Progress in Precision Agriculture

Series editor
Margaret A. Oliver, Soil Research Centre, University of Reading,
Berkshire, Berkshire, United Kingdom

This book series aims to provide a coherent framework to cover the multidisciplinary subject of Precision Agriculture (PA), including technological, agronomic, economic and sustainability issues of this subject. The target audience is varied and will be aimed at many groups working within PA including agricultural design engineers, agricultural economists, sensor specialists and agricultural statisticians. All volumes will be peer reviewed by an international advisory board.

More information about this series at https://link.springer.com/bookseries/13782

Ruth Kerry • Alexandre Escolà
Editors

Sensing Approaches for Precision Agriculture

Editors
Ruth Kerry
Department of Geography
Brigham Young University
Provo, UT, USA

Alexandre Escolà
Department of Agricultural and Forest Engineering Research Group on AgroICT & Precision Agriculture – GRAP
Universitat de Lleida/Agrotecnio-CERCA Center
Lleida, Catalunya, Spain

ISSN 2511-2260 ISSN 2511-2279 (electronic)
Progress in Precision Agriculture
ISBN 978-3-030-78433-1 ISBN 978-3-030-78431-7 (eBook)
https://doi.org/10.1007/978-3-030-78431-7

This Springer imprint is published by the registered company Springer Nature Switzerland AG
The registered company address is: Gewerbestrasse 11, 6330 Cham, Switzerland

Preface

Many endeavours in precision agriculture use some kind of sensor to gain relatively inexpensive information on the spatial and temporal variation in crops, soil, weeds, diseases, and so on. However, information about sensors is scattered throughout the literature. This text fills an important niche by bringing together information on a wide range of sensors that are used in precision agriculture in one book. Included are sensors that are well-established and regularly used in commercial precision agriculture as well as those that are currently being developed and researched. The book contains review chapters, case study chapters and chapters that include both review and case study sections. The full range of sensors used in precision agriculture is considered in the review chapters: (2) Satellite Remote Sensing for Precision Agriculture, (3) Sensing Crop Geometry and Structure, (4) Soil Sensing, (5) Sensing with Wireless Sensor Networks, (6) Sensing for Health, Vigour and Disease Detection in Row and Grain Crops, (7) On-Combine Sensing Techniques in Arable Crops, (8) Sensing in Precision Horticulture, (9) Sensing from Unmanned Aerial Vehicles, and (10) Sensing for Weed Detection. These chapters provide a logical and thorough review of the types of sensors that have been used to observe different phenomena within precision agriculture as well as the delivery platforms that have been used for sensing. Readers are provided with a rapid overview of the sensing solutions currently adopted and the trends in research towards developing new applications. In addition, the pros and cons of particular sensing approaches are considered, the standard best practices for using such sensors are discussed, and in some cases indications of current relative costs are given. The chapters with case studies: 4) Soil Sensing, (10) Sensing for Weed Detection, (11) Applications of Sensing to Precision Irrigation, (12) Applications of Optical Sensing of Crop Health and Vigour, and (13) Applications of Sensing for Disease Detection give detailed examples of some typical and cutting-edge applications of sensors in precision agriculture and the kinds of insights that the sensors used can provide to the sub-fields of precision agriculture. The book ends with an evaluation of potential future directions in sensor research for precision agriculture, which sensors show most promise for certain applications and the areas where increased research emphasis should be applied. The text provides sufficient detail to act as a handbook for practitioners trying to determine the best sensing approaches to use in a given situation. The target audiences of this book are upper-level undergraduate and graduate students, new professionals, scientists and practitioners of precision agriculture, and agricultural engineers. The book could be used in general agriculture and precision agriculture courses and also in courses on environmental monitoring and policy making at universities.

Provo, UT, USA Ruth Kerry

Lleida, Catalunya, Spain Àlex Escolà

Acknowledgements

We express thanks to each of the chapter authors for their valuable contributions to this book, which was produced during difficult times for agricultural research. We are also grateful to the anonymous reviewers who rigorously reviewed every chapter and provided feedback which helped improve the structure, clarity and quality of each chapter. Finally, we would like to thank Professor Margaret Oliver, the Springer Precision Agriculture Series editor, for her sage advice and guidance throughout our work on this engaging book project.

Contents

About the Editors

Ruth Kerry Trained in the UK at the Universities of Oxford (BA, MA Oxon.) and Reading (MSc and PhD) in geography and soil spatial analysis for precision agriculture, respectively, Ruth has been based in the Geography Department at Brigham Young University in Utah, USA, since receiving her PhD in 2004. She was hired as an associate professor at Brigham Young University in January 2021. She is also an affiliate assistant Professor in the Department of Crop, Soil and Environmental Sciences at Auburn University, Alabama, USA (2017–2022). Her research has involved mapping and spatial statistics for precision agriculture, soil studies and other applications. She has collaborated on research projects with academics in the UK, the USA, Europe, Iran and Chile across a range of precision agriculture and soil science topics. Along with her own research publications, she has a broad overview of the subjects of precision agriculture and spatial analysis having served as guest editor of two special issues of the journal *Precision Agriculture* (2008 and 2019), featuring articles from ECPA conferences, and a double special issue of *Geographical Analysis* on geostatistical applications. She has attended ECPA conferences since 2003 and has served on the scientific review committee for the last several meetings. She serves on the editorial board for *Precision Agriculture* (Springer) and is currently the treasurer of the International Society of Precision Agriculture (ISPA).

Alexandre Escolà Alex studied technical agricultural engineering at undergraduate level and agronomical engineering at master's level, both at the School of Agrifood and Forestry Science and Engineering of the Universitat de Lleida, Catalonia. His PhD and master's theses were related to the use of sensors to enhance pesticide spray applications in fruit orchards, and he received his PhD in 2010. In 2001, he was employed as part-time equivalent to assistant professor in the same centre and since 2006 as an equivalent to assistant professor. Since 2013, he has been hired as an equivalent to associate professor. His research is carried out within the GRAP, the Research Group on AgroICT & Precision Agriculture at the Universitat de Lleida/Agrotecnio-CERCA Centre, in which he is currently serving as the coordinator. His research lines focus on electronic 3D characterization of vegetation, especially in fructiculture and viticulture, and on developing precision agriculture techniques and technologies in general. He is member of the Department of Agricultural and Forest Engineering and teaches at the School of Agrifood and Forestry Science and Engineering and at the Polytechnic School of the Universitat de Lleida. He has attended ECPA and ICPA conferences since 2007, and in 2013, he chaired the 9th ECPA, held in Lleida. Since then, he has been a member of the ECPA programme committee and has served on the scientific review committee. He assisted the ISPA in the process of producing a modern definition of precision agriculture and participated in its translation into Catalan and Spanish. He serves on the editorial boards for *Precision Agriculture* (Springer) and *Sensors* (MDPI).

Contributors

Viacheslav I. Adamchuk Department of Bioresource Engineering, McGill University, Ste-Anne-de-Bellevue, QC, Canada

Elnaz Akbari University of California, Merced, CA, USA

Evangelos Anastasiou Agricultural University of Athens, Athens, Greece

Silvia Arazuri School of Agricultural Engineering and Biosciences, Public University of Navarre (UPNA), Pamplona, Spain

Asim Biswas School of Environmental Sciences, University of Guelph, Guelph, ON, Canada

Carlos Campillo Centro de Investigaciones Científicas y Tecnológicas de Extremadura (CICYTEX), Finca La Orden, Junta de Extremadura, Guadajira, Badajoz, Spain

Jaume Casadesús Institut de Recerca i Tecnologia Agroalimentàries (IRTA), Centre UdL IRTA, Lleida, Catalonia, Spain

Hongyan Chen School of Natural and Environmental Sciences, University of Newcastle, Newcastle-upon-Tyne, UK

Sven Christensen Department of Plant and Environmental Science, University of Copenhagen, Copenhagen, Denmark

Yafit Cohen Agricultural Engineering Institute, Agricultural Research Organization, Volcani Centre, Rishon LeZion, Israel

Curtis Cribben Bridgestone Americas, Inc., Burlington, NC, USA

Ana Isabel de Castro Megías National Institute for Agricultural and Food Research and Technology (INIA-CSIC), Madrid, Spain

Mads Dyrmann Department of Engineering, Aarhus University, Aarhus, Denmark

Reza Ehsani University of California, Merced, CA, USA

Alexandre Escolà Department of Agricultural and Forest Engineering, Research Group on AgroICT & Precision Agriculture – GRAP, Universitat de Lleida/Agrotecnio-CERCA Center, Lleida, Catalunya, Spain

Chris Fidelis Cocoa board of Papua New Guinea, Rabaul, Papua New Guinea

Damien J. Field Sydney Institute of Agriculture, the University of Sydney, Eveleigh, NSW, Australia

Spyros Fountas Agricultural University of Athens, Athens, Greece

David W. Franzen North Dakota State University, Fargo, ND, USA

Eduardo Galilea Department of Environmental and Soil Sciences, Research Group on AgroICT & Precision Agriculture – GRAP, Universitat de Lleida/Agrotecnio-CERCA Centre, Lleida, Catalonia, Spain

Nitsan Graff Agricultural Engineering Institute, Agricultural Research Organization, Volcani Institute, Rishon LeZion, Israel

The Robert H. Smith Faculty of Agriculture, Food and Environment, The Hebrew University of Jerusalem, Rehovot, Israel

Eduard Gregorio Department of Agricultural and Forest Engineering, Research Group on AgroICT & Precision Agriculture – GRAP, Universitat de Lleida/Agrotecnio-CERCA Center, Lleida, Catalonia, Spain

Perry J. Hardin Department of Geography, Brigham Young University, Provo, UT, USA

Jonathan E. Holland James Hutton Institute, Invergowrie, Dundee, UK

Austin Hopkins Department of Plant and Wildlife Science, Brigham Young University, Provo, UT, USA

Hsin-Hui Huang Department of Bioresource Engineering, McGill University, Ste-Anne-de-Bellevue, QC, Canada

Thomas Isakeit Texas A&M University, College Station, TX, USA

Carmen Jarén School of Agricultural Engineering and Biosciences, Public University of Navarre (UPNA), Pamplona, Spain

Ryan R. Jensen Department of Geography, Brigham Young University, Provo, UT, USA

Rasmus N. Jørgensen Department of Engineering, Aarhus University, Aarhus, Denmark

Ruth Kerry Department of Geography, Brigham Young University, Provo, UT, USA

Newell R. Kitchen USDA-ARS, Columbia, MO, USA

Morten S. Laursen Department of Engineering, Aarhus University, Aarhus, Denmark

Vasileios Liakos Crop and Soil Sciences Department, University of Georgia, Athens, GA, USA

PEARL Lab, Department of Agriculture - Agrotechnology, University of Thessaly, Larissa, Greece

Jordi Llorens Department of Agricultural and Forest Engineering, Research Group on AgroICT & Precision Agriculture – GRAP, Universitat de Lleida/ Agrotecnio-CERCA Center, Lleida, Catalonia, Spain

Dan S. Long USDA-Agricultural Research Service, Pendleton, OR, USA

Ainara López-Maestresalas School of Agricultural Engineering and Biosciences, Public University of Navarre (UPNA), Pamplona, Spain

Carlos Lopez-Molina School of Agricultural Engineering and Biosciences, Public University of Navarre (UPNA), Pamplona, Spain

Diana Marin School of Agricultural Engineering and Biosciences, Public University of Navarre (UPNA), Pamplona, Spain

José A. Martínez-Casasnovas Department of Environmental and Soil Sciences, Research Group on AgroICT & Precision Agriculture – GRAP, Universitat de Lleida/Agrotecnio-CERCA Centre, Lleida, Catalonia, Spain

John D. McCallum USDA-Agricultural Research Service, Pendleton, OR, USA

Yuxin Miao University of Minnesota, St. Paul, MN, USA

Sandra Millán Centro de Investigaciones Científicas y Tecnológicas de Extremadura (CICYTEX), Finca La Orden, Junta de Extremadura, Guadajira, Badajoz, Spain

Budiman Minasny Sydney Institute of Agriculture, the University of Sydney, Eveleigh, NSW, Australia

Jose P. Molin USP, Piracicaba, SP, Brazil

David J. Mulla Department of Soil, Water & Climate, University of Minnesota, St. Paul, MN, USA

Robert L. Nichols Cotton Incorporated, Cary, NC, USA

José Manuel Peña Institute of Agricultural Sciences (ICA-CSIC), Madrid, Spain

Claudia Pérez-Roncal School of Agricultural Engineering and Biosciences, Public University of Navarre (UPNA), Pamplona, Spain

Maria del Henar Prieto Centro de Investigaciones Científicas y Tecnológicas de Extremadura (CICYTEX), Finca La Orden, Junta de Extremadura, Guadajira, Badajoz, Spain

Vasilis Psiroukis Agricultural University of Athens, Athens, Greece

Jesper Rasmussen Department of Plant and Environmental Science, University of Copenhagen, Copenhagen, Denmark

Todd Sanderson Commonwealth Scientific and Industrial Research Organisation, Canberra, Australia

Gonzaga Santesteban School of Agricultural Engineering and Biosciences, Public University of Navarre (UPNA), Pamplona, Spain

Yehoshua Saranga The Robert H. Smith Faculty of Agriculture, Food and Environment, The Hebrew University of Jerusalem, Rehovot, Israel

Peter C. Scharf University of Missouri, Columbia, MO, USA

James S. Schepers USDA-ARS, Lincoln, NE, USA

Kanika Singh Sydney Institute of Agriculture, the University of Sydney, Eveleigh, NSW, Australia

John L. Snider Crop and Soil Sciences Department, University of Georgia, Athens, GA, USA

Didier Snoeck CIRAD, Tree Crop Based Systems Research Unit, Montpellier Cedex 5, France

Bo Stenberg Department of Soil and Environment, Swedish University of Agricultural Sciences, Skara, Sweden

James A. Taylor ITAP, University of Montpellier, INRAE, Institut Agro, Montpellier, France

School of Natural and Environmental Sciences, University of Newcastle, Newcastle-upon-Tyne, UK

J. Alex Thomasson Mississippi State University, Mississippi State, MS, USA

Bruno Tisseyre ITAP, Univ Montpellier, INRAE, Institut Agro, Montpellier, France

Marcus Travers Soil Essentials, Hilton of Fern, Brechin, Scotland, UK

Rodrigo G. Trevisan USP, Piracicaba, SP, Brazil

Nikos Tsoulias Leibniz Institute for Agricultural Engineering and Bioeconomy (ATB), Potsdam, Germany

Jorge Urrestarazu School of Agricultural Engineering and Biosciences, Public University of Navarre (UPNA), Pamplona, Spain

George Vellidis Crop and Soil Sciences Department, University of Georgia, Athens, GA, USA

Tianyi Wang Texas A&M AgriLife Research, Dallas, TX, USA

Xiwei Wang Nanjing Forestry University, Xuanwu, China

Johanna Wetterlind Department of Soil and Environment, Swedish University of Agricultural Sciences, Skara, Sweden

Chenghai Yang USDA Agricultural Research Service, College Station, TX, USA

David Yinil Cocoa board of Papua New Guinea, Rabaul, Papua New Guinea

Manuela Zude-Sasse Leibniz Institute for Agricultural Engineering and Bioeconomy (ATB), Potsdam, Germany

Chapter 1
Introduction and Basic Sensing Concepts

Alexandre Escolà and Ruth Kerry

Abstract This chapter provides the background about the need to characterize the spatial and temporal variation of crop status, soils, diseases, weeds and pests within fields, and the need to have data about the variation of these variables in precision agriculture (PA). The role of sensors is to provide quantifiable, objective, repeatable and cost-effective data in a simple way for the farmer to make more informed management decisions. Chapters 2–10 identify the most important specific sensing techniques used in PA in the form of reviews that cover the following topics: (2) Satellite Remote Sensing, (3) Sensing Crop Geometry and Structure, (4) Soil Sensing, (5) Sensing with Wireless Sensor Networks, (6) Sensing for Health, Vigour and Disease Detection in Row and Grain Crops, (7) On-Combine Sensing Techniques in Arable Crops, (8) Sensing in Precision Horticulture, (9) Sensing from Unmanned Aerial Vehicles and (10) Sensing for Weed Detection. Chapters 11–13 focus on small case studies: (11) Applications of Sensing to Precision Irrigation, (12) Applications of Optical Sensing of Crop Health and Vigour and (13) Applications of Sensing for Disease Detection. In the conclusion to the book, there is a section on how we expect sensors and analysis to develop. At the end of the introductory chapter some basic concepts are explained to facilitate the reading of the book, the use of sensors and the data they produce.

Keywords Precision agriculture · Sensors · Sensing systems · Sensor data · GNSS · Metrology

A. Escolà (✉)
Department of Agricultural and Forest Engineering, Research Group on AgroICT & Precision Agriculture – GRAP, Universitat de Lleida/Agrotecnio-CERCA Center, Lleida, Catalunya, Spain
e-mail: alex.escola@udl.cat

R. Kerry
Department of Geography, Brigham Young University, Provo, UT, USA

R. Kerry, A. Escolà (eds.), *Sensing Approaches for Precision Agriculture*, Progress in Precision Agriculture, https://doi.org/10.1007/978-3-030-78431-7_1

1.1 Background

Early definitions of precision agriculture (PA) emphasized the management of spatial variation within agricultural fields (Blackmore 1994) to maximize profits and reduce environment pollution (Fenton 1998). More recently, the current (2019) official definition of PA recognized by the International Society for Precision Agriculture (ISPA 2019) was developed in consultation with 46 PA experts coordinated by Dr. Nicolas Tremblay, Dr. Àlex Escolà and Dr. Viacheslav Adamchuk: "Precision agriculture is a management strategy that gathers, processes and analyzes temporal, spatial and individual data and combines it with other information to support management decisions according to estimated variability for improved resource use efficiency, productivity, quality, profitability and sustainability of agricultural production."

Inherent in the current and early definitions of PA is the need to characterize the spatial variation of crop status, soils, diseases, weeds and pests within fields and the need for data about the variation of these variables. Observing these elements and status by traditional methods usually requires sampling, which is often inconsistent, biased, destructive, time-consuming and expensive to complete. According to the ISPA definition, PA could be practiced without any technological help, as temporal, spatial or individual plant or animal data could be gathered, processed and analyzed by simply using human senses, a paper and a pencil. Most farmers are very aware of the variation within their fields and orchards, and some may be managing them with a simple site-specific approach. However, to make PA economically viable, sensing approaches have been used from the outset and are now increasingly used to obtain dense spatial data sets at a greatly reduced cost compared to traditional sampling and laboratory analysis. Sensors are key in obtaining data about crops, soil, and so on, in a quantifiable, objective, repeatable, cost-effective and simple way, although the last two criteria might not always be satisfied.

Some of the first sensors to be used in PA were yield monitors for cereal crops to characterize the spatial variation in yield, which was seen as a first step to determining the causes of yield variation and managing them. Yield monitors enabled the collection of high resolution spatial data on the variation in yield and then interpolate results into map form without having to do time-consuming spatial sampling at harvest. Use of yield sensors has not been without its problems and much research has been devoted to investigating sources of error and developing methods to pre-process such data to reduce error or to refine the sensors (see Chap. 7). Key to the use of these first sensors in PA, and the use of many sensors today, has been global navigation satellite systems (GNSS), beginning with the global positioning system (GPS), which were able to reduce considerably the time to obtain accurate positioning data. Although it is currently feasible to practice PA without the need for any GNSS receiver, GPS is considered to be one of the main triggers in the development and adoption of PA as many sensing approaches could not be implemented without accurate positioning data. Some basic concepts on GNSS will be covered at the end of this introductory chapter.

Over time, the number of sensors available for observing various phenomena in PA has increased almost exponentially and there is now a wide variety of sensing approaches to characterize spatial variation in various agronomically important phenomena. The 2019 definition of PA emphasizes the importance of changes in spatial variation over time, between years or even within individual growing seasons. This increasing interest in temporal analysis for PA means that repeated sampling of many phenomena is needed, sometimes numerous times within a season. Such temporal analysis would be economically untenable with traditional sampling and laboratory approaches. Current requirements in agriculture are leading to the almost exponential increase of interest in accurate yet inexpensive sensing approaches.

1.2 Purpose, Aims, Structure and Audience of this Book

A large proportion of the research in the PA literature uses sensors, but the output is scattered in various journals and reports. The idea for this book was proposed at the European Conference on Precision Agriculture held in 2017 in Edinburgh, UK. We noticed there, and at other PA conferences, that a large proportion of presentations involved some sort of sensing application. Sessions are usually based on the fields of application, however, such as weed studies, precision viticulture, precision irrigation or soil studies and only rarely are there occasional coherent sessions on particular sensing approaches. We saw the need for a text that brings together the variety of sensing approaches currently used in PA in one document to discuss properly the pros and cons of the different approaches. This book aims to bring together research based on the types of sensor and sensing systems commercially available to monitor crop characteristics, status and their environment. It also considers sensing systems in development or testing phases and their applications in PA. The book aims to bring together the 'state of the art' of the most popular sensing techniques and the current state of research of where sensors are applied in PA. The book will provide a broad overview of sensing in PA and a coherent introduction for new professionals and research scientists. However, the book does not cover proprioceptive or interoceptive sensors mounted on PA machinery or robots, as they are designed to measure the interaction of such equipment with the environment or to get internal data for their operation, respectively. Those sensors are already covered in another book of the series titled "Innovation in Agricultural Robotics for Precision Agriculture - A Roadmap for Integrating Robots into Precision Agriculture". Chapters on specific topics and case studies will provide depth and enable implementation of the methods by users. Readers will be introduced to the potential applications of a range of different sensors, how they should be used properly, their limitations for use in PA and accuracy assessments as well as current relative prices of some sensors compared to rates charged for standard sampling and laboratory analyses. Sensing is of great value in PA because it usually provides cost-effective and often near real-time data for making more informed management decisions.

In addition to introducing the book, Chap. 1 provides some basic concepts related to sensors and sensing techniques that do not fall within the scope of the other chapters. These basic concepts will help readers to understand the book content better and make more appropriate use of sensors and their data. Chapters 2–10 of this book identify the most important specific sensing techniques used in PA. These form review chapters covering the following topics: (2) Satellite Remote Sensing, (3) Sensing Crop Geometry and Structure, (4) Soil Sensing, (5) Sensing with Wireless Sensor Networks, (6) Sensing for Health, Vigour and Disease Detection in Row and Grain Crops, (7) On-Combine Sensing Techniques in Arable Crops, (8) Sensing in Precision Horticulture, (9) Sensing from Unmanned Aerial Vehicles and (10) Sensing for Weed Detection. Some of these review chapters include case studies to illustrate the application of the various sensors. However, the focus of Chapters 11–13 is solely on small-case studies, each showing cutting edge applications of different sensing methods: (11) Applications of Sensing to Precision Irrigation, (12) Applications of Optical Sensing of Crop Health and Vigour and (13) Applications of Sensing for Disease Detection. In conclusion to the book, there is a section on how we expect sensors and analysis to develop. The reference sections of each chapter alone, particularly the review chapters, have a wealth of information for readers who wish to explore the application of sensors in PA further.

This text provides sufficient detail to act as a handbook for practitioners. It will also be relevant to the wider field of digital agriculture, the adoption of which and some of its principals is increasing at an exponential pace. The theme of the ASA-CSSA-SSSA 2019 annual meeting, with about 600 sessions, 3000 research papers and 4600 presenters, was “Embracing the Digital Environment” which reflects the digital revolution within agriculture to which sensors have served as a keystone.

The target audiences of this book are upper level undergraduate and graduate students, new professionals, scientists and practitioners of PA and agricultural engineers. Readers are provided with a rapid overview of the sensing solutions currently adopted and the trends in research towards developing new applications. The book could be used in general agriculture and PA courses and also in courses on environmental monitoring and policy making.

1.3 Sensing Approaches

A wide range of sensing approaches is covered in this book. Two broad groups of approach are remote and proximal sensing approaches. Although remote sensing implies any measurement done without direct contact with the medium or object being measured, in this book the traditional PA approach is adopted. Thus, remote sensing involves the observation of the earth’s surface from satellite or airborne systems whereas proximal sensing systems collect information near the earth’s surface, from ground-based platforms.

1.3.1 Remote Sensing Systems

Remote sensing approaches typically observe reflectance of different parts of the electromagnetic spectrum from spaceborne or airborne sensors. Satellite remote sensing approaches have been used in PA from the outset investigating basic vegetation indices such as the normalized difference vegetation index (NDVI), a chlorophyll content-based index which is related to crop health. However, more recently, the increase in the spatial, spectral and temporal resolution of imagery from some satellites has enabled greater use of satellite remote sensing platforms in PA. The spatial resolution in remote sensing imaging refers to the size of image pixels in terms of the area captured on the ground or the footprint, that is the smallest area of observation. This is sometimes called ground sample distance (GSD) although it may differ slightly if interpolation is performed in the final image product. The reflectance characteristics of each pixel indicate the average reflectance of the surface over the area of the pixel. Traditional remote sensing platforms like Landsat have pixel sizes of 30 m for most bands whereas the relatively new Sentinel-2 imagery has a pixel size of 10 m for RGB and near-infrared (NIR) bands. Satellites such as GeoEye-1, Worldview-3 and Pleiades 1A, B now produce imagery pixel sizes of <2 m (see Chap. 2 for more detail). Some newer satellites also have shorter revisit times, some revisit as often as daily, which makes temporal studies using satellite remote sensing increasingly viable. The increased spectral resolution of some satellites in recent years has revolutionized the study of phenomena such as drought (see Chap. 11) and disease detection within individual growing seasons. Increased spectral resolution has been of particular help in identifying certain diseases before their effects are visible to the naked eye (see Chap. 13 for specific examples).

Chapter 2 provides an exhaustive review of satellite remote sensing platforms and their characteristics in terms of spatial, temporal and spectral resolution as well as the swath width and cost per unit area. A wide range of vegetation and soil indices that have been calculated for PA research is reviewed. Its summary tables provide an excellent quick reference guide. Also reviewed are hyperspectral and sun-induced fluorescence (SIF) satellite specifications and narrow band vegetation indices that the former have been used to calculate. The discussion of how synthetic aperture radar satellites and satellite-based digital surface model products have been used is very useful. Indeed, the former has proved particularly useful for estimating soil surface moisture and has the advantage of being able to collect data at night and when there is cloud cover. The chapter concludes by identifying future satellite remote sensing needs for PA. The need for more hyperspectral imagery in the future is particularly important and the economic advantage of freely available data is addressed. Sentinel-2 is identified as a particularly viable option for future studies given its relatively smaller pixel size, good spectral resolution and more frequent revisit times than Landsat 8, other freely available data.

Satellite remote sensing techniques are applied and mentioned in other chapters of the book. For example, Chapter 6 discusses the use of satellite and other remote sensing products and derived vegetation indices to identify plant nutrient deficiencies and diseases for row and grain crops, while Chap. 8 considers the use of remote sensing products by type of band and the diseases that have been identified with those bands. The use of remote sensing approaches to estimate yield and fruit maturity is important for a wide range of crops in precision horticulture (Chap. 8). Chapter 8 also considers the complex numerical analysis techniques that are often required to deal with the hyperspectral imagery needed for disease detection. Chapter 10, on weed detection, notes the low spatial resolution of most freely available satellite imagery as an issue for weed detection in PA, therefore, other remote sensing platforms have been favoured for this activity. If some of the new, high spatial resolution satellite imagery becomes available without charge, satellite remote sensing will still be a cost-effective option for weed detection. Chapter 11, on precision irrigation, includes a case study that used NDVI from Sentinel-2 imagery to characterize the spatial variation in different varieties of fruit trees and different irrigation sectors. Finally, in Chap. 13, case study 1 used 2 m resolution visible and near-infrared (VIS-NIR) GeoEye imagery to identify areas with Cotton Root Rot disease and determine a precision fungicide application protocol.

Early in PA, aerial photographs, in particular from standard surveys, were a relatively frequently used remote sensing product (Robert 2002; Kerry and Oliver 2003) because of the potentially small GSD represented by pixels (usually ~1–3 m) compared to early freely available satellite imagery (~20–30 m). However, with the limited spectral resolution of photographs and the high cost of custom flights where there is no nearby airstrip or farmer routinely using light aircraft to spray crops, and so on, the use of aerial imagery captured by manned aircraft has decreased. However, it is still economically viable for those with easy access to a plane, and Chap. 10 uses manned aircraft to acquire imagery for weed detection. The aircraft need to be flown at low altitudes to ensure sufficient spatial accuracy; nevertheless, such data are mainly useful for identifying large patches of weeds only. In case study 2 of Chap. 12, aircraft captured aerial imagery in the VIS-NIR wavelengths was used to classify NDVI for vineyard blocks to indicate vine water status. In case study 1 of Chap. 13 aerial imagery captured by manned aircraft in the VIS-NIR wavelengths was used to identify areas with Cotton Root Rot disease. The case study compared the ability of this imagery to identify the disease with high resolution (2 m) GeoEye data and unmanned aerial vehicle (UAV) captured imagery. The use of UAVs has revolutionized remote sensing for PA. Chapter 9 reviews the use of UAVs in PA together with a case study that investigates UAV imagery to determine side-dress fertilization of winter cereals. The chapter discusses the various pros and cons of imagery from UAVs compared to satellite and manned flight-based aerial photograph imagery. The UAVs and some associated cameras are relatively inexpensive and provide images with very high spatial resolution (pixels as small as 1 cm). Case study 1 of Chap. 13 illustrates how relatively high resolution (1–2 m pixels) satellite and manned aircraft imagery can both guide precision fungicide applications successfully. The study also shows how the higher spatial resolution from UAV imagery makes plant by plant fungicide applications a possibility for the future. Having said this, data storage and processing issues can be important drawbacks of using UAV images unless standard automated imagery processing services are used. The case study in Chap. 9 illustrates how Sentinel-2 and UAV imagery used to direct fertilizer side-dressing can increase profits, but the latter increased profits slightly more.

The cheapest cameras or sensors that can be mounted on UAVs have limited spectral resolution, therefore, several vegetation indices using only wavelengths in the visible range have been developed. The finer temporal resolution of UAV data over satellite remote sensing products makes such data much more amenable to within season temporal analysis of crops. Flights can be made regularly apart from when winds are high. The low altitudes of UAV flights means that cloud cover is less of a problem than for satellite imagery. This temporal flexibility for drone flights has resulted in a great increase in temporal studies which observe the crop throughout the growing season. In addition to the review and case study of UAVs in Chap. 9, Chap. 6 briefly discusses the use of UAV imagery to determine N deficiency. Case study 1 in Chap. 10 discusses the use of UAV imagery for mission planning to spray weed patches with commercial sprayers and case study 4 in Chap. 12 uses UAV imagery to improve the functioning of a potato crop model and to spatialize it within a field. Although the use of UAVs is interesting in research, it needs to be improved further to upscale for regular use on large farms.

1.3.2 Proximal Sensing Systems

There are many passive and active optical or spectroscopic sensors currently in use in PA, both in remote and proximal sensing. The main limitation of passive sensors is the variation in lighting in the open environment, therefore, they are sometimes applied to plant and soil samples in the laboratory where lighting conditions can be controlled. The sample collection and preparation adds considerably to the cost, but improves the accuracy of any values derived from the approach. Optical or spectroscopic proximal sensors are often distinguished from one another by the platform on which they are mounted. When the crop is being monitored such sensors are frequently mounted on all-terrain vehicles (ATVs) or on farm machinery, sensing the crop while routine operations are being carried out. At other times, they are attached to masts within a field. This can be a good way to get almost continual temporal coverage of crop status. This is a potentially economic approach if wanting to sense a limited area of crop frequently because no fuel or labour are needed compared with when sensors are mounted on mobile equipment. Case study 1 of Chap. 12 investigates the use of an optical Crop Circle sensor mounted 1.5 m above the ground to sense the side curtain canopy areas of grape vines. The data from this sensor were used to calculate several vegetation indices to monitor berry characteristics such as size and pH throughout the growing season. Case study 2 of Chap. 11 captured oblique thermal images of cotton fields in Israel using infrared thermal cameras mounted on a 20 m vertical mast. The images were used to determine leaf water potential and derive rates of drip irrigation.

Examples of optical sensors mounted on farm machinery or ATV vehicles are quite common in PA. Chapter 7 focuses solely on sensors that are mounted on combines. There is a detailed section on combine-mounted near-infrared spectroscopy instruments to sense different aspects of grain quality. In case study 2 of Chap. 10, a machine vision camera was mounted on an ATV to obtain dense imagery of the vegetation in the field which was analyzed by machine learning software to distinguish monocot and dicot weeds from the crop. In addition to spectroscopy sensors mounted on farm machinery (Chap. 7) or masts, there are handheld devices that can perform spectroscopy in the field or laboratory. One of the earliest examples and most commonly used of these handheld devices are soil plant analysis development (SPAD) or NDVI meters, although such instruments are also frequently mounted on farm machinery (Chaps. 6 and 7).

Proximal sensors have been particularly favoured over remote sensing approaches for sensing soil characteristics because they allow closer proximity to the medium being studied than remote sensors. Case studies 1, 3 and 4 in Chap. 4 all use NIR spectroscopy either alone or together with other sensing approaches. Case study 1 used on-the-go NIR spectroscopy, while case studies 3 and 4 used the method in the laboratory on air-dried soils. These case studies note that the accuracy and reliability of results from proximal soil sensors often differs considerably when spectroscopic approaches are applied in the laboratory under constant moisture and lighting conditions, *in situ* in the field or are measured on-the-go. Case studies 2 and 3 in Chap. 13 use hyperspectral cameras under laboratory conditions with constant lighting to detect Laurel wilt and Esca in avocado leaves and vine leaves, respectively, before the effects of these diseases on the leaves are visible to the human eye. Hyperspectral imaging has also been used to determine the degree of fruit ripeness (Chap. 8).

At the far end of electromagnetic spectrum are gamma rays. Gamma-ray radiometry has proved useful for mapping soil-derived plant nutrients and for soil mapping at various scales as illustrated by case study 2 of Chap. 4. There are also many active optical sensors that have been used for various aspects of sensing. Stereo cameras and light detection and ranging (LiDAR) have been used for three dimensional 3D modelling of plants, for autonomous robot guidance and to determine the size of fruit (Chaps. 3 and 8). Obtaining 3D data from crops may help farmers and advisors to understand the development and variability of the crop better throughout the growing season to make more informed decisions on canopy management, applications of plant protection products and other operations. Chapter 7 discusses the use of laser and LiDAR-based sensors on combines to detect crop height, density and variability in biomass. Chapter 6 mentions the use of LiDAR data from a UAV digital surface model being subtracted from a digital elevation model to model crop height and yield.

Probably, the most commonly used types of proximal sensor for soil sensing are geophysical sensors and soil moisture sensors. Soil moisture is probably the most temporally variable soil property, therefore, such sensors are often left in place to determine irrigation timing and rates. Chapters 5 and 11 examine the use of wireless sensor networks (WSN) with soil moisture sensors in detail. Geophysical sensors that measure EC_a or resistivity of the soil have been used or calibrated to infer various soil properties or to determine suitable management zones within fields. However, some of the best uses of such geophysical data are when they are combined with other sensor data. Case study 1 of Chap. 4 investigates the use of geophysical instruments on a multi-sensor platform and case study 2 of Chap. 4 uses geophysical data combined with gamma-ray radiometry. Chapter 6 also mentions situations where plant health is assessed through soil sensing or where geophysical instruments are used to define management zones or areas with markedly different soils. Most proximal soil sensors are related to a range of soil properties and do not measure one individual soil property. Consequently, most proximal soil sensing approaches need to be calibrated to reflect values of a given soil property. The case studies in Chap. 4 discuss this issue of calibration of sensed data in detail and investigate the relative expense and sampling effort required for good calibration. To date, more success has been achieved with proximal sensors for mapping the more permanent properties of the soil such as texture and therefore water content and organic matter content compared to nutrients. However, geophysical sensors usually reflect problems with soil salinity well.

Data from sensors are usually dense and increasingly complex numerical methods are needed to analyze these large data sets. Practitioners of PA cannot be expert in all of these methods so there is increasing demand for automated machine learning approaches so that the labour costs for sampling and analysis are not replaced by labour costs for data analysis. Although some techniques are described in the chapters of this book, they do not cover the full range of analysis and numerical techniques in detail because there are limits to the scope within a single book and modelling will be the focus of another forthcoming book in the Springer Precision Agriculture Series.

1.4 Basic Sensing Concepts for Precision Agriculture

Although the readers of this book are not expected to design sensors from scratch, their use in PA requires some basic understanding of sensor principles, how to express the recorded data and what the possible applications in PA are. The aim of this section is to clarify some basic concepts used throughout the book to help the reader understand its content better and make better use of their sensors and data.

1.4.1 When Sensors Are Used in Precision Agriculture

Precision agriculture can be practiced by following a 4-stage cycle. The first stage is data gathering about the crop and its environment. The second is data processing and information extraction. The third stage is decision making and the last is operation in the field. Sensors are used in the first and fourth stages with two main purposes: (1) capturing data about the crop and its environment, but also (2) monitoring the equipment which is carrying the sensors themselves (agricultural machinery or robots). As mentioned, this book covers only those sensors that record data about the crop and its environment.

The PA cycle described above may be followed using two different approaches in what is referred to as map-based precision agriculture and real-time precision agriculture. In the former, sensors are used in the first stage of the PA cycle to gather georeferenced data which needs to be processed (filtered, normalized, etc.) and interpolated to create maps representing the spatial distribution of the variables measured throughout a field or orchard (stage 2). Those maps, together with other georeferenced or mapped ancillary data, are used to make a decision (stage 3) about what specific management operations should be carried out and with what specifications. The output of stage 3 is usually another map, the so-called *prescription map*, representing the field or orchard and what to do in it following a site-specific management approach (rates to apply, intensity of a specific operation, etc.). Finally, in stage 4 it is time to go to the field and execute the prescription map with either manually operated conventional equipment or with variable rate technology (VRT) equipment. From the moment data are captured until the management operation is carried out, several hours, days or even weeks can pass. This time delay can have a negative effect if a problem needs to be addressed rapidly due to changing conditions (i.e. a treatment against a moving pest), but it can give the farmer or advisors time to integrate several information sources and supervize and validate the decision made, whether the decision is made by a human or is automated.

The second approach for PA practitioners is the so-called real-time approach. In this approach, crop or soil sensors are mounted on VRT equipment and the PA cycle is executed on-the-go, on a nearly real-time basis, so that the time between sensing and acting is a matter of some milliseconds only, depending on the distance between the sensor and the VRT equipment and on the forward speed of the agricultural machinery. Although this approach also follows the four stages in the cycle, there is no time for the controller to integrate many information sources or for the farmer or operator to supervize and validate the decision made. In this case, even though a GNSS receiver is not required, it is very convenient to have it for three different purposes: (1) to avoid overlap between already treated areas and those currently being treated, (2) to record the sensor readings with their locations to create a map subsequently to show the spatial distribution of the measured quantity, and (3) to record and geo-reference what the VRT equipment did in the field to create what is called an *as-applied map*. This will not be a map about the crop, soil or environmental readings, but about the internal sensors of the VRT equipment, called proprioceptive and interoceptive sensors, to monitor the rates actually applied. Thus, the farmer will be able, at least, to supervize what was actually done in the field *a posteriori*.

1.4.2 How Sensors Are Used in Precision Agriculture

Both remote and proximal sensors may be active or passive and that will determine the way in which they are used and under what conditions. An active sensor emits its own energy towards the object being measured (usually some form of radiation but also other forms of energy such as electricity) and captures the returned energy to infer a property or status. A passive sensor does not emit any energy and simply captures the reflected energy from an external source. A simple example is a red, green, blue (RGB) camera. When lighting conditions are sufficient, the camera simply records the amount of red, green and blue light captured by the optics, operating as a passive sensor. However, when ambient light is not sufficient, some cameras use a flash. That is, they emit their own light and capture the returning scattered light, turning it into an active sensing system.

Passive sensors rely on external energy sources. In agriculture, usually the sun. That is why they can only operate during daytime (or in some cases at night with artificial light) and their readings are affected by sunlight intensity. In addition, the further the sensor is from the target, the larger the negative effects on the energy captured will be. This is of particular relevance in satelliteborne sensors, which means that their readings need to be corrected according to atmospheric status to obtain absolute values that can be compared to data captured on different dates. Passive sensors that record visible light from satellites are useful only during daytime and under cloudless conditions.

Alternatively, proximal and remote active sensors can be used during night time and under cloudy conditions, extending the sensing time window or even the working time when the sensor can be used in a real-time PA approach. That is the case of active N proximal sensors and of synthetic aperture radar sensing devices mounted on satellites such as Sentinel-1.

Regardless of whether they are active or passive, proximal or remote, sensors can be operated manually or fully automated. In addition, the time required to complete a measurement also needs to be considered. This characteristic affects the time invested in sensing and may influence the sensing approach, therefore affecting the final spatial resolution of the measurements. In manually operated sensors, the usual approach is to design a discrete sampling strategy, either systematic or stratified. With automated sensors, the readings may require a trigger signal or may be continuous at a specific rate of update or output frequency. Both the trigger signal or the output frequency of the sensors will condition the spatial resolution of the data together with the forward speed of the sensor. If the sensor does not need to stop for a specific time to make a measurement, the usual approach is the on-the-go sensing strategy. In that case, the sensor will be mounted on a ground, aerial or space platform and will capture data at a specific temporal rate or at pre-determined distance intervals, resulting in much higher spatial resolution data than those obtained by discrete sensors.

1.4.3 Sensing Resolutions

Most common concerns when using sensors are associated with the resolution of one type or another. According to the International Vocabulary of Metrology (BIPM, 2012), resolution is the *smallest change in a quantity being measured that causes a perceptible change in the corresponding indication*. Such an indication is the *quantity value provided by a measuring instrument or a measuring system*. In analogue devices, the resolution of a display device is to be considered, which is the *smallest difference between displayed indications that can be meaningfully distinguished*. Those resolutions are expressed in the same units as the quantity being measured.

Regarding imaging sensors, their resolution is related to the detail perceivable in the acquired images. The higher is the resolution, the more detailed are the images and the smaller the objects that can be distinguished. In digital imaging sensors, such resolution is frequently expressed by the number of pixels of the sensor. That is, the number of elementary units or samples of the whole image. The resolution can be expressed as a round number, obtained from multiplying the number of pixel columns by the number of rows of the sensor, or by simply mentioning the count of columns and rows. Thus, a 10 megapixel resolution sensor is a device producing digital images of 3648 columns and 2736 rows, or 9 980 928 pixels. That resolution, together with the distance of the sensor from the target will determine the spatial resolution of the acquired images.

The main types of resolution considered in PA are spatial, temporal, spectral and radiometric resolutions. Spatial resolution of imaging sensors may be considered as the size of the captured target represented in each pixel or, alternatively, its inversion. The resolution of digital images is usually expressed as the equivalent size of a side of a pixel (e.g. pixel size of 1 m) or as the count of pixels per unit target area (e.g. 5 pixels cm^{-2} of leaf or 9 pixels m^{-2} of ground surface). It is similar to point clouds where resolution is usually expressed as the number of points per unit ground area or per unit of the target area both for airborne and terrestrial laser scanners. Depending on many factors, such resolutions may range from less than one to some tens of points per square metre of ground in the former and from hundreds to thousands of points per square metre of ground in the latter. For other aspects, it is also common to express spatial resolution as the number of samples per unit of ground area or per grid size. Thus, taking one soil sample per hectare is also a way to express spatial resolution. A technical term to express such spatial resolutions is the ground sample distance or GSD, which expresses the equivalent distance between samples on the ground. In this case, the pixels in digital images would be considered samples as well. The advantage of this way of expressing resolution is that the quantity is a length and it can be expressed using the international system base unit. Another thing to consider in relation to sampling and spatial resolution is the sample spatial support. For a satellite image with a GSD of 30 m the spatial support is areal giving the average reflectance characteristics of the 30 m square on the ground. Other data have punctual support as they are taken at specific points on the ground separated by a given GSD. Still, other data have a pseudo-point support. Using a pseudo-point support is standard practice in soil sampling and involves collecting five or so samples within a metre or a few metres squared at the nodes of a 100 m square grid and mixing them prior to analysis. This mixing or averaging tries to reduce local noise in the data. This procedure is often referred to as the bulking strategy. Similar approaches are also used in plant analyses such as taking SPAD meter measurements of several leaves from a small area around the nodes of a sampling grid.

Temporal resolution is related to the time between two consecutive measurements of the same area or target. It is usually expressed in time units when time is long and in frequency units when it is short. In remote sensing from satellites, the temporal resolution is set by the platform (satellite) and it is usual to refer to it as return time, revisit time or revisit frequency, commonly expressed in days. However, other sampling techniques with higher temporal resolutions tend to express it as one sample per time unit (1 sample h^{-1}) or as a number of samples per time unit when resolution is higher (10 Hz or samples s^{-1}). That also applies to video cameras, where the number of images or frames per time unit is an important specification. Thus, the usual temporal resolution for videos is 24 Hz or images per second or frames per second but this resolution may vary up to several hundreds or even thousands of hertz in high speed cameras for slow motion recording.

It is difficult to obtain data at both high spatial and temporal resolutions. It is usually a trade-off when deciding what sensing approach to use. In proximal sensing, when sensors are mounted on mobile ground platforms, the spatial resolution depends on the forward speed of the platform and it tends to be as fast as possible when hourly costs are applied. A way to increase spatial resolution without increasing hourly costs is with ground robotic platforms for scouting. Robots may work 24/7 and could take their time to reach the desired spatial resolution. The temporal resolution when using sensors mounted on machinery depends on the fuel costs, the ability to enter the field regularly and the damage that could be done to the crop. A way to increase the temporal resolution is by installing sensors permanently in the field. This way, very high temporal resolutions can be achieved but, on the other hand, spatial resolution tends to be low due to the cost of requiring several sensors over the farm. Satellite remote sensing runs on a pre-set schedule and provides users with pre-set spatial resolution data, whereas manned and unmanned airborne platforms are more flexible but may be more expensive and or the data more complex to process. As mentioned previously, the key is finding a balance between requirements, costs and complexity in processing the data.

Finally, spectral and radiometric resolutions refer to radiometric sensors, which capture energy from different bands of the electromagnetic spectrum; the bands are the fraction of the spectrum to which the sensor is sensitive. Spectral resolution is defined as the ability to resolve features in the electromagnetic spectrum and is usually expressed as bandwidth. Bands can be broader or narrower depending on the sensor and their size or width is expressed in length units, usually nanometres or micrometres. However, spectral resolution is sometimes expressed as the number of spectral bands measured by the sensor. For example, the Sentinel-2 multispectral instrument measures up to 13 bands with bandwidths ranging from 15 nm to 175 nm. Radiometric resolution has to do with how finely an imaging sensor records the different levels of brightness in each band. The available levels depend on the radiometric resolution and are expressed in bits, in a similar way to the A/D converters. The radiometric resolution of Sentinel-2 is 12 bit, allowing up to 4096 levels to be recorded.

1.4.4 Global Navigation Satellite Systems

Global navigation satellite systems (GNSS) receivers are not precisely sensors, but in PA they are strongly connected to them because of the importance of obtaining both data on the measured property and the location of the measurement. They crop up many times throughout this book, therefore we consider them here.

First, notice that the general term should be GNSS receiver rather than GPS as the latter is only one out of the four currently available navigation satellite-based systems available globally. Some years ago, most receivers in the western hemisphere would certainly be only GPS, but that is no longer the case as most professional receivers are currently compatible with GPS (USA), Glonass (Russia), Galileo (Europe) and Beidou (China). The GNSS receivers account for location, navigation and last, but not least, timing. The time obtained from GNSS receivers is one of the most accurate times available to common users and it can be of interest to synchronize the readings of several sensors scattered across wide areas.

What is important when choosing and using a GNSS receiver is the required accuracy to register sensor data with location. Stand-alone receivers may obtain coordinates with errors of up to several metres. Ground-based and satellite-based augmentation systems are available ranging from sub-metre to centimetre accuracies and precision. Service fees range from free to several hundred Euros per year.

Another important aspect is how to include the coordinates properly in technical documents and presentations. It is important to clarify that any coordinate should be accompanied by their units and by a reference to identify them uniquely on the Earth. That is done by adding to the coordinate tuple the coordinate reference system (CRS) they are referred to. A CRS is a coordinate system that is referenced through a datum to the Earth (OGP, 2012). There are several CRS sub-types that depend on how the Earth's curvature is dealt with. The most common in PA are geographic and projected CRS sub-types. The former uses latitude and longitude expressed in angular units to locate geographic features on Earth. Height is usually given relative to a reference ellipsoid (ellipsoid height) or, if available, to a geoid (orthometric height, equivalent to expressing heights above mean sea level). Projected CRSs are based on geographic ones but use a projection to transfer curved coordinates to a plane. The most used projection in PA is the universal transverse Mercator (UTM) where locations are expressed as Cartesian coordinates (x,y,z) in metres in different zones of a specific UTM grid, where columns are expressed by numbers and rows by letters. To make the coordinates unambiguous, the datum specifies the mathematical model of an ellipsoid to represent the earth and the primer or zero meridian to which longitudes are referred. Thus, when using the GPS system, coordinates provided by receivers are expressed in a geographic CRS sub-type with the WGS84 datum; the correct way to cite them would be as follows, always including the datum:

0.596158° E, 41.629470 N, WGS84 (using dd.dddddd° for latitude and longitude)
0° 35.76948′ E, 41° 37.7682′ N, WGS84 (using dd° mm.mmmmm')
0° 35′ 46.1688″ E, 41° 37′ 46.092″ N, WGS84 (using dd° mm' ss.ssssss")

Care is required with augmentation systems because they sometimes convert the coordinates to a local CRS. For example, the official datum in Europe is the ETRS89 and the one in the USA is NAD83, either with geographic or projected coordinates. In geographic information system software, it is usual to refer to the CRS using the EPSG codes (OGP, 2021). Each code specifies the CRS including its sub-type and the datum. For example, the EPSG code 4326 represents geographic coordinates for the datum WGS84, and the EPSG code 25831 represents projected UTM coordinates for the ETRS89 datum in zone 31.

When projected CRS sub-types are used, it is important to specify what the projection is and the zone that the coordinates refer to. In the UTM projection, the earth's surface is divided into a grid with 60 columns (numbered from 1 to 60) and 24 rows (named from south to north A to Z). There are singularities around the north and south poles and in other areas regarding rows and columns. Thus, with projected coordinates in an official project in Europe, the previous coordinates should be cited as follows:

x or easting: 299761 m, y or northing: 4611429 m, UTM 31 T, ETRS89

where 31 T is the UTM grid zone. It is important to include the UTM zone because the x or eating coordinates might occur in each of the 60 × 24 cells of the UTM grid. The coordinates of the examples define the location of the School of Agrifood and Forestry Science and Engineering of the Universitat de Lleida, in Lleida, Catalonia.

1.4.5 Concepts Related to Metrology

In any book related to sensors and their applications there should be some reference to metrology and how to record and express properly the data sensors produce. In addition to the above definitions of resolution, there are some other basic concepts on sensing to be considered. In this section, we have extracted concepts from the International Vocabulary of Metrology, hereafter VIM (BIPM, 2012) and from the International System of Units brochure, hereafter SI (BIPM, 2019). Both documents are published and updated regularly by the Bureau International des Poids et Mesures (BIPM), which is an international organization with 63 member states and 40 associate states and economies (in January 2021) working cooperatively on measurement science and standards.

Sensing systems provide users with magnitudes of specific properties of a phenomenon, object or substance. The properties are represented by quantities and their magnitudes can be expressed with a number and a reference, usually, a measurement unit. At this point, the authors would like to encourage the use of the International System of Units to ease the scientific use and communication of data in technical or scientific publications (BIPM, 2019).

According to the VIM (VIM, 2012), the term *sensor* refers to one of the elements in a measuring instrument or measuring system. In this book terms such as measuring instrument, measuring system, sensor, transducer and sensing system are used as synonyms for the sake of simplicity and to reach a broader audience.

Once a measuring instrument or system makes a measurement, it provides an indication or reading. Sensor users may be interested in the errors involved and how to minimize them. The error or measurement error is the measurement obtained minus a reference value, resulting in a numerical quantity. The accuracy is the closeness of the measurement to a true value of the quantity being measured. The smaller is the measurement error, the more accurate is the measurement. However, accuracy is not a quantity and should not be given a numerical value. Another important term is precision. According to the VIM, precision is the *closeness between measurements obtained by replicate measurements on the same or similar objects under specified conditions*. Sometimes precision is misused when accuracy is meant; precision represents the dispersion of repeated measurements. Moreover, precision is expressed numerically in terms of standard deviation, variance or coefficient of variation of the data.

To improve the accuracy of measurements, a good calibration is required. A calibration establishes a relation between measurements obtained by measuring systems and those provided by measurement standards. Such relation can be expressed by a calibration diagram or a calibration curve and then used to obtain quantity values from instrument measurements. Once a calibration is obtained, the measurement systems require adjustment to *provide prescribed indications corresponding to given values of the quantity to be measured* (BIPM, 2012). Calibration and adjustment should not be confused.

Finally, as many of the measurement principles, techniques and systems used in PA are not standardized yet, it is always important to include an accurate and detailed description of the techniques and measurement systems used for readers to understand the published results better.

References

BIPM (2012) International vocabulary of metrology (VIM 3rd edn). Bureau International des Poids et Mesures. Available at https://www.bipm.org/en/publications/guides. Accessed 1 Feb 2021

BIPM (2019) SI brochure: the international system of units (SI 9th edn). Bureau International des Poids et Mesures. Available at. https://www.bipm.org/en/publications/si-brochure. Accessed 1 Feb 2021

Blackmore S (1994) Precision farming; an introduction. Outlook Agric 23:275–280

Fenton JP (1998) On-farm experience of precision agriculture. The International Fertiliser Society, York

ISPA (2019) Precision Ag definition. International Society of Precision Agriculture.. Available at http://www.ispag.org/about/definition. Accessed 1 Feb 2021

Kerry R, Oliver MA (2003) Variograms of ancillary data to aid sampling for soil surveys. Precis Agric 4:261–278

OGP (2012) Geomatics guidance note number 7, part 1 (OGP publicactions 373-7-1). International Association of Oil & Gas Producers. Available at. https://epsg.org/guidance-notes.html. Accessed 1 Feb 2021

OGP (2021) Geodetic database. International Association of Oil & Gas Producers.. Available at https://epsg.org/search/by-name. Accessed 1 Feb 2021

Robert PC (2002) Precision agriculture: a challenge for crop nutrition management. Plant Soil 247:143–149

Chapter 2
Satellite Remote Sensing for Precision Agriculture

David J. Mulla

Abstract The objective of this chapter is to review a wide range of satellite sensors that have potential for application in precision agriculture (PA). Classes of satellite sensors included are multispectral, hyperspectral and synthetic aperture radar satellite sensors. Multispectral and radar sensors for digital surface models are also reviewed. For each class of sensor, years of satellite operation, and wavelengths, bandwidths, spectral resolution, revisit frequencies and spatial resolutions are considered. These factors, along with signal to noise ratio and cost, determine the applicability of satellite remote sensing to PA. For each satellite, examples are given regarding the use and potential applications in PA. One of the most promising new satellites for PA applications is the Sentinel-2 platform provided by the European Space Agency (ESA) at no cost. This is an excellent option with its good spatial resolution ranging from 10 to 20 m, high spectral resolution (12 bits per pixel), and a relatively quick revisit frequency of 5 days. Narrow wavebands useful for diagnosing specific soil conditions or crop stressors with hyperspectral satellite imagery are reviewed. Potential applications in PA are discussed involving synthetic aperture radar (SAR), sun-induced fluorescence (SIF) and digital surface model (DSM) satellite products.

Keywords Multispectral · Hyperspectral · Fluorescence · Radar · Digital surface model

2.1 Introduction

Remote sensing has always played an important role in precision agriculture (PA) by providing information about spatial and temporal variation in soil, landscape or crop characteristics that affect management decisions (Bhatti et al. 1991; Moran

---

D. J. Mulla (✉)
Department of Soil, Water & Climate, University of Minnesota, St. Paul, MN, USA
e-mail: mulla003@umn.edu

R. Kerry, A. Escolà (eds.), *Sensing Approaches for Precision Agriculture*, Progress in Precision Agriculture, https://doi.org/10.1007/978-3-030-78431-7_2

et al. 1997; Mulla 2013), including seeding, fertilizing, spraying, irrigating or tilling (i.e. identifying the right input, at the right place, right time and right rate – the 4R approach; Yost et al. 2019).

Satellite remote sensing in PA initially relied on wideband (> 35 nm bandwidth) multispectral remote sensing of bare soils in the blue (B), green (G), red (R) and near-infrared (NIR) bands using Landsat 5 imagery (Bhatti et al. 1991). The spatial resolution (30 m) and return frequency (16 days) of Landsat 5 were acceptable for these initial efforts. Subsequently, satellite remote sensing platforms expanded rapidly to include higher resolution multispectral satellites with quicker return frequencies (e.g. IKONOS and QuickBird) that are more suited for PA applications (Bausch and Khosla 2010). Another innovation was the launch of the Earth Observing (EO) 1 satellite with the Hyperion sensor (Apan et al. 2004) that recorded narrow band (< 11 nm bandwidth) hyperspectral reflectance in the visible (VIS), NIR and short wave infrared (SWIR) wavelengths.

There is a wide variety of other satellite sensors including sun-induced fluorescence (SIF), synthetic aperture radar (SAR) backscatter sensing and satellite-based sources of digital surface models (DSMs) that have been studied infrequently in PA. They have a wide range of wavelengths (VIS, X-, C- and L-band radar), spatial resolutions (5 m to 40 km) and potential applications (photosynthetic rates, soil moisture contents, soil roughness and surface elevation).

Satellite-based imagery has many advantages and disadvantages (Sozzi et al. 2018). Advantages include repeating global spatial coverage, allowing practitioners of PA in any part of the world to download and process imagery for their area of interest on multiple dates. Other advantages include the potential ability of satellite imagery to identify areas of a field with poor soil or unhealthy crop. This advantage is the key feature that makes satellite imagery useful for PA so that management decisions can be customized and problems affecting these regions addressed in a timely fashion. As will be shown here, multispectral and hyperspectral satellite remote sensing have been widely studied for applications in PA, including detection of crop nutrient deficiencies (Söderström et al. 2017), crop diseases and pest activity (Li et al. 2015), crop water stress (Jackson et al. 2004), crop yield and biomass (Thenkabail et al. 2013), soil characteristics (Demattê et al. 2007), and for delineation of management zones (Nawar et al. 2017).

The primary disadvantage of satellite imagery that relies on reflectance is interference from cloud cover. Whitcraft et al. (2015) studied the impact of cloud cover for croplands across the globe on the satellite revisit frequency needed to provide reasonably clear imagery during eight-day periods in different months. When reasonably clear imagery is defined as having at least a 95 % chance of satellite pixels that are unobstructed by clouds, the required satellite revisit frequency during the month of July in the 'corn belt' region of the Midwestern US would range from 1 to 3 days. Landsat 7 satellite (with a revisit frequency of 16 days) is thus not capable of providing cloud free imagery reliably during any 8-day period in July for this region. This disadvantage can be overcome using multispectral satellite imagery with higher revisit frequencies (e.g. WorldView) or SAR satellites that can obtain imagery through clouds.

Other disadvantages of satellite imagery include the need for significant amounts of data processing. Processing could include, for example, mosaicking, orthorectification, masking of clouds or regions of no interest, atmospheric corrections for haze and dust, conversion of digital numbers (DNs) to calibrated reflectance or backscatter values, corrections for angle of incidence of sensor or time of day, and top of atmosphere or top of crop canopy corrections for reflectance (Moran et al. 1997). In addition, there is often a need to determine what wavelengths are of most interest, and to compute appropriate spectral or vegetation indices (Thenkabail et al. 2013), or statistical analyses such as regression, principal component analysis (PCA) or machine learning to extract usable information. Processing takes time, requires advanced computing and geo-spatial skills, and limits the ability to make timely management recommendations.

The objective of this chapter is to review a wide range of satellite sensors that have potential application in PA. Classes of satellite sensors reviewed include multispectral, hyperspectral, sun-induced fluorescence and synthetic aperture radar satellite sensors. In addition, the use of multispectral and radar sensors for digital surface models is reviewed. For each class of sensor, the review discusses years of satellite operation, and wavelengths, bandwidths, spectral resolution, revisit frequencies and spatial resolutions involved. Examples are provided showing how information from each satellite sensor has been used in PA, or in applications that have potential for use in PA.

Many satellites included in the review have been decommissioned and no longer actively collect imagery. Nevertheless, these sources have been used for important PA research, and could continue to provide useful archival information for PA over the next decade in combination with historical yield maps, soil maps, and other auxiliary information (e.g. for delineation of management zones).

2.2 Multispectral Satellites

Multispectral remote sensing typically involves measuring broadband (>35 nm bandwidth) reflectance in the VIS and NIR bands from 430–950 nm (Table 2.1). Reflectance data may be correlated directly with soil or crop characteristics of interest in the VIS (B, G, R) and NIR wavelengths, or may be converted into one of many vegetation indices prior to correlation (Mulla 2013). Bare soil reflectance can indicate spatial or temporal variation in soil organic matter, moisture, texture, bulk density, carbonate or iron oxide content (Mzuku et al. 2005). Reflectance from crops can indicate spatial and temporal variation in the concentration of plant pigments or crop biomass characteristics (Pinter Jr et al. 2003). Crop pigments include chlorophyll a and b, anthocyanin and carotenoids. Each absorbs visible radiation preferentially at specific wavelengths, including B and R wavelengths for chlorophyll a (430 and 650 nm) and b (450 and 650 nm), and violet to G (550 nm) for carotenoids. Anthocyanins have relatively weak absorption at all VIS wavelengths. As crop nitrogen deficiency increases, reflectance in the B and R wavelengths typically

Table 2.1 Multispectral satellite years of operation, revisit frequency, spatial resolution, wavelength, bandwidth and radiometric resolution (bits) specifications

Satellite	Years	Revisit Time (d)	Spatial Resolution (m)	Wavelengths (nm)	Bandwidths (nm)	Bits
Landsat 7 (ETM+)	1999–present	16	30	450–520 (B), 520–600 (G), 630–690 (R), 770–900 (NIR), 1550–1750 (SWIR 1), 2090–2350 (SWIR 2), 10400–12500 (TIR)	60–80 (VIS), 130–200 (NIR), 160–200 (SWIR), 2100 (TIR)	8
Ikonos-2	2000–2015	3	3.2	430–535 (B), 466–620 (G), 590–710 (R), 715–918 (NIR)	66–89 (VIS), 96 (NIR)	11
QuickBird	2001–2015	1–3.5	2.9	450–520 (B), 520–600 (G), 630–690 (R), 760–900 (NIR)	60–80 (VIS), 140 (NIR)	12
RapidEye	2009–present	5.5	5	440–510 (B), 520–590 (G), 630–685 (R), 690–735 (RE), 760–850 (NIR)	45–70 (VIS), 90 (NIR)	12
GeoEye-1	2009–present	3	1.84	450–510 (B), 510–580 (G), 655–690 (R), 780–920 (NIR)	35–70 (VIS), 140 (NIR)	11
WorldView 2	2009–present	1.1	0.46	450–510 (B), 510–580 (G), 585–625 (Y), 630–690 (R), 705–740 (RE), 770–895 (NIR 1), 860–1040 (NIR 2)	35–70 (VIS), 125–180 (NIR)	11
WorldView 3	2014–present	<1	1.24–3.7	450–510 (B), 510–580 (G), 585–625 (Y), 630–690 (R), 705–740 (RE), 770–895 (NIR 1), 860–1040 (NIR 2), 1195–2365 (SWIR 1-8)	35–70 (VIS), 125–180 (NIR), 30–70 (SWIR)	11
Pleiades 1A,B	2011–present	1	2	430–550 (B),490–610 (G), 600–720 (R), 750–950 (NIR)	120 (VIS), 200 (NIR)	12
SPOT 6,7	2012–present	1–5	6	450–520 (B), 530–600 (G), 620–690 (R), 780–920(NIR)	70–80 (VIS), 120 (NIR)	12
ResourceSat-2	2012–present	26	5.8	520–590 (G), 620–680 (R), 770–860 (NIR)	60–70 (VIS), 90 (NIR)	10

(continued)

Table 2.1 (continued)

Satellite	Years	Revisit Time (d)	Spatial Resolution (m)	Wavelengths (nm)	Bandwidths (nm)	Bits
Landsat 8	2013–present	16	30	452–512 (B), 533–590 (G), 636–673 (R), 851–879 (NIR), 1566–1651 (SWIR 1), 2107–2294 (SWIR 2), 1363–1384 (Cirrus), 10600–11190 (TIR 1), 11500–12510 (TIR 2)	37–60 (VIS), 28 (NIR), 85–187 (SWIR), 590–1010 (TIR)	12
KompSat-3	2013–present	1.4	2.8	450–520 (B), 520–600 (G), 630–690 (R), 760–900 (NIR)	60–80 (VIS), 140 (NIR)	14
Flock Dove	2016–present	1	3–4	455–515 (B), 500–590 (G), 590–670 (R), 780–860 (NIR)	60–90 (VIS), 80 (NIR)	16
Sentinel 2A, 2B	2015–present	5	10–20	492.1* (B), 559* (G), 665* (R), 703.8* (RE 1), 739.1* (RE 2), 779.7* (RE 3), 833* (NIR 1), 864* (NIR 2), 1376.5* (Cirrus), 1610.4*(SWIR 1), 2185.7* (SWIR 2)	18–98 (VIS), 32–45 (NIR), 141–238 (SWIR)	12

Multispectral bands include blue (B), green (G), red (R), and near-infrared (NIR), as well as short wave infrared (SWIR) and thermal infrared (TIR) wavelengths

increases as a result of reduced absorption by nitrogen containing chlorophyll pigments. Reflectance at G wavelengths also increases because chlorophyll pigments do not absorb radiation strongly at G wavelengths. Crop canopy characteristics such as biomass, leaf area index (LAI), leaf orientation, size and density are typically expressed in reflectance at NIR wavelengths. As canopy biomass increases, NIR reflectance generally increases.

Changes in multispectral reflectance in satellite imagery across a field and on different dates can indicate a wide range of conditions that are amenable to precision management. These include soil fertility status, crop stresses due to nutrients, disease, insects and water, and differences in growth, biomass and yield of crops from a variety of factors (Mulla and Miao 2016).

Wavelengths and bandwidths for reflectance data are relevant for PA applications. Nearly all multispectral satellites collect imagery at B, G, R and NIR wavelengths, and the bandwidths are generally between 35 and 200 nm (Table 2.1). Narrower bandwidths (higher spectral resolution), in general, provide greater ability to discriminate between different types of factors that cause crop stress (e.g. crop deficiencies due to nitrogen or phosphorus). Multispectral satellites may also collect reflectance at yellow, red edge (RE) and other NIR wavelengths. Red edge reflectance occurs from 680 to 750 nm between the R and NIR wavelengths. It is the region where reflectance increases dramatically as chlorophyll absorption decreases.

The SWIR and thermal IR (TIR) capabilities also exist on some multispectral remote sensing satellites, and these bands are often useful for estimating crop residue cover or crop water stress, although the utility of TIR bands is diminished because of coarse spatial resolution (120 m for Landsat 8). Another important aspect of remote sensing imagery is radiometric resolution. This refers to the number of individual grey scale values measured between black and white (Al-Wassai and Kalyankar 2013). As the number of grey scale values (DNs) increases, more detail can be distinguished in the remote sensing image. Radiometric resolution is expressed in terms of the number of bits per pixel. Radiometric resolution for Sentinel-2A imagery with 12 bits provides 4096 DNs, while for ResourceSat 2 imagery there are 10 bits or 1024 DNs (four times fewer individual grey scale values).

Spatial resolution and revisit frequency of satellite-based multispectral imagery are of primary relevance to applications involving PA (Fig. 2.1). Spatial resolution of multispectral satellites varies from 30 to 0.5 m. Pixels with coarser spatial resolution are more likely to be heterogeneous, while finer spatial resolution generally leads to more homogeneous pixels. At coarse spatial resolution, reflectance values for a single pixel may represent an average for bare soil, healthy or diseased crop and sunlit or shaded crop, making it difficult to identify specific locations accurately where management interventions are needed. At fine spatial resolution, these limitations become less. Revisit frequency of multispectral satellites varies from 1 to 15 days. Satellites with revisit frequencies of from 1 to 3 days are optimum in areas of the world with frequent cloud cover, but unfortunately satellites with this capacity (e.g. WorldView and Planet) only provide user access at rather substantial costs. Landsat 8 and Sentinel-2 provide multispectral imagery at no cost, and of these two,

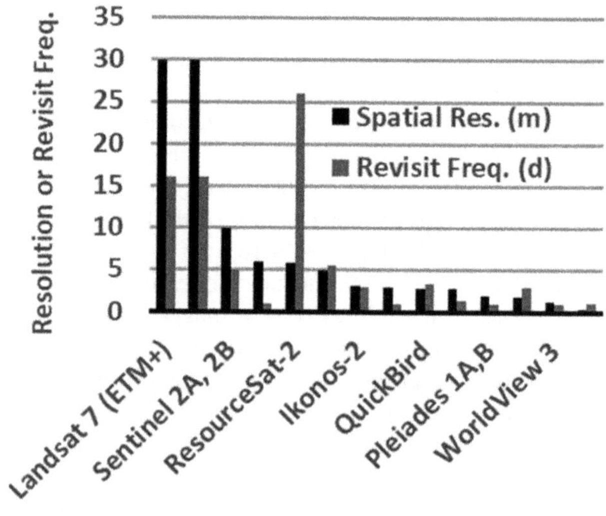

Fig. 2.1 Spatial resolution (m) and revisit frequency (d) for multispectral satellites

the 5-day revisit frequency of Sentinel-2 is reasonably useful for PA applications in large regions of the world.

The cost of acquiring satellite imagery is an important consideration affecting the applicability to PA. The European Space Agency provides Sentinel-2 multispectral satellite imagery at no cost. This is an excellent option for PA applications with its good spatial resolution ranging from 10 to 20 m, high spectral resolution (12 bits per pixel) and a relatively quick revisit frequency of 5 days. Landsat 8 multispectral imagery is also free, but this source is less appealing for PA applications (especially in cloudy areas of the world) because of a relatively long revisit frequency (16 days) and a coarser spatial resolution (30 m).

2.2.1 *Landsat 7 and 8*

Landsat 7 and 8 are US-based public satellite remote sensing systems that have provided data since 1999 and 2013, respectively. Revisit time, spatial and spectral resolution, and wavebands for Landsat 7 and 8 are summarized in Table 2.1, while swath width, cost and minimum area for purchased data are summarized in Table 2.2. Landsat 8 imagery is free to use, which, together with the moderate spatial resolution, makes it attractive for applications involving PA. A typical Landsat 7 or 8 scene comprises an area of 37,000 km^2, which is an extremely large area or region and hence the imagery must be re-sampled for precision agricultural applications.

Numerous applications of Landsat imagery exist in PA (Table 2.3). Scudiero et al. (2015) used Landsat 7 imagery to estimate regional variation in soil salinity across agricultural fields in the west San Joaquin Valley. A canopy response salinity index (CRSI) based on NIR, R, G and B reflectance was calibrated and validated

Table 2.2 Multispectral satellite swath width, cost and minimum area purchased

Satellite	Swath Width (km)	Cost (US $/km^2)	Minimum Area Purchased (km^2)
Landsat 7 (ETM+)	185	Free after 2008	37,000
Ikonos 2	11.3	$10.00	100
QuickBird	16.4	$17.50	100
RapidEye	77	$1.42	100
GeoEye-1	15.2	$27.50	100
WorldView-2	16.4	$29.00	100
WorldView-3	13.1	$24.00	100
Pleiades 1A, B	20	$21.25	100
SPOT-6,7	60	$5.75	3600
ResourceSat 2	70	$147*	4900
Landsat 8	185	Free	37,000
KompSat 3	15	$24.00	25
Flock dove	20–24.6	$1.22	100
Sentinel-2A, 2B	290	Free	12,000

Table 2.3 Satellite based multispectral broad-band vegetation indices used in precision agriculture. B, G, R, RE, NIR and SWIR refer to blue, green, red, red edge, near-infrared and shortwave infrared reflectance. LE, Rn, H and φ_q are latent heat, net radiation, sensible heat and soil heat flux

Index and Satellite	Name of Index and Example Applications	Wavebands and References
NDVI Ikonos GeoEye-1 WorldView-2 Landsat 8 KompSat 3	Normalized difference vegetation index Variable-rate N and fungicide in sugar beets N deficiency in turf grass Citrus greening disease Purple spot disease in asparagus Area of rice paddy fields	(NIR-R)/(NIR ± R) Seelan et al. (2003), Caturegli et al. (2015), Li et al. (2015), Navrozidis et al. (2018) and Yeom et al. (2013)
GNDVI QuickBird	Green NDVI N deficiency in maize	(NIR-G)/(NIR ± G) Bausch and Khosla (2010)
GRVI Planet dove	Green relative vegetation index Biomass of alfalfa and maize	(G-R)/(G ± R) Houborg and McCabe (2016)
NDRE RapidEye WorldView-3	Normalized difference red edge N uptake in wheat Avocado fruit size and yield	(NIR-RE)/(NIR ± RE) Magney et al. (2017) and Robson et al. (2017)
MCARI RapidEye	Modified chlorophyll absorption in reflectance index Leaf N status in wheat	((RE-R)-0.2)(RE-R)/(RE) Eitel et al. (2007)
CIred edge Sentinel-2 Sentinel-2	Chlorophyll index red edge Canopy chlorophyll content in potatoes N uptake in wheat	(RE1/R)-1 Clevers et al. (2017) and Söderström et al. (2017)
GRVI QuickBird	Green ratio vegetation index Sorghum yield	NIR/G Yang et al. (2006)
SAVI Landsat 7 Landsat 8	Soil adjusted vegetation index Soil properties Potato yield	1.5*(NIR-R)/(NIR ± R ± 0.5) Demattê et al. (2007) and Al-Gaadi et al. 2016
TVI RapidEye SPOT-6	Triangular vegetation index Leaf N status in wheat Powdery mildew disease in wheat	0.5*[120(NIR-R)-200(R-G)] Eitel et al. (2007) and Yuan et al. (2016)
ARVI SPOT-6	Atmospherically resistant vegetation index Powdery mildew disease in wheat	NIR-(2R-B)/NIR+(2R-B) Yuan et al. (2016)
NDWI Landsat 7	Normalized difference water index Crop water stress in maize and soya beans	(NIR-SWIR)/(NIR ± SWIR) Jackson et al. (2004)
CRSI Landsat 7	Canopy response salinity index Soil salinity	(NIR*R)-(G*B)/(NIR*R) ± G*B) Scudiero et al. (2015)
METRIC Landsat 8	Mapping ET at high resolution with internalized calibration ET in maize and cotton	$LE = R_n - H - \varphi_q$ Gowda et al. (2008)

using ground truth measurements of soil electrical conductivity in 22 agricultural fields with an EM38 Dual Dipole sensor (Geonics Ltd., Mississauga, Ontario, Canada). A wide range in soil salinity characteristics was discovered, ranging from no salinity to extremely saline (>25 dS m^{-1}) areas.

Demattê et al. (2007) used Landsat 7 to estimate spatial variation in soil properties across large agricultural regions of Brazil. Ground truth soil sampling was conducted across an area of 43,000 ha. For calibration, only remotely sensed pixels from bare soil were used, based on a soil adjusted vegetative index (SAVI) and a soil line developed by plotting red reflectance on the x-axis against NIR reflectance on the y-axis. Soil texture and organic matter content were accurately predicted using regression equations developed using Landsat 8 reflectance in the R, NIR, and SWIR 1 and 2 bands. Cation exchange capacity was accurately predicted using a regression equation based on Landsat 8 reflectance in the G, NIR, and SWIR 1 and 2 bands. Soil fertility characteristics, including soil potassium, phosphorus and pH were only poorly predicted using Landsat 8 reflectance. Likewise, Pongpattananurak et al. (2012) used Landsat 7 imagery together with geo-referenced ground truth soil samples, to model the spatial distribution of soil texture for the state of Jalisco, Mexico, an area of approximately 8 million ha.

Al-Gaadi et al. (Al-Gaadi et al. 2016) used Landsat 8 imagery to estimate spatial patterns in potato yield for three centre pivot irrigated fields in eastern Saudi Arabia. Spectral indices that had strong predictive power for potato yield include the normalized difference vegetation index (NDVI), estimated from the ratio of NIR and R reflectance values (NIR–R)/(NIR + R), and the soil adjusted vegetative index (SAVI), estimated by (NIR–R)/(NIR + R + L)(1 + L), where L is a canopy background adjustment factor. Differences in potato yield across each field were as large as 10 t ha^{-1}, indicating that variable management of irrigation and fertilizer applications are warranted.

Jackson et al. (2004) estimated crop water stress for maize (*Zea mays*) and soya bean (*Glycine max*) crops in Iowa, USA using Landsat 7 imagery based on a normalized difference water index (NDWI = (NIR–SWIR)/(NIR + SWIR)). The NDWI was closely correlated with ground truth measurements of vegetative water content. Regression equations for vegetative water content as a function of NDWI were used to develop regional maps showing vegetative water content on five dates during the 2002 growing season with clear sky satellite imagery.

Remotely sensed information is useful for monitoring crop water stress and making management decisions relevant to variable-rate irrigation (Khanal et al. 2017; de Lara et al. 2019; Seigfried et al. 2019). Evapotranspiration (ET) is one of the most important components of the water balance in semi-arid areas; it is a key factor for optimizing irrigation water management. Allen et al. (2007) pioneered the use of a satellite-based energy balance method for mapping evapotranspiration with internalized calibration (METRIC). Gowda et al. (Gowda et al. 2008) estimated ET with good accuracy based on Landsat satellite imagery for irrigated maize and cotton (*Gossypium spp.*) in Texas, USA, relative to soil moisture balance estimates of ET (Gowda et al. 2008). The METRIC approach for estimating ET is available worldwide using Landsat 8 satellite imagery via Google Earth's EEFlux engine (Allen et al. 2015).

2.2.2 Ikonos 2

Ikonos 2 was a private satellite operated by DigitalGlobe providing data from 2000 to 2015 (Table 2.1) before being decommissioned (Table 2.1). Seelan et al. (2003) used Ikonos imagery over four growing seasons across the North Central US region to help agricultural producers improve management decisions related to variable-rate nitrogen fertilizer applications, effectiveness of variable-rate fungicide applications, delineation of management zones and estimation of crop damage caused by spray drift, flooding and Rhizoctonia infestation in sugar beet (*Beta vulgaris*) (Table 2.3).

Soil properties such as clay and total carbon content were estimated by Sullivan et al. (2005) using Ikonos imagery for two Alabama fields in a cotton–peanut (*Arachis hypogaea*) rotation. A combination of Ikonos G and NIR reflectance values was able to estimate soil clay or organic matter contents at accuracies ranging from 34 to 61 % in these fields (prior to tillage).

2.2.3 QuickBird

QuickBird was a private satellite operated by DigitalGlobe providing data from 2001 to 2015, when it re-entered the Earth's atmosphere (Table 2.1). Bausch and Khosla (2010) used QuickBird imagery to assess spatial patterns in nitrogen deficiency across an irrigated maize field in eastern Colorado (Table 2.3). The field had several nitrogen strip treatments, including a reference strip receiving sufficient nitrogen. QuickBird Green NDVI (GNDVI) values were calculated using (NIR–G)/(NIR + G), and these values were normalized relative to the nitrogen reference strip. By comparison with ground truth data, it was determined that QuickBird values for normalized GNDVI of <0.96 indicated nitrogen deficiency. Maps showing the spatial distribution of nitrogen deficiency were developed so that the farmer could spread additional nitrogen fertilizer variably in these areas to correct nitrogen deficiencies.

Yang et al. (2006) evaluated the feasibility of QuickBird imagery to estimate spatial patterns in sorghum (*Sorghum bicolor*) yield for two agricultural fields in Texas (Table 2.3). QuickBird green ratio vegetation index (GRVI=(NIR/G)) reflectance values were moderately correlated with yield monitor data for one field, and strongly correlated in the second field. Unsupervized classification of QuickBird imagery was used to prepare maps for each study field showing regions with low, moderate or high sorghum yield. These maps could provide a basis for dividing each field into management zones.

2.2.4 RapidEye

RapidEye launched a constellation of five satellites in 2009 with technical specifications summarized in Tables 2.1–2.2. RapidEye was the first satellite to collect red edge (RE) wavelength reflectance data at wavelengths of 690–735 nm. The RE reflectance is known to be sensitive to nitrogen deficiency in crops.

RapidEye imagery has been widely used in PA studies to estimate crop nitrogen status. Eitel et al. (2007) reported that the ratio of RapidEye estimated modified chlorophyll absorption reflection index (MCARI = [((RE–R) –0.2)(RE–R)](RE–G)] and modified triangular vegetation index (TVI) spectral indices were significantly correlated with nitrogen content of flag leaves in three Idaho wheat (*Triticum aestivum*) fields (Table 2.3). The TVI is estimated from NIR and G reflectance (0.5*[120(NIR–R) –200(R–G)]). Whereas MCARI is very sensitive to chlorophyll content of leaves, TVI is more affected by structural features of wheat plants. The ratio of MCARI/MTVI was a better predictor of leaf N content than either spectral index alone. Magney et al. (2017) expanded on the Eitel et al. (2007) study by collecting RapidEye imagery from four Idaho and Washington wheat farms over three seasons. The normalized difference red edge (NDRE=(NIR–RE)/(NIR + RE)) spectral index explained 81 % of the variability in end-of-season cumulative wheat N. The NDRE index was superior to sixteen other vegetative indices studied, including NDVI, MCARI, GNDVI and TVI.

2.2.5 GeoEye-1

The GeoEye-1 satellite, operated by DigitalGlobe, provides multispectral reflectance data from 2009 to the present (Tables 2.1–2.2). Caturegli et al. (2015) evaluated the feasibility of using GeoEye-1 imagery to estimate spatial patterns in N deficiency for three warm-season and two cool-season varieties of turfgrass in the Tuscany region of Italy (Table 2.3). Each experimental turfgrass variety received a wide range of N fertilizer applications, and two weeks later GeoEye-1 satellite-based remote sensing NDVI data were collected over the experimental sites on the same day that turfgrass clippings were collected for laboratory analysis of tissue N content. Satellite NDVI values explained from 87–94 % of the variability in tissue N content for warm- and cool-season turfgrass varieties. GeoEye-1 satellite imagery could be used to estimate spatial variations in N deficiency for turfgrass grown on golf courses, parks and turfgrass production farms.

2.2.6 WorldView-2 and 3

Launched in 2009, WorldView-2 provides multispectral imagery across a wide range of wavebands at a revisit frequency of 1.1 days with a spatial resolution of 0.46 m (Table 2.1). Because of the very high cost (Table 2.2), only large corporations and other wealthy clients use WorldView-2 or 3 for specialized data collection.

Li et al. (2015) used WorldView-2 imagery collected in December, 2010 to identify orange trees in a Florida test grove that were afflicted with the citrus greening disease (Table 2.3). Trees were classified as being either healthy or diseased based on ground truth observations taken with a handheld spectrometer. Tree canopies were segmented from background soil or grass, and further masked to eliminate portions of the canopy in shade or full sun. The remaining part of the canopy was retained for analysis using WorldView-2 NDVI and GNDVI spectral indices. On average, WorldView-2 spectral indices could distinguish between healthy and diseased trees 76 % of the time. This approach could help citrus producers to identify diseased trees rapidly in their orchards.

Robson et al. (2017) used WorldView-3 imagery collected in two commercial avocado (*Persea americana*) groves in the Queensland, Australia region to estimate spatial and temporal variation across years in yield and fruit size (Table 2.3). The NDRE spectral indices from WorldView-3 imagery had consistently positive correlation with fruit yield and size from hand harvested trees in the two groves. Correlation coefficients were generally weak, however, ranging from 0.29 to 0.37 for fruit yield and size, respectively. WorldView satellite estimates were greatly improved, however, compared with visual estimates of fruit yield and size by avocado producers. Field-scale maps showing spatial variation in avocado fruit yield could be used by producers for decisions concerning the rate of fertilizer, herbicides and irrigation to apply in each portion of the orchard.

2.2.7 Pleiades 1A and 1B

Pleiades 1A and 1B were satellites launched by the French Space Agency in 2011 and 2012, respectively. Technical specifications and costs of Pleiades imagery are summarized in Tables 2.1 and 2.2 Navrozidis et al. (Navrozidis et al. 2018) compared the feasibility of using Pleiades 1A and Landsat 8 imagery to estimate infestation of asparagus with purple spot disease in southern France (Table 2.3). Ground truth estimates were estimated by experienced crop scouts in three agricultural fields using a numerical infestation severity scale (0–100). Pleiades 1A imagery for the same fields was used to estimate NDVI and SAVI at spatial resolutions of 2 m. Pleiades 1A NDVI values were moderately well correlated with ground truth measurements of disease infestation severity. Regression models for disease severity based on Pleiades 1A imagery were used to develop maps showing the spatial variation in disease severity for each of the three study fields.

2.2.8 SPOT-6 and 7

The satellite pour l'observation de la terre (SPOT) 6 and 7 satellites were privately operated by SPOT Image and have provided multispectral data since 2012 with specifications summarized in Tables 2.1 and 2.2. The SPOT satellites orbit the Earth in coordination with the Pleiades 1A and 1B satellites. Yuan et al. (2016) used SPOT-6 imagery to map powdery mildew infestation of winter wheat in Shaanxi Province of central China (Table 2.3). A FieldSpec spectrometer (ASD Inc., Boulder, CO) was used to collect ground truth data in wheat canopies with no, light or heavy disease severity. The G and R SPOT-6 reflectance features, and SPOT-6 estimates of NDVI, TVI, and atmospherically resistant vegetation index (ARVI) features were strongly correlated with ground truth data for disease incidence. The ARVI is estimated using NIR, R and B reflectance [NIR–(2R–B)/NIR+(2R–B)]. Estimation of disease incidence was achieved using spectral angle papper (SAM) with all five SPOT-6 reflectance features, after first normalizing the reflectance features relative to reflectance of a healthy crop. Maps showing disease incidence for two regions of Shaanxi Province had prediction accuracies of 86 and 68 %, respectively, for healthy and diseased crops.

2.2.9 ResourceSat 2

ResourceSat 2 is an Indian satellite providing multispectral data from 2012 to the present (Tables 2.1 and 2.2). Kumar et al. (2015) used ResourceSat 2 LISS IV imagery from a single date in spring of 2013 to classify crops in the Varanasi region of India. support vector machines (SVM) could classify crops with an overall accuracy of over 93 % relative to ground truth data collected from the region by trained observers. Classification accuracy was tested for all possible combinations of pairs of different crops. Crop classification included wheat, maize, linseed (*Linum usitatissimum*), lentil (*Lens culinaris*), mustard (*Brassica juncea*) and barley (*Hordeum vulgare*), as well as minor crops lumped in a single category, including gram (*Cicer aerietinum*), sugarcane (*Saccharum officinarum*), pigeon pea (*Cajanus cajan*) and vegetables.

2.2.10 KompSat 3

KompSat 3 (also known as Arirang 3) is a Korean satellite launched in 2013 with technical and cost specifications summarized in Tables 2.1 and 2.2. KompSat 3 imagery has been used infrequently for agricultural applications. Yeom et al. (2013) used NDVI values obtained from KompSat 3 imagery taken at two distinct phenological growth stages to distinguish rice (*Oryza sativa*) paddies in two districts of

South Korea from other landuse classes (Table 2.3). Aerial imagery from three years was obtained over known rice paddies to develop a dataset that could be used to train the rice paddy maximum likelihood classification algorithm. Accuracies of maps correctly identifying rice paddies ranged from 80–83 % in the two agricultural districts. The study indicated that the maps were useful for PA, because they could be used to estimate the area of each rice paddy. This information is needed to determine how much fertilizer the producer needs to purchase for each rice paddy.

2.2.11 Planet Lab Flock CubeSats

Planet Labs developed an innovative approach to satellite imagery with the development and deployment of miniature Flock CubeSats weighing 4 kg each, with dimensions of 0.003 m^3. Miniaturization allowed a single rocket to deploy multiple Dove satellites (a Flock of Doves) simultaneously. These were deployed in 10 flocks launched between 2016 and 2019 with 3–88 doves per flock. Each Flock of Doves is placed in a sun synchronous orbit around Earth, with successive CubeSats in a given launch following slightly different offset (but somewhat overlapping) orbital tracks to ensure wider swath coverage in each orbit. Imagery from each satellite must be orthorectified and combined with images from all other CubeSat imagery in the same Flock to produce a scene. Specifications for Dove CubeSats are summarized in Tables 2.1 and 2.2.

Houborg and McCabe (2016) were among the first scientists to explore the use of Planet Dove CubeSats for PA (Table 2.3). At that time, Dove CubeSats followed the International Space Station orbit, had relatively low signal to noise ratios, lacked consistency in reflectance from different CubeSats in the Flock, and lacked corrections for atmospheric interference or top of the atmosphere radiance. To overcome these limitations, reflectance from CubeSats was calibrated using Landsat 8 imagery (after atmospheric corrections), which is characterized by large signal to noise ratios and high quality sensors. Planet CubeSat and Landsat 8 imagery were obtained in four successive months for 47 centre pivot irrigated agricultural fields near Riyadh, Saudi Arabia. The difference in timing of imagery from CubeSat and Landsat 8 was no greater than 4 days. Planet imagery was degraded to a spatial resolution of 30 m and spatially co-registered to match Landsat 8 spatial resolution and spatial location, and a series of data mining operations was applied to develop ten rules for predicting Planet NDVI based on Landsat NDVI and Planet RGB reflectance and Green–Red VI (GRVI=(G–R)/(G + R)) spectral index values. These relationships (leading to an accuracy of 97 % with the spatially degraded Planet NDVI values) were then used to estimate Planet NDVI values at a spatial resolution of 3 m. Resulting maps of NDVI showed high spatial resolution in identifying spatial variation in biomass for irrigated alfalfa (*Medicago sativa*), grass (*Dactylis glomerata*) and maize (*Zea mays*) fields.

McCabe et al. (2017) conducted a follow-up study of spatial patterns in LAI and crop evapotranspiration (Table 2.3) with more recently launched Dove CubeSats for the 47 centre pivot irrigated fields near Riyadh evaluated by Houborg and McCabe (2016). By this time, Dove CubeSats were being launched into sun synchronous orbits and sensors on different Dove satellites were cross-calibrated using fixed ground calibration sites based on imagery from RapidEye, Landsat 8 and Sentinel-2 satellites, which all have high quality sensors and radiometric calibration. Landsat 8 NDVI data together with other VI data were used to extract LAI values from Dove imagery at spatial resolutions of 30 m, and these relationships were used to estimate LAI from 3 m Dove reflectance data. In addition, a Priestly–Taylor algorithm was used to estimate crop water use from spatial patterns in LAI.

2.2.12 Sentinel-2A and 2B

The European Space Agency launched Sentinel-2A and 2B in 2015 and 2017, respectively. Each satellite orbits the Earth with a phase difference of 180°, with technical specifications and swath width summarized in Tables 2.1 and 2.2. Unlike other satellites, Sentinel collects imagery in three distinct RE wavebands and two distinct NIR wavebands. Sentinel bandwidths are narrower than those collected by other multispectral satellites, offering greater diagnostic capability of specific crop stressors, e.g. stresses from nutrient deficiencies, disease or water. Significantly, the European Space Agency provides Sentinel-2 imagery at no charge.

Clevers et al. (2017) studied canopy chlorophyll content (CCC) of potato crops (Table 2.3) grown in the Netherlands using Sentinel-2 imagery collected from May to October, 2016. Ground truth data were collected for CCC using a SPAD meter and a CROPSCAN multispectral radiometer (Rochester, MN). Potatoes received 10 N treatments differing in amount of initial and split fertilizer application rates. A Sentinel-2 chlorophyll index based on RE1 reflectance ($CI_{red\ edge}$ = ((RE1/R)–1)) explained 58 % of the spatial variation in measured CCC values. In general, CCC values and N deficiencies were most pronounced in regions of the field receiving the lowest rates of N fertilizer.

Söderström et al. (Söderström et al. 2017) found that Sentinel-2 $CI_{red\ edge}$ values were useful in predicting crop N uptake for winter wheat crops across southern Sweden (Table 2.3). Ground truth data were collected weekly from mid-April to mid-June from 26–36 winter wheat fields using a handheld Yara N sensor (Yara Gmbh, Hanninghof, Germany). Sentinel-2 $CI_{red\ edge}$ values explained 88 % of the spatial variation in measured above-ground crop N uptake. This approach shows the feasibility of having technical service providers collect ground truth data on crop N uptake at different growth stages for selected fields to develop regional maps of crop N uptake based on Sentinel-2 imagery. These regional maps could be used to make management decisions on in-season variable-rate N fertilizer applications.

2.3 Hyperspectral Satellites

The primary advantage of hyperspectral remote sensing relative to multispectral sensing is that it involves narrower bandwidths (<10 nm). Reflectance spectra from hyperspectral imagery show excellent continuity from one band to another, typically starting at B wavelengths and ending in the NIR or SWIR wavelengths. Continuous spectra are amenable to more sophisticated and wavelength specific interpretations than multispectral imagery. Sophisticated processing techniques for hyperspectral imagery include lambda–lambda plots to determine orthogonal pairs of wavelengths that contain information that is highly correlated with soil or crop characteristics. In general, it is not useful to construct narrow band hyperspectral vegetation indices using pairs of closely spaced wavelengths because reflectance at those wavelengths respond similarly (redundancy) to soil or crop characteristics. It is more instructive to construct vegetation indices using wavelengths that are orthogonal to maximize the information content. For example, Thenkabail et al. (2013) used EO 1 Hyperion hyperspectral narrow band vegetation indices to study wheat, rice, cotton, alfalfa and maize crop characteristics (biomass, LAI and crop height) in Uzbekistan. Lambda–lambda graphs for two band NDVI-like vegetation indices showed that the optimum wavelengths involved a combination of red and NIR bands (e.g. $(NIR_{855}—R_{687})/(NIR_{855} + R_{687})$). The best four bands for crop biomass, LAI and crop height were at 550, 687, 855 and 1180 nm.

Hyperspectral imagery can be interpreted using a variety of methods other than lambda–lambda comparisons. These include partial least squares regression (PLSR) and PCA. Both of these techniques attempt to determine which wavelengths are most strongly correlated with reflectance characteristics of the soil or crop being studied. In contrast to PCA, PLSR eliminates wavelengths that contain redundant information. Casa et al. (2013) used PLSR to estimate spatial variation in soil texture from two fields near Rome, Italy based on Proba 1 CHRIS hyperspectral reflectance data collected in 37 narrow bands. The PLSR model relied primarily on four bands at 489, 631, 683 and 987 nm to estimate clay, silt or sand content.

Hyperspectral remote sensing is much more powerful than multispectral imagery in identifying specific soil or crop characteristics of interest, such as distinguishing spatial variation in soil organic carbon content (Gomez et al. 2008) from variation in soil texture (Casa et al. 2013), or distinguishing crop characteristics such as plant pigment concentrations from crop biomass characteristics, crop residue cover or crop water stress (Thenkabail et al. 2013). Crop pigments have specific absorption or reflectance bands at 450, 515, 550, 570 and 650 nm. In contrast, crop biomass, LAI and crop height affect reflectance at 687, 760, 855, 1045 and 1100 nm. Crop residue and the associated content of lignin or cellulose affects reflectance at 1548, 1620, 1690, 2025, 2133 and 2205 nm. Finally, crop water or water stress affects reflectance at 970, 1180, 1245, 1450, 1760, 1950, 2050, 2145 and 2173 nm. The narrow bands of hyperspectral remote sensing allow for numerous combinations of two and three band vegetation indices (Thenkabail et al. 2013; Mulla 2013). Selection of relevant wavelengths and their combinations through vegetation indices has an important effect on the ability to discern specific soil or crop features of interest.

2.3.1 Earth Observing (EO) 1 Hyperion

Earth Observing (EO) 1 with a Hyperion imaging spectrometer was the first satellite to record hyperspectral reflectance data. The EO 1 Hyperion was launched in 2001 and decommissioned in 2016. The revisit frequency was 16 days, and spatial resolution was 30 m (Table 2.4). Hyperspectral reflectance data were collected at wavelengths ranging from 400–2500 nm in 10-nm bandwidths.

Wu et al. (2010) evaluated the accuracy of predicting leaf chlorophyll content and LAI for eight types of vegetation (including flax (*Linum usitatissimum*), maize (*Zea mays*), potato (*Solamum tuberosum*) and tea (*Camellia sinensis*) crops) in thirty sites in China using ten vegetation indices calculated from Hyperion imagery (Table 2.5) collected on March 28, 2004. Ground truth data were obtained using laboratory analysis of leaf tissue samples, whereas LAI was estimated using an LAI-2000 Plant Canopy Analyzer.

Vegetation indices evaluated using linear regression with ground truth measurements included narrow band versions of NDVI, a modified NDVI, SR (=R_{750}/R_{705}), TCARI (=3(R_{750}–R_{705}) –0:2(R_{750}–R_{550})(R_{750}/R_{705})), MCARI (=TCARI/3), OSAVI (=(1 + 0:16)(R_{750}–R_{705})(R_{750} + R_{705} + 0.16)) and TVI, as well as ratios between TCARI/OSAVI and MCARI/OSAVI (Wu et al. 2010). The ratio MCARI/OSAVI explained 71 % of the variability in leaf chlorophyll content measurements, performing better than the TCARI/OSAVI ratio (R^2 = 0.66), the modified NDVI (R^2 = 0.66) and the modified SR (R^2 = 0.68) vegetation indices. Estimates of chlorophyll content were moderately good for the SR (R^2 = 0.54) and NDVI (R^2 = 0.63) indices. Inclusion of OSAVI helped to reduce interference from bare soil, while MCARI was sensitive to green and red edge absorption differences caused by variation in concentration of chlorophyll a pigments. Measured values for LAI were

Table 2.4 Hyperspectral and sun-induced fluorescence (SIF) satellite specifications

Satellite	Years	Revisit Time (d)	Pixel Length (km)	Spectral Imagery	Wavelengths (nm)	Bandwidth (nm)
EO-1 Hyperion	2001–2016	16 days	0.030	Hyperspectral	400–2500	10
Proba 1 CHRIS	2001–present	1–8 days	0.017–0.034	Hyperspectral	400–1050	1.25–11
EnviSat Sciamachy	2002–2012	3 days	30	SIF (O_2-A band)	650–790	0.48
MetOp-B GOME-2	2007–present	1 day	40	SIF (O_2-A band)	650–790	0.5
GoSat TANSO FTS	2009–present	3 days	10	SIF (O_2-A band)	757–775	0.022
Sentinel-5P TROPOMI	2017–present	17 days	7	SIF (O_2-A band)	675–775	0.5
TanSat	2017–present	<16 days	2	SIF (O_2-A band)	758–778	0.044

Table 2.5 Hyperspectral satellite-based narrow-band vegetation indices used in precision agriculture. R refers to reflectance at the wavelength (nm) in subscript. NIR refers to near-infrared reflectance

Index and Satellite	Wavebands and Example Applications	Name of Index and References
NDVI EO 1 Hyperion	(NIR855-R687)/(NIR855 ± R687) Crop biomass, LAI and height	Normalized difference vegetation index Thenkabail et al. (2013)
SR EO 1 Hyperion	(R750/R705) Crop leaf chlorophyll content and LAI	Simple ratio Wu et al. (2010)
TCARI EO 1 Hyperion	3*[(R750-R705)-0.2*(R750-R550)(R750/R705)] Crop leaf chlorophyll content and LAI	Transformed chlorophyll absorption in reflectance index Wu et al. (2010)
OSAVI EO 1 Hyperion	(1 ± 0.16)(R750-R705)(R750 ± R705 ± 0.16) Crop leaf chlorophyll content and LAI	Optimized soil adjusted vegetation index Wu et al. (2010)
TVI Proba 1 CHRIS	0.5*[120(NIR750-R550)-200(R670-G550)] Leaf chlorophyll and LAI in maize	Triangular vegetation index Vincini et al. (2006)
DWSI EO 1 Hyperion	(R800 ± R550/R1660 ± R680) Orange rust disease in sugar cane	Disease water stress index Apan et al. (2004)

predicted most accurately ($R^2 = 0.67$) with the MCARI vegetation index, which tends not to saturate at large chlorophyll concentrations and reduces interference from bare soil. The TVI was also reasonably accurate ($R^2 = 0.63$), but responded more to changes in chlorophyll content than MCARI. Other vegetation indices performed worse than MCARI and TVI, including SR ($R^2 = 0.47$) which was strongly affected by interference from bare soil. Estimation of LAI was more challenging than estimation of leaf chlorophyll content, due in part to the interaction between LAI and chlorophyll content.

Apan et al. (2004) assessed the feasibility of detecting sugar cane orange rust using forty narrow band vegetation indices derived from Hyperion imagery (Table 2.5) collected over Queensland, Australia on April 2, 2002. Orange rust disease symptoms can be expressed as changes in leaf colour, leaf structure or leaf moisture content. Presence or absence of orange rust was identified in 142 or 159 locations, respectively, within commercial sugar cane fields. Thirty percent of these locations were used for validation of prediction accuracy using linear discriminant function models calibrated using data from the remaining locations. The greatest accuracy ($R^2 > 61$ %) at predicting locations with infected sugar cane were disease water stress (DWSI) indices (R_{800}/R_{1660}, R_{1660}/R_{550} or ($R_{800} + R_{550}/R_{1660} + R_{680}$)) that included ratios of reflectance at shortwave IR 1660 nm and either NIR (800 nm) or G (550 nm) wavelengths. Onset of orange rust could be detected accurately with the DWSI indices as an indirect result of lesions and ruptured leaves that enhanced loss of moisture. Vegetation indices based on red edge inflection point (REIP) reflectance at wavelengths of 690–730 nm had poor accuracy in identifying diseased locations. In this particular study, orange rust disease was more closely associated with loss of cellular moisture than changes in leaf pigmentation.

2.3.2 *Proba 1 CHRIS*

The Project for Onboard Autonomy satellite (Proba 1) is a Belgian satellite launched in 2001. A hyperspectral compact high resolution imaging spectrometer (CHRIS) is mounted on Proba 1 with technical specifications summarized in Table 2.4.

Vincini et al. (2006) used CHRIS reflectance data collected on June 28, 2005 at three sensor view angles 36, 0 and – 36 °) with a spatial resolution of 17 m to study LAI and leaf chlorophyll content in maize (Table 2.5) at two planting dates (separated by 14 days) and sugar beet crops subject to three fertilizer application rates (0, 90 and 180 kg N ha^{-1}). Ground truth measurements of LAI and leaf chlorophyll content were obtained using a Licor LAI 2000 meter and a Minolta SPAD 502 meter, respectively. Narrow band vegetation indices such as Simple Ratio ($SR=NIR_{800}/R_{680}$) and TVI $=0.5*[120(NIR_{750}--R_{550}) -200(R_{670}-G_{550})]$, were evaluated, where the subscripted numbers are wavelengths in nm. The SR, TVI and LAI were larger for early planted maize than for that planted late. The SR was larger for maize at a view angle of 36 ° than at angles of 0 or – 36°, while TVI was larger at an angle of –36° than at angles of 0 or 36°. The SR and TVI increased most at different maize planting dates with a view angle of –36°. SR and TVI increased as N application rate increased in sugar beet, with the magnitude of increases being largest for an angle of –36°. For greatest sensitivity in assessing differences in maize LAI or sugar beet response to N application rate, vegetation indices based on a viewing angle of –36° performed better than at 0 or 36°.

Delegido et al. (2008) developed an innovative spectral regression approach to remove the effect of view angle on hyperspectral response of the CHRIS sensor to LAI and leaf chlorophyll content. The CHRIS images were collected from 54 pixels in 2003–2004 at a spatial resolution of 34 m at a nadir view angle corresponding to locations with ground truth data. Images were collected for each pixel at five view angles, namely; nadir, ±36 and ± 55°.

Ground truth data for these findings were based on measurements in mid-July of 2003–2004 for LAI with a Licor LAI 2000 and for leaf chlorophyll content with an Opti-Sciences CCM 200 meter in a 5 x 10 km area of Spain. A diverse mix of rainfed and irrigated crops were studied, including garlic (*Allium sativum*), alfalfa (*Medicago sativa*), onion (*Allium cepa*), maize (*Zea mays*), potato (*Solamum tuberosum*) and sugar beet (*Beta vulgaris*). In general, there was a moderately strong positive correlation between ground truth measurements for LAI and leaf chlorophyll content, which theoretically poses a challenge for estimating either of these variables independently by satellite remote sensing techniques.

A third degree polynomial regression equation with four coefficients was developed for CHRIS hyperspectral reflectance from the observed crops between wavelengths of 500 and 750 nm. The second coefficient (B) was positively correlated with ground truth measurements of LAI, and explained 71 % of the spatial variability in observed LAI measurements. This relationship was not affected by view angle or leaf chlorophyll content. Separately, the area under the hyperspectral reflectance curve of the observed crops between 600 to 700 nm explained 66 % of the variability in measured leaf chlorophyll content. This relationship was almost independent of viewing angle, and was not affected by LAI.

2.4 Sun-Induced Fluorescence Satellites

Sun-induced fluorescence (SIF) refers to the release of excess energy absorbed by chlorophyll pigments in photosystems I and II of plant leaves. The SIF is closely correlated with rates of photosynthesis that are controlled by these pigments. It is distinct from and of much smaller magnitude than reflectance of energy by chlorophyll pigments. The source of fluorescence is energy received from the sun at wavelengths of 430–650 nm during daylight hours by the plant. Fluorescence is emitted at longer wavelengths than the absorbed energy, typically at wavelengths ranging from 650–790 nm with peaks at 685 and 740 nm (Aasen et al. 2019).

Oxygen molecules in the Earth's atmosphere absorb most of the incoming solar radiation at wavelengths of 687 (O_2-B) and 761 (O_2-A) nm. In these two telluric absorption bands, the magnitude of SIF is larger than the magnitude of reflected energy from chlorophyll pigments (Joiner et al. 2011). Many satellite-based sensors described here have been developed to measure SIF in the O_2-A band (Table 2.4). To obtain these measurements, the sensors must measure fluorescence energy at high (>1000) signal to noise ratio (SNR) in narrow bands, typically with a bandwidth less than 0.5 nm. Extracting SIF from satellite sensor measurements in the O2-A band is typically accomplished using one of two methods. The first is radiative transfer modelling and principal component analysis (Joiner et al. 2011), and the second is either Fraunhofer line discrimination (FLD) or singular vector decomposition (SVD) as discussed by Liu et al. (2015) and Du et al., (Du et al. 2018). The second method is more accurate for satellite sensors with narrower spectral resolution (Du et al. 2018) and larger SNR. While the sensitivity of SIF to crop productivity and photosynthetic rates is impressive, applications of SIF to PA are limited by the coarse spatial resolution (2–40 km) of satellite-based SIF measurements (Table 2.4).

2.4.1 *EnviSat Schiamachy*

The European Space Agency launched EnviSat in 2009 and decommissioned it in 2012. EnviSat carried a scanning imaging absorption spectrometer for atmospheric cartography (Sciamachy) with technical specifications summarized in Table 2.4. Maier et al. (2004) first suggested the use of satellite-based estimates of SIF as an alternative to NDVI for estimating spatial and temporal variation in crop photosynthesis for PA. The NDVI is unsuited for estimates of photosynthetic rates, because it is affected by bare ground, leaf pigments, crop biomass and tends to saturate with large biomass.

2.4.2 MetOp B Gome 2

Launched in 2009 by the European Space Agency, MetOp B carries a global ozone monitoring experiment 2 (GOME 2) sensor with technical specifications summarized in Table 2.4. Guanter et al. (2014) compared GOME 2 SIF measurements across the globe from 2007–2011 with photosynthetic gross primary productivity (GPP) measurements from flux towers in the US Midwestern 'corn belt' (CB) and western Europe (WE). Temporal averages of SIF estimates from these two regions at a spatial resolution of 40-km were strongly correlated with GPP measurements from flux towers at a resolution of 1-km. Monthly mean GPP estimates were much larger in the CB with a large proportion of maize (a C4 crop) than WE grasslands.

Sun et al. (2015) showed that GOME 2 estimates of SIF anomalies from long-term average values were strongly correlated with onset of drought in Texas and the Great Plains regions during 2011–2012. Crop photosynthesis declined dramatically in Texas during the hot and dry summer of 2011. This decline was closely correlated with patterns in soil moisture depletion. This study shows that satellite-based estimates of SIF could be used to identify areas experiencing crop stress in near real time.

2.4.3 GoSat TANSO FTS

Launched in 2009 by Japan, GoSat carries a thermal and near-infrared sensor for carbon observation–fourier Transform spectrometer (TANSO–FTS). TANSO FTS with technical specifications summarized in Table 2.4.

Joiner et al. (2011) were among the first to estimate SIF using global imagery from the GoSat TANSO FTS sensor. Spatial patterns in SIF were estimated globally for July 2009, showing large rates of photosynthesis instantaneously in the US Midwestern CB and the southern China rice growing regions. As would be expected for agricultural regions, maximum fluorescence was observed in summer months for northern latitudes, and winter months for southern latitudes.

2.4.4 *Sentinel-5P TROPOMI*

Launched in 2016 by the European Space Agency, Sentinel-5P carries the tropospheric monitoring instrument (TROPOMI). TROPOMI collects SIF data according to technical specifications summarized in Table 2.4. Köhler et al. (2018) developed a high resolution global scale map of SIF from TROPOMI data collected during April 8–15, 2018. This map shows areas with active photosynthesis in crops in the San Joaquin Valley of California, the rice and wheat growing areas south of Beijing, China, and the Nile Delta region of Egypt. In contrast, photosynthesis occurring in the US Midwestern CB was minimal during this time period. Köhler et al. (2018) also showed that TROPOMI has much better spatial resolution (7-km) than older satellite based SIF sensors, such as the GOME 2 sensor (40-km).

2.4.5 *TanSat*

TanSat is a Chinese satellite launched in 2017 with a carbon dioxide sensor (CDS). The CDS measures SIF in the narrow wavelength range from 758–778 nm at increments of 0.044 nm (Table 2.4). While TanSat has the best spatial resolution of all satellite SIF sensors (2-km), SNR is relatively low at 360.

Du et al. (2018) compared monthly average global SIF measurements from March 2017 to March 2018 from TanSat with Enhanced Vegetation Index (EVI) and gross primary production (GPP) measurements from MODIS. For better comparison between datasets, MODIS data were aggregated to a spatial resolution of 2-km, the spatial resolution of TanSat. TanSat SIF measurements were generally larger in the US Midwestern CB than elsewhere globally. TanSat SIF measurements agreed closely in magnitude and temporal pattern with MODIS GPP estimates. A linear relation developed between TanSat SIF and MODIS GPP measurements showed that SIF could be multiplied roughly by 5.37 to obtain GPP estimates.

2.5 Synthetic Aperture Radar Satellites

The SAR satellite sensors actively emit microwave signals at short (3 cm), medium (5 cm) or long (23 cm) wavelengths. Short, medium and long wavelength sensors are referred to as X-, C- or L-band SAR sensors, respectively. Microwave energy is emitted from SAR satellites during the day and night, and passes through clouds without interference. Thus, SAR imagery is available when skies are cloudy or at night when the sun is not shining (Rosen et al. 2000), an advantage over imagery from multispectral or hyperspectral reflectance-based imagery.

Radar signals are typically emitted from SAR satellites at an oblique incidence or view angle. Incoming microwave energy interacts with surface elements such as topography, vegetation, soil surface roughness and soil moisture (Quesney et al. 2000), resulting in variable magnitudes of energy back-scattered to receiving antennae. The magnitude of back-scattered energy is affected by wavelength and incidence angle of the signal, as well as by polarization. Older SAR satellites such as JERS 1 (see below) transmitted and received microwave energy at a single wavelength and single polarization (McNairn and Brisco 2004). Newer SAR satellites can transmit and receive at a range of polarization modes, including transmitting in horizontal (H) or vertical (V) polarization and receiving in horizontal or vertical polarization, resulting in modes such as linear polarization HH, VV, HV or VH (McNairn and Brisco 2004) dual polarization (HH and HV or VV and VH), or quad polarization (HH, VV, HV and VH). Horizontally and vertically polarized signals interact differently with standing crops. In general, horizontally polarized signals penetrate crop canopies better than vertically polarized signals, providing more information about soil characteristics.

Extracting soil surface characteristics from SAR backscatter involves several issues. Variation in topography affect backscatter through variation in the local incidence angle of incoming radar signals (Baghdadi et al. 2008). These effects can be removed by radiometric modelling based on an accurate digital surface model. Radiometric corrections are also required to estimate radar backscatter coefficients from the digital number (DN) for each pixel. Finally, speckle noise, resulting from multipath backscatter and interference should be reduced by smoothing or filtering algorithms.

Estimates of soil surface roughness tend to be more accurate using HH or HV polarization at longer wavelengths (e.g. L-band radar) than with shorter wavelengths (C- or X-band), and at higher sensor view angles than lower angles (Baghdadi et al. 2008). In contrast, estimating soil moisture is best when backscatter is measured at both lower and higher view angles in multiple polarization modes, in order to remove the effects of soil roughness. The SAR is not very accurate at distinguishing moisture contents in soil with volumetric moisture contents greater than 35–40 % (Baghdadi et al. 2008).

Precision agriculture could benefit from the information provided by SAR satellites as illustrated by example applications summarized for each SAR satellite below. The high spatial resolution and good revisit frequencies (Fig. 2.2) of the Sentinel-1, CosmoSkyMed and TerraSAR-X/TanDEM-X satellites, combined with their ability to see through clouds are attractive attributes. The SAR imagery from these satellites may be combined with high resolution multispectral satellite imagery to develop useful PA applications involving spatial and temporal variation in crop height (or biomass), soil roughness or soil moisture. The SAR satellites with coarse spatial resolution (e.g. SMOS) may also be useful in PA, particularly if their products can be downscaled by fusion with one of the finer resolution SAR satellites in Fig. 2.2.

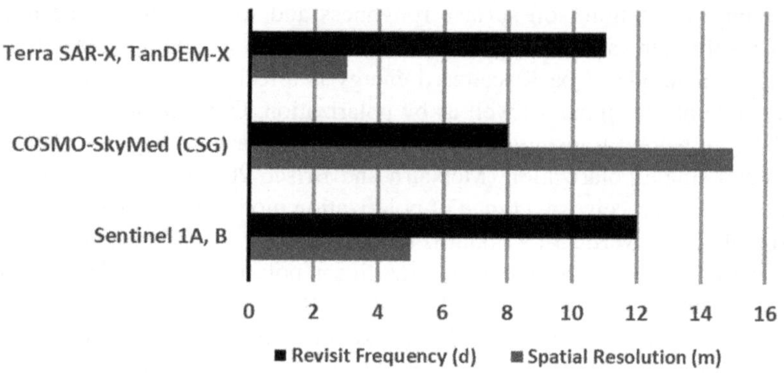

Fig. 2.2 Revisit frequency (d) and spatial resolution (m) for C-band and X-band synthetic aperture radar (SAR) satellites

2.5.1 Japanese Earth Resources Satellite 1

The Japanese Earth Resources Satellite (JERS) collected L-band SAR data from 1992–1998 (Table 2.6). Metternicht (1998) studied soil salinity in Bolivia using JERS 1 SAR data collected in May of 1994. Speckle in the radar image was removed based on moving window filtering techniques leading to spatial averaging of pixels into pixels no smaller than 37.5 m. Radar backscatter was greater for rougher ploughed alkaline soils (>3.6 cm roughness height) than for smoother unploughed saline soils.

2.5.2 Advanced Land Observing Satellites 1 and 2

The Advanced Land Observing Satellites (ALOS-1 and 2) were launched by Japan in 2006 and 2015, respectively. The ALOS-1 was decommissioned in 2011, but ALOS-2 continues to collect data. The ALOS satellites transmit L-band SAR using a phased array L-band synthetic aperture radar (PALSAR) instrument, with specifications provided in Table 2.6.

Torbick et al. (2011) acquired twenty-five fine beam and three wide beam ALOS PALSAR data sets (each with HH polarization) for the Sacramento Valley rice growing region of California in the period from December 2006 to April 2007. The objective of the study was to identify which fields were planted to rice and identify when each field was flooded. Orthorectified National Agricultural Imagery Program (NAIP) data with a spatial resolution of 1 m were used to select 250 pixels with rice and 225 pixels with other crops. Rice fields were classified with an accuracy of 97 % relative to 1 m spatial resolution National Agricultural Imagery Program (NAIP) data for over 500 sites with known crop type. Flooding condition was estimated accurately using ScanSAR in 96 % of ground truth fields, of which roughly

Table 2.6 Synthetic aperture radar (SAR) Satellite specifications

Satellite	Class	Years	Wavelength (GHz, cm)	Spatial Resolution (m)	Revisit Time (d)	Cost/Scene Area km^2
JERS 1	L-Band SAR	1992–1998	1.275, 23.5	18	44	Free
ALOS-1 (PALSAR)	L-Band SAR	2006–2011	1.275, 23.5	10–30	46	Free
ALOS-2 (PALSAR)	L-Band SAR	2015–present	1.2365–1.2785, 22.6	3–10	14	$2187/ 2450
SMOS (MIRAS)	L-Band SAR	2010–present	1.41, 21	10,000	1–3	Free
SMAP	L-Band passive radiometer	2015–present	1.41, 21.3	30,000	2–3	Free
ERS 1, 2 (AMI)	C-Band SAR	1992–2000, 1996–2011	5.3, 5.66	25	35	Free
RadarSat 1	C-Band SAR	1997–2013	5.3, 5.66	12.5	2–3	$2707/2500
RadarSat 2	C-Band SAR	2008–present	5.45, 5.55	3–30	12	$2707/2500
EnviSat (ASAR)	C-Band SAR	2002–2012	5.331, 5.63	1000	35	Free
Sentinel-1A, B	C-Band SAR	2014– or 2016–present	5.405, 5.55	5	12	Free
COSMO-SkyMed (CSG)	X-Band SAR	2007–2018	9.6, 3.1	3–15	1–8	$3400/1600
Terra SAR-X, TanDEM-X	X-Band SAR	2007– or 2011–2016	9.65, 3.11	1–3	11	$1190/1500 and $1587/1500

half were flooded. These data could be used to estimate water usage patterns in different seasons.

2.5.3 Soil Moisture and Ocean Salinity

The Soil Moisture and Ocean Salinity (SMOS) satellite, launched in late 2009, is a joint venture of France and Spain through the European Space Agency. The SMOS carries a microwave imaging radiometer using SAR (MIRAS) which transmits L-band radar pulses in H + V polarization (Table 2.6). The L-band radar backscattering coefficients are affected by soil moisture, soil surface roughness and vegetation. Effects of vegetative moisture can be removed based on estimates of LAI (Champagne et al., 2015). The MIRAS data are used to estimate soil surface moisture content at depths of 3–5 cm whenever soil is not frozen, although estimates are at a coarse spatial resolution of 10-km.

Martínez-Fernández et al. (2016) compared SMOS estimates of soil moisture from 2010–2014 with measurements of soil moisture from a network of soil sensors installed at shallow depths across an agricultural region in the Duero Basin of Spain. Both satellite and experimental measurements of soil moisture were converted into a soil water deficit index (SWDI) based on estimates of field capacity and available moisture capacity. The SMOS estimates for SWDI explained 72 % of the variation in experimentally measured SWDI. The SMOS SWDI estimates showed that on average, there were 35 weeks of soil moisture deficit (drought) from 2010–2014.

For applications involving agriculture, the spatial resolution of SMOS estimates for soil moisture are too coarse (10 km). Methods for downscaling SMOS soil moisture estimates to finer scales (e.g. 1 km) have been reviewed by Peng et al. (2017). Two major approaches for downscaling include fusion of finer scale radiometric satellite remote sensing with SMOS estimates or use of statistical or deterministic hydrologic models.

2.5.4 *Soil Moisture Active Passive*

The Soil Moisture Active/Passive Mission (SMAP) satellite was launched in 2015 as an effort by NASA to monitor surface soil moisture and soil frost. The SMAP transmitted L-band SAR pulses in VV, HH and HV polarizations at a frequency of 1.26 GHz (23.8 cm). Unfortunately, the L-band SAR instrument failed after 6 months in orbit. A passive microwave radiometer on the SMAP satellite collects L-band data (Table 2.6), which are used to provide global estimates of soil surface moisture. The SMAP has a revisit frequency of 2–3 days, with a very coarse spatial resolution of 36-km, although resolutions of 9-km can be achieved by estimating soil moisture in areas of overlapping SMAP imagery.

El Hajj et al. (2017) compared SMAP and SMOS estimates for soil moisture with measured values from a regional monitoring network in southern France. The SMAP estimates of soil moisture under-estimated measured volumetric soil moisture content by 0.045 %. In contrast, SMOS estimates had poorer accuracy, under-estimating measured volumetric soil moisture content by 0.095 %.

2.5.5 *European Remote Sensing 1 and 2*

The European Space Agency launched European Remote Sensing (ERS) satellites ERS 1 and 2 in 1992 and 1996, respectively. The ERS 1 was decommissioned in 2000, while ERS 2 was decommissioned in 2011. These satellites transmitted

C-band synthetic aperture radar data (Table 2.6) using an active microwave instrument (AMI).

Radar backscatter from satellites is collected at incidence angles greater than 20°. Under these conditions, backscatter is affected by soil dielectric properties (controlled by soil moisture), soil roughness and vegetation cover (Quesney et al. 2000). Quesney et al. (2000) developed a method for estimating soil moisture contents at the watershed scale in a large wheat growing region of France using high resolution (12.5 m) AMI data from ERS 1 and 2. Vegetation effects were removed from backscatter data based on optical thickness of the wheat canopy. Soil roughness effects were removed by studying the effect of tillage furrow directions on backscatter at different soil moisture contents relative to radar view angle. Soil moisture contents could not be estimated during May and June because of interference from crop biomass, or during periods in late autumn when harvested crop residue covered the soil. These results show the complexity and challenges of estimating soil moisture accurately from satellite radar backscatter data.

2.5.6 RadarSat 1 and 2

The Canadian Space Agency launched RadarSat 1 and 2 in 1997 and 2008, respectively. RadarSat 1 was decommissioned in 2013, while RadarSat 2 continues to collect data at present. The C-band SAR specifications for RadarSat 2 are summarized in Table 2.6.

RadarSat 1 transmitted radar signals at a single frequency and a single polarization mode, whereas RadarSat 2 transmits radar signals in three different polarization modes simultaneously. The VV polarization has poor ability to penetrate vegetation with vertical canopy structure (e.g. wheat crops), but can penetrate maize canopies more easily. In contrast, HH polarization penetrates crop canopies more readily, leading to information about soil characteristics. The HV or VH cross-polarization (generally assumed to be equal) provides important information about canopy structure or crop residue cover (McNairn and Brisco 2004). In addition, radar backscatter is less sensitive to tillage row direction in HV or VH cross-polarization mode than in HH polarization mode. Quad polarization mode involves collecting VV, HH, HV and VH radar backscatter data simultaneously.

Gherboudj et al. (2011) acquired RadarSat 2 images from July, 2008 for an agricultural region of Saskatchewan, Canada to estimate crop height and vegetation moisture content, soil surface roughness and soil moisture content. Ground truth data for these variables were collected from sixteen large agricultural fields growing wheat (*Triticum aestivum*), lentils (*Lens culinaris*), canola (*Brassica rapa*), peas (*Pisum sativum*) and alfalfa (*Medicago sativa*), together with some fallow fields. A co-polarization correlation coefficient based on back-scattered radar signals in the VV and HH modes was used to develop an empirical relationship to estimate crop height (ranging from 10 to 100 cm) at different sensor incidence angles (ranging from 30 to 45°). Soil roughness was then calculated based on a depolarization factor using the difference between radar backscatter coefficients in the VH and VV modes, vegetation height and incidence angle of the sensor. Finally, soil moisture is estimated by inverting an empirical model developed by Oh (2004) based on estimated values for surface roughness. Calibration accuracy was very good for crop height, vegetation water content and soil moisture, with accuracies of 75, 65 and 60 %, respectively. Surface roughness was estimated with poor accuracy (15 %).

2.5.7 EnviSat

EnviSat was launched in 2002 by the European Space Agency, and was decommissioned in 2012. The EnviSat Advanced SAR (ASAR) instrument transmitted C-band radar at a frequency of 5.331 GHz or 5.63 cm (Table 2.6).

Pathe et al. (2009) studied ASAR GM data acquired over Oklahoma, USA in HH mode from 2004–2006. Soil surface moisture data were measured at 75 locations in the Oklahoma MESONET. The ASAR data from multiple passes were used with a change detection approach for reference areas to estimate the backscatter coefficients for dry and wet soil at a radar incidence angle of 30°. These data were used to estimate the sensitivity coefficient to soil moisture. Backscatter coefficients and the sensitivity coefficient were very sensitive to vegetative cover (e.g. forested land versus agricultural land). These three coefficients were then used to estimate ASAR based soil moisture contents. The ASAR pixels at 1-km spatial resolution were degraded to a spatial resolution of 3-km to improve the accuracy of estimating soil moisture. The average Pearson correlation coefficient between 3-km scale ASAR estimated and ground truth measurements of soil moisture was about 0.5.

2.5.8 Sentinel-1A and -1B

The European Space Agency launched Sentinel-1A and -1B in 2013 and 2015, respectively. The Sentinel-1A and -1B satellites operate 180° out of phase in the same orbit for a revisit frequency of 12 days. Sentinel-1A and -1B transmit C- band SAR pulses with dual polarization of HH and HV at a frequency of 5.405 GHz or 5.55 cm (Table 2.6).

Ferrant et al. (2017) acquired twenty-five Sentinel-1 images over India from 2016–2017 to identify irrigated or rainfed agricultural fields and to estimate rates of irrigation application during the dry and monsoon seasons. Sentinel-1 imagery was collected in interferometric wide swath mode with VH and VV polarization at a spatial resolution of 10 m. Ratios of VV/VH backscattering were computed for image classification. Ground truth data for classification were obtained from field surveys conducted for 192 rice paddies, 286 irrigated fields with maize, vegetables or mangos (*Mangifera indica*), and 428 rainfed locations with cotton or natural vegetation.

Classification of Sentinel-1 pixels into flooded rice paddies, irrigated maize, irrigated vegetables and rainfed cotton was achieved using Random Forest (RF) supervized classification with half the ground truth data. Classification was accurate for flooded rice paddies either in the dry or monsoon season, due to the sensitivity of Sentinel-1 for flooded soils. Sentinel-1 imagery accurately identifed rainfed cotton planted before the monsoon season.

2.5.9 Cosmo SkyMed

Cosmo SkyMed is a constellation consisting of two Italian satellites transmitting X-band SAR pulses at a frequency of 9.6 GHz or 3.1 cm (Table 2.6). Revisit frequency is 1–8 days. Cosmo SkyMed data were available from 2007–2018.

Gorrab et al. (2015) studied the feasibility of using Cosmo SkyMed SAR radar imagery to estimate soil roughness and soil moisture content for agricultural fields with different roughness near Kairouan, Tunisia. Four images were collected in Ping Pong mode (spatial resolution of 7.9 m) with dual (HV/HH) polarization at sensor incidence angles of 26 or 36° during November and December of 2013. Seven images were also collected from Terra SAR-X radar during the same time period with dual (HH/VV) polarization at a sensor incidence angle of 36° in Spotlight mode (spatial resolution of 1.8 m). Terra SAR-X backscatter values were used at reference field locations described below to calibrate for differences in radar backscatter from different satellites in the Cosmo SkyMed constellation.

Ground truth data were collected from fifteen bare agricultural fields for soil surface roughness, moisture content and soil bulk density. Roughness measurements were converted to root mean square (RMS) and correlation length (l). The RMS values varied from 0.24 to 3.4 cm for smooth to rough fields, respectively. Volumetric moisture contents ranged from about 5 to 32 %.

Cosmo SkyMed radar backscatter in HH polarization mode at a sensor incidence angle of 36° explained 62 and 53 % of the variability in measured RMS for wet and dry soils, respectively. Dryer soils were generally rougher than wetter soils because of differences in tillage and fallow practices. Accuracy of radar backscatter at predicting soil roughness was poor ($R^2 = 0.4$) for a sensor incidence angle of 26°. In general, radar backscatter is more sensitive to changes in soil roughness at larger sensor incidence angles. Cosmo SkyMed HH polarization backscatter predicted variability in soil volumetric moisture content at an accuracy of 64 % for a sensor incidence angle of 36°.

2.5.10 Terra SAR-X and TanDEM-X

Germany launched Terra SAR-X in 2007 and a second nearly identical satellite TanDEM-X in 2011. The two satellites fly in close proximity, using a helix shaped flight pattern that allows the satellites to collect radar imagery data with considerable spatial overlap. This allows image processing using interferometric techniques. Both satellites transmit X-band SAR pulses at a frequency of 9.65 GHz, or 3.11 cm (Table 2.6).

Baghdadi et al. (2008) studied soil surface roughness and moisture content using images collected from Terra SAR-X for two agricultural watersheds in France. Images were collected at low and high sensor incidence angles (26 to 52°) Spotlight mode with HH polarization at a spatial resolution of m. A single ALOS PALSAR image was also collected in one watershed using HH polarization at a sensor incidence angle of 38°.

Ground truth measurements of soil moisture, surface roughness and soil bulk density were obtained from fields in each of the two watersheds on each of three dates in early 2008. Soils were very wet on all dates of field data collection, ranging from 27 to 41 % volumetric water content. A profilometer was used for measurements of surface roughness, and these measurements were transformed into RMS roughness and correlation length (l) using the autocorrelation function of surface roughness.

Terra SAR-X radar backscatter increased with surface roughness up to an RMS value of 1.5 cm, and thereafter leveled off. Backscatter was larger at high sensor angles of incidence. The difference in backscatter between smooth and rough fields was greater for high sensor angles of incidence than for low angles of incidence. The wettest soil reduced backscatter more than reductions in backscatter on moderately wet soil, regardless of sensor angle of incidence. Measurements of surface roughness with Terra SAR-X were most accurate for dryer soil with little emerged crop vegetation using high sensor angles of incidence. The longer wavelength ALOS PALSAR sensor showed a greater sensitivity to differences in backscatter between smooth and rough fields than the shorter wavelength Terra SAR-X sensor.

2.6 Satellite-Based Digital Surface Model Products

Global DSMs include at least a fraction of the heights of vegetation canopies or buildings, and are not equivalent to surface digital elevation models (DEMs) where the heights of vegetation and buildings have been removed. However, in agricultural fields without dense vegetation, DSMs are equivalent to DEMs. Two primary approaches for developing global DSMs include analysis of interferometry from SAR satellites (Rosen et al. 2000) or analysis of stereo images from multispectral satellites (Deilami and Hashim 2011).

The SAR-based DSMs can be obtained using cross-track interferometry (XTI) or repeat track interferometry (RTI) (Rosen et al. 2000). The SAR antennae transmit energy to the Earth's surface perpendicular to the satellite orbital track, at incidence angles typically greater than 20°. The interferometric phase difference between the signals received by each antenna can be analyzed to infer the elevation of the surface target, while location in the orbital track and angle of incidence identify the x, y coordinates of the target. In XTI, the satellite has two antennae. In standard XTI mode one antenna transmits radar signals, while each antenna can receive backscatter radar signals from the surface. In XTI ping pong mode, each antenna transmits and then receives its own signals. In RTI, the satellite returns to approximately the same target location in successive orbital paths that may be separated by seconds, days, weeks or years. Elevation data are extracted using interferometric analyses of the phase differences between radar backscatter on successive passes.

Panchromatic stereo images obtained from multispectral satellites can be used to develop global DSMs (Deilami and Hashim 2011). Two methods are used to collect stereo images. In the first method, images from successive orbital paths are georeferenced, then target locations in each image are identified using signal, feature or structural matching operations. Target elevations are estimated based on comparisons with ground truth elevation measurements at benchmark locations. The second method involves analysis of panchromatic imagery collected in rapid succession along the same orbital path by viewing the target before, at and after nadir. Accuracy of DSMs derived from stereo images involving panchromatic multispectral imagery is generally better for the second method than the first. However, accuracy of DSMs derived from panchromatic imagery are generally worse than DSMs derived from interferometric SAR techniques.

The DEMs are of great value in PA (Bishop and McBratney 2002; Nawar et al. 2017). They can be used to help delineate management zones, and predict spatial patterns in soil properties and crop yield. They can also be used to derive terrain attributes such as slope, aspect, a topographic wetness index, and so on, which influence soil properties and crop growth. High quality DEMs are available globally using one of four satellite-based platforms described below. The spatial resolution of these DEMs ranges from 5 to 30 m (Fig. 2.3), while the accuracy measured as root mean square error (RMSE) varies from 2 to 20 m. These DEMs have particular value for PA applications in developing countries, where low resolution DEMs have hindered the adoption of PA for decades.

2.6.1 *Shuttle Radar Topography Mission*

The Shuttle Radar Topography Mission (SRTM) was based on interferometric Synthetic Aperture Radar in the C (5.3 GHz) and X (9.6 GHz) bands collected simultaneously from the eleven day Endeavor Shuttle flight between 56° S and 60° N latitudes during February of 2000. Originally, global DEMs were made available at a spatial resolution of 90 m (Table 2.7). In 2015, the USGS provided a global

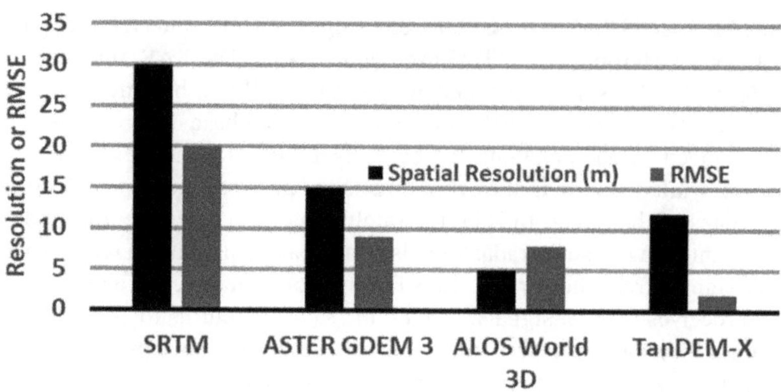

Fig. 2.3 Spatial resolution (m) and root mean square error (RMSE) for satellites that provide globally available digital elevation model (DEM) data

Table 2.7 Global sources of space derived Digital Surface Models (DSM) and their attributes

Attribute	SRTM	ASTER GDEM v3	ALOS World 3D	TanDEM-X
Years	February, 2000	1999–present	2006–2011	2010–present
Imagery	C-Band SAR, X-Band SAR	Panchromatic Stereo	Panchromatic Stereo	X-Band SAR
Wavelength	5.3, 9.6 GHz	520–600, 630–690 nm	520–770 nm	9.65 GHz
Pixel Resolution (m)	30, 90	15	5, 30	12, 30
RMSE Vertical Accuracy (m)	<20 m	<9 m	<8 m	<2 m
Cost	Free	Free	Free (30 m DEMs)	Free to scientists (12 m DEMs)

DEM based on SRTM data at a spatial resolution of 30 m (SRTMGL1). Vertical accuracies (RMSE) of this product were 5.9–12.4 m, depending on which area of the globe was evaluated (Alganci et al. 2018).

2.6.2 *Advanced Spaceborne Thermal Emission and Reflection Radiometer*

The advanced spaceborne thermal emission and reflection radiometer (ASTER) sensor was launched into orbit on the Terra satellite during 1999. It has been collecting panchromatic stereo imagery at wavelengths of 520–600 (G/Y) and 630–690 (R) nm through to the present (Table 2.7). Global DEMs (GDEM) covering 83° N to 83° S were released by NASA and the Japanese Ministry of Economy, Trade and

Industry (METI) in 2009 (GDEM v1), 2011 (GDEM v2) and 2015 (GDEM v3). Vertical accuracies (RMSE) of GDEM v1 were < 20 m, while accuracies of GDEM v2 and v3 were < 15 m and < 8.52 m, respectively (Gesch et al. 2016; Alganci et al. 2018).

2.6.3 *Panchromatic Remote-Sensing Instrument for Stereo Mapping*

The Advanced Land Observing Satellite launched in 2006 and decommissioned in 2011 carried a panchromatic remote sensing instrument for stereo mapping (PRISM). The PRISM included three forward, nadir or backward looking sensors that collected reflectance at wavelengths ranging from 520 (G) to 770 (NIR) nm with a spatial resolution of 2.5 m (Table 2.7). Multiple images for any given location across the globe were collected across the 2006–2011 time-frame over which PRISM operated. These images were oriented, matched and masked to eliminate clouds, snow, ice and water. They were then stacked across time to calculate and correct elevation bias and fill gaps that had been masked. Mosaics of DSMs were produced at a 5 m spatial resolution. This global DSM product was released in 2014 under the name ALOS World 3D. Vertical accuracy (RMSE) of the ALOS 3D DEM averaged over study regions in Cambodia, Senegal, Nepal and Sri Lanka ranged from 1.91–14.43 m (Tadono et al. 2014). In May 2015, the DSM was degraded to a spatial resolution of 30 m, renamed ALOS World 3D 30 (AW3D30), and provided for download without cost. Further updates of AW3D30 continued through April, 2019. The vertical accuracy (RMSE) of AW3D30 ranges from 2.94–7.15 m (Alganci et al. 2018; Li and Zhao 2018), depending on ground slope and other factors.

2.6.4 *TanDEM-X*

Terra SAR-X and TanDEM-X fly in a helical path around each other separated by less than 500 m while they orbit the Earth. One of the satellites transmits X-band radar at 9.65 GHz, while both measure back-scattered radar signals simultaneously. Global DSMs were developed using interferometric backscatter collected in this manner from December 2010 to January 2015. Spatial resolution of the DSMs was originally 12 m, but a spatially degraded DSM is also available at a spatial resolution of 30 m (Table 2.7).

Wessel et al. (2018) evaluated the accuracy of TanDEM-X DSMs using over 3 million kinematic GPS elevations from around the globe, as well as elevation data from 23,000 benchmark locations supplied by the US National Geodetic Survey for a variety of land cover types. Vertical accuracy (RMSE) across six continents compared with kinematic GPS elevations averaged 1.29 m. In the US, vertical accuracies (RMSE) across three general land cover types were 1.44 m for developed areas, 1.12 m for low cover areas including agricultural lands and 1.84 m for forested lands. The larger RMSE values in developed and forested lands (relative to agricultural land) resulted from backscatter returns from the top of buildings or trees, which were not representative of backscatter from the ground surface.

2.7 Conclusions for the Chapter

The applicability of satellite remote sensing to PA depends on many factors. Revisit frequency is important, not only to increase the chances of acquiring imagery on clear days, but also to detect and diagnose locations experiencing crop stress or changes in soil conditions in a timely fashion. Revisit frequencies of one day are widely available for the WorldView, Planet Lab Flock of Doves, and SMOS satellites. Satellites with revisit frequencies that are more poorly suited for PA include Landsat 7/8 and Sentinel-5P. Availability of imagery from commercial satellites may also be limited by competition from other users, particularly with WorldView.

Spatial resolution of multispectral imagery has an important influence on the homogeneity of remote sensing pixels and applicability to PA. In general, spatial resolution of imagery that has the best potential for use in PA ranges between 0.5- and 30 m. This capability is available in imagery from many of the existing multispectral satellites, with Sentinel-2 having the advantage of being provided for download at no cost. Coarser spatial resolution is less applicable for use in small fields than in large fields, while fine spatial resolution is generally applicable across all scales of farming.

In general, diagnosis of specific soil conditions or crop stressors is enhanced with narrow band spectral or vegetation indices available with satellite-based hyperspectral imagery. At present, the only source of satellite-based hyperspectral imagery is Proba 1 CHRIS. There is a pressing need for additional satellites with hyperspectral sensors, particularly those with high signal to noise ratios, fine spatial resolution and short revisit frequencies. Launch of the German Environmental Mapping and Analysis (EnMAP) hyperspectral satellite with a spatial resolution of 30 m is currently planned for 2021. Another option for future satellite hyperspectral imaging is the US development of HyperSat, with a planned spatial resolution of 10 m. Many satellites have recently been launched to measure sun-induced fluorescence (SIF) in the narrow hyperspectral O_2-A band, and research has shown that SIF is a powerful tool for detecting crop stress due to a variety of factors (Sun et al. 2015; Köhler et al. 2018). However, spatial resolution of satellite based SIF imagery (> 2 km) is unsuited for PA, unless SIF imagery is fused and downscaled using finer resolution imagery from multispectral satellites.

Traditionally, PA has focused on management of fertilizer inputs (such as nitrogen and phosphorus). There is an increasing awareness of the need for precision management of crops based on soil moisture. Synthetic aperture radar (SAR) satellites have proved to be effective at estimating soil moisture at spatial scales ranging from 1 to 30,000 m with revisit frequencies ranging from 1 to 45 days. The SAR imagery at finer spatial scales and quicker revisit frequencies consistent with the requirements for application to PA are available with the ALOS-2, Sentinel-1 and Terra SAR-X satellites. A major advantage of SAR satellites is their ability to collect data at night or in the presence of cloud cover. Additional research is needed to determine the feasibility of fusing SAR satellite imagery relating to soil moisture, with multispectral imagery relating to soil conditions and crop stresses for new innovations in PA.

Satellite imagery offers the possibility to facilitate greater adoption of PA in developing countries, where requisite information resources such as soil and topographic maps are available at only very coarse scales. Hyperspectral and SAR satellites have each shown potential for mapping soil characteristics useful in PA (Gomez et al. 2008; Baghdadi et al. 2008). Digital Surface Models are available from a variety of satellite platforms, the most useful for PA being ALOS World 3D (RMSE <8 m) and TanDEM-X (RMSE <2 m).

Cost of acquiring satellite imagery is an important consideration affecting the applicability to PA. The European Space Agency provides Sentinel-2 multispectral satellite imagery at no cost. This is an excellent option for PA with its good spatial resolution ranging from 10–20 m, high spectral resolution (12 bits per pixel), and a relatively quick revisit frequency of 5 day. Landsat 8 multispectral imagery is also free, but this source is less appealing for PA (especially in cloudy areas of the world) because of its relatively long revisit frequency (16 days) and a coarser spatial resolution (30 m). Sentinel-1 C-band SAR imagery at a spatial resolution of 5 m and a revisit frequency of 12 days is provided free of charge by the European Space Agency. This is a viable option for PA involving spatial and temporal variation of soil moisture, especially in irrigated crops. Global Digital Surface Models are available at no cost from multiple sources, the most applicable to PA are those involving TanDEM-X (12 m spatial resolution) and ALOS World 3D (30 m spatial resolution).

References

Aasen H, Van Wittenberghe S, Medina NS, Damm A, Goulas Y, Wieneke S, Hueni A, Malenovský Z, Alonso L, Pacheco-Labrador J, Cendrero-Mateo MP (2019) Sun-induced chlorophyll fluorescence II: review of passive measurement setups, protocols, and their application at the leaf to canopy level. Remote Sens 11:927

Al-Gaadi KA, Hassaballa AA, Tola E, Kayad AG, Madugundu R, Alblewi B, Assiri F (2016) Prediction of potato crop yield using precision agriculture techniques. PLoS One 11:e0162219

Alganci U, Besol B, Sertel E (2018) Accuracy assessment of different digital surface models. ISPRS Int J Geo-Inf 7:114

Allen RG, Tasumi M, Trezza R (2007) Satellite-based energy balance for mapping evapotranspiration with internalized calibration (METRIC)—model. J Irrig Drain Eng 133:380–394

Allen RG, Morton C, Kamble B, Kilic Huntington J, Thau D, Gorelick N, Erickson T, Moore R, Trezza R, Ratcliffe I (2015) EEFlux: a landsat-based evapotranspiration mapping tool on the Google Earth Engine. In: Proceedings Am. Soc. Agric. Biol. Eng. Symposium on Emerging

Technologies for Sustainable Irrigation-A Tribute to the Career of Terry Howell, Sr., Am. Soc. Agric. Biol. Eng., St Joseph, pp 1–11
Al-Wassai FA, Kalyankar NV (2013) Major limitations of satellite images. arXiv preprint arXiv:1307.2434
Apan A, Held A, Phinn S, Markley J (2004) Detecting sugarcane 'orange rust'disease using EO-1 Hyperion hyperspectral imagery. Int J Remote Sens 25:489–498
Baghdadi N, Cerdan O, Zribi M, Auzet V, Darboux F, El Hajj M, Bou Kheir R (2008) Operational performance of current synthetic aperture radar sensors in mapping soil surface characteristics in agricultural environments: application to hydrological and erosion modelling. Hydrol Proc 22:9–20
Bausch WC, Khosla R (2010) QuickBird satellite versus ground-based multi-spectral data for estimating nitrogen status of irrigated maize. Precis Agric 11:274–290
Bhatti AU, Mulla DJ, Frazier BE (1991) Estimation of soil properties and wheat yields on complex eroded hills using geostatistics and thematic mapper images. Remote Sens Environ 37:181–191
Bishop TFA, McBratney AB (2002) Creating field extent digital elevation models for precision agriculture. Precis Agric 3:37–46
Casa R, Castaldi F, Pascucci S, Palombo A, Pignatti S (2013) A comparison of sensor resolution and calibration strategies for soil texture estimation from hyperspectral remote sensing. Geoderma 197:17–26
Caturegli L, Casucci M, Lulli F, Grossi N, Gaetani M, Magni S, Bonari E, Volterrani M (2015) GeoEye-1 satellite versus ground-based multispectral data for estimating nitrogen status of turfgrasses. Int J Remote Sens 36:2238–2251
Champagne C, Davidson A, Cherneski P, L'Heureux J, Hadwen T (2015) Monitoring agricultural risk in Canada using L-band passive microwave soil moisture from SMOS. J Hydromet 16:5–18
Clevers J, Kooistra L, Van Den Brande M (2017) Using Sentinel-2 data for retrieving LAI and leaf and canopy chlorophyll content of a potato crop. Remote Sens 9:405
de Lara A, Khosla R, Longchamps L (2019) Soil water content and high resolution imagery: maize yield. Agronomy 9:174
Deilami K, Hashim M (2011) Very high resolution optical satellites for DEM generation: a review. Eur J Sci Res 49:542–554
Delegido J, Fernandez G, Gandia S, Moreno J (2008) Retrieval of chlorophyll content and LAI of crops using hyperspectral techniques: application to PROBA/CHRIS data. Int J Remote Sens 29:7107–7127
Demattê JAM, Galdos MV, Guimarães RV, Genú AM, Nanni MR, Zullo J Jr (2007) Quantification of tropical soil attributes from ETM+/LANDSAT-7 data. Int J Remote Sens 28:3813–3829
Du S, Liu L, Liu X, Zhang X, Zhang X, Bi Y, Zhang L (2018) Retrieval of global terrestrial solar-induced chlorophyll fluorescence from TanSat satellite. Sci Bull 63:1502–1512
Eitel JUH, Long DS, Gessler PE, Smith AMS (2007) Using in-situ measurements to evaluate the new RapidEye™ satellite series for prediction of wheat nitrogen status. Int J Remote Sens 28:4183–4190
El Hajj M, Baghdadi N, Zribi M, Rodríguez-Fernández N, Wigneron J, Al-Yaari A, Al Bitar A, Albergel C, Calvet JC (2017) Evaluation of SMOS, SMAP, ASCAT and Sentinel-1 soil moisture products at sites in southwestern France. Remote Sens 10:569
Ferrant S, Selles A, Le Page M, Herrault PA, Pelletier C, Al-Bitar A, Mermoz S, Gascoin S, Bouvet A, Saqalli M, Dewandel B (2017) Detection of irrigated crops from Sentinel-1 and Sentinel-2 data to estimate seasonal groundwater use in South India. Remote Sens 9:1119
Gesch D, Oimoen M, Danielson J, Meyer D (2016) Validation of the ASTER global digital elevation model version 3 over the conterminous United States. Int Archives Photogramm Remote Sens Spat Inf Sci 41:143
Gherboudj I, Magagi R, Berg AA, Toth B (2011) Soil moisture retrieval over agricultural fields from multi-polarized and multi-angular RADARSAT-2 SAR data. Remote Sens Environ 115:33–43

Gomez C, Rossel RAV, McBratney AB (2008) Soil organic carbon prediction by hyperspectral remote sensing and field Vis-NIR spectroscopy: an Australian case study. Geoderma 146:403–411

Gorrab A, Zribi M, Baghdadi N, Mougenot B, Chabaane Z (2015) Potential of X-band TerraSAR-X and COSMO-SkyMed SAR data for the assessment of physical soil parameters. Remote Sens 7:747–766

Gowda P, Chávez J, Howell T, Marek T, New L (2008) Surface energy balance based evapotranspiration mapping in the Texas high plains. Sensors 8:5186–5201

Guanter L, Zhang Y, Jung M, Joiner J, Voigt M, Berry JA, Frankenberg C, Huete AR, Zarco-Tejada P, Lee JE, Moran MS (2014) Global and time-resolved monitoring of crop photosynthesis with chlorophyll fluorescence. Proc Natl Acad Sci 111:E1327–E1333

Houborg R, McCabe M (2016) High-resolution NDVI from Planet's constellation of earth observing nano-satellites: a new data source for precision agriculture. Remote Sens 8:768

Jackson TJ, Chen D, Cosh M, Li F, Anderson M, Walthall C, Doriaswamy P, Hunt ER (2004) Vegetation water content mapping using Landsat data derived normalized difference water index for corn and soybeans. Remote Sens Environ 92:475–482

Joiner J, Yoshida Y, Vasilkov AP, Middleton EM (2011) First observations of global and seasonal terrestrial chlorophyll fluorescence from space. Biogeosciences 8:637–651

Khanal S, Fulton J, Shearer S (2017) An overview of current and potential applications of thermal remote sensing in precision agriculture. Comput Electron Agric 139:22–32

Köhler P, Frankenberg C, Magney TS, Guanter L, Joiner J, Landgraf J (2018) Global retrievals of solar-induced chlorophyll fluorescence with TROPOMI: first results and intersensor comparison to OCO-2. Geophys Res Lett 45:10–456

Kumar P, Gupta DK, Mishra VN, Prasad R (2015) Comparison of support vector machine, artificial neural network, and spectral angle mapper algorithms for crop classification using LISS IV data. Int J Remote Sens 36:1604–1617

Li H, Zhao J (2018) Evaluation of the newly released worldwide AW3D30 DEM over typical landforms of China using two global DEMs and ICESat/GLAS data. IEEE J Sel Top Appl Earth Observ Remote Sens 11:4430–4440

Li X, Lee WS, Li M, Ehsani R, Mishra AR, Yang C, Mangan RL (2015) Feasibility study on Huanglongbing (citrus greening) detection based on WorldView-2 satellite imagery. Biosyst Eng 132:28–38

Liu L, Liu X, Hu J (2015) Effects of spectral resolution and SNR on the vegetation solar-induced fluorescence retrieval using FLD-based methods at canopy level. Eur J Remote Sens 48:743–762

Magney TS, Eitel JU, Vierling LA (2017) Mapping wheat nitrogen uptake from RapidEye vegetation indices. Precis Agric 18:429–451

Maier SW, Günther KP, Stellmes M (2004) Sun-induced fluorescence: A new tool for precision farming. In: VanToai T, Major D, McDonald M, Schepers J., Tarpley L (eds) Digital imaging and spectral techniques: applications to precision agriculture and crop physiology, Am. Soc. Agron. Spec. Pub. 66, Madison, pp 209–222

Martínez-Fernández J, González-Zamora A, Sánchez N, Gumuzzio A, Herrero-Jiménez CM (2016) Satellite soil moisture for agricultural drought monitoring: assessment of the SMOS derived soil water deficit index. Remote Sens Environ 177:277–286

McCabe MF, Aragon B, Houborg R, Mascaro J (2017) CubeSats in hydrology: ultrahigh-resolution insights into vegetation dynamics and terrestrial evaporation. Water Resour Res 53:10017–10024

McNairn H, Brisco B (2004) The application of C-band polarimetric SAR for agriculture: a review. Can J Remote Sens 30:525–542

Metternicht GI (1998) Fuzzy classification of JERS-1 SAR data: an evaluation of its performance for soil salinity mapping. Ecol Model 111:61–74

Moran MS, Inoue Y, Barnes EM (1997) Opportunities and limitations for image-based remote sensing in precision crop management. Remote Sens Environ 61:319–346

Mulla DJ (2013) Twenty five years of remote sensing in precision agriculture: key advances and remaining knowledge gaps. Biosyst Eng 114:358–371
Mulla DJ, Miao Y (2016) Precision farming. In: Thenkabail PS (ed) Remote sensing handbook, Land resources monitoring, modeling and mapping with remote sensing, vol II. Taylor & Francis Publ., CRC Press, Boca Raton, pp 161–178
Mzuku M, Khosla R, Reich R, Inman D, Smith F, MacDonald L (2005) Spatial variability of measured soil properties across site specific management zones. Soil Sci Soc Am J 69:1572–1579
Navrozidis I, Alexandridis TK, Dimitrakos A, Lagopodi AL, Moshou D, Zalidis G (2018) Identification of purple spot disease on asparagus crops across spatial and spectral scales. Comput Electron Agric 148:322–329
Nawar S, Corstanje R, Halcro G, Mulla D, Mouazen AM (2017) Delineation of soil management zones for variable rate fertilization: a review. In: Sparks DL (ed) Advances in agronomy, vol 143, Academic, pp 175–245
Oh Y (2004) Quantitative retrieval of soil moisture content and surface roughness from multipolarized radar observations of bare soil surfaces. IEEE Trans Geosci Remote Sens 42:596–601
Pathe C, Wagner W, Sabel D, Doubkova M, Basara JB (2009) Using ENVISAT ASAR global mode data for surface soil moisture retrieval over Oklahoma. USA IEEE Trans Geosci Remote Sens 47:468–480
Peng J, Loew A, Merlin O, Verhoest NE (2017) A review of spatial downscaling of satellite remotely sensed soil moisture. Rev Geophys 55:341–366
Pinter PJ Jr, Hatfield JL, Schepers JS, Barnes EM, Moran M, Daughtry CS, Upchurch DR (2003) Remote sensing for crop management. Photogramm Eng Remote Sens 69:647–664
Pongpattananurak N, Reich R, Khosla R, Aguirre-Bravo C (2012) Modeling the spatial distribution of soil texture in the state of Jalisco. Mexico Soil Sci Soc Am J 76:199–209
Quesney A, Le Hégarat-Mascle S, Taconet O, Vidal-Madjar D, Wigneron JP, Loumagne C, Normand M (2000) Estimation of watershed soil moisture index from ERS/SAR data. Remote Sens Environ 72:290–303
Robson A, Rahman M, Muir J (2017) Using worldview satellite imagery to map yield in avocado (Persea americana): a case study in Bundaberg, Australia. Remote Sens 9:1223
Rosen PA, Hensley S, Joughin I, Li FK, Madsen S, Rodríguez E, Goldstein R (2000) Synthetic aperture radar interferometry. Proc IEEE 88:333–382
Scudiero E, Skaggs TH, Corwin DL (2015) Regional-scale soil salinity assessment using landsat ETM+ canopy reflectance. Remote Sens Environ 169:335–343
Seelan SK, Laguette S, Casady GM, Seielstad GA (2003) Remote sensing applications for precision agriculture: a learning community approach. Remote Sens Environ 88:157–169
Seigfried J, Khosla R, Longchamps L (2019) Multispectral satellite imagery to quantify in-field soil moisture variability. J Soil Water Conserv 74:33–40
Söderström M, Piikki K, Stenberg M, Stadig H, Martinsson J (2017) Producing nitrogen (N) uptake maps in winter wheat by combining proximal crop measurements with Sentinel-2 and DMC satellite images in a decision support system for farmers. Acta Agric Scand Sec B—Soil Plant Sci 67:637–650
Sozzi M, Marinello F, Pezzuolo A, Sartori L (2018) Benchmark of satellites image services for precision agricultural use. In: Groot Koerkamp PWG, Lokhorst C, Ipema AH, Kempenaar C, Groenestein CM, van Oostrum C, Ros N (eds) Proceedings Ag. Eng. Conf. on New Engineering Concepts for a Valued Agriculture, Wageningen, The Netherlands pp 8–11
Sullivan DG, Shaw JN, Rickman D (2005) IKONOS imagery to estimate surface soil property variability in two Alabama physiographies. Soil Sci Soc Am J 69:1789–1798
Sun Y, Fu R, Dickinson R, Joiner J, Frankenberg C, Gu L, Xia Y, Fernando N (2015) Drought onset mechanisms revealed by satellite solar-induced chlorophyll fluorescence: insights from two contrasting extreme events. J Geophys Res Biogeosci 120:2427–2440
Tadono T, Ishida H, Oda F, Naito S, Minakawa K, Iwamoto H (2014) Precise global DEM generation by ALOS PRISM. ISPRS Ann Photogramm Remote Sens Spat Inf Sci 2:71

Thenkabail PS, Mariotto I, Gumma MK, Middleton EM, Landis DM, Huemmrich KF (2013) Selection of hyperspectral narrowbands (HNBs) and composition of hyperspectral twoband vegetation indices (HVIs) for biophysical characterization and discrimination of crop types using field reflectance and Hyperion/EO-1 data. IEEE J Sel Top Appl Earth Observ Remote Sens 6:427–439

Torbick N, Salas WA, Hagen S, Xiao X (2011) Monitoring rice agriculture in the Sacramento Valley, USA with multitemporal PALSAR and MODIS imagery. IEEE J Sel Top Appl Earth Observ Remote Sens 4:451–457

Vincini M, Frazzi E, D'Alessio P (2006) Angular dependence of maize and sugar beet VIs from directional CHRIS/Proba data. In: Proceedings 4th ESA CHRIS PROBA workshop, Esrin, Fracati, Italy, pp 19–21.. http://earth.esa.int/workshops/4th_chris_proba/index.html. Accessed 11 Aug 2020

Wessel B, Huber M, Wohlfart C, Marschalk U, Kosmann D, Roth A (2018) Accuracy assessment of the global TanDEM-X digital elevation model with GPS data. ISPRS J Photogramm Remote Sens 139:171–182

Whitcraft AK, Becker-Reshef I, Justice CO (2015) A framework for defining spatially explicit earth observation requirements for a global agricultural monitoring initiative (GEOGLAM). Remote Sens 7:1461–1481

Wu C, Han X, Niu Z, Dong J (2010) An evaluation of EO-1 hyperspectral hyperion data for chlorophyll content and leaf area index estimation. Int J Remote Sens 31:1079–1086

Yang C, Everitt JH, Bradford JM (2006) Comparison of QuickBird satellite imagery and airborne imagery for mapping grain sorghum yield patterns. Precis Agric 7:33–44

Yeom J, Han Y, Kim Y (2013) Separability analysis and classification of rice fields using KOMPSAT-2 high resolution satellite imagery. Res J Chem Environ 17:136–144

Yost MA, Sudduth KA, Walthall CL, Kitchen NR (2019) Public–private collaboration toward research, education and innovation opportunities in precision agriculture. Precis Agric 20:4–18

Yuan L, Pu R, Zhang J, Wang J, Yang H (2016) Using high spatial resolution satellite imagery for mapping powdery mildew at a regional scale. Precis Agric 17:332–348

Chapter 3
Sensing Crop Geometry and Structure

Eduard Gregorio and Jordi Llorens

Abstract This chapter provides an introduction to the most common 3D sensing systems used for the extraction of crop geometry and structure information. These systems have been grouped into four families based on their principle mode of operation and performance: photogrammetric techniques, ultrasound sensors, optical sensors and depth cameras. The uses and applications of these systems vary considerably, from the simple indication of the presence or absence of vegetation to the generation of high-density point clouds or 3D models of crops and the ability to provide simultaneous colour and depth data. Throughout this chapter, the operating principles and main applications of all these systems are presented, comparing their advantages and limitations. We concluded that it is not possible to select an optimal sensor for use in the broad field of precision agriculture, but rather that the most suitable device should be chosen for each particular application. The future of crop geometry sensing lies in the development of automated methods to analyse the huge amount of data provided, an area in which the application of artificial intelligence algorithms seems to have promising prospects.

Keywords Crop structure · Photogrammetry · Ultrasonic sensor · LiDAR sensor · Depth camera

3.1 Introduction

The geometric characterization of crops was and sometimes still is usually based on manual measurements that are prone to errors and have large associated labour and time costs. Consequently, the development of automated methods to obtain plant features in a quick, accurate and efficient way has attracted much interest. The simplest technique is to use digital cameras, based on a charge-coupled device (CCD)

E. Gregorio · J. Llorens (✉)
Department of Agricultural and Forest Engineering, Research Group on AgroICT & Precision Agriculture – GRAP, Universitat de Lleida/Agrotecnio-CERCA Center, Lleida, Catalonia, Spain
e-mail: eduard.gregorio@udl.cat; jordi.llorens@udl.cat

R. Kerry, A. Escolà (eds.), *Sensing Approaches for Precision Agriculture*, Progress in Precision Agriculture, https://doi.org/10.1007/978-3-030-78431-7_3

or complementary metal-oxide-semiconductor (CMOS) sensors, to estimate properties such as plant height or diameter. However, this technique has limited ability to obtain geometric characteristics due to the overlap between vegetative organs. Three-dimensional modelling is needed to acquire an in-depth knowledge of crop geometry. A myriad of photogrammetric techniques has been developed to generate three-dimensional models from digital photographs. Section 3.2 of this chapter reviews the two that are most often applied in crop structural characterization: stereo vision and structure from motion.

As an alternative to photogrammetry, a set of active sensors can be used based on the emission of a signal and the detection of its return after interaction with the target under study (i.e. the crop). Unlike photogrammetric techniques, these measurements are not affected by lighting conditions, although they do not provide colour information. Depending on the signal type, it is possible to distinguish between ultrasound sensors, which emit high-frequency mechanical waves, and optical sensors, most notably used in light detection and ranging (LiDAR) systems. These two typologies are presented in Sects. 3.3 and 3.4, respectively.

A new family of sensors, known as depth cameras, has emerged in recent years. These cameras can take pictures with range (depth) data on a pixel basis, as described in Sect. 3.5. Although the resolution is lower than in photogrammetric or LiDAR-based techniques, these are low-cost sensors which, in many cases, also provide simultaneous colour information (red, green, blue-depth – RGB-D – sensors). Finally, Sect. 3.6 ends this chapter with a brief discussion on future trends in crop geometry sensing.

As an introduction to each specific technology, Table 3.1 offers a comparison of the main specifications (range, wavelength, resolution, price, and so on) of several commercial sensors used for the electronic characterization of crops. The technologies behind these sensors and some examples of successful research studies are discussed in this chapter.

3.2 Photogrammetric Techniques

This section presents the principle of operation of stereo vision and structure from motion techniques, and reviews their application in crop geometry characterization. Other photogrammetric techniques, such as shape-from-silhouette (Shlyakhter et al. 2001), shape-from-focus (Billiot et al. 2013) and photometric stereo (Bernotas et al. 2019) are also used in agriculture, although to a lesser extent.

3.2.1 Stereo Vision

Stereo vision (SV) is a photogrammetric technique inspired by human vision which allows 3D information of a scene or object to be obtained from images taken simultaneously by two monocular cameras, although multifocal cameras can also be used

Table 3.1 Specifications of different commercial sensors used for crop geometry and structure characterization

Sensor	Siemens PXS400 3RG61 .5	Hokuyo 30LX-EW	Velodyne VLP-16	Leica P201
Principle	Ultrasound (Time-of-flight, ToF)	LiDAR (ToF)	LiDAR (ToF)	LiDAR (ToF)
Range (m)	0.4–3.0	0.1–30	100	120
Wavelength (nm)	Ultrasound 120 kHz	905 (IR)	903 (IR)	808 (IR) 658 (VIS)
Channels /beams	1	1	16 (±15°)	1 + rotating mirror + rotating base
Multi-echo/return (max returns)	–	3	2 (strongest and last)	1
Accuracy (mm)	Depends on environmental conditions*	±50	±30	6 at 100 m
Scanning angular range (°)	–	270	360	H 360 V 270
Angular resolution (°)	–	0.25	0.1–0.4	0.002***
Measurements (points/s)	200 ms**	43 240	300 000	100 0000
Interface	Analogue	Ethernet	Ethernet	Ethernet/Wireless
More Info	Siemens AG (2008)	Hokuyo LTD (2014)	Velodyne Inc. (2020)	Leica Geosystems (2013)
Sensor	**CamCube2.0**	**MS Kinect v1 (K1)**	**MS Kinect v2 (K2)**	**Intel D415**
Principle	ToF camera	Structured-light + RGB camera	ToF camera + RGB camera	Active IR stereo vision
Range (m)	0.3–7.0	0.8–4.0	0.5–4.5	0.3–10
Wavelength (nm)	870 (IR)	IR	IR	850 (IR)
Depth FoV	40° × 40°	57° × 43°	70° × 60°	65° × 40°
RGB Resolution	–	640 px × 480 px	1920 px × 1080 px	1920 px × 1080 px
Depth Resolution	204 px × 204 px	640 px × 480 px	512 px × 424 px	1280 px × 720 px
Frame Rate	25 fps (typical)	30 fps	30 fps	up to 30 fps (RGB) up to 90 fps (depth)
Interface	USB 2.0	USB 2.0	USB 3.0	USB-C
More Info	PMD Technologies GmbH (2009)	Pagliari and Pinto (2015)	Pagliari and Pinto (2015)	Intel Corporation (2020)

*Accuracy of an ultrasonic sensor depends on several environmental properties such as the air temperature, humidity, atmospheric pressure and air movement, among others
**Response delay
***Angular accuracy

(Rovira-Más et al. 2009). Depth information is obtained by identifying homologous (common) pixels in both images and measuring the variation in their position, a parameter known as disparity. The following steps need to be completed to obtain 3D point clouds: (1) camera calibration, (2) stereo rectification, (3) stereo matching and (4) point cloud reconstruction. Calibration consists of obtaining both intrinsic camera parameters (e.g. focal length, lens aberration) and extrinsic parameters (position and orientation of images). Stereo rectification aims to align the epipolar lines of both images (i.e. the lines that connect the real point being measured and the centre of projection of both images). This simplifies the search for homologous pixels in the two images since these are always located in the corresponding epipolar lines. The typical arrangement of an SV system is shown in Fig. 3.1, where $P(x, y, z)$ is a real point in the scene, while x_L and x_R are their horizontal projections on the left and right images, respectively. The third step of the process is to find homologous pixels, also known as stereo matching, and aims to determine the disparity between the two views. Finally, a 3D point cloud reconstruction is performed based on the disparity image and applying geometric triangulation. Following the example of Fig. 3.1, the depth R is computed using the following expression (Rovira-Más et al. 2011):

$$R = \frac{b \cdot f}{d}, \tag{3.1}$$

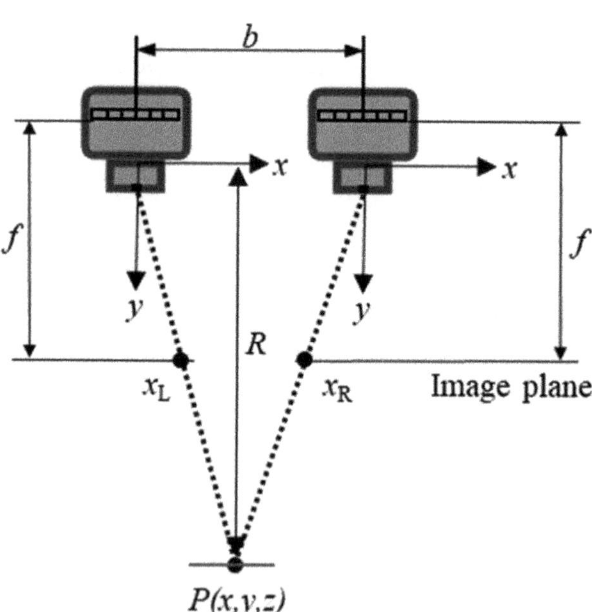

Fig. 3.1 Binocular stereo vision system

where b is the baseline or distance between the lenses of both cameras, f is the focal length (the same is assumed for both cameras) and $d = x_L - x_R$ is the disparity.

An important advantage of SV is its ability to generate high resolution 3D point clouds. During the processing of the point cloud, the availability of colour information allows more efficient segmentation. In addition, in an SV-based system both images are captured simultaneously using cameras attached to a fixed frame, thus minimizing the risk of failure associated with moving elements. However, SV systems require prior calibration and the results are affected by the robustness of the stereo matching algorithms, as well as by the prevailing lighting conditions. The need to process large amounts of information has traditionally been a major difficulty associated with SV systems. In recent years, the development of increasingly fast processors, together with the introduction of low-cost stereo cameras (Li et al. 2017; Oliveira et al. 2018) has allowed the use of SV in real-time applications.

In the field of geometric characterization of crops, Ivanov et al. (1995) were pioneers in using SV to obtain 3D models of maize (*Zea mays* L.) plants. In their study, the stereo matching was done manually, labelling the leaves and identifying the homologous points in their contours. Estimates of the horizontal profile of the leaf area index (LAI), vertical profile of the leaf area density (LAD), as well as of leaf position and orientation, were made. Many of the SV studies have been carried out under controlled laboratory conditions. In this way, He et al. (2003) developed an SV system for transplant growth analysis. A sweet potato transplant population was measured, and strong correlations were obtained between the destructively measured mass and the estimated mass volume with a coefficient of determination of $R^2 = 0.88$. For their part, Andersen et al. (2005) obtained 3D images of ten wheat plants using simulated annealing during the stereo matching process. These authors determined the size of the plant and its leaf area, with good agreement with actual values. The SV was also applied by Biskup et al. (2007) in a laboratory study of the daytime and nocturnal movement of a soya bean (*Glycine max* L. merr.) plant and to quantify its drought stress from the zenith leaf angle distribution. The system was also used to monitor a real soya bean plantation, verifying its robustness under field conditions. Müller-Linow et al. (2015) continued the previous work by developing more robust image processing routines to generate 3D reconstructions of the plants and to determine the LAI and the leaf angle distribution. Trials were conducted with sugar beet plants, and it was observed how leaf angles vary throughout the season and differ depending on the variety studied. The latest advances in matching algorithms have allowed Bao et al. (2019a) to develop a robotic platform for field phenotyping of sorghum (*Sorghum bicolor* L. Moench) plants. The system comprises six side-viewing camera pairs and can be used for the measurement of multiple properties such as plant height, plant width, convex hull volume, plant surface area and stem diameter. Other SV applications include 3D tree reconstruction for automated blossom thinning (Nielsen et al. 2012), geometric characterization during the plant seedling stage (Xiong et al. 2017), and the development of systems to measure plant growth automatically, both under controlled indoor conditions (Yeh et al. 2014) and in greenhouses and outdoor fields (Lati et al. 2013).

In addition, SV has been widely used in autonomous navigation systems of agricultural vehicles (Reid and Searcy 1987; Rovira-Más et al. 2004; Kise et al. 2005).

In this context, it is worth noting the SV system developed by Kise and Zhang (2008a), which had the double goal of locating crop rows for tractor guidance and 3D crop mapping. This system was validated in a soya bean field with three different plots according to the days elapsed since planting. The height differences were clearly detected. In another project, Kise and Zhang (2008b) developed an SV sensor comprising two multispectral cameras placed on a terrestrial vehicle. The spectral information was combined with 3D data to generate multispectral 3D field images of the soya bean field. In this vein, Zhao et al. (2018) recently presented a prototype that combines SV and a concave grating spectrometer (450–790 nm). This allows simultaneous monitoring of the biochemical properties and morphological features of crops.

3.2.2 Structure from Motion and Multi-View Stereo

The combination of structure from motion (SfM) and multi-view stereo (MVS) techniques enables 3D point clouds to be generated from a set of images taken from different viewing angles (Ullman 1979; Seitz et al. 2006). Unlike SV, a single RGB camera is moved to different positions taking a large number of pictures with a large degree of overlap between them. In the SfM process, invariant feature points have to be identified in several images. For this purpose, numerous feature detector algorithms have been developed (Tareen and Saleem 2018), most notably the scale-invariant feature transform, SIFT, (Lowe 1999) and the speeded-up robust features, SURF (Bay et al. 2006). A sparse 3D point cloud is then generated, the intrinsic and extrinsic parameters of the camera are determined, and the camera position and orientation estimated. Camera projective matrices computed with SfM are used by the MVS approach to provide a dense 3D point cloud, increasing the number of points by two or three orders of magnitude (James and Robson 2012).

In the last ten years, the application of SfM–MVS has revolutionized 3D plant modelling, particularly when a high level of detail is required (Quan et al. 2006; Pound et al. 2014). The development of these techniques has been made possible by the increase in computing capacity and the development of commercial photogrammetric software that automatically generate 3D reconstructions (Probst et al. 2018). The SfM–MVS is a low-cost image-based technique and, as can be seen in Fig. 3.2a, provides very high resolution, accurate and realistic 3D representations. In addition, unlike SV, prior camera calibration is not required, it is not necessary to know its position or orientation. However, the acquisition of multiple images, as well as their processing, are important limitations in real-time applications.

The SfM–MVS has been widely applied to plant phenotyping in indoor conditions. Rose et al. (2015) used this technique to monitor the growth of five tomato plants for six days. Between 40 and 70 photographs per plant were taken to generate the 3D point cloud. From this, the leaf area and the main stem height were determined, obtaining a strong correlation with a coefficient of determination of $R^2 > 0.96$ with the reference measurements made with a laser sensor. Similarly, Santos and

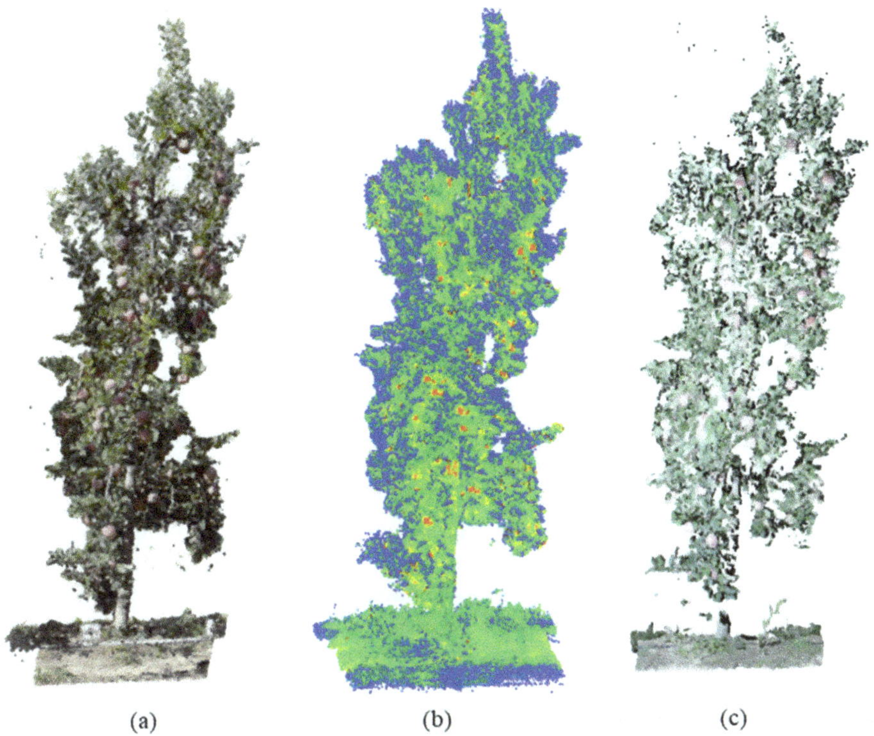

Fig. 3.2 The 3D point clouds of an apple (*Malus domestica* Borkh.) tree generated using: (**a**) structure from motion and multi-view stereo, (**b**) a LiDAR-based system (colour denotes intensity of returns) and (**c**) a depth camera

Rodrigues (2016) used a low resolution camera (1.3 MP) and MVS for 3D reconstruction of sunflower (*Helianthus annuus* L.) and maize plants, both in the initial and in the final stage of development (2 m high). For the maize plant, the leaf lengths were determined with errors below 9 %. In an experiment with cucumber (*Cucumis sativus* L.), pepper (*Capsicum annuum* L.) and eggplant (*Solanum melongena* L.), Hui et al. (2018) concluded that a minimum of 60–80 images were required to reconstruct 95 % of the leaves. Duan et al. (2016) also demonstrated the possibility of using SfM–MVS to study plant growth (in this case, wheat) in greenhouse conditions.

Recently, SfM–MVS has been applied in crop phenotyping under field conditions. Jay et al. (2015) generated 3D models of five plant species (sunflower (*Helianthus annuus* L.), savoy cabbage (*Brassica oleracea var. sabauda* L.), cauliflower (*Brassica oleracea var. botrytis* L.), Brussel sprouts (*Brassica oleracea var. gemmifera* L.) and sugar beet (*Beta vulgaris* L.)). The RGB images were acquired by displacing a camera along the row. The plants were discriminated from the environment on the basis of colour and plant height. Other authors used SfM–MVS to estimate crop growth in the field. In this context, Zhang Y et al. (2018b) monitored the evolution of sweet potato plants over four months and under six different fertilization conditions. Good estimations for plant height, leaf number and area were obtained ($R^2 > 0.97$). For their part, Mortensen et al. (2018) mounted a stereo camera on an agricultural robot and monitored several Cos (*Lactuca sativa* L. var. Cos) and Iceberg lettuces (*Lactuca sativa* L. var. Iceberg) over five weeks. Correlations were strong between the estimated surface areas and the actual fresh weight with coefficients of determination of $R^2 = 0.84–0.94$). Likewise, Andújar et al. (2018) detected and classified different weed species in a maize field. This work demonstrated the use of the SfM–MVS technique to obtain dense point clouds from which reliable plant models can be generated. Their results agreed with those obtained by Martinez-Guanter et al. (2019), who compared 3D crop models (maize, sugar beet and sunflower) using SfM and RGB-D systems.

Fruit detection, including of wine grapes (Herrero-Huerta et al. 2015) or apple (Gené-Mola et al. 2020b), is another area where SfM–MVS has a strong potential. In this context, Rose et al. (2016) developed a platform for field vineyard phenotyping with RGB cameras and an on-board real-time kinematic-global navigation satellite system receiver (RTK-GNSS). The MVS was applied to georeferenced images, obtaining 3D point clouds of the vineyard with which it was possible to detect the number and size of berries.

As has been shown, SfM–MVS has the potential to create 3D plant reconstructions at the organ level. However, its large-scale field application currently faces several challenges, including the development of automatic image acquisition and processing systems and the need to resolve wind-related limitations (plant position can vary from one image to another).

3.3 Ultrasonic Sensors

Ultrasonic sensors use the transmission and reception of an ultrasound wave signal (a graphical description of a wave emission can be seen in Fig. 3.3) to detect objects and some of their properties at certain distances. The distance to a target is found by measuring the time interval between wave emission and reception, a principle known as time-of-flight measurement. The properties of the target can be interpreted by analysing the echo wave that returns from the target (echo wave in Fig. 3.3). Using a piezo-electric or di-electric membrane device, the sensor is able to generate ultrasonic sound wave packages that bounce back from the target and, after a very short time interval, are usually received and read by the same device or using a specific membrane to act as receiver.

The simplest analysis technique is the 'pulse-echo' method, where a short burst of ultrasound is emitted and the time taken by the signal to return is measured. The range resolution, the smallest distance that can be measured, is determined by the resolution of the time measurement. The time is found considering the phase of the returned signal, and the higher the frequency the better is the resolution. However, it is not possible to indefinitely increase the frequency to improve distance

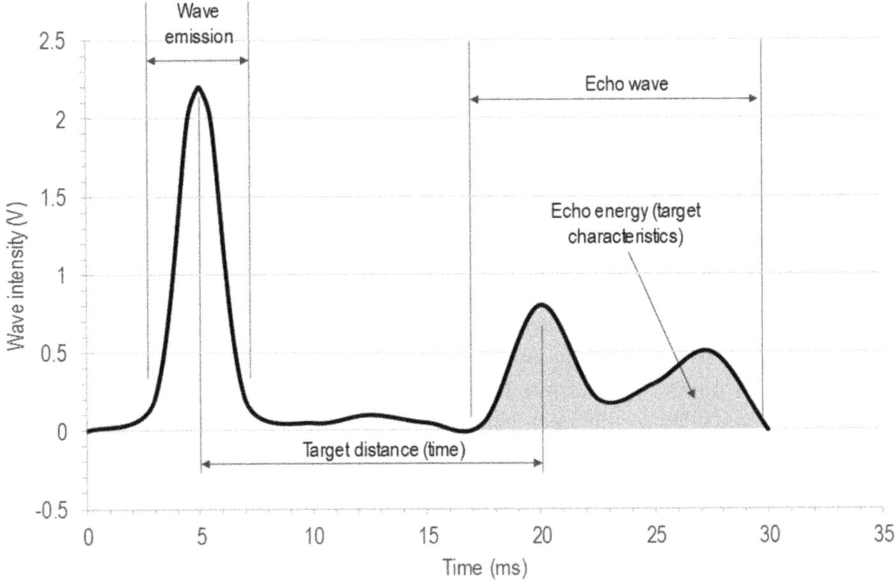

Fig. 3.3 Typical acoustic waveform: both emitted and returned wave are shown

resolution because, as is well known, ultrasound is attenuated as it travels through a medium and this attenuation increases with frequency (Wykes et al. 1994).

Sprayer control systems have been developed, using different techniques, which sense the presence or absence of target plants and control the sprayer output in an on–off or selective manner (Reichard and Ladd 1981; Ladd et al. 1981). At the time of their development, these systems detected only the presence of targets using photoelectric systems but not actual target characteristics. In 1987, a commercially available orchard air-blast sprayer control system (Roper Grower's Cooperative, Winter Garden, FL) represented a refinement of the selective control approach. Five ultrasonic transducers on either side of the sprayer could detect the presence or absence of tree foliage, and spray nozzles were then activated accordingly (Giles et al. 1987).

With respect to the performance and accuracy of this technology, different scientific studies have been carried out to verify the feasibility of using this kind of sensor for crop measurement. In 2007, Zaman et al. proposed a study to investigate the errors in citrus canopy volume measured with a 10-transducer ultrasonic orchard measurement array. The factors studied were ground speed accuracy, uncalibrated air temperature and the effect of ultrasonic transducer crosstalk. Deviations in the driving path from the centreline between two rows and the height error in the transducer array due to improper tyre inflation and uneven ground were also estimated. A few years later, Jeon et al. (2011) presented further research in which ultrasonic sensors were subjected to simulated environmental and operating conditions to determine their durability and accuracy. Conditions tested included exposure to extended cold, outdoor temperatures, cross-winds, temperature change, dust clouds, travel speeds and spray effect. They verified that these conditions have to be taken into account in the process to determine canopy distance in real and simulated plant canopies.

Similar studies were carried out on apple and olive (*Olea europaea* L.) trees by Escolà et al. (2011) and Gamarra-Diezma et al. (2015), respectively. In those cases, the main objective was to obtain a good distance measurement to estimate crop volume at different tree heights. In the apple tree study, they found that the sensors under study were required to be at a minimum distance of 60 cm apart because of errors from interference at <10 cm. In this case, the narrow distance between crop rows (<4 m) contributed to ensuring operation with short inter-sensor distance because the interference effect caused by the reflection of the subsequent sensor's wave was less important. In olive tree plantations, where there are wide row distances (>10 m), distance between sensors was increased to 130 cm. Accurate volume measurement is an important aspect in the process of tree crop characterization. By using ultrasonic sensors, it is possible to obtain this measurement if the distance between sensors is taken into account.

Ultrasound can also be used in field and bush crop measurements. For example, an ultrasonic sensor is a simple way to measure distance and can be mounted on agricultural machinery to improve harvesting procedures. Zhao et al. (2010) used two ultrasonic sensors mounted on either side of a combine harvesting header to detect presence of the crop and to measure width in the header. With that information, they could determine when the system needed to start and stop collecting data, as well as the real cutting width which is an important parameter in a yield monitor.

With its ability to measure plant height, the ultrasonic sensor can be used as a weed sensor detector, especially when differences between crop or soil and weeds are evident. Andújar et al. (2011, 2012) proposed different studies in weed detection using ultrasound sensors and applied the technique to two important field crops, maize and winter wheat (*Triticum aestivum* L.). In maize, where the aim was to differentiate grasses and broad-leaved weeds, the sensor's static readings permitted discrimination of 81 % of pure grasses and up to 99 % of pure stands of broad-leaved weeds. With dynamic measurements, its potential to detect weed infestations was confirmed. In the winter wheat application, the authors considered the hypothesis that weed-infested zones have more biomass than non-infested areas. Among the main results, they found that ultrasound readings could discriminate areas with weed infestation with a success rate of about 92.8 %. After detecting a weed infestation, a method to reduce it can be applied. In this respect, Zaman et al. (2011) proposed a weed spraying system using eight individual nozzles with an ultrasonic sensor on each to detect the weeds and activate the spraying. The selective application was adjusted depending on weed height above the wild blueberry (*Vaccinium angustifolium*) crop. In this case, weed height was between 35 cm and 55 cm higher than that of the crop, which ranged from 12 cm to 30 cm.

For the non-destructive assessment of forage mass in legume-grass mixtures, Fricke et al. (2011) proposed the use of an ultrasonic sensor yield mapping system in two field experiments. Important conclusions were drawn from stationary and on-the-go experiments on pure swards and mixtures of legumes and perennial ryegrass (*Lolium perenne* L.). In this research, they georeferenced the ultrasonic sensor readings using a high-accuracy differential global positioning system (DGNSS). Forage mass, calculated on the basis of information gathered from ultrasonic height measurements, could be explained at an acceptable range of accuracies (R^2 from 0.6 to 0.86). The relationship between ultrasonic sward height and forage mass was affected by various factors: sward type, weed abundance, abrupt forage mass changes and growth periods. At the end of their project, they proposed further lines of research to improve the technique, including combining information from image sensors.

In this particular respect, sensor fusion could be an interesting option when ultrasonic sensors on their own cannot extract all the information required for a specific purpose from the environment. With this in mind, Farooque et al. (2013) integrated a system comprising an ultrasonic sensor, a digital colour camera and a slope sensor. All the data from the sensors were georeferenced with an RTK-GNSS receiver. This integrated system was applied in a commercially managed wild blueberry field, with the sensors mounted on a specially adapted harvester. In this research, as in previous studies, the ultrasonic sensor was used to measure plant height which was then correlated with fruit yield determined by counting blue pixels captured by the colour camera. All the data were processed and represented on a map in which five differentiated zones were defined. In a study by Fricke and Wachendorf (2013), ultrasonic height measurements were also taken, but in this case the information was combined with spectral-radiometric reflection measurements. To determine whether biomass predictions made using only the ultrasonic sward height could be improved by complementing the measurements with vegetation indices VIs, various hyperspectral VIs were evaluated. Spectral reflections measured on the canopy appeared to explain differences detected in biomass measurements. In a more recent paper, Yuan et al. (2018) proposed the objective of detecting and measuring wheat plant height by comparing ultrasonic sensor, LiDAR and data from unmanned aerial vehicles (UAVs), and the results were compared with manual measurements. The ultrasonic and laser sensors were mounted on a ground-based multi-sensor phenotyping system. The ultrasonic sensor performed worse than the other systems.

Upchurch et al. (1993) designed an intelligent tree trunk diameter calliper. The system was calibrated to work with objects (tree trunks in this case) with diameters from 1.6 cm to 19 cm with a small mean error of 0.04 cm. Tree trunk diameter is an interesting property to measure in orchard production, and in this case the measurement technique was improved and automated by means of ultrasonic sensors.

The use of ultrasonic sensors in orchard crops has facilitated the gathering of important data to improve specific field tasks. In the 1980s, two interesting studies by McConnell et al. (1983) and Giles and Delwiche (1988) were published with similar titles: "*Electronic measurement of tree-row volume*" and "*Electronic measurement of tree canopy volume*", respectively. Their objective was to measure the distance to the foliage of the crop from the position of the sensor; this can be done from the centre of two rows in row measurement or from the tree side in measurements of individual trees. In McConnell et al. (1983), a mast with a range of transducers was mounted to read the shape of the crop at different heights. In Giles and Delwiche (1988), three ultrasonic sensors were mounted on each side of the sprayer to characterize the crop as the target of a spray application. Subsequently, this process was applied in citrus crops by other authors. Tumbo et al. (2002), for example, mounted a total of 20 ultrasound transducers, 10 on either side of a mast mounted behind a tractor, to read the crop shape. With this number of sensors, cross-talk interference was evident, so the system was designed to trigger the sensors in sequential groups of three to prevent this effect. When the effect of canopy foliage density and the ground speed of the system was tested in citrus trees (Zaman and Salyani 2004), it was found that canopy foliage density had a significant effect on ultrasonic volume estimation, with greater differences obtained in low density canopies. However, there were no differences in volume measurements at ground speeds of 1.6 km h^{-1} to 4.7 km h^{-1}. Later, the information provided by the ultrasonic sensors was mapped using DGPS (Schumann and Zaman 2005). For that study, interpretation of tree height and volume was made using the data from 10 ultrasound sensors mounted on a vertical mast at 0.6 m increments, all reading the same side of the row. That required two passes on each alleyway to cover all of the field. The mast was mounted on a trailer pulled by a car. Measurement accuracy was reduced given the short distance between sensors, but it was sufficient to provide maps of application rates to adjust sprayers.

A year later, an electronic sprayer control system was proposed by Solanelles et al. (2006) to adjust a variable-rate spray to different crops: olive, pear (*Pyrus communis* L.) and apple. In this case, information from the ultrasonic sensor was used to estimate tree canopy width which was then compared to the maximum tree width of the whole orchard. On the basis of the ratio obtained, different solenoid valves were adjusted to supply the appropriate amount of liquid spray to the canopy.

Other studies have since been performed to adapt the principle of canopy measurements for variable-rate spray application to grapevines (Gil et al. 2007, 2013; Llorens et al. 2010). The objective of the ultrasonic measurements was to calculate vine volume, then using an application coefficient, defined as litres of liquid per cubic metre of crop, the sprayer output was adjusted using proportional solenoid valves.

As can be seen in Fig. 3.3, it is also possible to extract information from the analysis of the energy of the echo wave after it bounces off the target. This type of analysis allows prediction of another interesting property, the canopy density. This variable can be used to improve estimates of volume on the basis of information about how the crop is organized internally within the estimated volume. Stajnko et al. (2012) proposed an analysis of the envelope on 15-cm ultrasound bands to measure canopy size and density, and then adjust spray application accordingly. Palleja and Landers (2015) analyzed the returned ultrasound wave in a more in-depth and accurate way. For their research, they used four ultrasound sensors for each side of the crop (grapevine and apple) which were activated synchronously to prevent interference. With this system, they were able to provide an estimate of canopy density specific to each side of the crop, with an error of 14 % in vineyards and 3.8 % in apple trees.

Since 2015, little research using ultrasonic sensors has been published mainly because of the decrease in use of this kind of sensor because of advances in other technologies, for example LiDAR sensors, which provide better and faster data for the electronic characterization of crops. Some of these sensors are described in the following section of this chapter.

3.4 Optical Sensors

The main optical sensors used in canopy characterization are described below. First, photoelectric sensors are introduced as a simple technology to detect the presence of and the distance to a target. Second, the LiDAR principle is presented as a well-known but complex technology that is used to extract interesting and useful information from the surroundings; in this case, surroundings composed of different crop structures.

3.4.1 Photoelectric Sensors

Photoelectric sensors are active sensors able to detect objects from the interruption or reflection of light, often infrared, within a defined sensing range. Depending on how the sensor is used and the type that is employed, they permit the presence or absence of one or various objects to be determined. In the case of reflection type sensors, (Fig. 3.4), the objects located beyond the sensing cut-off point are disregarded. These sensors are usually low-cost, compact and rugged.

These photoelectric sensors are commonly used in industrial environments to detect the position of physical elements through the interruption of a fixed optical light beam. They are usually applied for near-range detection and have a specific range that depends on the type used (Fig. 3.4). Wide ranges of detection are not possible. There are three main types of these easy-to-mount sensors. In a through-beam system (Fig. 3.4a), separate sensor, emitter and receiver devices are installed on either side of the area of work. In a retroreflective system (Fig. 3.4b), the light source and receiver unit are in the same housing and the sensor is aimed at a reflector which sends the signal back to the receiving unit when there is no object in between. Finally, in a diffused system (Fig. 3.4c) the light source and receiver are housed in the same device and the light beam is reflected back by the target.

Although used far less than in industrial environments, some studies used these sensor types or adapted versions of them, to detect vegetative elements. Some examples of these studies are described below.

Hooper et al. (1976) designed a photoelectric sensor to distinguish between plant material and soil. They analyzed the differences in plant vs. soil reflectivity to detect the plant structure. The operation of the system depended on measuring the ratio of visible to near-infrared reflected radiation, which is less for leaves than for soil due

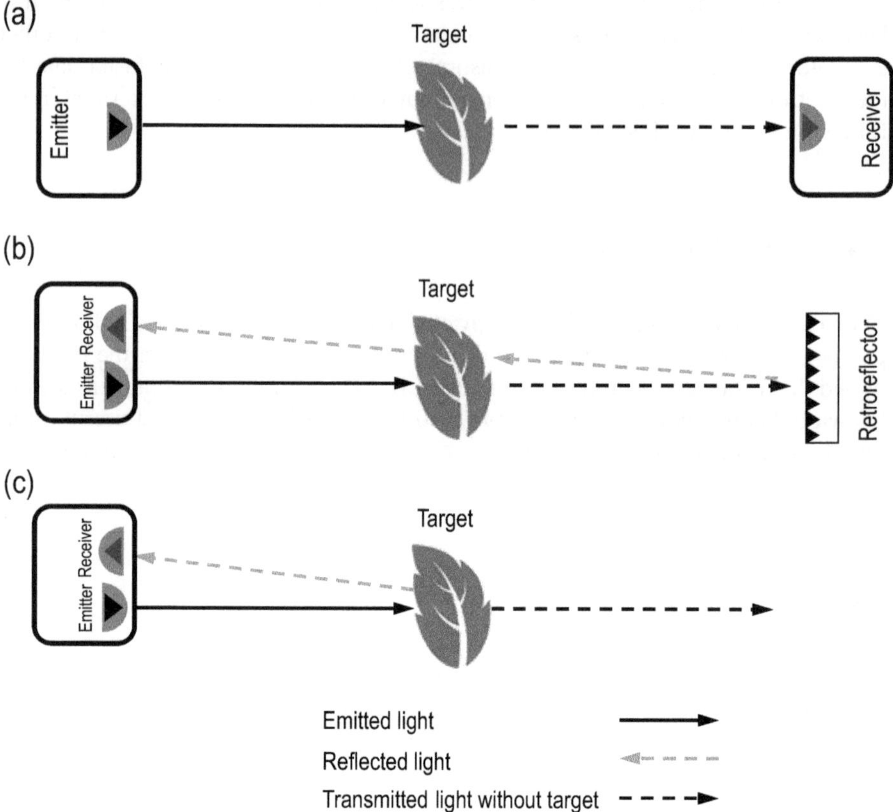

Fig. 3.4 Types of photoelectric sensors. (**a**) through-beam system, (**b**) retroreflective system and (**c**) diffused system

to the absorption of visible red radiation by chlorophyll. For this specific sensor, the working principle is somewhat different from the principle described in Fig. 3.4. In this case, the sensor needs to be able to read the reflected radiation. To do this, the receiver measures the intensities of near-infrared and visible radiation reflected from a surface illuminated by the emitter.

The photoelectric sensor has also commonly been used to adjust machinery output. A selective applicator for post-emergence herbicides in row crops was developed and tested by Shearer and Jones (1991). In this study, the presence of weeds was sensed using a modulated NIR light source and phototransistor receiver. A solenoid valve was activated, or deactivated as required, through a logic circuit to release herbicide through the spray nozzle. The results showed herbicide savings of 15 % with no adverse effect on weed control.

Ladd et al. (1981) proposed a selective sprayer using pulse-modulated solenoid valves. In this case, IR sensors were used to detect plants and trigger pesticide spraying. The crops detected with this sensor were cabbage, cauliflower and pepper.

Field testing of the system confirmed that pest control was maintained while pesticide use was reduced by 24–51 %. In this case, the field crop system developed required the target to be able to pass through the emitter-detector sensor pair.

Miller et al. (2003) tested a variable-rate granular fertilizer spreader with both GPS-guided prescription mapping and real-time tree presence measurement with photoelectric sensors in citrus groves. The variable-rate fertilizer unit performed well for site-specific fertilizer application with large-scale variability within a grove. They used six units of Banner QMT42 (Banner Engineering, Minneapolis, Minn.) long-range diffuse photocells to measure the citrus trees and to adjust the prescription rate output from the prescription map.

Some more recent research is a robotic system presented by Hayashi et al. (2010). The authors used the sensor to detect the presence of a piece of fruit that is manipulated by an arm of the robotic system. In the system, the photoelectric sensor works on a gripper that is responsible for grasping and cutting the peduncle of a strawberry while a suction device holds the fruit.

3.4.2 LiDAR Sensors

As an advancement of the use of the simple photoelectric directional sensors, LiDAR technology can aim the light beam in different directions. The LiDAR sensors use light, usually laser in infrared mode, to measure the distance to the sensor surroundings.

A general classification of LiDAR commercial solutions can be made according to the dimensions (directions of axis) scanned. Following this criterion, it is possible to distinguish 1D, 2D and 3D LiDAR sensors. The first type can only provide information in one direction, while 2D LiDAR sensors can provide information from two directions. To generate a whole 3D point cloud, the LiDAR system needs either to use two internal rotary mirrors to direct the laser beam to different angles, or to have the ability to rotate its head 360° to scan the surroundings. If the LiDAR sensor has neither of these capabilities, the sensor needs to be moved to different positions, for which a positioning system is required. The 3D LiDAR sensors can provide information from the surroundings in three directions when the sensor is stationary in one position. According to the operation system, there are solid-state sensors without mechanically moving parts and sensors with moving parts. Usually, those moving parts are rotating mirrors that are able to redirect a single laser beam to different directions.

The first step in the working process of the sensor is to generate a beam of infrared light that is redirected to a different direction using a mirror. Multiple sensor formats are possible, from a limited angular range of detection (short field of view) to a 360° angular range of detection. In all cases, the raw data received from the sensors are stored in polar format (angle and distance for each distance measurement) with the origin in the centre of the sensor. Then, using a basic trigonometry transformation, each measured point can be located in a Cartesian coordinate system. To generate a whole point cloud of a target the LiDAR system needs to be moved to different positions, for which a precise positioning system is required. Simultaneous location and mapping (SLAM), inertial measurement units (IMU) and GNSS receivers are used to register the different scans and to geo-reference the position of the sensor. The SLAM algorithms enable the LiDAR sensor position to be estimated in relation to the points that are captured instantly and in previous scans. In this process, the movement is estimated sequentially by comparing point clouds. This allows a precise map of 3D point clouds to be created using identified features captured and located in the scanning process. The IMU sensor is an electronic device that enables inertial information (orientation, angular rate, force and acceleration) of the LiDAR sensor movement to be recorded during the scanning process. With that information, it is possible to correct or compensate the LiDAR positioning. The SLAM+IMU systems are combined with GNSS systems to provide better accuracy of the point cloud captured with LiDAR sensors.

This section will focus on terrestrial laser scanners (TLS), laser-based systems that can be used statically (with 3D capture capability) or in motion after being mounted on a mobile or portable system (either with 2D or 3D capture capabilities). This chapter does not cover aerial LiDAR, which is commonly used in electronic forest characterization where large areas are scanned and processed. In our case, for close-range terrestrial crop characterization, the technique is called mobile terrestrial LiDAR or laser scanning (MTLS), with the LiDAR sensors deployed using a terrestrial vehicle or platform to scan a crop. This system can be designed using all manner of terrestrial vehicles or indoor equipment, including all-terrain vehicles, cars, bicycles, movable platforms, specific engine-equipped prototypes, manual trolleys that are pushed alongside the crop or even backpacks to be carried by a person on foot. An example of a MTLS using a robotic platform is shown in Fig. 3.5a.

A brief description is provided below of the most important research studies that have been published to date on the use of LiDAR to extract information about agricultural crops.

An early system design in one of the first studies to use a LiDAR system was presented by Vanderbilt et al. (1979). They applied what was then a new laser measurement technique for the first time to measure the irradiance distribution in a vegetative canopy of wheat. The new technique was conceptually similar to the traditional manual point quadrat methodology, but using the capabilities of an electronic system. The laser light source was mounted on a tripod and moved at an angle (pan), measuring the height where each light beam first hit the foliage. According to the authors, in comparison with other available methods the laser technique was amenable to automation and applicable to canopies of various heights and at different measurement distances. This was in effect the beginning of laser-based crop measurements. A decade later, Walklate (1989) used a similar optical instrument for measuring crop geometry based on the probability of a light beam intersecting plant surfaces within a crop. In this case, the laser emitter was based on a helium-neon laser and the return from the target was detected with a backscattered light collection system. The system was applied to a barley (*Hordeum vulgare* L.) crop and the results compared with crop area density measurements obtained in destructive

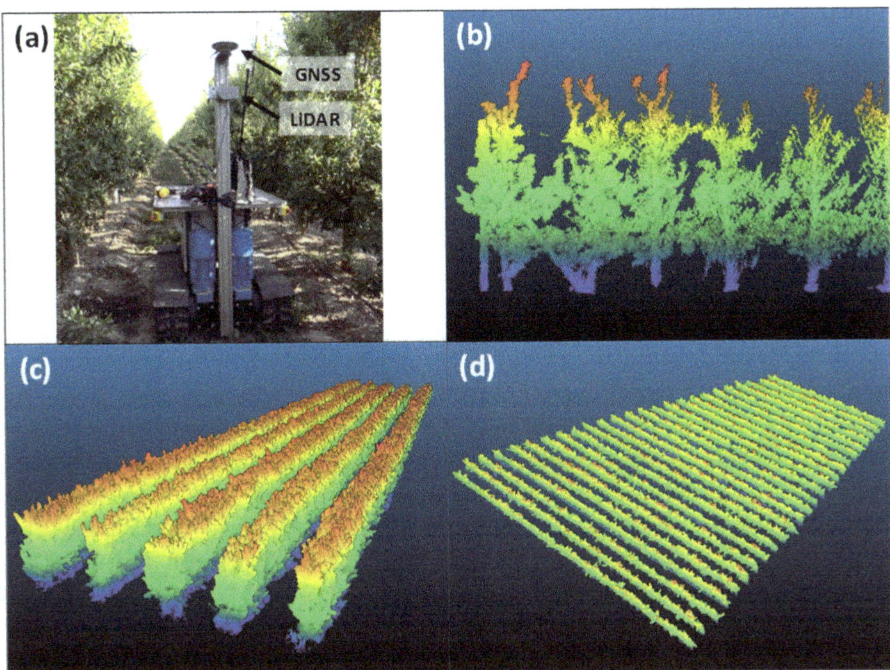

Fig. 3.5 Procedure and point cloud results from terrestrial LiDAR scanning. (**a**) mobile terrestrial robotic platform with sensor and positioning system, (**b**) close-range point cloud of a scanned apple tree row section, (**c**) five-row almond orchard and (**d**) vineyard point cloud. All LiDAR point clouds are represented using a colour scale representing height

manual procedures. In that study, Walklate proposed that the LiDAR data processing methodology could be understood using the Poisson law, where the probability of a LiDAR laser beam intersecting a plant surface was taken as similar to that of a beam of sunlight penetrating a canopy.

As with other sensor types, when LiDAR sensor measurements are proposed for use in crop characterization, accuracy and performance are factors of fundamental importance that need to be taken into account. With this in mind, several authors have undertaken specific studies on the behaviour of these measurements under the effect of different interactions. When used to measure distances, LiDAR systems are faced with the problem that, due to its size, the laser beam may intercept more than one object, with partial interception on multiple targets. In such cases, the distance measurement could be affected by the way this interception is produced. This effect, known as mixed pixels or edge effects, was assessed by Sanz-Cortiella et al. (2011). In their study, the effect on distance measurement was analyzed when LiDAR LMS200 laser beams were blocked by different surface shapes. Mixed pixels usually appear when a canopy is scanned with LiDAR sensors. Their main conclusion was that in a partial beam blockage scenario, the distance measured with the sensor depends more on the blocked energy than the blocked surface area. Their research showed that it is important to know the dimensions of the laser beam cross-section in sensor simulations, and these also need to be known when selecting the LiDAR sensor.

Another clear example of the effect on LiDAR sensor performance relates to factors that can affect volume measurement in tree crops when no GNSS system is used, as described in Palleją et al. (2010). According to the authors, the distance to the crop and the angular movement of the sensor are crucial to obtain accurate volume estimates. They concluded that all procedures for tree volume estimation should incorporate additional devices such as inertial measurement units (IMUs) or methods to control or estimate and correct the trajectory.

The LiDAR position and point of view are also very important, especially when the shape of the crop can create shadows that make it impossible to scan some parts of it. This effect can be considerable and difficult to solve with very dense crops. The effect of the position of the sensor in relation to the crops was studied in Bietresato et al. (2016), where they used a scanning system with two LiDAR sensors working at different heights and at the same time as a LiDAR stereovision system. Another important effect in crop characterization is when the scanning is incomplete (only one side of the crop row). When tree crops are characterized electronically, the sensor operator can decide whether the scanning system collects information from each side of the crop or not. In fact, according to Auat Cheein et al. (2015), it can represent a reduction in accuracy of up to 30 % in estimates of tree crown volume. Today, an automated crop characterization service does not need to cover all of the field, but a discontinuous systematic sampling procedure can be applied to extract a raster map of the field. For example, in del-Moral-Martínez et al. (2016) two LAI characterization strategies were proposed in vineyard crops: continuous on-the-go scanning with computing of 1 m LAI values along the rows, or discontinuous systematic scanning of specific sampling sections separated by up to 15 m along the rows. The resulting raster map of the field was unaffected by the method employed.

As in the ultrasonic sensors section, the following research studies in which LiDAR sensors were used to extract information, focus first on field crops and then on orchard crops.

LiDAR sensors have been used to measure and numerically describe the architecture of field crops. In this case, the sensor needs to be moved or placed in a high position over the ground to be able to read the top and interior of a crop that is normally extended over the soil surface. One early study is described in Vanderbilt et al. (1990), in which a LiDAR sensor was used to extract crop density in maize. In this case, the method used provided an interception coefficient for classes of vegetation in layers of the canopy viewed in various directions. Sometimes, the main crop is not the target of the laser readings, as for example with weeds that can grow in inter-row areas of maize fields (Andújar et al. 2013). In that study, the authors proposed and tested a procedure for weed characterization. The weeds were detected by measuring their height, which had a strong correlation with manually measured height values that had a coefficient of determination of $R^2 = 0.88$. The ability of the system to discriminate between weeds was also carried out, with a success rate of 77.7 % in the classification of Johnson grass *(Sorgum halepense*).

In Lumme et al. (2008), other field crops (namely barley, oat (*Avena sativa* L.) and wheat) were scanned to extract information about growth heights, which was then used to study the effect of different nitrogen fertilization strategies on yield for each plot studied. In this particular study, a movable rack about 3 m high was used to have a better point of view and to move the sensor over the crop. In a similar study, a static 3D LiDAR scanner was used to scan a wheat crop (Hosoi and Omasa 2009). In this case, the sensor was mounted in different positions around the field to extract a complete three-dimensional model of the crop. Using LiDAR data and a profiling voxel-based method, the authors determined the plant area index and plant area density profiles, with the latter property was then used to estimate the actual dry weight of each organ type (ears, stems and leaves).

Sometimes, a particular crop characterization study is undertaken to improve harvest operations. This was the case in Saeys et al. (2009), where two different models of Sick LMS LiDAR sensors were used to scan small grain crop plots. In this research, the interesting option of using the sensor on a combine harvester to scan the amount of crop harvested was explored at the same time as the combine was harvesting the cereal.

Llop et al. (2016) showed that LiDAR sensors can also be used in greenhouses. In their study, they used the sensor to measure tomato (*Solanum lycopersicum* L.) crop properties. Crop layout, with narrow alleyways and tall and thin plants usually planted in pairs, made the scanning process difficult and meant that the sensor could see only one side of the plant. Nonetheless, despite this drawback they found a good correlation between manual and electronic measurements in volume and density variables.

An important factor in the case of orchards is the ability of the sensor system to generate a point cloud that defines the whole geometry of the crop, especially if the scanning process is carried out from different points of view. In general, as this kind of crop is organized in rows, the points of view are limited to two sides of the crop. With this limitation, all the information provided from each location needs to be synchronized correctly, and the whole point cloud can subsequently be processed as a unique element. One of the pioneers in this technique was the team led by Peter Walklate (Walklate et al. 2002) who used a tractor-mounted LiDAR system to obtain structural details of an apple orchard and compare the performance of different spray deposition models. The goal was to adjust the pesticide output from an axial fan sprayer based on the estimated crop structure properties.

Similar LiDAR sensors were subsequently used for characterization of a wide range of tree crops, including pear, apple, wine grape, citrus, avocado (*Persea americana*) and olive (Palacín et al. 2007; Rosell et al. 2009; Llorens et al. 2011; Moorthy et al. 2011; Pfeiffer et al. 2018). In those studies, LiDAR systems were developed to obtain 3D point clouds from which large amounts of plant information could be extracted to determine variables such as height, width, volume, LAI and LAD. For some scanned trees, the coefficient of determination between manually measured volumes and those obtained from the processed point cloud was as high as 0.976. In one study on the characterization of an olive tree crop (Escolà et al. 2017), the first and second returns from the readings were analyzed, while data from the third return were not considered reliable and were discarded. Then, to present crop property measurements, they used interpolation to represent the status of the crop as a digital map. These maps, available during the growing season of the crop, can be used to extract differences in growth rate between two crop stages, an important factor for analyses of the time dynamics of a field. To improve and speed up the process, some authors have proposed scan aggregation to obtain better LAI measurements, or using information from only one side to obtain the crop values instead of scanning the crop from both sides. This simplification was shown to give good correlations in north–south oriented vineyards, an important factor if this process needs to be applied in different situations (Arnó et al. 2013, 2015).

Given the potential of LiDAR technology to extract crop properties, large field areas can be scanned for different purposes. Underwood et al. (2016) undertook an extensive study in tree detection, and flower and fruit mapping in almond orchards. They measured 580 trees three times each season for two years. For this, they used a LiDAR system plus an imaging system. With the LiDAR sensor they located the trees and measured their volumes and with the imaging system they counted the flowers and fruits. Scanning large areas and combining these two different technologies is an interesting option for implementation in the near future.

With the aim of improving crop management operation, Escolà et al. (2013) carried out a real-time characterization of an orchard canopy using a LiDAR sensor to adjust a variable-rate algorithm for spray application. An application coefficient was used to convert canopy volume into the required spray flow rate. Good correlations with coefficients of determination of $R^2 > 0.9$ between spray flow rate and canopy cross-sectional area were obtained. Siebers et al. (2018) also considered crop management processes in their study. After scanning vineyards grown under different management systems (single cordon and spur with pruned or minimally pruned vineyards) in two different locations, they obtained strong correlation coefficients between pruning weight and vine wood (trunk and cordon), with volume extracted from LiDAR point clouds.

A new and interesting use of LiDAR data can be seen in recent studies (Gené-Mola et al. 2019a; Tsoulias et al. 2020). They used the information provided by reflectance or intensity values (amount of infrared light returned from the target) to detect and locate apples. Differences in reflectance values between leaves, branches and apples can be sufficient to detect the position of the fruits, and using a clustering process the apples can then be counted. In a subsequent study, these results were improved significantly by applying forced air flow and using multi-view approaches (Gené-Mola et al. 2020a). This detection process obtained similar results to those obtained by imaging systems, but with the advantage of providing the coordinates of the fruit location. With the same objective, but targeting oranges, Méndez et al. (2019) used a static 3D LiDAR sensor with colour capture of each point to locate and count oranges in a commercial field. They tested the system on two different crop management strategies, pruned and unpruned trees. In the pruned trees they found a significant regression between actual and modelled fruit number ($R^2 = 0.63$,

$p = 0.01$), but because of problems with leaf occlusion the coefficient of determination was not significant in the unpruned trees.

The 3D LiDAR systems have also been used in electronic orchard characterization. In grapevine, for example, wood measurements were used to obtain the real crop volume (trunks and cordons) (Keightley and Bawden 2010), and in tomato the plants were scanned from three different positions (Hosoi et al. 2011). Using a post-processing procedure, it was possible to convert leaf points into polygons which were in turn used to estimate leaf properties including LAD, LAI and the leaf inclination angle. Similar to this process, but using intensity values at the same time, a system was proposed to extract the leaf inclination angle in pear trees (Balduzzi et al. 2011). The main key in the analysis was that the intensity information provided by TLS systems depends on the local inclination of the measured surface.

A particular case of the use of LiDAR sensors to improve combine harvester capabilities was presented in Zhang and Grift (2012). Using a Sick LiDAR, they evaluated its ability in static and dynamic mode to measure stem height in *Miscanthus giganteus* (*Miscanthus* x *giganteus*), a crop used for bioenergy production. The stem height measurement system that the authors were developing was intended for future use as a component of the so called 'look-ahead yield monitor' of a combine harvester. Mean errors of just 5.08 and 3.8 % were obtained in the static and dynamic approaches, respectively.

As ever larger crop areas have begun to be scanned using LiDAR sensors, automatic procedures to extract crop properties have become increasingly necessary. In their study, Colaço et al. (2017) scanned a 25-ha field using LiDAR with a GNSS receiver to geo-reference all the points obtained. After delimiting the individual trees along a row, they used convex-hull and alpha-shape algorithms to reproduce the shape of the crowns and then calculate the canopy volume and height. The estimated canopy properties were similar using the two algorithmic methods.

Finally, in relation to the recent development of full-waveform hyperspectral LiDAR systems which allow spectral and spatial information to be obtained of each measuring point, Zhang et al. (2019) proposed the use of a practical calibration approach to resolve the system's problem of a lack of consistency in the pulse-eco arrival times of multiple spectral channels.

3.5 Depth Cameras

Since the 2000s, low-cost depth cameras have become available on the market. These devices can generate real-time 3D images without the need for mobile components. Sometimes the term RGB-D camera is also used, although this term should strictly be reserved for depth cameras that simultaneously provide colour and depth information. Current depth cameras are based on one of the following operating principles (Giancola et al. 2018): structured-light (SL), time-of-flight (ToF) or stereo vision (SV).

3.5.1 Structured-Light Sensors

Structured-light is an active stereo vision technique based on the emission of a known light pattern by a projector (Khoshelham and Elberink 2012) that is distorted when hitting a target, generating a disparity map that is captured by a camera (Fig. 3.6). The distance is determined by applying the same expression as in stereo vision (Eq. 3.1) except that, in this case, the baseline is equal to the separation between the projector and the camera. The most popular SL camera is the Microsoft Kinect v1 (referred to hereafter as the K1), launched in 2010. This camera comprises an IR laser projector, an IR CMOS sensor and an RGB camera, and provides synchronized colour and depth images (Table 3.1). Although the K1 was originally developed for the gaming market, it quickly attained popularity in different research fields due to its ability to provide 3D colour images in real time (30 fps), its compact design and its low cost. Another advantage of SL cameras is that they require less power than ToF cameras and so active cooling is not necessary (Sarbolandi et al. 2015). The main limitations of the K1 are its short range (0.8–4 m) and that it was designed for indoor conditions, with the result that some depth data are lost if the camera is operated under direct sunlight. Conversely, deep images are available at night but colour information degrades. The best performance outdoors is achieved under low light conditions (e.g. sunrise, sunset or cloudy days) (Rosell-Polo et al. 2015).

The K1 has been widely used for indoor plant phenotyping; Chéné et al. (2012) were the first to study this application. These authors segmented leaves of a rosebush (*Rosa* L.) from RGB-D data and measured the leaf curvature and orientation of a yucca (*Yucca* L.) plant. They also demonstrated the possibility of merging 3D plant images with thermal images to improve the identification of infected leaves. Xia et al. (2015) presented a complete methodology for 3D segmentation of plant leaves using K1 measurements that was validated under greenhouse conditions. Paulus et al. (2014) compared the K1 with two higher-cost laser scanning systems to measure plant geometry. They concluded that the K1 could replace these more expensive laser scanners in numerous phenotyping applications, particularly for characterizing volumetric objects.

With respect to field studies, Azzari et al. (2013) used the K1 at night to measure different plants (wild artichoke (*Cynara cardunculus* L.), bristly ox-tongue (*Picris echioides* L.)), estimating properties such as the basal diameter, plant height and volume. Rosell-Polo et al. (2015) combined the K1 with an RTK-GNSS positioning system to generate georeferenced 3D point clouds of apple tree rows, obtaining good correlations between the estimated tree heights and the physical values. These authors also applied the K1 to plant organ classification (leaves, flowers and branches) in apple and pear trees. As demonstrated by Andújar et al. (2016b), the K1 can also be used for plant growth monitoring.

Characterization of woody plants and estimation of biomass is another important application of SL sensors. Nock et al. (2013) tested the Asus Xtion camera to measure woody plant stems, concluding that it was capable of measuring branches with

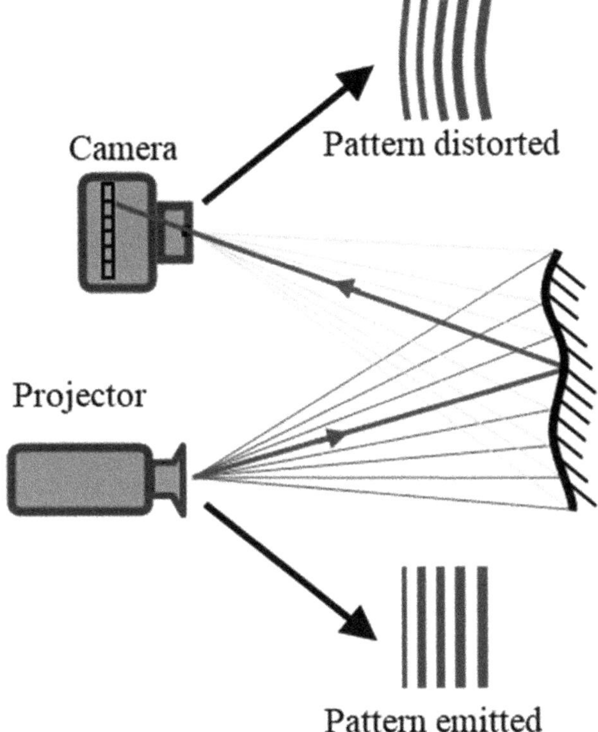

Fig. 3.6 Typical set-up of a structured-light system. The projected pattern is distorted by the object's surface and then it is captured by the camera for 3D reconstruction

a diameter >6 mm. In this context, Andújar et al. (2015) analyzed what the best viewing angle was for biomass estimation in poplar (*Populus* L.) seedings. They concluded that the angle depends on the growth stage, with a top view appropriate for one-month-old plants and a frontal view for one-year-old plants.

3.5.2 *Time-of-Flight Cameras*

The ToF cameras are active imaging sensors that emit continuous-wave (CW) modulated light in the near-infrared (NIR) with light-emitting diodes (LEDs). As shown in Fig. 3.7, they are based on measuring the phase shift φ between the emitted and the received signal for each pixel (CMOS sensor). The depth R is computed as (Foix et al. 2011):

$$R = \frac{c}{4\pi f_{\mathrm{m}}}\varphi, \tag{3.2}$$

where c is the speed of light and f_m is the modulation frequency of the emitted signal.

Conventional ToF cameras provide depth data (depth image) and the intensity of the return signal (amplitude signal), but they do not give colour information. Some ToF cameras (e.g. Mesa Imaging SR4000 or PMD CamCube) also provide a confidence matrix to indicate the reliability of the depth data.

The main advantage of ToF cameras over LiDAR and SV systems is their ability to provide depth data at high rates of sampling. Limitations include the low resolution of the depth images (Table 3.1) and noise associated to depth data. Kazmi et al. (2014) compared several ToF cameras with an SV sensor both indoors and outdoors. They concluded that, although the SV provides higher resolution images and is more robust under sunlight, the ToF cameras have the advantages of lower computational costs and the ability to measure targets with non-uniform textures.

In crop characterization, the successful application of ToF cameras for automatic plant probing and phenotyping has been demonstrated (Klose et al. 2009; Alenya et al. 2013). Plant phenotyping is usually performed manually, and so its automation using ToF cameras reduces labour hours and human error. In this context, Chaivivatrakul et al. (2014) developed an automatic phenotyping system where a ToF camera is integrated in a rotating table. Maize plant measurements were made, estimating the leaf geometry (length, area, angle) and the stem diameter. Likewise, van der Heijden et al. (2012) developed a system for the phenotyping of greenhouse pepper plants by combining RGB and ToF cameras. Strong correlations were obtained between estimated plant heights and manual measurements. For the phenotyping of small grain cereals under field conditions, Busemeyer et al. (2013) presented a multi-sensor platform that integrated various optical sensors (ToF cameras, light curtain imaging, laser distance sensors, hyperspectral and RGB cameras) with a modular architecture adaptable to different species.

Another application of ToF cameras is the automatic sensing of inter-plant spacing at initial growth stages (Nakarmi and Tang 2012). Accurate measurements of inter-plant spacing can contribute to distributing the inputs evenly to different plants. Karkee and Adhikari (2015) used ToF cameras to generate apple tree

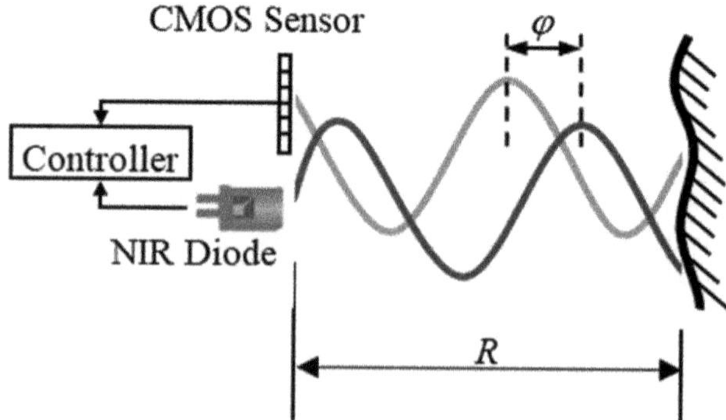

Fig. 3.7 Principle of operation of a ToF camera. (Adapted from Vázquez-Arellano et al. 2016)

skeletons. From these 3D reconstructions, a branch identification accuracy of 77 % was achieved. This is the first step in the development of an automatic pruning system.

In 2014, Microsoft launched a second version of the Kinect sensor (hereafter referred to as K2) based on the ToF principle instead of structured-light (Table 3.1). The main difference between the K2 sensor and conventional ToF cameras is the presence of an RGB camera, which provides colour information together with the depth data (like the K1). In experimental tests of the K2, Sarbolandi et al. (2015) concluded that, unlike the K1, it can be used for operation in daylight conditions. This feature is particularly interesting for agricultural applications, although it should be noted that the number of measured points is reduced as the sunlight increases. In addition, the K2 is more accurate and precise than the K1. Fig. 3.2c shows the 3D point cloud of an apple tree obtained using a K2 sensor. Although the point density is less than that provided by LiDAR measurements (Fig. 3.2b), the K2 provides colour information. Compared to SfM (Fig. 3.2a), K2 point clouds are less realistic, but real-time acquisition is possible.

These factors explain why the K2 has become very popular in 3D crop characterization as a low-cost alternative to LiDAR systems. For instance, the K2 was used as the basis for the development of plant-to-plant phenotyping platforms under controlled conditions (McCormick et al. 2016; Sun and Wang, 2019). With respect to field phenotyping, several studies have used the K2 to determine multiple geometric properties of maize plants (Hämmerle and Höfle 2016; Vázquez-Arellano et al. 2018; Bao et al. 2019b), including plant height and orientation, leaf angle, and stem diameter and position, among others.

The K2 was also used for weed detection and control by Andújar et al. (2016a). In this study, 3D measurements were used to separate weeds from maize plants and subsequently obtain weed and maize estimates of biomass. With the ultimate goal of developing a system for robotic weed control, Gai et al. (2020) used the K2 to detect broccoli (*Brassica oleracea* L. var. botrytis L.) and lettuce (*Lactuca* L.) at different growth stages. They achieved detection rates of more than 90 % for both crops.

The K2 has also been proposed by Rosell-Polo et al. (2017) as a cost-effective alternative to MTLS for the 3D characterization of orchard crops. These authors developed an MTLS based on the combination of a K2 with an RTK-GNSS system. The system was validated experimentally in a vineyard, obtaining the most accurate 3D point clouds when the field of view was adjusted to one vertical pixel column. Similarly, Bengochea-Guevara et al. (2017) used a mobile platform with an onboard K2 sensor to enable the easy acquisition of information in the field. The approach used was also applied in vineyards and included a method to correct the drift that 3D reconstruction of large crop rows present.

3.5.3 Active Stereo Vision

Recently, new low-cost compact stereo vision sensors, like the Intel RealSense series, have appeared on the market. These devices include a laser projector that emits a known pattern that helps to improve the image matching process. Vit and Shani (2018) compared these sensors with other current SL and ToF cameras and demonstrated that the new active stereo sensors are suitable for outdoor agricultural phenotyping, with the advantage of lower power requirements. Likewise, Milella et al. (2019) developed an automatic grapevine phenotyping system based on a RealSense camera and showed its capacity for grape bunch detection and to estimate the canopy volume.

3.6 Conclusions for the chapter

As with all data acquisition technologies, the expectation is that crop geometry sensing in the future will be cheaper and faster, and that larger data volumes and more accurate results will be possible. For now, as there is no single sensor that can meet all the requirements, the most suitable sensor needs to be chosen for each particular situation. In this respect, an important limitation of photogrammetric sensors is the effect of changes in light conditions. In contrast, LiDAR systems are more reliable but their cost is significantly higher and they do not provide colour information. The RGB-D camera sensors are a low-cost solution that record colour and depth images, but their resolution is lower than that obtained with LiDAR systems or photogrammetry.

At the time of writing, only a limited amount of equipment fitted with ultrasonic sensors, and in some cases LiDAR, is available on the market for application with commercial agricultural machinery. It is currently used only for high value crops or on machines such as those for pesticide application or harvesting extensive field crops The logical next step that would lead to the large-scale incorporation of sensors in agricultural applications is for crop geometry characterization to be sufficiently accurate and fast enough to allow the real-time adjustment of operating parameters of the machine (spray rate, cutting height, etc.).

The other technologies (ToF cameras, SL sensors, active SV, and so on) continue in the research phase, although some commercial applications are possible in the near future. Their practical implementation will generally be determined on a cost-benefit basis. Photogrammetry, for example, has an interesting future because it can be applied with cameras with low resolution and with fewer camera specifications. If the cost of the sensor exceeds the cost of the equipment or machine to which a new capacity must be added, it is clear that it will not be implemented commercially. If, on the other hand, the sensor is used to generate maps to provide information for future field operations, that is something that could be implemented in the near future. In other words, the real-time adjustment of an agricultural operation will be more difficult than map-based applications. Thus, we can anticipate that complex equipment (with the best sensors and processing systems) will be available to produce accurate maps of crops. Then, agricultural machinery that performs specific operations will be able to read these maps and adapt their operations to the needs of the crop. Sensors with less complexity than those initially mounted can be installed to adjust machine operations further, improving the resolution offered by the maps.

In relation to the data obtained from all the sensors described in this chapter, another issue that needs to be improved is the volume of data that can be processed. Dealing with large amounts of data from continuous scanning is a major problem, especially when manual processes are required. For this reason, automated systems to process large and accurate point cloud files are necessary (Colaço et al. 2017). In this context, another aspect that needs to be improved is automatic on-board processing of data through which the final result can be made available as soon as the scanning process ends (Underwood et al. 2016).

Powerful tools to deal with this type of problem exist in the fields of artificial intelligence and big data. In this regard, deep learning for computer vision and data processing has emerged as a family of techniques with great potential in agriculture (Kamilaris and Prenafeta-Boldú 2018). For example, convolutional neural networks (CNNs) can be used to detect, classify or segment vegetative organs (trunks, leaves, fruits, etc.) in 2D images (Jiang and Li 2020). In addition, the simultaneous availability of colour and depth information (multi-modal data), obtained with RGB-D cameras or photogrammetry, can be used to improve the performance of neural networks, as demonstrated for branch and fruit detection applications (Zhang et al. 2018a; Gené-Mola et al. 2019b). Depth data can also be used to project 2D detections on to 3D point clouds (Shi et al. 2019; Gené-Mola et al. 2020b).

A pending challenge is the development of CNN models that allow the direct processing of 3D point clouds. Jin et al. (2020), for example, have implemented a voxel-based convolutional network to separate maize stems and leaves based on the measurements of a terrestrial LiDAR system. Their results outperformed those obtained with traditional clustering methods. However, the application of deep learning techniques in crop geometric characterization needs more publicly-available datasets with multi-modal images of different crops to train the models.

Deciphering the complexity of plant and crop structures has been made possible through the availability of advanced data extraction technologies. As the volumes of data increase and the information becomes ever more complex, increasing the speed at which the data can be extracted and analyzed is the next big challenge for the present and future generations of precision agricultural engineers.

References

Alenya G, Dellen B, Foix S et al (2013) Robotized plant probing: leaf segmentation utilizing time-of-flight data. IEEE Robot Autom Mag 20:50–59

Andersen HJ, Reng L, Kirk K (2005) Geometric plant properties by relaxed stereo vision using simulated annealing. Comput Electron Agric 49:219–232

Andújar D, Escolà A, Dorado J et al (2011) Weed discrimination using ultrasonic sensors. Weed Res 51:543–547

Andújar D, Weis M, Gerhards R (2012) An ultrasonic system for weed detection in cereal crops. Sensors 12:17343–17357

Andújar D, Escolà A, Rosell-Polo JR et al (2013) Potential of a terrestrial LiDAR-based system to characterise weed vegetation in maize crops. Comput Electron Agric 92:11–15

Andújar D, Fernández-Quintanilla C, Dorado J (2015) Matching the best viewing angle in depth cameras for biomass estimation based on poplar seedling geometry. Sensors 15:12999–13011

Andújar D, Dorado J, Fernández-Quintanilla C et al (2016a) An approach to the use of depth cameras for weed volume estimation. Sensors 16:1–11

Andújar D, Ribeiro A, Fernández-Quintanilla C et al (2016b) Using depth cameras to extract structural parameters to assess the growth state and yield of cauliflower crops. Comput Electron Agric 122:67–73

Andújar D, Calle M, Fernández-Quintanilla C et al (2018) Three-dimensional modeling of weed plants using low-cost photogrammetry. Sensors 18:1077

Arnó J, Escolà A, Vallès JM et al (2013) Leaf area index estimation in vineyards using a ground-based LiDAR scanner. Precis Agric 14:290–306

Arnó J, Escolà A, Masip J et al (2015) Influence of the scanned side of the row in terrestrial laser sensor applications in vineyards: practical consequences. Precis Agric 16:119–128

Auat Cheein FA, Guivant J, Sanz R et al (2015) Real-time approaches for characterization of fully and partially scanned canopies in groves. Comput Electron Agric 118:361–371

Azzari G, Goulden M, Rusu R (2013) Rapid characterization of vegetation structure with a Microsoft Kinect sensor. Sensors 13:2384–2398

Balduzzi MAF, Van der Zande D, Stuckens J et al (2011) The properties of terrestrial laser system intensity for measuring leaf geometries: a case study with Conference pear trees (Pyrus Communis). Sensors 11:1657–1681

Bao Y, Tang L, Breitzman MW et al (2019a) Field-based robotic phenotyping of sorghum plant architecture using stereo vision. J F Robot 36:397–415

Bao Y, Tang L, Srinivasan S et al (2019b) Field-based architectural traits characterisation of maize plant using time-of-flight 3D imaging. Biosyst Eng 178:86–101

Bay H, Tuytelaars T, Van Gool L (2006) SURF: speeded up robust features. In: Leonardis A, Bischof H, Pinz A (eds) Computer vision – ECCV 2006. ECCV 2006. Lec Notes Comput Sci 3951:404–417. Springer, Berlin/Heidelberg

Bengochea-Guevara J, Andújar D, Sanchez-Sardana F et al (2017) A low-cost approach to automatically obtain accurate 3D models of woody crops. Sensors 18:30

Bernotas G, Scorza LCT, Hansen MF et al (2019) A photometric stereo-based 3D imaging system using computer vision and deep learning for tracking plant growth. Gigascience 8:1–15

Bietresato M, Carabin G, Vidoni R et al (2016) Evaluation of a LiDAR-based 3D-stereoscopic vision system for crop-monitoring applications. Comput Electron Agric 124:1–13

Billiot B, Cointault F, Journaux L et al (2013) 3D image acquisition system based on shape from focus technique. Sensors 13:5040–5053

Biskup B, Scharr H, Schurr U et al (2007) A stereo imaging system for measuring structural parameters of plant canopies. Plant Cell Environ 30:1299–1308

Busemeyer L, Mentrup D, Möller K et al (2013) BreedVision – a multi-sensor platform for non-destructive field-based phenotyping in plant breeding. Sensors 13:2830–2847

Chaivivatrakul S, Tang L, Dailey MN et al (2014) Automatic morphological trait characterization for corn plants via 3D holographic reconstruction. Comput Electron Agric 109:109–123

Chéné Y, Rousseau D, Lucidarme P et al (2012) On the use of depth camera for 3D phenotyping of entire plants. Comput Electron Agric 82:122–127
Colaço AF, Trevisan RG, Molin JP et al (2017) A method to obtain orange crop geometry information using a mobile terrestrial laser scanner and 3D modeling. Remote Sens 9:10–13
del-Moral-Martínez I, Rosell-Polo JR, Company J et al (2016) Mapping vineyard leaf area using mobile terrestrial laser scanners: should rows be scanned on-the-go or discontinuously sampled? Sensors 16:1–13
Duan T, Chapman SC, Holland E et al (2016) Dynamic quantification of canopy structure to characterize early plant vigour in wheat genotypes. J Exp Bot 67:4523–4534
Escolà A, Planas S, Rosell JR et al (2011) Performance of an ultrasonic ranging sensor in apple tree canopies. Sensors 11:2459–2477
Escolà A, Rosell-Polo JR, Planas S et al (2013) Variable rate sprayer. Part 1 – orchard prototype: design, implementation and validation. Comput Electron Agric 95:122–135
Escolà A, Martínez-Casasnovas JA, Rufat J et al (2017) Mobile terrestrial laser scanner applications in precision fruticulture/horticulture and tools to extract information from canopy point clouds. Precis Agric 18:111–132
Farooque AA, Chang YK, Zaman QU et al (2013) Performance evaluation of multiple ground based sensors mounted on a commercial wild blueberry harvester to sense plant height, fruit yield and topographic features in real-time. Comput Electron Agric 91:135–144
Foix S, Alenya G, Torras C (2011) Lock-in time-of-flight (ToF) cameras: a survey. IEEE Sensors J 11:1917–1926
Fricke T, Wachendorf M (2013) Combining ultrasonic sward height and spectral signatures to assess the biomass of legume–grass swards. Comput Electron Agric 99:236–247
Fricke T, Richter F, Wachendorf M (2011) Assessment of forage mass from grassland swards by height measurement using an ultrasonic sensor. Comput Electron Agric 79:142–152
Gai J, Tang L, Steward BL (2020) Automated crop plant detection based on the fusion of color and depth images for robotic weed control. J F Robot 37:35–52
Gamarra-Diezma JL, Miranda-Fuentes A, Llorens J et al (2015) Testing accuracy of long-range ultrasonic sensors for olive tree canopy measurements. Sensors 15:2902–2919
Gené-Mola J, Gregorio E, Guevara J et al (2019a) Fruit detection in an apple orchard using a mobile terrestrial laser scanner. Biosyst Eng 187:171–184
Gené-Mola J, Vilaplana V, Rosell-Polo JR et al (2019b) Multi-modal deep learning for Fuji apple detection using RGB-D cameras and their radiometric capabilities. Comput Electron Agric 162:689–698
Gené-Mola J, Gregorio E, Auat Cheein F et al (2020a) Fruit detection, yield prediction and canopy geometric characterization using LiDAR with forced air flow. Comput Electron Agric 168:105121
Gené-Mola J, Sanz-Cortiella R, Rosell-Polo JR et al (2020b) Fruit detection and 3D location using instance segmentation neural networks and structure-from-motion photogrammetry. Comput Electron Agric 169:105165
Giancola S, Valenti M, Sala R (2018) A survey on 3D cameras: metrological comparison of time-of-flight, structured-light and active stereoscopy technologies, Springer Briefs in Computer Science. Springer
Gil E, Escolà A, Rosell JR et al (2007) Variable rate application of plant protection products in vineyard using ultrasonic sensors. Crop Prot 26:1287–1297
Gil E, Llorens J, Llop J et al (2013) Variable rate sprayer. Part 2 – vineyard prototype: design, implementation, and validation. Comput Electron Agric 95:136–150
Giles DK, Delwiche MJ (1988) Electronic measurement of tree canopy volume. Trans ASAE 31:264–272
Giles DK, Delwiche MJ, Dodd RB (1987) Control of orchard spraying based on electronic sensing of target characteristics. Trans ASAE 30:1624–1636
Hämmerle M, Höfle B (2016) Direct derivation of maize plant and crop height from low-cost time-of-flight camera measurements. Plant Methods 12:50

Hayashi S, Shigematsu K, Yamamoto S et al (2010) Evaluation of a strawberry-harvesting robot in a field test. Biosyst Eng 105:160–171

He DX, Matsuura Y, Kozai T et al (2003) A binocular stereovision system for transplant growth variables analysis. Appl Eng Agric 19:611–617

Herrero-Huerta M, González-Aguilera D, Rodriguez-Gonzalvez P et al (2015) Vineyard yield estimation by automatic 3D bunch modelling in field conditions. Comput Electron Agric 110:17–26

Hokuyo LTD (2014) Distance data output/UTM-30LX-EW [WWW Document]. URL: https://www.hokuyo-aut.jp/search/single.php?serial=170. Accessed 1 May 2020

Hooper AW, Harries GO, Ambler B (1976) A photoelectric sensor for distinguishing between plant material and soil. J Agric Eng Res 21:145–155

Hosoi F, Omasa K (2009) Estimating vertical plant area density profile and growth parameters of a wheat canopy at different growth stages using three-dimensional portable lidar imaging. ISPRS J Photogramm Remote Sens 64:151–158

Hosoi F, Nakabayashi K, Omasa K (2011) 3-D modeling of tomato canopies using a high-resolution portable scanning lidar for extracting structural information. Sensors 11:2166–2174

Hui F, Zhu J, Hu P et al (2018) Image-based dynamic quantification and high-accuracy 3D evaluation of canopy structure of plant populations. Ann Bot 121:1079–1088

Intel Corporation (2020) Intel® RealSense™ Product Family D400 Series [WWW Document]. URL: https://www.intelrealsense.com/wp-content/uploads/2020/06/Intel-RealSense-D400-Series-Datasheet-June-2020.pdf. Accessed 10 Feb 2020

Ivanov N, Boissard P, Chapron M et al (1995) Computer stereo plotting for 3-D reconstruction of a maize canopy. Agric For Meteorol 75:85–102

James M, Robson S (2012) Straightforward reconstruction of 3D surfaces and topography with a camera: accuracy and geoscience application. J Geophys Res Earth Surf 2003–2012:117

Jay S, Rabatel G, Hadoux X et al (2015) In-field crop row phenotyping from 3D modeling performed using structure from motion. Comput Electron Agric 110:70–77

Jeon HY, Zhu H, Derksen R et al (2011) Evaluation of ultrasonic sensor for variable-rate spray applications. Comput Electron Agric 75:213–221

Jiang Y, Li C (2020) Convolutional neural networks for image-based high-throughput plant phenotyping: a review. Plant Phenomics 2020:1–22

Jin S, Su Y, Gao S et al (2020) Separating the structural components of maize for field phenotyping using terrestrial LiDAR data and deep convolutional neural networks. IEEE Trans Geosci Remote Sens 58:2644–2658

Kamilaris A, Prenafeta-Boldú FX (2018) Deep learning in agriculture: a survey. Comput Electron Agric 147:70–90

Karkee M, Adhikari B (2015) A method for three-dimensional reconstruction of apple trees for automated pruning. Trans ASABE 58:565–574

Kazmi W, Foix S, Alenyà G et al (2014) Indoor and outdoor depth imaging of leaves with time-of-flight and stereo vision sensors: analysis and comparison. ISPRS J Photogramm Remote Sens 88:128–146

Keightley KE, Bawden GW (2010) 3D volumetric modeling of grapevine biomass using Tripod LiDAR. Comput Electron Agric 74:305–312

Khoshelham K, Elberink SO (2012) Accuracy and resolution of Kinect depth data for indoor mapping applications. Sensors 12:1437–1454

Kise M, Zhang Q (2008a) Development of a stereovision sensing system for 3D crop row structure mapping and tractor guidance. Biosyst Eng 101:191–198

Kise M, Zhang Q (2008b) Creating a panoramic field image using multi-spectral stereovision system. Comput Electron Agric 60:67–75

Kise M, Zhang Q, Rovira Más F (2005) A stereovision-based crop row detection method for tractor-automated guidance. Biosyst Eng 90:357–367

Klose R, Penlington J, Ruckelshausen A (2009) Usability of 3D time-of-flight cameras for automatic plant phenotyping. Bornimer Agrartech Berichte 69:93–105

Ladd TL, Reichard DL, Simonet DE (1981) Integration of a photoelectrically operated intermittent sprayer with action level thresholds for control of lepidopteran pests of cabbage. J Econ Entomol 74:698–700
Lati RN, Filin S, Eizenberg H (2013) Estimating plant growth parameters using an energy minimization-based stereovision model. Comput Electron Agric 98:260–271
Leica Geosystems (2013) Leica ScanStation P20 [WWW Document]. URL: https://w3.leica--geosystems.com/downloads123/hds/hds/scanstation_p20/brochures-datasheet/leica_scanstation_p20_dat_en.pdf. Accessed 10 Jan 2020
Li D, Xu L, Tang X et al (2017) 3D Imaging of greenhouse plants with an inexpensive binocular stereo vision system. Remote Sens 9:508
Llop J, Gil E, Llorens J et al (2016) Testing the suitability of a terrestrial 2D LiDAR scanner for canopy characterization of greenhouse tomato crops. Sensors 16(9):1435
Llorens J, Gil E, Llop J et al (2010) Variable rate dosing in precision viticulture: Use of electronic devices to improve application efficiency. Crop Prot 29:239–248
Llorens J, Gil E, Llop J et al (2011) Ultrasonic and LIDAR sensors for electronic canopy characterization in vineyards: Advances to improve pesticide application methods. Sensors 11(2):2177–2194
Lowe DG (1999) Object recognition from local scale-invariant features. In: Proceedings of the Seventh IEEE International Conference on Computer Vision. IEEE, vol 2, pp 1150–1157
Lumme J, Karjalainen M, Kaartinen H, et al (2008) Terrestrial laser scanning of agricultural crops. Int Arch Photogramm Remote Sens Spat Inf Sci XXXVII. Pa:563–566
Martinez-Guanter J, Ribeiro Á, Peteinatos GG et al (2019) Low-cost three-dimensional modeling of crop plants. Sensors 19:2883
McConnell RL, Elliot KC, Blizzard SH et al (1983) Electronic measurement of tree-row volume. In: ASAE Annual International Meeting. St. Joseph, MI
McCormick RF, Truong SK, Mullet JE (2016) 3D sorghum reconstructions from depth images identify QTL regulating shoot architecture. Plant Physiol 172:823–834
Méndez V, Pérez-Romero A, Sola-Guirado R et al (2019) In-field estimation of orange number and size by 3D laser scanning. Agronomy 9:885
Milella A, Marani R, Petitti A et al (2019) In-field high throughput grapevine phenotyping with a consumer-grade depth camera. Comput Electron Agric 156:293–306
Miller WM, Whitney JD, Schumann A et al (2003) A test program to assess VRT granular fertilizer applications for citrus. In: ASAE Annual International Meeting. American Society of Agricultural and Biological Engineers, Las Vegas, Nevada, USA
Moorthy I, Miller JR, Jimenez Berni JA et al (2011) Field characterization of olive (Olea europaea L.) tree crown architecture using terrestrial laser scanning data. Agric For Meteorol 151:204–214
Mortensen AK, Bender A, Whelan B et al (2018) Segmentation of lettuce in coloured 3D point clouds for fresh weight estimation. Comput Electron Agric 154:373–381
Müller-Linow M, Pinto-Espinosa F, Scharr H et al (2015) The leaf angle distribution of natural plant populations: assessing the canopy with a novel software tool. Plant Methods 11:11
Nakarmi AD, Tang L (2012) Automatic inter-plant spacing sensing at early growth stages using a 3D vision sensor. Comput Electron Agric 82:23–31
Nielsen M, Slaughter DC, Gliever C (2012) Vision-based 3D peach tree reconstruction for automated blossom thinning. IEEE Trans Ind Inf 8:188–196
Nock C, Taugourdeau O, Delagrange S et al (2013) Assessing the potential of low-cost 3D cameras for the rapid measurement of plant woody structure. Sensors 13:16216–16233
Oliveira F, Souza A, Fernandes M et al (2018) Efficient 3D objects recognition using multifoveated point clouds. Sensors 18:2302
Pagliari D, Pinto L (2015) Calibration of Kinect for Xbox One and comparison between the two generations of Microsoft sensors. Sensors 15:27569–27589
Palacín J, Pallejà T, Tresanchez M et al (2007) Real-time tree-foliage surface estimation using a ground laser scanner. IEEE Trans Instrum Meas 56:1377–1383

Palleja T, Landers AJ (2015) Real time canopy density estimation using ultrasonic envelope signals in the orchard and vineyard. Comput Electron Agric 115:108–117
Palleja T, Tresanchez M, Teixido M et al (2010) Sensitivity of tree volume measurement to trajectory errors from a terrestrial LIDAR scanner. Agric For Meteorol 150:1420–1427
Paulus S, Behmann J, Mahlein A-K et al (2014) Low-cost 3D systems: suitable tools for plant phenotyping. Sensors 14:3001–3018
Pfeiffer SA, Guevara J, Cheein FA et al (2018) Mechatronic terrestrial LiDAR for canopy porosity and crown surface estimation. Comput Electron Agric 146:104–113
PMD Technologies GmbH (2009) PMD[vision] CamCube 2.0 Datasheet V. No. 20090601
Pound MP, French AP, Murchie EH et al (2014) Automated recovery of three-dimensional models of plant shoots from multiple color images. Plant Physiol 166:1688–1698
Probst A, Gatziolis D, Strigul N (2018) Intercomparison of photogrammetry software for three-dimensional vegetation modelling. R Soc Open Sci 5:172192
Quan L, Tan P, Zeng G et al (2006) Image-based plant modeling. ACM Trans Graph 25:599
Reichard DL, Ladd TL (1981) An automatic intermittent sprayer. Trans ASABE 24:893–896
Reid J, Searcy S (1987) Vision-based guidance of an agriculture tractor. IEEE Control Syst Mag 7:39–43
Rose J, Paulus S, Kuhlmann H (2015) Accuracy analysis of a multi-view stereo approach for phenotyping of tomato plants at the organ level. Sensors 15:9651–9665
Rose J, Kicherer A, Wieland M et al (2016) Towards automated large-scale 3D phenotyping of vineyards under field conditions. Sensors 16:2136
Rosell JR, Llorens J, Sanz R et al (2009) Obtaining the three-dimensional structure of tree orchards from remote 2D terrestrial LIDAR scanning. Agric For Meteorol 149(9):1505–1515
Rosell-Polo JR, Auat Cheein F, Gregorio E et al (2015) Advances in structured light sensors applications in precision agriculture and livestock farming. In: Advances in agronomy, pp 71–112.
Rosell-Polo JR, Gregorio E, Gene J et al (2017) Kinect v2 sensor-based mobile terrestrial laser scanner for agricultural outdoor applications. IEEE/ASME Trans Mech 22:2420–2427
Rovira-Más F, Zhang Q, Reid JF (2004) Automated agricultural equipment navigation using stereo disparity images. Trans ASAE 47:1289–1300
Rovira-Más F, Wang Q, Zhang Q (2009) Bifocal stereoscopic vision for intelligent vehicles. Int J Veh Technol 2009:123231
Rovira-Más F, Zhang Q, Hansen AC (2011) Mechatronics and intelligent systems for off-road vehicles. Springer, London
Saeys W, Lenaerts B, Craessaerts G et al (2009) Estimation of the crop density of small grains using LiDAR sensors. Biosyst Eng 102:22–30
Santos TT, Rodrigues GC (2016) Flexible three-dimensional modeling of plants using low- resolution cameras and visual odometry. Mach Vis Appl 27:695–707
Sanz-Cortiella R, Llorens-Calveras J, Rosell-Polo JR et al (2011) Characterisation of the LMS200 laser beam under the influence of blockage surfaces. Influence on 3D scanning of tree orchards. Sensors 11:2751–2772
Sarbolandi H, Lefloch D, Kolb A (2015) Kinect range sensing: structured-light versus time-of-flight Kinect. Comput Vis Image Underst
Schumann AW, Zaman QU (2005) Software development for real-time ultrasonic mapping of tree canopy size. Comput Electron Agric 47:25–40
Seitz SM, Curless B, Diebel J et al (2006) A comparison and evaluation of multi-view stereo reconstruction algorithms. In: 2006 IEEE Computer Society Conference on Computer Vision and Pattern Recognition (CVPR'06), New York, 17–22 June
Shearer SA, Jones PT (1991) Selective application of post-emergence herbicides using photoelectrics. Trans ASAE 34:1661–1666
Shi W, van de Zedde R, Jiang H et al (2019) Plant-part segmentation using deep learning and multi-view vision. Biosyst Eng 187:81–95
Shlyakhter I, Rozenoer M, Dorsey J et al (2001) Reconstructing 3D tree models from instrumented photographs. IEEE Comput Graph Appl 21:53–61

Siebers M, Edwards E, Jimenez-Berni J et al (2018) Fast phenomics in vineyards: development of GRover, the grapevine rover, and LiDAR for assessing grapevine traits in the field. Sensors 18:2924
Siemens AG (2008) Simatic sensors catalog: sensor technology for factory automation FS 10 2009
Solanelles F, Escolà A, Planas S et al (2006) An electronic control system for pesticide application proportional to the canopy width of tree crops. Biosyst Eng 95:473–481
Stajnko D, Berk P, Lešnik M et al (2012) Programmable ultrasonic sensing system for targeted spraying in orchards. Sensors 12:15500–15519
Sun G, Wang X (2019) Three-dimensional point cloud reconstruction and morphology measurement method for greenhouse plants based on the Kinect sensor self-calibration. Agronomy 9:596
Tareen SAK, Saleem Z (2018) A comparative analysis of SIFT, SURF, KAZE, AKAZE, ORB, and BRISK. In: 2018 International Conference on Computing, Mathematics and Engineering Technologies (iCoMET). IEEE, pp 1–10
Tsoulias N, Paraforos DS, Xanthopoulos G et al (2020) Apple shape detection based on geometric and radiometric features using a LiDAR laser scanner. Remote Sens 12:2481
Tumbo SD, Salyani M, Whitney JD et al (2002) Investigation of laser and ultrasonic ranging sensors for measurements of citrus canopy volume. Appl Eng Agric 18:367–372
Ullman S (1979) The interpretation of structure from motion. Proc R Soc Lond Ser B Biol Sci 203:405–426
Underwood JP, Hung C, Whelan B et al (2016) Mapping almond orchard canopy volume, flowers, fruit and yield using lidar and vision sensors. Comput Electron Agric 130:83–96
Upchurch BL, Glenn DM, Vass G et al (1993) An ultrasonic tree trunk diameter caliper. HortTechnology 3:89–91
van der Heijden G, Song Y, Horgan G et al (2012) SPICY: towards automated phenotyping of large pepper plants in the greenhouse. Funct Plant Biol 39:870
Vanderbilt VC, Bauer ME, Silva LF (1979) Prediction of solar irradiance distribution in a wheat canopy using a laser technique. Agric Meteorol 20:147–160
Vanderbilt VC, Silva LF, Bauer ME (1990) Canopy architecture measured with a laser. Appl Opt 29:99
Vázquez-Arellano M, Griepentrog H, Reiser D et al (2016) 3-D Imaging Systems for Agricultural Applications—A Review. Sensors 16:618
Vázquez-Arellano M, Paraforos DS, Reiser D et al (2018) Determination of stem position and height of reconstructed maize plants using a time-of-flight camera. Comput Electron Agric 154:276–288
Velodyne Inc (2020) Puck [WWW Document]. URL: https://velodynelidar.com/products/puck/. Accessed 10 Jan 2020
Vit A, Shani G (2018) Comparing RGB-D sensors for close range outdoor agricultural phenotyping. Sensors 18:4413
Walklate PJ (1989) A laser scanning instrument for measuring crop geometry. Agric For Meteorol 46:275–284
Walklate PJ, Cross JV, Richardson GM et al (2002) Comparison of different spray volume deposition models using LIDAR measurements of apple orchards. Biosyst Eng 82:253–267
Wykes C, Webb P, Nagi F (1994) Ultrasonics arrays for automatic vehicle guidance. Control Eng Pract 2:164
Xia C, Wang L, Chung B-K et al (2015) In situ 3D segmentation of individual plant leaves using a RGB-D camera for agricultural automation. Sensors 15:20463–20479
Xiong X, Yu L, Yang W et al (2017) A high-throughput stereo-imaging system for quantifying rape leaf traits during the seedling stage. Plant Methods 13:7
Yeh Y-HF, Lai T-C, Liu T-Y et al (2014) An automated growth measurement system for leafy vegetables. Biosyst Eng 117:43–50
Yuan W, Li J, Bhatta M et al (2018) Wheat height estimation using LiDAR in comparison to ultrasonic sensor and UAS. Sensors 18:3731

Zaman QU, Salyani M (2004) Effects of foliage density and ground speed on ultrasonic measurement of citrus tree volume. Appl Eng Agric 20:173–178

Zaman QU, Schumann AW, Hostler HK (2007) Quantifying sources of error in ultrasonic measurements of citrus orchards. Appl Eng Agric 23:449–453

Zaman QU, Esau TJ, Schumann AW et al (2011) Development of prototype automated variable rate sprayer for real-time spot-application of agrochemicals in wild blueberry fields. Comput Electron Agric 76:175–182

Zhang L, Grift TE (2012) A LIDAR-based crop height measurement system for Miscanthus giganteus. Comput Electron Agric 85:70–76

Zhang J, He L, Karkee M et al (2018a) Branch detection for apple trees trained in fruiting wall architecture using depth features and Regions-Convolutional Neural Network (R-CNN). Comput Electron Agric 155:386–393

Zhang Y, Teng P, Aono M et al (2018b) 3D monitoring for plant growth parameters in field with a single camera by multi-view approach. J Agric Meteorol 74:129–139

Zhang C, Gao S, Niu Z et al (2019) Calibration of the pulse signal decay effect of full-waveform hyperspectral LiDAR. Sensors 19:5263

Zhao C, Huang W, Chen L et al (2010) A harvest area measurement system based on ultrasonic sensors and DGPS for yield map correction. Precis Agric 11:163–180

Zhao H, Xu L, Shi S et al (2018) A high throughput integrated hyperspectral imaging and 3D measurement system. Sensors 18:1068

Chapter 4
Soil Sensing

Viacheslav I. Adamchuk, Asim Biswas, Hsin-Hui Huang, Jonathan E. Holland, James A. Taylor, Bo Stenberg, Johanna Wetterlind, Kanika Singh, Budiman Minasny, Chris Fidelis, David Yinil, Todd Sanderson, Didier Snoeck, and Damien J. Field

Abstract In addition to the overview of diversity in soil sensing technologies, this chapter presents four case studies to illustrate the practical use of these technologies to enhance precision agriculture in Canada, the United Kingdom, Sweden and Papua New Guinea. These studies represent investigations of different instruments, field conditions and targeted soil properties. However, in all four cases, proximal soil sensing was used to predict selected soil properties to generate relatively accurate maps that could help to implement site-specific crop management successfully.

Asim Biswas: Introduction
Viacheslav I. Adamchuk and Hsin-Hui Huang: Introduction and Case Study 4.1
Jonathan E. Holland and James A. Taylor: Case Study 4.2
Bo Stenberg and Johanna Wetterlind: Case Study 4.3
Kanika Singh, Budiman Minasny, Chris Fidelis, David Yinil, Todd Sanderson, Didier Snoeck and Damien J. Field: Case Study 4.4

V. I. Adamchuk (✉) · H.-H. Huang
Department of Bioresource Engineering, McGill University,
Ste-Anne-de-Bellevue, QC, Canada
e-mail: viacheslav.adamchuk@mcgill.ca

A. Biswas
School of Environmental Sciences, University of Guelph, Guelph, ON, Canada
e-mail: biswas@uoguelph.ca

J. E. Holland
James Hutton Institute, Invergowrie, Dundee, UK

J. A. Taylor
ITAP, University of Montpellier, INRAE, Institut Agro, Montpellier, France

School of Natural and Environmental Sciences, University of Newcastle,
Newcastle-upon-Tyne, UK

B. Stenberg · J. Wetterlind
Department of Soil and Environment, Swedish University of Agricultural Sciences,
Skara, Sweden
e-mail: Bo.Stenberg@slu.se; Johanna.Wetterlind@slu.se

R. Kerry, A. Escolà (eds.), *Sensing Approaches for Precision Agriculture*,
Progress in Precision Agriculture, https://doi.org/10.1007/978-3-030-78431-7_4

Keywords Proximal sensing · Soil properties · Soil EC_a · Soil spectrometry · Sensor fusion · Thematic maps

4.1 Introduction

Over the last three decades, a key area of sensor development to support precision agriculture (PA) has been soil sensing. In addition to stationary soil sensors dedicated to monitoring dynamic soil properties over time, there continues to be a persistent need to characterize spatial soil heterogeneity accurately. Until recently, intensive soil sampling with follow-up standardized laboratory analysis has been an important strategy to obtain thematic soil maps representing soil properties that can be used to implement differentiated management of agricultural fields in response to spatially variable local needs. Unfortunately, the relatively high cost of traditional soil sampling and analysis by commercial laboratories make it economically infeasible to achieve the sampling density needed to model the spatial variation of many soil properties accurately. Thus, a popular approach is to collect one sample per hectare using a 100 m × 100 m grid. However, some soil characteristics do not exhibit reliable spatial dependencies at such long separation distances. This is especially the case in conditions with relatively high natural soil variability or due to anthropogenetic factors that affect soil attributes (Adamchuk et al. 2010). Certainly, further reduction of sampling density to accommodate financial constraints of farm operations reduces the accuracy of interpolated soil maps. In addition, deployment of geostatistical tools to interpolate soil maps using a relatively small number of samples frequently causes misleading results; to avoid this, careful analysis of sampling data needs to be done (Webster and Oliver 2007).

One way to reduce per sample costs of soil property measurements is to deploy automatic sampling implements and transition laboratory analytics to less laborious techniques. The latter means shifting from the traditional use of chemical extraction methods for different soil properties, which were combined with conventional analytical instruments, to the use of new equipment with increased capacity for chemical extraction, or spectral-based sensor systems to avoid the wet-chemistry phase altogether. Extensive development of new analytical methods relying on sensor techniques, such as laser-induced breakdown spectroscopy, X-ray fluorescence, and/or diffuse

K. Singh · B. Minasny · D. J. Field
Sydney Institute of Agriculture, the University of Sydney, Eveleigh, NSW, Australia
e-mail: kanika.singh@sydney.edu.au; budiman.minasny@sydney.edu.au

C. Fidelis · D. Yinil
Cocoa board of Papua New Guinea, Rabaul, Papua New Guinea

T. Sanderson
Commonwealth Scientific and Industrial Research Organisation, Canberra, Australia

D. Snoeck
CIRAD, Tree Crop Based Systems Research Unit, Montpellier Cedex 5, France

reflectance of light in the ultraviolet, visible, near-infrared and mid-infrared parts of the spectrum make it feasible, not only to reduce the cost of laboratory analysis, but to bring sensor technologies closer to the sampled fields to optimize sample analysis logistics. However, many chemical soil properties that are used to assess nutrient availability to plants cannot be measured directly without proper sample preparation and chemical extraction. To address this issue, increasing use of deep learning algorithms and other artificial intelligence tools make it feasible to predict the results of conventional soil tests based on the indirect relationships among multiple soil attributes.

Improving the quality of thematic soil maps while attaining affordable costs can be accomplished by deploying the previously mentioned spectral and other sensing techniques directly in the field. For example, by obtaining intensive *in situ* point-based soil measurements, or conducting soil mapping when moving mobile sensor instruments across the field (i.e. using on-the-go sensors). The quest to enable measurements of soil properties in field conditions has resulted in a new area of research and development called proximal soil sensing (PSS). In combination with remote sensing of the soil discussed in Chap. 2, PSS technologies enable high-density data representing the spatial distribution of sensor measurements across agricultural fields.

With rapid developments in electronics, a wide array of measurement principles and deployment approaches can be used to obtain rapid and reliable signals suitable to support PA practices. Thus, some PSS tools can be added to a specific agricultural machine or implement and provide sensor-based control of these operations in real time. In principle, this approach is similar to the widely accepted global navigation satellite system (GNSS) based automatic guidance of tractors as well as self-propelled machinery and variable-rate management of agricultural inputs in response to plant characteristics (using proximal plant sensors). Alternatively, soil mapping can be conducted as a stand-alone field operation to produce high-density PSS measurements that can be analyzed together with other data layers to develop management prescription maps (e.g., map-based approach). While real-time sensor signals are important for highly dynamic soil properties, such as water content, a map-based approach is suitable for soil properties that are more stable over time, such as texture or organic matter content.

4.1.1 Proximal Soil Sensing Technologies

Hummel et al. (1996), Sudduth et al. (1997), Adamchuk et al. (2004, 2017, 2018), Shibusawa (2006), and Viscarra Rossel et al. (2011) provide overviews of proximal sensing systems that have been used to sense soil properties that change spatially and with time. With recent improvements in many aspects of sensing technologies and, more importantly, data interpretation, additional new commercial tools as well as research projects extend the list of available solutions. Although agriculture is a key area of deployment for most PSS solutions, some provide valuable tools for other industries, such as archaeology, mining, ecology and natural resource sciences.

Sensors differ in terms of the mode of operation and inherence of measured values. Thus, mobile (i.e., on-the-go) platforms can be used to obtain high-density soil data at a specific point in time. Nevertheless, monitoring stations enable the observation of changes in soil conditions over time. Some sensors provide measurements that pertain to soil surface or specific depths, while others provide signals integrated through the soil profile according to the depth sensitivity curve and/or produce an array of values that correspond to different depths within a specified operational range.

4.1.2 Topography

When mapping agricultural fields, geographic coordinates are recorded together with other measurements. These coordinates frequently include field elevation that reveals accurate representation of field topography at the centimetre level, when real-time kinematic (RTK) GNSS equipment is used. Similar to light detection and ranging (LiDAR) maps obtained through remote sensing, these field topography data can serve as a primary layer of spatially intensive data to assess soil growing conditions. In addition to the elevation data, topography derivatives, such as slope, aspect ratio, topography wetness index, and so on, can be used to define spatially variable water accessibility.

4.1.3 Geophysical Sensors

Several near surface geophysics instruments have been used historically to produce soil maps for PA. The most popular of these are the maps of apparent soil electrical conductivity (EC_a), measurement of soil's ability to conduct electrical charge. Such maps can be obtained using galvanic contact resistivity, electromagnetic induction (EMI), or capacitively-coupled electrical resistivity principles to induce an electrical current in the soil and measure the change of potential as a factor of distance between transmitting and receiving elements of measurement circuits. Since geometrical relationships between these elements define the shape of depth sensitivity curves, multiple receiving elements have been used to obtain data representing different soil depths. With EMI instruments, effective measurement depth can also change when lifting the instrument above the soil surface or when changing the frequency of the alternating current. Furthermore, EMI sensors have been used to create less popular maps of magnetic susceptibility, while galvanic contact electrodes have been deployed to measure soil capacitance (ability of soil to store electrical charge) which is directly related to the volumetric water content. For more detailed mapping of layering within the soil profile, another type of geophysical instrument, ground penetrating radar (GPR), has been used for certain agricultural practices.

In many cases, geophysical data is collected using transects, by travelling across an agricultural field at a constant distance between passes. The resulting maps reveal the spatial variation of several soil attributes such as: soil salinity, texture, organic matter content, moisture, bulk density, etc. These maps have been used to define relatively homogeneous areas of the field, frequently called 'management zones' that have distinct water and nutrient storage potential and could be linked to changes in some chemical soil properties caused through natural processes under uniform field management.

4.1.4 Spectral Sensors

Many PSS instruments are used to measure reflected, or emitted, energy in different parts of the electromagnetic spectrum. Because of accessibility, diffuse reflectance of visible and near-infrared light has been used to predict soil properties, such as organic matter, texture, water content and even some chemical characteristics using the entire range of soil preparation from *in situ* measurements to laboratory tests on prepared soil samples. To simplify optical measurements, combinations of light emitting diodes and photoresistors have been used for specific wavelength measurements that could be conducted on-the-go. Unlike remote sensing, PSS technology allows sensing light reflectance below the soil surface at a specific depth, while avoiding plant residues and other surficial effects. Mid-infrared reflectance (MIR), X-ray fluorescence (XRF), microwave and other spectroscopic tools have been increasingly used to enhance the ability to predict certain soil characteristics without chemical extraction. Despite their potential to measure the concentration of important chemicals, using these for *in situ* measurements has presented many challenges. This is not the case with the relatively popular mapping of gamma-ray radioactivity emitted by several isotopes typically present in agricultural soils. The maps of gamma-ray count can also be linked to soil texture and an array of other properties defined by soil material.

4.1.5 Mechanistic Sensors

Mechanical means to map soil structure have been used as well as spectral tools. There is an entire family of prototyped mobile sensors to map the soil's ability to resist mechanical penetration. These sensors can successfully delineate potentially compacted areas of the field, and, in certain cases, even the depth of a hardpan (compacted layer of soil). Some sensors also measured soil-sensor friction related to cohesive soil characteristics. Furthermore, sensor vibration affected by interaction with the soil has been measured using acoustic sensors. In addition to mechanical impedance, soil tilth can be assessed using air permeability through measured volume and/or pressure of air injected into soil. Seismic sensing is another mechanistic approach used in geophysics that may be used for agricultural soils in the future.

4.1.6 Electrochemical Sensors

The direct measurement of chemical soil properties, such as pH or the activity of selected ions other than hydrogen, can be achieved using electrochemical sensors, such as ion-selective electrodes or ion-selective field effect transistors. Unlike most sensors mentioned above, electrochemical sensors require stabilization time and the presence of soil solution, which makes deployment for high-density field mapping relatively challenging. The same can be said of non-dispersive infrared sensors (NDIR) used to measure biological activity of soil microorganisms through the detection of the dynamics and concentration of certain gases, such as CO_2.

4.1.7 *Soil Sensor Data Interpretation*

4.1.7.1 Sampling Design

Proximal soil sensors are well known for their generation of high-resolution data through indirect measurement of soil properties and thus require calibration or development of mathematical or statistical relationships with laboratory measured data. While, field collection and laboratory measurement of more samples can provide a stronger and often better relationship between sensor and laboratory measured soil data, this requires a considerable budget and may limit the advantages of proximal soil sensing. Selecting a proper sampling design (Brus and de Gruijter 1997), optimization of sampling locations in geographical and/or in feature space, is the only way to provide a thorough assessment of the relationship between sensor and laboratory measured data. For example, geostatistical sampling optimizes the sampling pattern in geographical space (Vašát et al. 2010), while feature space, a virtual space is constructed based on a set of environmental covariates (Hengl et al. 2003). In developing sampling designs, high-resolution data collected by proximal soil sensors together with other data such as remote sensing and drone imagery and physiographic data, can be used as environmental covariate layers (Fig. 4.1).

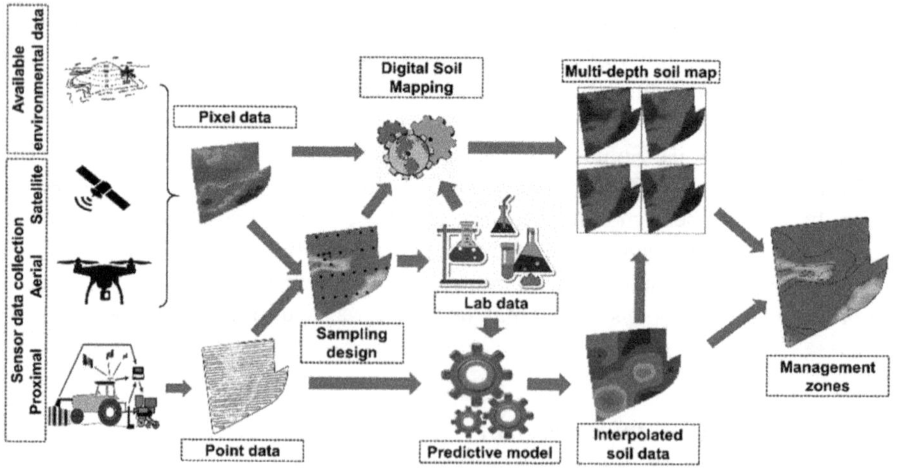

Fig. 4.1 Generalized framework for sensor data analytics

4.1.7.2 Sensor Calibration and Validation

Laboratory measurement of soil samples at optimized locations can then be used to develop calibration relationships with sensor data using various statistical and mathematical methods at those locations and then the relation can be extrapolated to develop a high-resolution soil data layer. These calibration strategies may range from simple linear regression to multivariate chemometrics and machine learning to data mining approaches (Ji et al. 2019), which are often validated using an internal (e.g., repeated, k-fold, leave-one-out cross-validation) and/or an external validation dataset. While many such methods have been investigated (Brungard et al. 2015; Kim et al. 2012; Smith et al. 2012; Triantifilis et al. 2012), machine learning, a general term for a broad set of models used to discover patterns in data and to make predictions (Witten et al. 2017), has shown strong potential. Although machine learning is most often applied to large databases, it is an attractive tool for learning about and making spatial predictions of soil classes or properties because knowledge about relationships between soil and environmental covariates is often poorly understood (Grunwald et al. 2012). However, as the relationship between soil characteristics of agronomic importance and soil sensor data may be limited to a set of conditions, wider application of calibration relations or models might not always be appropriate. Moreover, often with a limited amount of laboratory soil data, available global or local models such as spectral models can be used to predict soil properties at high resolutions based on sensor measurements. More often than not, however, there is a lack of consensus on the most appropriate methods for developing relationships with a particular sensor as the choice of these methods may determine a sensor's predictive ability.

4.1.7.3 Sensor Data Fusion

Although research has been conducted with individual proximal soil sensors (Mahmood et al. 2013; Stenberg and Viscarra Rossel 2010; Weindorf et al. 2014) or has been combined to a limited degree (Ji et al. 2019; Saifuzzaman et al. 2019; Xu et al. 2019), the integration or fusion of data from multiple PSS systems might help to complete characterization of soil because of the complementary information provided by these sensors. Soil sensor data fusion integrates data and information from multiple sensors through various methods, including statistical and machine learning or data mining (Adamchuk et al. 2011c; Castanedo 2013), to predict different soil properties. Other advanced methods have been used for data fusion in engineering applications, such as robotics, and represent the future for applications in soil science (Dong et al. 2009). Therefore, it is critical to choose the right combination of sensors and appropriate numerical methods to integrate data for the success of soil sensor data fusion.

4.1.8 Digital Soil Mapping

The capability of proximal soil sensors to measure soil properties at depths can be used to characterize and quantify the horizontal and vertical variability in soil properties using 3D digital soil mapping techniques (Hengl et al. 2014). While the calibration can allow detailed characterization of soil properties, maps of spatial variation can be used for developing management strategies including site-specific crop input applications. Commonly used spatial interpolation techniques of point data, such as inverse distance weighing or geostatistical analysis-based methods, can be used to develop maps of soil properties from proximal soil sensors. While these interpolations can show the variability in the measured properties, processes controlling the variability of these properties can be very complex and interconnected and should be taken into consideration when quantifying spatial variation and developing maps. Digital soil mapping reflects the soil-landscape relationship (McBratney et al. 2003) and detailed maps can be created by analyzing sensor data statistically. For example, sensors can allow us to review soil profiles at various depths. Thus, sensors substantially increase the efficiency of mapping and allow more accurate and quantitative prediction of soil classes and/or soil properties at any location. This can be a powerful approach in assisting optimal decisions on environmental and agricultural management as digital soil mapping can provide information on continuous and quantitative soil properties by incorporating high-resolution digital soil sensing and mapping techniques (Fig. 4.1). The availability of soil information at depths, however, remains one of the major challenges for full characterization and mapping of soil properties in 3D.

4.1.9 Management Zone Delineation

Integrated multi-layer data from on-the-go sensors can be converted into depth-wise data by various inversion techniques (Jiang et al. 2019). On-the-spot sensors can also uniquely provide soil information at multiple depths (Zhang et al. 2017). Multi-layer information collected using the proximal sensors can enable the mapping of soil properties in 3D (Fig. 4.1). These sensor-based maps can be used to fragment the study fields in accordance with spatially constrained data clusters, commonly known as management zones (MZs). Clustering algorithms, either partitioning, or hierarchical, such as k-means (Zalik 2008) have been used to delineate MZs (Fridgen et al. 2004; Dhawale et al. 2016). However, the complexity and frequently occurring discontinuities of certain MZs make the management operations difficult and complex, and the technology less appealing to potential users. These MZs can be used throughout the decision-making process for optimized management of our land resources including variable-rate and depth fertilizer application, variable depth seeding, variable-rate irrigation based on the total available water in the soil profile and recommendation on tillage regimes.

4.2 Introduction of Case Studies

In the first study on "Comparison of mapping soil properties in Canada using a multi-sensor platform and 1 ha grid sampling", a combination of three on-the-go sensors (galvanic contact resistivity, dual-wavelength subsurface soil reflectance, and direct contact soil pH) together with field elevation data were used to map soil properties in two Ontario fields. The study on "The value of management as a covariate in generating digital maps of soil chemical properties for northern UK arable fields" integrates apparent soil electrical conductivity measurements obtained using an electromagnetic induction instrument and gamma-ray radiometry to map two fields in Scotland. The study on "Validation of a NIR-based laboratory soil analysis procedure for less expensive farm-scale soil mapping" demonstrates the applicability of hyperspectral measurements of diffuse light reflectance to predict multiple soil properties for seven Swedish farms. Finally, the study on "Soil-based precision fertilisation using near-infrared spectroscopy for optimising cocoa production" further demonstrates the practical use of visible and near-infrared soil spectra for improving the management of cacao on four farms across Papua New Guinea.

4.2.1 Case Study 4.1. Comparison of Mapping Soil Properties in Canada Using a Multi-Sensor Platform and 1 ha Grid Sampling

4.2.1.1 Introduction

Characterizing field heterogeneity through proximal sensing systems (PSS) is a key to optimising crop production under PA. Proximal soil sensing-based soil mapping allows a more timely and cost-effective acquisition of large amounts of geospatial data than with standard grid-based sampling. The PSS sensors react to multiple soil properties, but no individual sensor can quantify all soil properties (Viscarra Rossel et al. 2011; Mahmood et al. 2012). To improve predictions of soil properties and to broaden the application of PSS technologies, sensor fusion seeks to incorporate the strengths of different sensors by assembling them on an on-the-go platform (Kuang et al. 2012; Mahmood et al. 2012; Adamchuk et al. 2011c). On-the-go sensing normally requires a set of soil samples collected at optimal field locations to calibrate sensor measurements to conventional soil properties. No approach has been standardized for how many and where samples should be taken for calibration (Peters 2006).

Currently, studies of optimal sampling approaches for sensor fusion calibration are rare, but several approaches have been proposed. These include approaches using grid sampling of various grid sizes (e.g., Mahmood et al. 2012). Grid sampling approaches ensure spatial coverage of the target area; however, the high sampling density required might not be cost effective for sensor fusion calibration beyond research activities. Other approaches have evaluated zone or cluster-based sampling algorithms, which differ from systematic grid sampling, in an effort to assign samples in both the geographic and attribute space for sensor-based mapping (Brus et al. 2006) or for sensor fusion calibration (Taylor et al. 2010; Adamchuk et al. 2011b). Finally, approaches to sample properly in the attribute space of multiple sensors have also been proposed, such as the latin hypercube sampling (LHS) method (Minasny and McBratney 2006). While LHS has been shown to perform better than random sampling, equal spatial strata, or principal component analysis in simulations, it does not consider local heterogeneity or geographic coverage to account for potential spatial factors that are not associated with the sensor's data.

Considering the advantages and accessibility of multi-sensor platforms that are capable of acquiring large amounts of soil sensing data across a landscape, this study compared two types of soil mapping: sensor-based (using a multi-sensor platform) and grid-based (using 1 ha grid sampling). Specifically, the aims were to (i) evaluate the effectiveness of the recently developed neighbourhood search algorithm (Dhawale et al. 2016) to determine calibration sampling locations for a multi-sensor platform, (ii) generate multiple linear regression calibration models between the soil sensor data and soil properties, and (iii) generate sensor-based soil maps and compare their prediction quality with those produced by spatial interpolation of 1 ha grid sampling data. The soil properties of interest were soil texture, cation exchange capacity (CEC), soil pH, buffer pH and soil organic matter (SOM).

4.2.2 *Materials and Methods*

Field data were collected in two agricultural fields (NX in 2015, ST in 2016) near Ottawa, Ontario, Canada. The NX and ST fields were ~ 40 ha and 45 ha, respectively. Data collection included a PSS survey and three types of soil sampling: grid, sensor calibration and validation sampling (Fig. 4.2).

4.2.2.1 The PSS Field Survey

The PSS system used was the on-the-go Veris Mobile Sensing Platform – MSP3 with the EC_a Surveyor, pH Manager and OpticMapper modules (Veris Technologies, Inc., Salina, Kansas, USA) operated by DuPont Pioneer (Richmond, Ontario, Canada). The MSP3 was used on transects spaced at ~15 m and at a maximum recommended speed of 13 km h^{-1}. The EC_a sensor measured two depths: shallow (0–0.30 m; $EC_{a\text{-}s}$) and deep (0–1.00 m; $EC_{a\text{-}d}$) at a 1-Hz logging rate. The pH sensor uses ion-selective electrodes (ISEs) and provided pH measurements (denoted pH_s) from an automated soil core collection procedure at 0.1 m depth at 0.1 Hz. The lower rate of sampling was necessary to allow electrode stabilization. The OpticMapper sensor collected spectral reflectance measurements at two specific

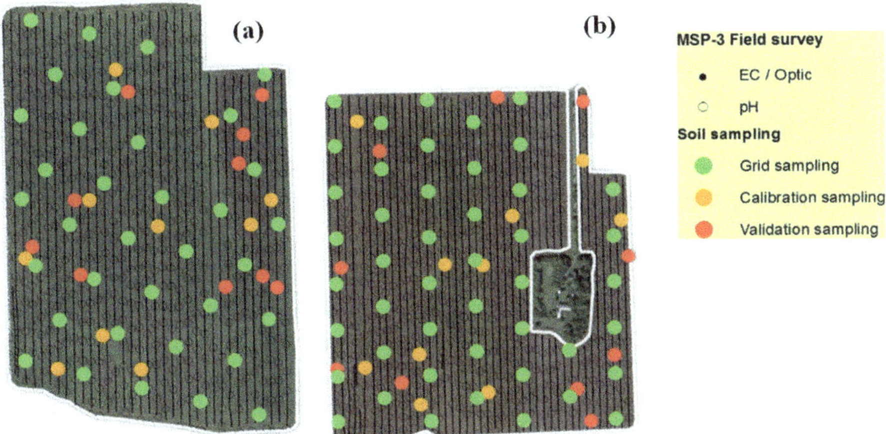

Fig. 4.2 Distribution of MSP3 field survey measurements and soil sampling locations from the two studied sites: (**a**) ST and (**b**) NX

wavelengths [660 nm, denoted as RED and 940 nm, denoted as NIR] at a depth of 0.05 m in the soil.

All sensor data were georeferenced using an external RTK-GNSS receiver (R9s, Modular GNSS Systems, Trimble, Inc., Sunnyvale, California, USA), which also recorded elevation to derive a digital elevation model (DEM) for each field on a 5 m grid. Slope, aspect and topographic wetness index (TWI) were derived from the DEM. Slope and aspect were modelled in ArcGIS. The TWI was generated in SAGA GIS version 2.4 (Hamburg, Hamburg, Germany).

4.2.2.2 Soil Sampling

Three sampling strategies were used:

(1) Grid-based sampling: Sampling for the two fields was on a 1 ha grid. Soil samples were bulked from multiple soil cores (0.15 m depth) collected around the centre of grid cells. In total, 35 and 46 samples were collected from the ST and NX fields respectively. These data were used to produce interpolated maps of soil properties.
(2) Sensor calibration sampling: For both fields, the four high-resolution sensor variables (pH_s, $EC_{a\text{-}s}$, $EC_{a\text{-}d}$ and elevation) were used for multi-sensor calibration sampling design (Dhawale et al. 2016). The algorithm transforms each data layer into 20 m × 20 m grid cells and assigns the average sensor value to the cells. A clustering procedure is used to identify the first cluster by isolating results with the greatest possible reduction in MSE. Neighbouring cells with relatively small variances are then aggregated across all data layers. A set of optimized spatial clusters that best delineates each data layer into multiple

homogenous areas is formed. Finally, 10 sampling locations in the centre of each zone were identified. These data were used to generate multiple linear regression (MLR) relationships between sensor data and measured soil properties at the sampling locations to produce soil maps by regression.

(3) Validation sampling: For each field, an additional 10 soil samples were randomly collected to validate the interpolation and prediction methods. Sampling and sensing in the NX field were not done within a 20 m buffer zone of the house and road to avoid potential edge effects.

At each field site (grid, calibration and validation) topsoil cores were extracted (depth – 0-0.15 m). Composite samples were prepared on site. Soil samples were sent to certified laboratories, A&L Canada Laboratories, Inc. (London, Ontario, Canada) for chemical analysis and Agro-Enviro-Lab (La Pocatière, Québec, Canada) for textural analysis and soil organic matter measurements. Laboratory soil properties available for prediction were soil pH, buffer pH (BpH), soil CEC, SOM, clay (%) and sand (%). Soil texture analysis was not performed for the NX field. For alkaline samples (pH > 7), BpH was not measured but estimated from pH using the method of Adamchuk et al. (2011a).

4.2.2.3 Quantitative Methods for Soil Mapping

(1) Mapping of the grid-based data: interpolation by ordinary kriging (OK) was used to map, pH, BpH, CEC, clay, sand and SOM, for the ST field and pH, BpH, CEC and SOM, for the NX field. Ordinary kriging and mapping were done in ArcGIS (ESRI, Redlands, California, USA). To emphasize the advantage of the OK approach in illustrating spatial heterogeneity of the dense sensor data, this study applied this interpolation method consistently throughout the study including for the relatively low-density grid sampling, which is not advised in practice because it is difficult to determine the variogram model accurately to describe the spatial structure.

(2) Sensor-based mapping: For sensor-based soil prediction, a multiple linear regression technique related one soil property to multi-sensors measurements at the calibration locations. All sensed data layers were interpolated by OK to the soil sampling locations for regression. The digital terrain attributes (DTAs) were also extracted at each soil sampling location. Table 4.1 lists the soil properties measured and expected relationships to the sensor data.

The best subset regression model was identified by calculating and comparing the Akaike information criterion (AIC) (Akaike 1973), mean square error (MSE), root mean square error of prediction ($RMSE_P$) and adjusted R^2_{adj} of potential models. The significance level of the selected model for predicting a specific soil chemical property was set at $\alpha = 0.05$. The models developed were used in ArcGIS to predict soil properties from the sensor data.

Table 4.1 Sensed variables expected to show relationships with soil properties

Response variable: Soil property of interest	Predictor variables: Multiple sensing data layers
pH	pH_s, EC_a and topography
BpH	pH_s, EC_a and topography
Clay content	EC_a, topography, reflectance and pH_s
Sand content	EC_a, topography and reflectance
SOM	Reflectance, EC_a, pH_s and topography
CEC	EC_a, topography and reflectance

4.2.2.4 Map Assessment

The quality of the interpolated and predicted soil maps was assessed by $RMSE_V$ using the interpolated or predicted values and the laboratory measured values at the 10 independent validation sites. To quantify the statistical difference between the accuracy of the two mapping approaches, the non-parametric Wilcoxon Signed-Ranks Test for paired samples was performed on the $RMSE_v$ at a significance level of 0.05.

4.2.3 Results and Discussion

4.2.3.1 Description of Soil Properties and Soil Sensor Responses

Both experimental fields had a similar range of pH, from strongly acidic (pH < 5.1) to neutral (6.6–7.3) and a similar standard deviation (SD ≈ 0.7). Overall, SOM was higher in field ST with 3.5–8.3% and 2.3–5.5% for field NX. The SD of SOM was greater in field ST. The CEC was slightly larger in ST than in field NX, which was consistent with higher SOM in field ST. In the latter, clay ranged from 11 to 40% (SD = 5.6%) and sand from 24 to 69% (SD = 12.1%).

Table 4.2 presents a summary of univariate statistics for sensing data layers from fields ST and NX. Both fields showed similar responses for most of the sensor data layers. Elevation differed, particularly the difference in elevation between fields (6 m for NX field and 10 m for the ST field). Maps of the available predictors for both fields are shown in Fig. 4.3.

4.2.3.2 Multiple Linear Regression Analysis

Best subset MLR models and fits are shown in Table 4.3. For both fields, good model fits were obtained (R^2_{adj} >0.87) for all properties. Selected predictors varied between properties but were very constant for a given property between the fields.

Table 4.2 Summary statistics of sensed measurements

Field	Sensor data layer	Number of measurements	Statistics[a]			
			Min.	Max.	Mean	SD
ST	pHs	611	4.9	7.9	6.4	0.67
	$EC_{a\text{-}s}$ (mS m^{-1})	9681	2	105	21	10.1
	$EC_{a\text{-}d}$ (mS m^{-1})	9681	0	287	25	16
	Elevation (m)	9429	57	67	60	2.1
	RED	8756	204	231	213	3.1
	NIR	8756	390	480	431	11.0
NX	pHs	728	5.2	7.8	6.2	0.44
	$EC_{a\text{-}s}$ (mS m^{-1})	11,655	2	85	24	7.8
	$EC_{a\text{-}d}$ (mS m^{-1})	11,655	0	46	24	6.7
	Elevation (m)	11,655	90	96	93	0.5
	RED	11,250	203	246	220	6.6
	NIR	11,250	371	545	441	23.6

[a]SD = standard deviation

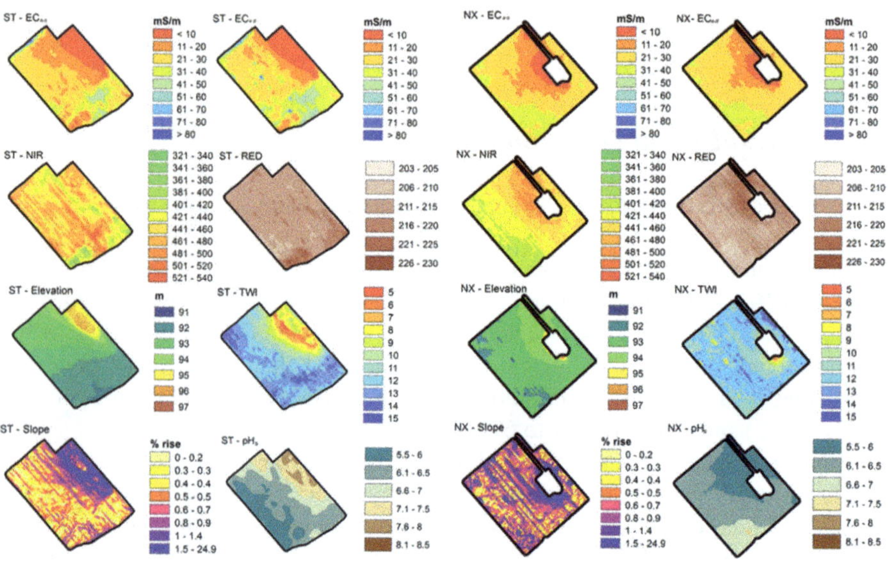

Fig. 4.3 Thematic maps of sensing measurements: $EC_{a\text{-}s}$, $EC_{a\text{-}d}$, NIR, RED, elevation, TWI, slope and pHs using ordinary kriging for ST (left) and NX (right) fields

4.2.3.3 Quality of Soil Mapping

The $RMSE_v$ for the two mapping approaches with the ten independent validation points for all available soil properties are shown in Table 4.4. Based on analysis of the map predictions at these validation points, the quality of sensor-based mapping was not significantly different ($p = 0.05$) from the grid-based mapping. Figure 4.4

Table 4.3 Regression analysis for sensor-based soil property prediction

Field	Soil property	Best multiple linear regression model	R^2_{adj}	$RMSE_p$
ST	pH	$-5.665 + 1.718\boldsymbol{pH_s} + 0.031\boldsymbol{EC_{a-s}}$[a]	0.92	0.26
	BpH	$7.759 - 0.507\boldsymbol{TWI} + 0.008\boldsymbol{EC_{a-s}} + 0.062(\boldsymbol{pH_s \cdot TWI})$	0.98	0.11
	Clay (%)	$-2.153 \cdot 10^2 + 1.711\boldsymbol{RED} + 2.251 \cdot 10^{-2}(\boldsymbol{EC_{a-s} \cdot TWI}) - 1.461 \cdot 10^{-3}(\boldsymbol{RED \cdot NIR})$	0.85	1.89
	Sand (%)	$110.432 - 5.461\boldsymbol{TWI} + 0.249(EC_{a-s} \cdot TWI) - 0.016(EC_{a-s} \cdot RED)$	0.92	4.57
	CEC (meq hg^{-1})	$12.442 + 0.033(\boldsymbol{EC_{a-s} \cdot TWI}) - 0.004(EC_{a-s} \cdot EC_{a-d})$	0.94	0.86
	SOM (%)	$67.123 - 0.195\boldsymbol{RED} - 2.659\boldsymbol{pH_s} + 0.003(\boldsymbol{EC_{a-s} \cdot EC_{a-d}}) - 0.032(\boldsymbol{pH_s \cdot EC_{a-d}})$	0.87	0.39
NX	pH	$4.731 + 0.010(\boldsymbol{pH_s \cdot EC_{a-s}})$	0.90	0.24
	BpH	$6.678 - 0.209\boldsymbol{TWI} + 0.028\boldsymbol{EC_{a-s}} + 0.025(\boldsymbol{pH_s \cdot TWI})$	0.92	0.11
	CEC (meq hg^{-1})	$12.359 - 0.035(\boldsymbol{EC_{a-d} \cdot TWI}) + 0.015(EC_{a-s} \cdot EC_{a-d})$	0.88	1.52
	SOM (%)	$59.314 - 0.189\boldsymbol{RED} - 1.620\boldsymbol{pH_s} - 0.296\boldsymbol{EC_{a-d}} + 0.005(\boldsymbol{EC_{a-s} \cdot EC_{a-d}})$	0.90	0.24

[a]Bold text indicates a significance level at 0.05, others are significant at 0.10

Table 4.4 Comparison of prediction quality between grid- and sensor-based soil property prediction methods using 10 validation samples

Field	Predicted soil property	Mapping approach	# of samples	$RMSE_V$
ST	pH	Grid	35	0.63
		Sensor	10	0.39
	BpH	Grid	35	0.37
		Sensor	10	0.30
	Clay (%)	Grid	35	1.93
		Sensor	10	2.75
	Sand (%)	Grid	35	8.74
		Sensor	10	6.39
	CEC (meq hg^{-1})	Grid	35	2.30
		Sensor	10	1.96
	SOM (%)	Grid	35	0.57
		Sensor	10	0.86
NX	BpH	Grid	46	0.21
		Sensor	10	0.25
	pH	Grid	46	0.39
		Sensor	10	0.38
	CEC (meq hg^{-1})	Grid	46	1.95
		Sensor	10	2.07
	SOM (%)	Grid	46	0.55
		Sensor	10	0.43

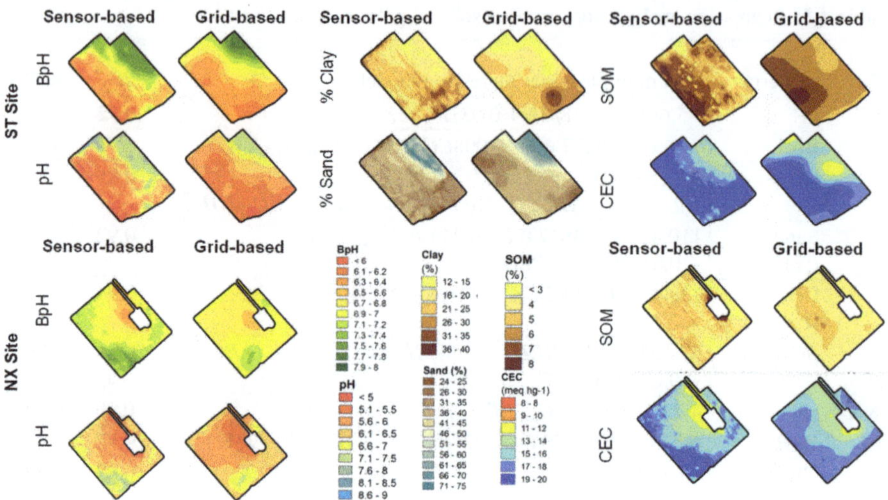

Fig. 4.4 Thematic maps for the ST site of buffer pH (BpH), pH, % clay, % sand, CEC and SOM, and buffer pH (BpH), pH, CEC and SOM for the NX site produced using grid-based and sensor-based prediction approaches

shows the grid-based and sensor-based maps of selected soil properties predicted for both field sites.

4.2.4 *General Discussion*

The two approaches produced very similar maps. However, the PSS-based maps showed more detail for all properties, which is associated with the higher spatial resolution of the sensor data. Neither of the two mapping approaches was always superior for soil properties across the two fields. The mapping of BpH and pH by the sensor-based approach was either comparable or better than the grid-based approach. Sensor-based CEC prediction was better than grid-based for ST (with $RMSE_V$ about 15% lower), but slightly poorer for NX. In contrast, sensor-based SOM mapping was better for NX, but poorer for ST.

Ten samples only were used to construct the regression model, therefore, the non-optimized sampling locations for the optic sensor could have resulted in poor prediction quality when incorporating an optic sensing data layer as one of the predictor variables. In theory, the optic sensing data layer should show some correlation with soil CEC which it did not in this study. Further investigations might assess whether optimizing the optic sensor's calibration locations might improve the prediction of % clay and SOM. Through the analysis of best subset regression for the prediction of soil properties of interest, $EC_{a\text{-}s}$ data were selected in all the regression models.

4.2.5 Conclusions

Compared to 1 ha grid soil mapping, multi-sensor-based mapping demonstrated an advantage for site-specific resource management in terms of comparable quality of soil maps, lower sampling density (i.e., 10 for sensor calibration compared with 35–46 for grid sampling) and consequently, less time and expense spent on field work and on laboratory testing. To deliver robust sensor-based soil characterization strategies, further research will be needed to investigate the integration of other sensor systems and to evaluate mapping quality for various soil types, climatic conditions, and agricultural practices. Because of similarities in soil series within a given geographic region, it might be feasible to generalize some parameters of soil prediction models, which will further reduce the number of calibration samples required within each agricultural field.

4.3 Case Study 4.2. Including Management as a Predictor Can Improve Subfield-Scale Digital Maps of Soil Chemical Properties

4.3.1 Introduction

Proximal sensors are increasingly being used to develop soil maps for PA to aid zone or site-specific management decisions. Most commercially available proximal soil sensors are based on sensing the apparent electrical conductivity (EC_a) of the soil. Another soil sensor that has gained some traction in agriculture is the gamma-ray (γ) radiometer, which records information on the low-level, natural emission of γ radiation from isotopes in the soil. Both EC_a and γ-radiometrics are general soil responses, and affected by multiple soil properties. For example, the EC_a response is strongly affected by clay content (%), clay mineralogy and soil moisture content (and their interactions), as well as being influenced by other soil properties. These proximal sensors have been used successfully to estimate individual soil properties with a local calibration. Physical soil properties, such as particle-size fractions, have been by far the most commonly predicted.

There are some dedicated sensors for soil chemistry, but limited commercialization has occurred to date. Research has demonstrated that soil chemical properties, such as cation exchange capacity (CEC) and pH, can be linked to EC_a (Triantafilis et al. 2009), to γ-radiometrics (Holland et al. 2017) or a combination of EC_a and γ-radiometrics (Taylor et al. 2010). However, there are limitations as soil predictions are site-specific, cannot be applied universally and require local calibrations to ensure accuracy.

Farm management practices (e.g. fertilizer and pesticide applications, cultivation, irrigation and drainage) change the soil. Despite this, management is not a widely used predictor within digital soil mapping models, and no examples of its use at an intra-field scale are known. The common assumption is that management is uniform across a field and is a non-spatial, constant effect. This poses an interesting conundrum for digital soil mapping for PA. If variable management is adopted within a field, will this influence the modelling and accuracy of future soil mapping within it? If, so, should management information be critical for digital soil mapping in PA?

The aim of this case study is to investigate the potential of management as a predictor in digital soil mapping in fields under varying management. This is done using a split field trial rather than a dedicated PA differential management approach. The interest here is placed on the generation of a digital soil mapping of soil chemical properties, as these are likely to be more strongly affected by management and are less reported than soil textural properties.

4.3.2 Materials and Methods

The study site was two fields at Balruddery Farm, James Hutton Institute, Dundee, UK (56.48° N; −3.13° W, WGS84) with sandy loam to sandy silt loam soils classified as Cambisols (WRB 2006). For both fields there was little difference in the topsoil (0–0.3 m) texture with only 3–4% variation in sand and silt contents. The two fields selected have a six-year rotation including potatoes (*Solanum tuberosum*), winter wheat (*Triticum spp.*), winter oilseed rape (*Brassica napus*), spring sown beans (*Phaseolus spp.*) and spring and winter sown barley (*Hordeum vulgare*). The fields are split into two half-field management treatments (conventional and integrated) (Fig. 4.5). The conventional treatment corresponds to local standard commercial practices for soil cultivation, fertilizer inputs and herbicide applications, while the integrated treatment includes over-winter cover crops, cereal straw incorporation, municipal green waste compost amendments and non-inversion tillage.

The EMI survey used a DUALEM 21S (DUALEM Inc., Mississauga, Ontario, Canada) sled-mounted at 0.25 m above the ground surface and surveyed ~20 m swath widths at a speed of ~5 km h^{-1}. The DUALEM-21S has horizontal co-planar (HCP) and perpendicular (PRP) arrays with 1 m and 2 m lengths. If depth of exploration (DOE) is the depth below an array within which the array accumulates 70% of its total sensitivity to conductivity, the DOE an HCP array is 1.6 array-lengths and the DOE of a PRP array is 0.5 array-lengths. Thus, the DOEs were approximately 0.25 m, 0.75 m, 1.35 m and 2.95 m for the 1 m PRP, 2 m PRP, 1 m HCP and 2 m HCP arrays respectively after allowance for the sled height. A fifth integrated EC_a variable (EARTH, with DOE ~1.3 m) was derived by finding, for each measurement point, a conductivity value for a uniform earth for which theoretical EC_a responses fit well to measured values (R. Taylor pers. comm.). Raw EMI data were checked for outliers and adjustments were made where sensor roll indicated a significant effect on values.

The γ-radiometrics survey was conducted with a SoilOptix sensor (SoilOptix Inc. Tavistock, ON, Canada) mounted on the front of an all-terrain vehicle (ATV) at

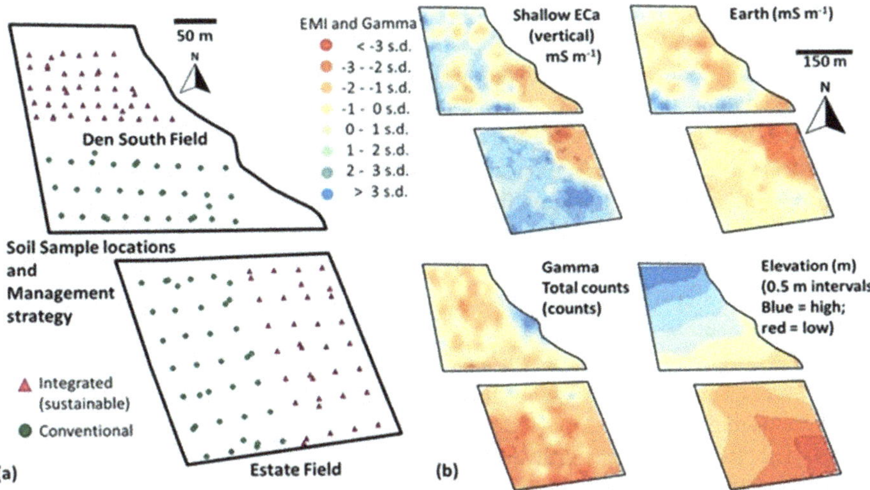

Fig. 4.5 The sampling strategy and delineation (**a**) of the two study fields (Den South is top and Estate is bottom) into integrated and conventional management and examples (**b**) of data layers obtained from EMI (PRP2 and EARTH layers), γ-radiometrics (Total counts) and elevation. The EMI and γ layers are mapped in half standard deviation increments and elevation is in 0.5 m intervals

a height of ~0.4 m. The SoilOptix sensor uses a CsI crystal and a scintillator to measure the amount of naturally occurring γ radiation. The γ-radiometer measures the energy in a spectrum, indicating the total amount of γ-radiation determined, termed the total count (TC).

The EMI sensor and γ-radiometer surveys were conducted on 22–23 March 2018 using both types of sensor coupled to a Trimble Ag-25 GNSS receiver (Trimble Inc., Sunnydale, CA, USA). All data were collected at 1 Hz. Maps of soil sensor variables were generated with local block kriging using the protocol of Taylor et al. (2007).

Soil cores were taken on a regular 25 m grid across both fields with additional cores sampled closer together at randomly selected grid points. There were 75 points sampled in the Den South field (11.2 samples ha^{-1}) and 72 points in the Estate field (9.0 samples ha^{-1}) (Fig. 4.5). The 0–0.3 m (topsoil) of the soil core was used for this analysis. Soil samples were air-dried and sieved (<2 mm) prior to chemical analysis. Soil pH was determined using a 1:2.5 water extraction (MAFF 1986). Soil organic carbon (%) was determined using a standard Dumas combustion technique. Soil exchangeable cations (Ca, Fe, K, Mg and Mn) were detected using a 1 M ammonium acetate extracting solution (Thomas 1983). Normality of the data was checked.

The effect of the management treatment on the soil chemical properties and on the proximal soil sensor properties was tested using the non-parametric Mann-Whitney U (MW) test due to low and unequal treatment sample numbers (n = 30–40) and unequal variances in the sensor data in the Estate Field. Only the interpolated soil sensor data at the sampling points was used for the MW test to avoid issues with spatial auto-correlation between data.

Prediction models of individual soil chemistry variables were constructed using the sensor data as the dependent variables (HCP, HCP2, PRP, PRP2, TC and Elev.) by MLR and regression forest analysis (RFA) with (+M) and without (−M) management included as a predictor for each modelling approach. For the MLR a stepwise approach was used to select the most parsimonious model from the full model and a leave-one-out cross-validation (LOOCV) was performed. The model fit from the LOOCV was assessed by R^2, RMSE and the three most dominant variables selected in the stepwise process were recorded. For the RFA, a bootstrap sampling method was used to avoid potential over-fitting compared to standard decision tree models. A single prediction was obtained from aggregation of the results of all individual trees. The predictions acquired from the regression prediction error out-of-bag were used to rank the importance of each predictor (Taghizadeh-Mehrjardi et al. 2016). The RFA was run with 300 regression trees and 30 iterations. The RFA model R^2, RMSE and the three most dominant variables were recorded.

4.3.3 Results and Discussion

The mean values for the topsoil chemical properties split by management treatments for each field are given in Table 4.5. The Fe and Mn results are not shown as the values were universally low and near the threshold of measurement. For the other soil chemical properties, there was a clear management effect in both fields with significantly larger values in the integrated treatment, especially for soil organic C and K. This accords with previously reported management effects for these fields (Hawes et al. 2018) and similar effects elsewhere (Zani et al. 2020).

There were some significant differences ($p < 0.001$) observed in the soil sensor data (data not shown). Since the fields were split simply N-S or E-W (Fig. 4.5) there was a possibility that the underlying soil variation may affect the sensor response. For both fields, TC was not significantly different while elevation was ($p < 0.01$). The latter can be associated with the splitting of the fields relative to the slope (Fig. 4.5). For the EMI survey, the Estate field had a lower response in the north-west quadrant that is only associated with the integrated treatment. Consequently, the range of EMI-derived variables in the integrated treatment was much greater, leading to a lower mean and greater variance, respectively. The spatial EC_a patterns were more evenly distributed between treatments in the Den South field, although there was a trend for slightly higher values (significant only for HCP2 and EARTH; $p < 0.05$) in the conventional treatment.

After modelling with and without management as a predictor, the soil chemical prediction models clearly indicated that there was a strong effect of management (Table 4.6) for selected soil variables. Overall, the soil variable with the best digital soil mapping model was carbon ($R^2 = 0.6$ to 0.7); however, there was no significant effect of management with both the MLR and RFA approaches. In contrast,

Table 4.5 Mean and standard deviation of topsoil (0–0.3 m) chemical properties for the conventional and integrated management treatments in two fields

Field	Soil variable	Conventional	Integrated
Den South	Carbon (%)	4.07 (0.38)	4.90 (0.55)
Den South	pH	5.66 (0.17)	5.82 (0.15)
Den South	Ca (mg kg^{-1})	9.34 (0.78)	10.95 (1.15)
Den South	Mg (mg kg^{-1})	1.33 (0.11)	1.57 (0.17)
Den South	K (mg kg^{-1})	0.37 (0.08)	0.68 (0.15)
Estate	Carbon (%)	2.71 (0.72)	3.31 (0.38)
Estate	pH	5.81 (0.18)	6.04 (0.17)
Estate	Ca (mg kg^{-1})	9.35 (0.99)	11.05 (0.88)
Estate	Mg (mg kg^{-1})	1.02 (0.10)	1.36 (0.11)
Estate	K (mg kg^{-1})	0.50 (0.12)	0.89 (0.22)

The mean is followed in each column by the standard deviation in brackets
Management was a significant factor for all soil chemical properties $p < 0.001$ using the MW test

Table 4.6 Statistical evaluation (R^2, RMSE) of the predicted soil variables (depth 0.3 m) and the main predictors (most important three where $p < 0.05$) using a multiple linear regression model (MLR) and random forest analysis (RFA) with (+M) and without (−M) the effect of conventional and integrated management treatments in the two study fields

		MLR		RFA		Main Predictors	
Soil variable	Management	R^2	RMSE	R^2	RMSE	MLR	RFA
Carbon	+ M	0.67	0.574	0.70	0.55	EL, H2, P2	EL, M, P
Carbon	− M	0.60	0.629	0.68	0.57	EL, P2, H	EL, P, EA
pH	+ M	0.38	0.167	0.32	0.17	M, EL, P2	EL, M, TC
pH	− M	0.18	0.192	0.25	0.18	H2, H, EL	EL, TC, EA
Ca	+ M	0.43	0.960	0.38	1.00	M, TC, P2	M, EL, P2
Ca	− M	0.11	1.196	0.11	1.20	H2, H	EL, EA, TC
Fe	+ M	0.62	0.002	0.49	0.00	EL, M	EL, P, EA
Fe	− M	0.58	0.002	0.46	0.00	EL, H2, H	EL, P, EA
K	+ M	0.58	0.155	0.57	0.16	M, EL	M, EL, EA
K	− M	0.19	0.216	0.33	0.20	H2, H	EK, EA, H2
Mg	+ M	0.64	0.143	0.69	0.14	M, EL, H2	EL, M, P
Mg	− M	0.31	0.200	0.49	0.17	EL	EL, P, TC
Mn	+ M	0.44	0.010	0.48	0.01	EL, H2, P2	EL, EA, H2
Mn	− M	0.43	0.010	0.55	0.13	EL, H2, P2	EL, EA, H2

Main predictors: M = Management, EL = ELEVATION, H = HCP, H2 = HCP2, P = PRP, P2 = PRP2, EA = EARTH, TC = total counts

management was a highly significant predictor for pH, Ca, K and Mg, especially for the MLR e.g. for K the R^2 was 0.58, but 0.19 without management. Modelling with a subset at a larger grid size (75 m, or 19% of points) also showed that including management improved model performance (data not shown). Assessment of the regression method used showed that when management is included, the MLR fit (R^2) becomes more similar to the RFA, but without management the prediction from

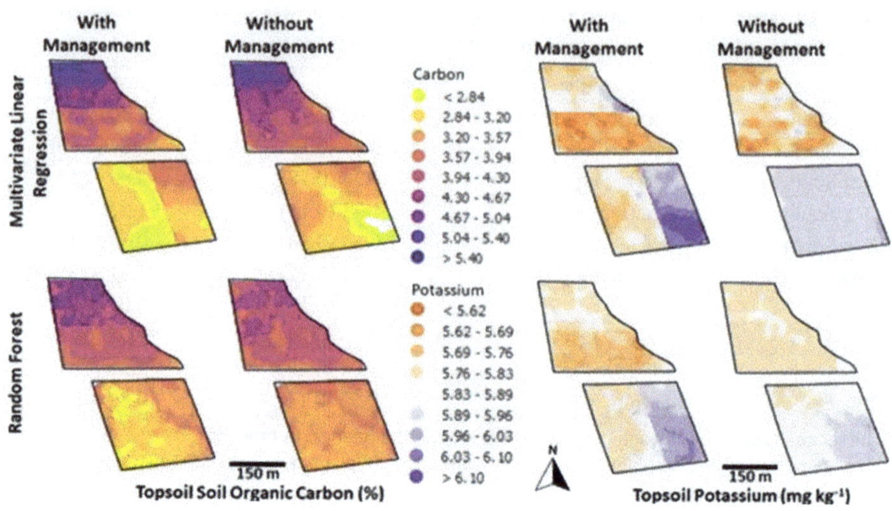

Fig. 4.6 Digital soil maps of the two study fields for topsoil (0–0.3 m) carbon (left) and potassium (right) from modelling with different approaches (Multivariate Linear Regression –top row; Random Forest Analysis – bottom row) with and without management (map columns) being included as a predictor

the RFA is better (larger R^2 and smaller RMSE) for all soil variables except Ca and Fe. This accords with other digital soil mapping research that has not used management information as a predictor and that advocates RFA over MLR. In this situation, the strongest predictor was generally elevation, which reflects the importance of slope and landscape position on pedogenesis for these two arable fields. The EMI-derived predictors were more significant than TC from the γ-radiometer. These results differed from those of Taylor et al. (2010) who found that the TC was the best predictor for pH and CEC and that elevation had little influence in the same region. These contrasting local results highlight the site-specific nature of many of these digital soil mapping models.

To visualize the effect of the modelling approach (MLR vs. RFA) or including (or excluding) management information, maps for topsoil C and K are shown in Fig.4.6. Even though the selected predictors differed (Table 4.6) and the maps show dissimilarities, the different approaches indicated a similar level of model fit (Table 4.6). The management predictor is not as strong for C in the RFA approach compared to the MLR, even though it was not one of the three strongest predictors selected in the MLR stepwise process. In contrast, management had a strong effect on topsoil K and model performance and its effect was evident in both MLR and RFA maps. Here again, the visual difference was weaker for RFA. The topsoil K maps with management appeared more coherent and reflected the raw data (not shown).

PA seeks to optimize production efficiency through site-specific input management. If PA is implemented in a continuous systematic manner over the medium to long term (5+ years) it seems reasonable to expect that differential management will spatially alter soil chemical properties (and perhaps soil physical properties in the longer term). If this is the case, then future digital soil mapping should routinely include management as a predictor. Indeed, Kaufmann et al. (2020) reported that management (different irrigation and fertilizer treatments) had a significant legacy effect on soil EC_a.

This case study also has immediate implications for the PA community. Management is a useful predictor for selected soil variables (Table 4.6; Fig. 4.6); thus, larger scale mapping such as at the farm scale could benefit from including field-specific management as a predictor in the modelling.

4.3.4 Conclusions

Using data from an established field-scale cropping systems experiment, it was shown that management had a significant and strong effect on the ability to predict most, but not all, topsoil chemical properties at the sub field-scale from proximally sensed data. In the future, the importance of management as a predictor is likely to increase with the increased adoption of site-specific management (e.g. variable application of inputs) according to high and low yielding zones within fields.

4.4 Case Study 4.3. Validation of a NIR-Based Laboratory Soil Analysis Procedure for Less Expensive Farm-scale Soil Mapping

4.4.1 Introduction

After more than two decades of increasing scientific interest, near-infrared (NIR) spectroscopy is recognized as a sensor technique well suited for several soil analyses, especially related to organic matter (OM) and texture (Soriano-Disla et al. 2014). Although cost-efficient, large-scale NIR calibration models have often lacked precision at the farm or field scale, a scale where the models are rarely validated. Adding a few local soil samples to the large models has significantly improved local predictions, but this is often no better than farm or field scale calibrations (Guerrero et al. 2016). Although good prediction models can be achieved with as few as 25 soil samples, the robustness of such models may be questioned (Wetterlind and Stenberg 2010).

Swedish farmers often analyze their soils for plant available nutrients and pH at a density of one soil sample per hectare (ha). However, the more expensive soil texture analyses are often only carried out on every second or third soil sample, if done at all. By adding NIR analysis to each sample, it would be possible to estimate soil texture and organic matter on all samples at considerably less cost than with the traditional laboratory methods. In this way the farmers would receive superior information on their fields to support decisions on site-specific management such as site-specific lime requirement, identifying draught sensitive areas and areas where the mineralisation of soil nitrogen makes additional fertilisation potentially superfluous.

With soil samples collected from several farms in the major agricultural regions of Sweden, analyzed by standard methods at a commercial laboratory, this case study aims to evaluate the possibility of determining a minimum number of calibration samples for reliable predictions of soil properties using local farm or field calibrations.

4.4.2 Method

Regular farm soil mapping samples from seven farms between 81 and 206 ha in size were selected at the analysis laboratory (Fig. 4.7, Table 4.7). Data for pH, plant available phosphorous, potassium and magnesium (P, K and Mg), soil texture and OM were available at a density of one sample per ha. The NIR spectra were also measured for all samples. Analyses were performed on the air-dried <2-mm fraction of the soil. Soil texture was analyzed through a sedimentation and sieving method and OM was determined through loss on ignition and corrected for structural water in clay minerals (Gee and Bauder 1986). Plant available nutrients were analyzed as ammonium acetate lactate-extractable nutrients and pH was measured in deionized water. The NIR spectra were measured with a FOSS NIR Systems 5000 between 1300 and 2400 nm with an analysis interval of 2 nm and a resolution of 5 nm.

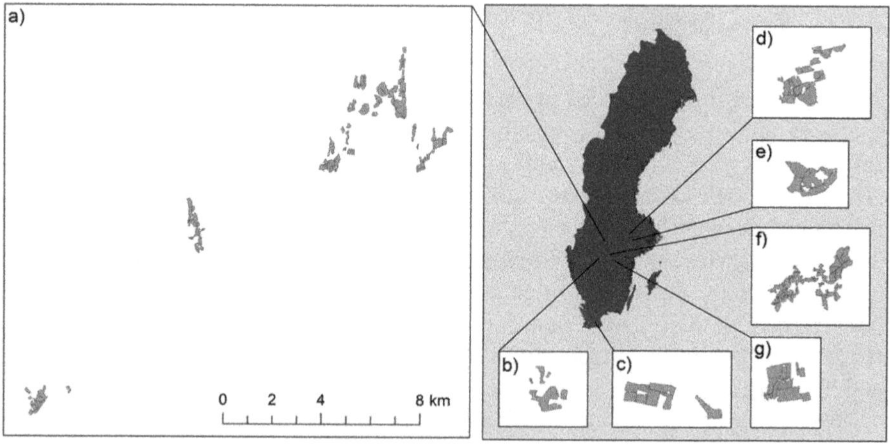

Fig. 4.7 Location in Sweden and land consolidation of the seven farms. From left to right: (**a**) Storfors (187 ha), (**b**) Mariestad (81 ha), (**c**) Sjöbo (169 ha), (**d**) Västerås (165 ha), (**e**) Eskilstuna (149 ha), (**f**) Askersund (206 ha) and (**g**) Linköping (175 ha). All farms are drawn in the same scale, according to the scale bar in (a)

Table 4.7 Ranges of soil OM, Clay, Sand, pH and plant available P at the seven farms

	n	OM (%)	Clay (%)	Sand (%)	pH	P (mg 100 g^{-1})
Askersund	206	1.6–19.0	1–26	1–88	5.5–7.7	1–39
Eskilstuna	149	1.5–11.5	41–70	1–12	6.0–7.4	2.3–12
Linköping	175	2.4–8.3	14–69	2–61	6.2–7.6	3–46
Mariestad	81	1.3–9.5	6–68	4–86	5.5–6.8	1–22
Sjöbo	169	2.8–9.2	11–47	9–60	5.7–7.7	1–34
Storfors	187	2.7–14.1	4–73	1–55	5.5–6.9	1–15
Västerås	165	1.6–8.7	8–56	1–63	5.3–7.1	2.1–47

Prediction models were calibrated on the absorbance spectrum transformed by the first derivative in the Unscrambler 10.3 software (Camo, Norway). For calibration, partial least squares (PLS) regression was applied to the entire spectrum. Calibrations were made for each farm individually on one-third of the samples for which the texture and OM were initially analyzed, i.e. every third sample in sequence. In addition, calibration samples were selected to represent the variations in the NIR spectrum following Kennard and Stone 1969, (KS). Up to half of the total number of samples in the interval of 10 were selected for calibration. The calibrations were subsequently used to determine the corresponding properties of the remaining samples. These samples are referred to as validation samples hereafter. The accuracy of the predictions was assessed with the RMSE of the model and by the coefficient of determination (R^2) of the model.

4.4.3 Results and Discussion

The RMSEs for OM and clay content for all seven farms as a function of the number of calibration samples are shown in Figs. 4.8 and 4.9. For clay and OM, the calibrations performed well with few calibration samples, and with 30–40 calibration samples the errors stabilized with RMSEs of 0.3–0.7% for OM and 2–4% for clay. This represents 5–15% of the mean value which should be compared with the uncertainty of the reference methods that is estimated at approximately 10%. Sand could also be predicted with good accuracy with 40 calibration samples. The Eskilstuna farm was an exception with its very small and variable sand content (Tables 4.7 and 4.8). Phosphorus, and in particular pH, correlated with the NIR spectrum for a few farms. They were, however, weak and very uncertain (Table 4.8). In terms of K and Mg, there were no useful relationships with NIR and the results are not reported in any further detail.

Clay and OM are the variables that are predicted best by NIR because clay minerals and organic matter have strong spectral features and have a large influence on the spectra, (Stenberg et al. 2010). Plant available nutrients or pH as such do not absorb in the near-infrared and any relationships that exist must be indirect. For

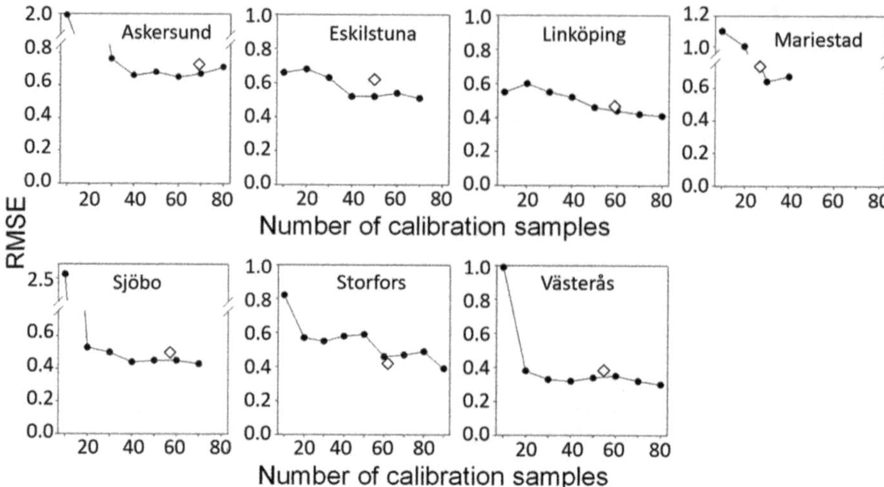

Fig. 4.8 The RMSE of predictions for OM content (%) as a function of number of KS-selected calibration samples. The open diamond represents a calibration model based on every third sample

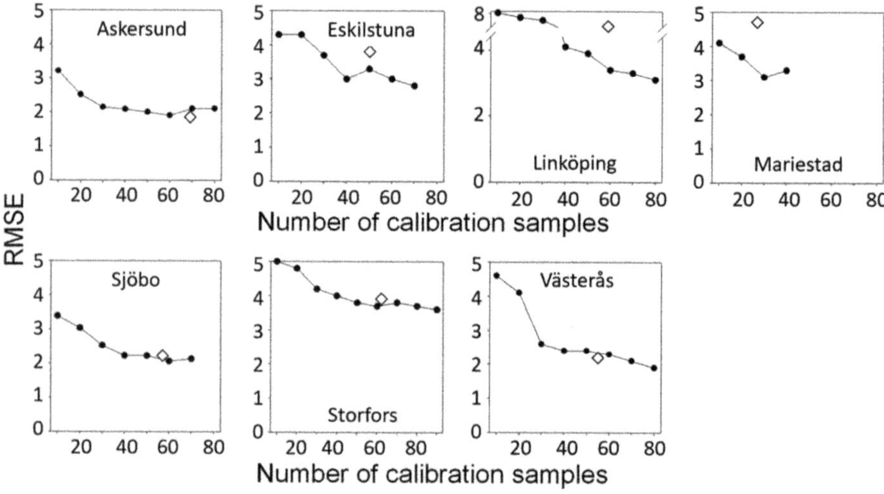

Fig. 4.9 The RMSE of predictions for clay content (%) as a function of number of KS-selected calibration samples. The open diamond represents a calibration model based on every third sample

example, clay and OM buffer pH. Clay and OM are also important for determining soil fertility, which in turn affects the removal of plant nutrients by crops. However, such relationships are highly unreliable and can be cancelled out by intensive fertilisation and liming. The results for pH and available plant nutrients correspond with the results from a previous study where the number of calibration samples needed for stable determinations of K, Mg, P-AL and pH were tested at four Swedish farms (Wetterlind 2012).

Table 4.8 Prediction results: R^2 and RMSE with the calibration based on 40 KS-selected samples

	OM (%)	Clay (%)	Sand (%)	pH	P (mg 100 g^{-1})
Askersund	0.59 (0.7)	0.85 (2)	0.79 (10)	0.22 (0.2)	0.04 (3.0)
Eskilstuna	0.58 (0.5)	0.65 (3)	0.10 (2)	0.47 (0.2)	0.00 (2.4)
Linköping	0.83 (0.5)	0.92 (4)	0.85 (5)	0.42 (0.2)	0.12 (8.5)
Mariestad	0.82 (0.7)	0.95 (3)	0.83 (7)	0.11 (0.2)	0.34 (1.7)
Sjöbo	0.79 (0.4)	0.91 (2)	0.80 (4)	0.39 (0.3)	0.00 (2.8)
Storfors	0.83 (0.6)	0.88 (4)	0.66 (4)	0.11 (0.3)	0.00 (2.7)
Västerås	0.95 (0.3)	0.81 (2)	0.33 (4)	0.03 (0.2)	0.42 (4.0)

As shown in Figs. 4.8 and 4.9, calibrations with 40 samples selected using the KS algorithm were either equally good or significantly better than the calibrations made on every third sample. This is because of a good distribution of calibration samples and the KS method also includes extreme samples in the calibration set and these do not need to be predicted. Extreme samples in the validation set mean that they are not represented by the model, which could mean a certain degree of extrapolation with more uncertain results as a consequence. With every third sample, it is coincidence that determines if the extreme samples end up in the calibration or validation sets. The KS method may involve moderate additional work and above all the NIR spectrum must be taken on all samples before the reference analyses can be made. Nevertheless, our conclusion is that the increased stability of calibrations, especially with few calibration samples, makes it worthwhile.

To estimate the value of the increased soil information a farmer would achieve by applying the proposed strategy with farm calibrations of 40 KS-selected samples, we used site-specific lime requirement as an example and compared the proposed strategy with using one sample per 3 ha which would be the 'business as usual case'. Lime requirement was used because it combines information on clay and OM content. To increase pH one-half unit, lime requirement (in the form of limestone flour, 50%CaO) was calculated as:

$$\text{Lime requirement } \left(\text{ton ha}^{?1}\right) = 0.5\left(1.9 + \left(\left(\left(3.5?\text{OM}\right) + \text{clay content}\right)/3.8\right)\right) \quad (4.1)$$

Figure 4.10 shows the effect of the analysis strategy on the lime requirement map at Västerås, a farm with one of the best NIR-calibration models. Table 4.9 reports the effect of analysis strategy for all farms as the difference with respect to a map interpolated on one sample per ha analyzed by the traditional method (reference map). It should be noted that this reference map is not an exact reflection of reality as one sample per ha is relatively sparse and may not capture all patterns found in the fields. All maps were interpolated using kriging.

Table 4.9 shows that lime requirement maps based on the NIR method reduces the difference from the reference map compared with a map based on one sample per 3 ha. The average difference did not change, but importantly, the most severe deviations were avoided. Where the differences between the two methods were small (e.g. at Sjöbo; Table 4.9), the NIR predictions were not necessarily poor

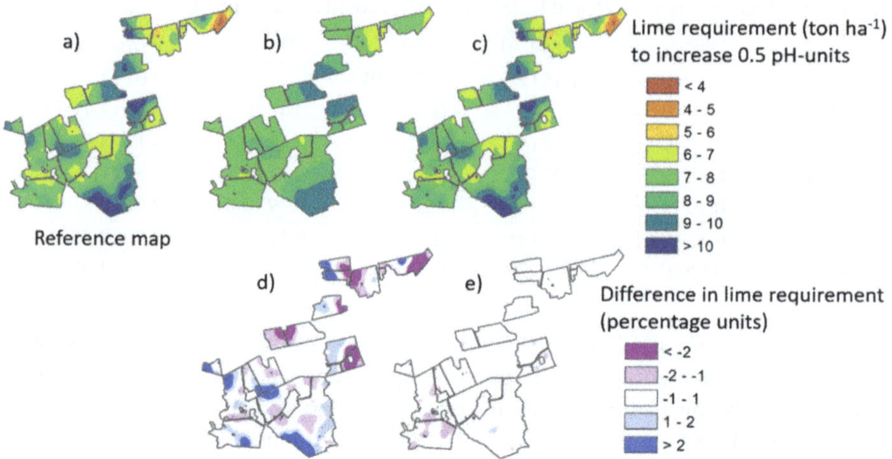

Fig. 4.10 Interpolated lime requirement maps for raising the pH 0.5 points at Västerås, calculated from clay and OM content based on (**a**) traditionally analyzed one sample per ha (165 samples; reference map), (**b**) traditionally analyzed one sample per 3 ha and (**c**) predictions from one sample per ha using 40 KS-selected calibration samples. Map (**d**) shows the difference between the reference map (**a**) and map (**b**), and (**e**) shows the difference between the reference map (**a**) and map (**c**)

(Table 4.8). However, the smaller soil variation meant that going from one sample per 3 ha to one sample per ha did not add much (Table 4.7).

4.4.4 Conclusions

For the seven farms used in this case study, the results show that soil texture and OM content can be analyzed with the same soil sample density as pH and plant available nutrients when the majority of samples are analyzed with the significantly faster and less expensive NIR technique instead of traditional methods. Depending on the variation in soil texture and OM, this method provides more detailed soil maps for site-specific liming or other soil-type related measures such as fertilisation, irrigation and seed rate determination. Even though sand content could not be predicted with the same accuracy and robustness as clay and OM content, the maps produced for sand content were still an improvement in most cases compared with the current situation with analysis on every third sample (maps not shown). Unfortunately, NIR spectroscopy could not produce reliable calibration models for pH and plant available nutrients.

Calculating the economic gain by using the NIR method is difficult because NIR based services are not available commercially. However, sampling and sample preparation have already been paid for via the pH and plant available nutrients analyses. The NIR analysis including sample preparation takes a few minutes and can be partially automated. The NIR instruments cost about €50,000 and are also used for

Table 4.9 Statistics of the difference in lime required (ton CaO ha^{-1} in 5 m × 5 m pixels) between the reference map (traditionally analyzed clay and OM with one sample per ha) and i) a map based on traditionally analyzed samples with one per 3 ha or ii) a map based on one sample per ha predicted with NIR. The smaller the difference the better. The NIR calibrations were based on 40 KS-selected samples

	Difference between the reference map and a map based on 1 sample per 3 ha^{-1}				Difference between the reference map and a map based on NIR predicted samples (KS40)			
	min.	max.	mean	stdev	min.	max.	mean	stdev
Askersund	−2.1	2.6	−0.1	0.49	−0.6	1.0	−0.0	0.22
Eskilstuna	−0.7	0.9	0.0	0.26	−0.5	0.6	0.0	0.18
Linköping	−3.5	2.6	−0.2	0.77	−1.1	1.5	0.2	0.41
Mariestad	−4.5	3.1	0.1	1.03	−1.7	1.0	0.0	0.34
Sjöbo	−1.1	1.7	−0.0	0.29	−0.9	0.5	−0.1	0.19
Storfors	−4.0	4.7	−0.2	0.96	−1.9	3.2	0.0	0.43
Västerås	−3.3	3.4	−0.0	0.86	−0.9	0.9	−0.1	0.24

animal feed and grain analyses, therefore they are often available in agricultural laboratories. The calibration and prediction procedure, including the selection of calibration samples, can be done in a highly standardized manner and should not take more than 5–15 minutes per farm. In addition, it should be possible to automate the process to a large extent.

4.5 Case Study 4.4. Soil-Based Precision Fertilisation Using Near-Infrared spectroscopy for Optimising Cocoa Production

4.5.1 Introduction

Precision soil fertilisation is required to optimize crop production. In Papua New Guinea (PNG), an eight-fold increase in cocoa production for export over the next few years is forecast (Singh et al. 2019). The recommended use of mineral fertilizers (i.e. single annual dosage of 120 g per tree of N: P: K: Mg at a 12:12:17:2 ratio) or organic fertilizers (5 kg $year^{-1}$ per tree - compost applied twice annually) has been proposed to reach the optimal cocoa yield (Konam et al. 2011). Cocoa is grown on soils with significant spatial variability in PNG which means that the 'one-size-fits-all' fertilizer recommendation is not efficient and cannot be used to achieve the target yield (Snoeck et al. 2016; Nelson et al. 2011). Precision fertilisation that ensures application is based on inherent soil properties to correct for deficiencies and replace nutrients that were exported through harvests is therefore desirable (Snoeck et al. 2016). To identify soil nutrient deficiencies, a suite of standard laboratory analyses, which is costly, is required. Another approach to gain soil information is based on near-infrared spectroscopy (NIRS), which can be an alternative to soil analysis to, reduce analysis cost and time (Henaka Arachchi et al. 2016).

The NIRS has been widely used in soil science, but using NIRS-based predictions in soil nutrient models is still unexplored. Therefore, this study's aim was to evaluate the potential of soil-based precision fertilisation using spectral data through a cocoa specific model called 'Soil Diagnosis'.

4.5.2 Materials and Method

4.5.2.1 Soil Sampling and Analysis

Soil samples from four smallholder cocoa farms (each, about 1 ha in area) were investigated in this study (Fig. 4.11 a and b): 1) the BOKA-PAN site is in the region of Bougainville. The soils are formed in recent volcanic ash from an active volcano (Mount Bagana). 2) the LAU-PAN site in the New Ireland Province, a high rainfall area. The soils are derived from limestone, shallow and reddish-brown clay. 3) the TSA-CCI site from the East New Britain Province, with soil derived from an ancient volcanic ash deposit (Andisol), and 4) the WAL-WIN site in the East Sepik Province (ESP). The soils are called 'meadow soils' of WAL-WIN, characterized as young, poorly drained alluvisols (Bleeker 1983; Nelson et al. 2011).

A total of 507 (128 at each site except LAU-PAN where due to shallowness of the soil there were only 123 samples) samples were collected, and typical soil testing procedures described by Blakemore et al. (1987) were conducted on the air-dried <2 mm fraction.

4.5.2.2 Spectral Data Acquisition and Pre-Processing

Spectral reflectance of soil samples was recorded in the laboratory in the visible to near-infrared (Vis-NIR) region (350–2500 nm) with 1-nm intervals using a portable spectroradiometer (Model: Field spec®3 FR, Analytical Spectral Devices Inc., USA).

4.5.2.3 Chemometric Modelling

The data were partitioned randomly into a calibration (80%) and a validation (20%) set using conditioned Latin hypercube sampling based on spectral data from each site. Cubist (a rule-based regression) was used to formulate the spectral models relating soil properties and nutrients with reflectance spectra. The accuracy of the predictions was assessed with RMSE (Bellon Maurel et al. 2011) and the concordance correlation coefficient (LCC), which measures the agreement between measured and predicted samples (Lin 1989).

Fig. 4.11 (**a**) The four sites marked in the main cocoa growing provinces of PNG and (**b**) soil profiles from the four sites (alphabetical order)

4.5.2.4 Fertilizer Recommendation Using Soil Diagnosis Software

Soil diagnosis is an MS Excel spreadsheet application used for computing the amount of nutrients needed for achieving optimum vigour and yield in a given soil (Snoeck et al. 2016). Soil test values for basic soil properties and nutrient contents and desired yield are provided as inputs to this software. The software computes the optimum values of total N, total P and available P by comparison with threshold levels and the (total exchangeable bases)/N ratios at the soil pH. Next, the software computes the required amounts of K, Ca, and Mg by comparing the actual levels in the soil with both the threshold levels and the optimum ratios between them (optimum ratios are: 8% K, 68% Ca and 24% Mg). Recommended doses also depend on the levels of base saturation (optimum: 60 to 100%). The user can accept the suggested data or can adjust them depending of specific local constraints. When the soil nutrient levels are corrected, additional amounts of fertilizers are computed to compensate for increased cocoa yields. The model was set to maintain an average yield of 1000 kg ha^{-1} of dry cocoa beans.

4.5.3 Results and Discussion

4.5.3.1 Descriptive Statistics

Overall, there was a large variation in soil properties over the four sites, particularly the texture and organic matter content which influenced cation exchange capacity (CEC). The WAL-WIN soil samples had larger CEC and SOC compared to soil samples from the other three sites (Table 4.10). With respect to available nitrogen (N), in general areas with higher rainfall are associated with higher N content (Bleeker 1983), particularly in the case of LAU-PAN. The large available phosphorous (P) content for the BOKA-PAN soils is linked to the freshly deposited and unweathered volcanic ash soil (Shoji and Takahashi 2002; Buytaert et al. 2007). Further, large exchangeable potassium (K) concentrations in the old volcanic soil of TSA-CCI are not surprising (Bleeker 1983). A clear soil textural association can be seen between clay values and calcium (Ca) content. These results suggest that soils are highly variable across provinces, which is of importance when creating soil spectral libraries.

4.5.3.2 Prediction of Soil Properties

Most soil properties were estimated accurately (LCC ≥ 0.7) using the Cubist model, except for pH, EC and Ex. K (Fig. 4.12). The predictions were excellent because (1) the calibration dataset was a good representation of the total dataset used (Shenk et al., 1991; Kuang and Mouazen 2012) and (2) the variability (Soriano-Disla et al. 2014; Ng et al. 2019) in properties of the four different soils was significant (Table 4.10).

Table 4.10 Mean and coefficient of variation (CV) expressed in percentage (shown in brackets) for selected soil properties of the four sites

Soil properties	BOKA-PAN	LAU-PAN	TSA-CCI	WAL-WIN
Sand, %	89 (5)	45 (53)	80 (9)	46 (28)
Clay, %	**5 (43)**	41 (60)	13 (45)	28 (35)
pH	6.63 (6)	6.48 (6)	7.10 (4)	6.61 (7)
EC, μ Scm^{-1}	113 (93)	169 (99)	132 (80)	112 (73)
SOC, %	**0.99 (99)**	1.38 (73)	1.22 (88)	1.94 (52)
CEC, cmolc kg^{-1}	19.0 (56)	34.0 (25)	28.3 (31)	42.3 (27)
BS, %	**27 (63)**	**31 (47)**	57 (29)	51 (31)
Avl. N, kg ha^{-1}	**132 (78)**	361 (60)	**174 (68)**	219 (35)
Avl. P, kg ha^{-1}	123 (74)	18 (61)	30 (62)	11 (89)
Ex. K, cmolc kg^{-1}	**0.15 (39)**	**0.19 (59)**	1.56 (71)	0.24 (33)
Ex. Ca, cmolc kg^{-1}	**3.20 (76)**	8.17 (53)	**1.68 (47)**	13.02 (32)
Ex. Mg cmolc kg^{-1}	**0.66 (30)**	1.73 (98)	11.70(39)	7.38 (57)

EC electrical conductivity, *SOC* soil organic carbon, *CEC* cation exchange capacity, *BS* base saturation, *Mg* magnesium, *Avl.* available, *Ex.* exchangeable. In bold are the values below lower thresholds (for thresholds, see Snoeck et al. 2016)

4.5.3.3 An Economically Viable Option for Measuring Soil Properties

The samples have captured variation within and across the four sites, which explains the large LCC values (Fig. 4.12). Therefore, the predicted soil values can be used in the fertilizer model to create a fertilization recommendation. Suppose only 100 samples were analyzed in the laboratory and used to create NIRS prediction models ($5400) compared to 500 soil samples analyzed in the laboratory for mapping of soil properties based on traditional methods alone ($27,000), there would be a saving of $21,600. This is a substantial sum in a developing country like PNG. Over time, this would pay for a Vis-NIR spectrometer (approximately $60,000).

4.5.3.4 Soil Fertilizer Recommendation

The soil diagnosis model was used to compute fertilizer requirements for each of the sites from the soil chemical analysis (Fig. 4.13a) and NIRS predicted values (Fig. 4.13b). The model calculates the nutrient requirements by the difference between the cocoa cultivar's needs and the amount that is available in the soil (Snoeck et al. 2016). Soil diagnosis computes the necessary amounts of nutrients by taking into account the thresholds and the balances between the nutrient ratios. For example, the need for N was calculated to be greater than the threshold and according to the "Sum of cations:N" ratio.

Similarly, P requirement was calculated to be greater than the threshold and dependent on the N:P ratio. The needs for K, Ca and Mg were calculated from the thresholds of each of the cations (as a function of the pH) and the optimum K:Ca:Mg

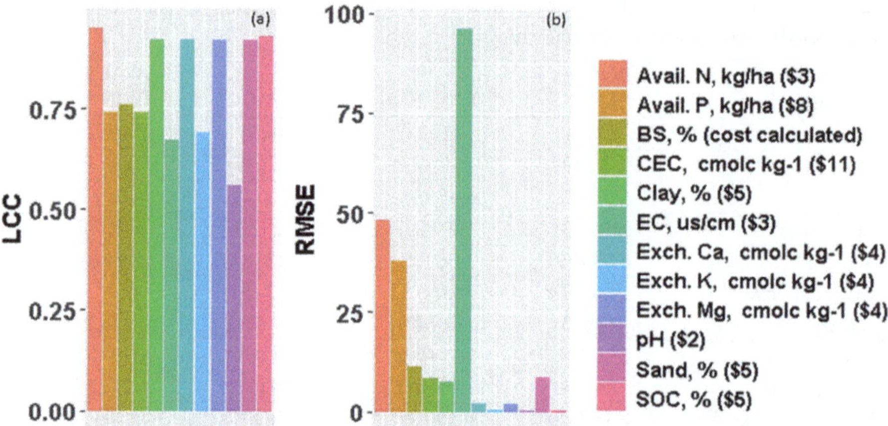

Fig. 4.12 Validation results for Cubist with (**a**) concordance correlation coefficient (LCC) for soil properties (in alphabetical order) and (**b**) root mean squared error (RMSE). The cost of analysis per sample in US $ is shown in brackets in the legend

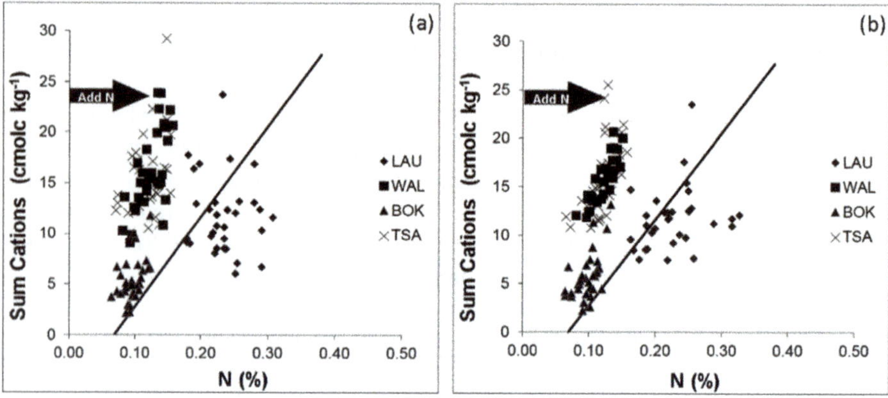

Fig. 4.13 The sum of cations: N ratio is important for computing fertilizer formulae for cocoa trees using the soil diagnosis method. Comparison of results obtained from the chemical (**a**) and spectral (**b**) data. Note: Soils above the diagonal line should respond positively to nitrogen fertilisation

ratio is 8:68:24, with a base saturation above 60 %. It can be seen that the points are grouped soil types with WAL-WIN and TSA-CCI having the closest nutrient balance (Fig. 4.13). This demonstrates the site-dependent nutrient availabilities, and that precision fertilisation is required for each site.

The young volcanic ash soils of BOKA-PAN had a low base saturation and nutrient content, yet large available P contents requiring the addition of most nutrients except P. The more weathered volcanic ash soils of ENB had a large available K, and medium to low P requiring Ca, Mg and P targeted fertilizer to balance optimum

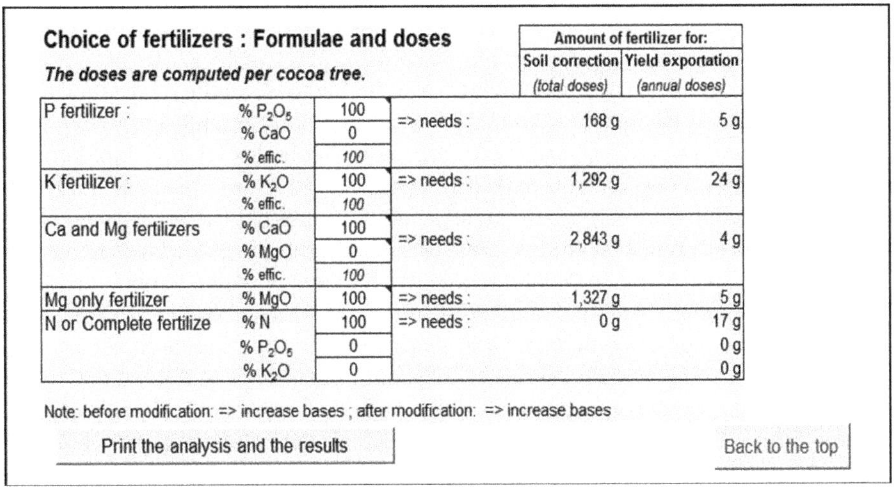

Choice of fertilizers : Formulae and doses

The doses are computed per cocoa tree.

				Amount of fertilizer for: Soil correction (total doses)	Amount of fertilizer for: Yield exportation (annual doses)
P fertilizer :	% P_2O_5	100	=> needs :	168 g	5 g
	% CaO	0			
	% effic.	*100*			
K fertilizer :	% K_2O	100	=> needs :	1,292 g	24 g
	% effic.	*100*			
Ca and Mg fertilizers	% CaO	100	=> needs :	2,843 g	4 g
	% MgO	0			
	% effic.	*100*			
Mg only fertilizer	% MgO	100	=> needs :	1,327 g	5 g
N or Complete fertilize	% N	100	=> needs :	0 g	17 g
	% P_2O_5	0			0 g
	% K_2O	0			0 g

Note: before modification: => increase bases ; after modification: => increase bases

Print the analysis and the results

Back to the top

Fig. 4.14 A fertilizer recommendation for LUA-PAN with low base saturation, and low K, Mg and Ca compared to N from the soil diagnosis model

uptake of nutrients. The shallow LAU-PAN soils with large N content required nutrients other than N (Fig. 4.13). Lastly, the poorly drained soils of WAL-WIN required soil physical correction first to target drainage issues, followed by chemical application.

Figure 4.14 demonstrates the fertilizer calculation for the LAU-PAN site, where the soils had low base saturation and low K, Mg and Ca compared to N. Therefore, N addition is not required for soil correction (Fig. 4.13), but a small annual dose was required because of nutrient export (Fig. 4.14). The recommendation was mainly for Ca, Mg, K and a small amount of P. The total doses are usually added every 5 years for soil correction. In contrast, annual doses are needed to replace the nutrients lost through yield export. Thus, if the average formula at a site is known related to its pedoclimatic conditions, the precision of NIRS prediction is sufficient to adjust the formula to the plot context where similar biophysical conditions exist.

4.5.4 Conclusion

This study demonstrated that NIRS can be used to predict soil properties on four different soils growing cocoa in PNG. A significant drop in the cost of laboratory analyses is possible if only a proportion of the samples are analyzed in the laboratory and the rest predicted using NIRS. This information can be fed into models such as 'Soil Diagnosis' for fertilizer recommendation and in future to test agronomy-based nutrient trials.

4.6 Conclusions for the Chapter

The overview of soil sensing technologies and the four case studies show that, available soil sensor systems allow both spatial and temporal monitoring of physical, chemical and or biological soil attributes. Because of complex relationships between individual sensor measurements and an array of soil properties, localized modelling of the relationship between sensor signals and conventional (laboratory) measurements of these attributes is required. Sensor fusion and the use of advanced analytical techniques, including artificial intelligence, have the potential to obtain accurate soil data with a relatively small number of soil samples for calibration. Today, many service providers produce thematic soil maps commercially based on sensor data. Thus, apparent soil electrical conductivity, gamma-ray spectroscopy, subsoil reflectance, and other on-the-go sensors have been used to map physical soil attributes successfully such as soil texture and organic matter content at a relatively fine field scale. More evolved radiometric techniques, including spectroscopy in different parts of the electromagnetic spectrum as well as electrochemical sensors have helped to improve the efficiency of determining changes in chemical soil attributes in both *in situ* and *ex situ* environments. Contemporary interest in soil health motivates researchers and technology developers to further explore the rapid measurement of biological soil properties. Telemetry monitoring stations have been efficient tools for time series analysis of many soil processes. In addition, field topography and vegetation sensing have become complementary elements of soil sensing practices. To deepen our understanding further of the entire spatio-temporal dynamics of key soil processes within the rooting zone and below, new sensor designs as well as new analytical tools are still being explored around the globe.

Acknowledgments Case study 4.1 was supported by funds from the Natural Science and Engineering Research Council of Canada (NSERC) Discovery Grant and through Ontario Ministry of Agriculture, Food and Rural Affairs (OMAFRA) New Directions Research Program (NDRP). Special thanks to Paul Hermans of DuPont Pioneer (Richmond, Ontario, Canada) for providing Veris MSP data and to Jeremy Nixon of J & H Nixon Farms and Michael Schouten of Schouten Dairy Farms. The authors acknowledge the contributions of the past and current PA and Sensor Systems (PASS) research team members Dr. Wenjun Ji, Dr. Long Qi, Maxime Leclerc, Dr. Yuanyuan Fu, and Matthieu Claustre for performing soil sampling.

The authors of case study 4.2 are very grateful for the assistance of Mr. M. Botha, Dr. D. Ronga and Dr. D. Cammarano with the data collection for this study. Dr. Philippe Lagacherie provided assistance with the RFA modelling and Mr. Rick Taylor (DUALEM Inc.) calculated the integrated ECa variable. Financial support was provided from the Scottish Society of Crop Research for soil sample analysis costs.

For Case study 4.3, Stiftelsen Lantbruksforskning is acknowledged for funding project H1033307 and Eurofins Agro, Kristianstad, Sweden, for providing data and selection of farms.

Case study 4.4 is supported by the Australian Centre for International Agricultural Research project 'Optimising soil management and health in PNG integrated Cocoa farming systems'; Field D; Australian Centre for International Agricultural Research (ACIAR) Research and Development Programs (R&D Programs).

References

Adamchuk VI, Hummel JW, Morgan MT et al (2004) On-the-go soil sensors for precision agriculture. Comput Electron Agric 44(1):71–91

Adamchuk VI, Ferguson RB, Hergert GW (2010) Soil heterogeneity and crop growth. In: Oerke EC, Gerhards R, Menz G, Sikora RA (eds) Precision crop protection – the challenge and use of heterogeneity. Springer, pp 3–16

Adamchuk VI, Jonjak AK, Wortmann CS et al (2011a) Case studies on the accuracy of soil pH and lime requirement maps. In: Stafford J (ed) Precision Agriculture: Papers from the 8th European Conference on Precision Agriculture, Prague, Czech Republic, 11–14 July 2011, pp 289–301

Adamchuk VI, Viscarra Rossel RA, Marx DB et al (2011b) Using targeted sampling to process multivariate soil sensing data. Geoderma 163(1):63–73

Adamchuk VI, Viscarra Rossel RA, Sudduth KA et al (2011c) Sensor fusion for precision agriculture. In: Thomas C (ed) Sensor fusion – foundation and applications. InTech, Rijeka, pp 27–40

Adamchuk VI, Allred B, Doolittle J et al (2017) Tools for proximal soil sensing. In: Ditzler C, Scheffe K, Monger HC (eds) Soil survey manual, USDA handbook 18. Government Printing Office, Washington, DC, pp 355–394

Adamchuk V, Ji W, Viscarra Rossel R et al (2018) Proximal soil and plant sensing. In: Shannon DK, Clay DE, Kitchen NR (eds) Precision agriculture basics. ASA-CSSA-SSSA, Madison, Wisconsin, pp 119–140

Akaike H (1973) Information theory and an extension of the maxi-mum likelihood principle. In: Petrov BN, Caski F (eds) Proceedings of the second international symposium on information theory. Akademiai Kiado, Budapest, pp 267–281

Bellon-Maurel V, McBratney A (2011) Near-infrared (NIR) and mid-infrared (MIR) spectroscopic techniques for assessing the amount of carbon stock in soils – critical review and research perspectives. Soil Biol Biochem 43(7):1398–1410

Blakemore LC, Searle PL, Daly BK (1987) Methods for chemical analysis of soils. NZ Soil Bur Sci Rep 80. Lower Hutt New Zealand

Bleeker P (1983) Soils of Papua New Guinea. Soils Papua New Guinea, Canberra

Brungard CW, Boettinger JL, Duniway MC et al (2015) Machine learning for predicting soil classes in three semi-arid landscapes. Geoderma 239:68–83

Brus DJ, de Gruijter JJ (1997) Random sampling or geostatistical modelling? Choosing between design-based and model-based sampling strategies for soil (with discussion). Geoderma 80(1–2):1–44

Brus DJ, de Gruijter JJ, Van Groenigen JW (2006) Designing spatial coverage samples using the k-means clustering algorithm. Dev Soil Sci 31:183–192

Buytaert W, Deckers J, Wyseure G (2007) Regional variability of volcanic ash soils in South Ecuador: the relation with parent material, climate and land use. Catena 70(2):143–154

Castanedo F (2013) A review of data fusion techniques. Scientific World Journal 2013:19 p

Dhawale N, Adamchuk VI, Huang HH et al (2016) Integrated analysis of multilayer proximal soil sensing data. In: Proceedings of the 13th International Conference on Precision Agriculture, St. Louis, Missouri, USA. July 31 – August 4, 2016

Dong J, Zhuang D, Huang Y et al (2009) Advances in multi-sensor data fusion: algorithms and applications. Sensors 9(10):7771–7784

Fridgen JJ, Kitchen NR, Sudduth KA et al (2004) Management Zone Analyst (MZA): software for subfield management zone delineation. Agron J 96(1):100–108

Gee GW, Bauder JW (1986) Particle-size analysis. In: Klute A (ed) Physical and mineralogical methods, 2nd edn. Soil Science Society of America, Madison, pp 383–411

Grunwald S, Thompson JA, Minasny B et al (2012) Digital soil mapping in a changing world. In: Digital soil assessments and beyond. CRC Press, pp 301–305

Guerrero C, Wetterlind J, Stenberg B et al (2016) Do we really need large spectral libraries for local scale SOC assessment with NIR spectroscopy? Soil Tillage Res 155:501–509

Hawes C, Alexander CJ, Begg GS et al (2018) Plant responses to an integrated cropping system designed to maintain yield whilst enhancing soil properties and biodiversity. Agronomy 8:229
Henaka Arachchi MPNK, Field DJ, McBratney AB (2016) Quantification of soil carbon from bulk soil samples to predict the aggregate-carbon fractions within using near- and mid-infrared spectroscopic techniques. Geoderma 267(1):207–214
Hengl T, Rossiter DG, Stein A (2003) Soil sampling strategies for spatial prediction by correlation with auxiliary maps. Aust J Soil Res 41:1403–1422
Hengl T, de Jesus JM, MacMillan RA et al (2014) SoilGrids1km? Global soil information based on automated mapping. PLoS One 9:e105992
Holland JE, Biswas A, Huang J et al (2017) Scoping for scale-dependent relationships between proximal gamma radiometrics and soil properties. Catena 154:40–49
Hummel JW, Gaultney LD, Sudduth KA (1996) Soil property sensing for site-specific crop management. Comput Electron Agric 14:121–136
Ji W, Adamchuk V, Chen S et al (2019) Simultaneous measurement of multiple soil properties through proximal sensor fusion: a case study. Geoderma 341:111–128
Jiang Q, Peng J, Biswas A et al (2019) Characterising dryland salinity in three dimensions. Sci Total Environ 682:190–199
Kaufmann MS, von Hebel C, Weihermüller L et al (2020) Effect of fertilizers and irrigation on multi-configuration electromagnetic induction measurements. Soil Use Manage 36:104–116
Kennard RW, Stone LA (1969) Computer aided design of experiments. Dent Tech 11(1):137–148
Kim J, Grunwald S, Rivero RG et al (2012) Multi-scale modeling of soil series using remote sensing in a wetland ecosystem. Soil Sci Soc Am J 76(6):2327–2341
Konam J, Namaliu Y, Daniel R et al (2011) Integrated pest and disease management for sustainable cocoa production: a training manual for farmers and extension workers, 2nd edn. ACIAR Monograph No. 131. Aust Cent Int Agric Res, Canberra, p 36
Kuang B, Mouazen AM (2012) Influence of the number of samples on prediction error of visible and near infrared spectroscopy of selected soil properties at the farm scale. Eur J Soil Sci
Kuang B, Mahmood HS, Quraishi MZ et al (2012) Sensing soil properties in the laboratory, in situ, and on-line. Adv Agron 114:155–223
Lin LI-K (1989) A concordance correlation coefficient to evaluate reproducibility. Biometrics
MAFF (1986) The analysis of agricultural materials, Reference Book 427. HMSO, London
Mahmood HS, Hoogmoed WB, Henten EJ (2012) Sensor data fusion to predict multiple soil properties. Precis Agric 13:628–645
Mahmood H, Hoogmoed W, van Henten E (2013) Proximal gamma-ray spectroscopy to predict soil properties using windows and full-spectrum analysis methods. Sensors 13(12):16263
McBratney AB, Santos MLM, Minasny B (2003) On digital soil mapping. Geoderma 117(1–2):3–52
Minasny B, McBratney AB (2006) Latin hypercube sampling as a tool for digital soil mapping. Dev Soil Sci 31:153–606
Nelson PN, Webb MJ, Berthelsen S et al (2011) Nutritional status of cocoa in Papua New Guinea. ACIAR Technical Reports No. 76. Aust Cent Int Agric Res, Canberra, p 67
Ng W, Minasny B, Montazerolghaem M et al (2019) Convolutional neural network for simultaneous prediction of several soil properties using visible/near-infrared, mid-infrared, and their combined spectra. Geoderma
Peters J (ed) (2006) Standards for electromagnetic induction mapping in the grains industry. Grains Research and Development Corporation, Australian Government
Saifuzzaman M, Adamchuk V, Buelvas R et al (2019) Clustering tools for integration of satellite remote sensing imagery and proximal soil sensing data. Remote Sens (Basel) 11(9):1036
Shenk JS, Westerhaus MO (1991) Population definition, sample selection, and calibration procedures for near infrared reflectance spectroscopy. Crop Sci
Shibusawa S (2006) Soil sensors for precision agriculture. In: Srinivasan A (ed) Handbook of precision agriculture: principles and applications. CRC Press, New York

Shoji S, Takahashi T (2002) Environmental and agricultural significance of volcanic ash soils. Glob J Environ Res 6:113–135
Singh K, Majeed I, Panigrahi N et al (2019) Near infrared diffuse reflectance spectroscopy for rapid and comprehensive soil condition assessment in smallholder cacao farming systems of Papua New Guinea. Catena 183:1–14
Smith CAS, Daneshfar B, Frank G (2012) Use of weights of evidence statistics to define inference rules to disaggregate soil survey maps. In: Minasny B, Malone BP, McBratney A (eds) Digital soil assessments and beyond. CRC Press, Sydney, pp 215–220
Snoeck D, Koko L, Joffre J et al (2016) In: Lichtfouse E (ed) Sustainable agriculture reviews. Springer, pp 155–202
Soriano-Disla JM, Janik LJ, Viscarra Rossel RA et al (2014) The performance of visible, near-, and mid-infrared reflectance spectroscopy for prediction of soil physical, chemical, and biological properties. Appl Spectrosc 49(2):139–186
Stenberg B, Viscarra Rossel RA (2010) Diffuse reflectance spectroscopy for high-resolution soil sensing. In: Proximal Soil Sensing. Springer, pp 29–47
Stenberg B, Viscarra Rossel RA, Mouazen AM et al (2010) Visible and near infrared spectroscopy in soil science. Adv Agron 107:163–215
Sudduth KA, Hummel JW, Birrell SJ (1997) Sensors for site-specific management. In: Pierce FJ, Sadler EJ (eds) The state of site-specific management for agriculture. ASA-CSSA-SSSA, Madson, pp 183–210
Taghizadeh-Mehrjardi R, Nabiollahi K, Kerry R (2016) Digital mapping of soil organic carbon at multiple depths using different data mining techniques in Baneh region, Iran. Geoderma 266:98–110
Taylor JA, Whelan BM, McBratney AB (2007) Establishing broadacre management classes. Agron J 99:1366–1376
Taylor JA, Short M, McBratney AB et al (2010) Comparison of the ability of multiple soil sensors to predict soil properties in a Scottish potato production system. In: Viscarra Rossel RA, McBratney AB, Minasny B (eds) Proximal soil sensing. Progress in Soil Science series. Springer. ISBN: 978-90-481-8858-1
Thomas GW (1983) Exchangeable cations. Methods of soil analysis: part 2 chemical and microbiological properties 9:159–165
Triantafilis J, Lesch SM, La Lau K et al (2009) Field level digital soil mapping of cation exchange capacity using electromagnetic induction and a hierarchical spatial regression model. Aust J Soil Res 47:651–663
Triantifilis J, Earl NY, Gibbs ID (2012) Digital soil-class mapping across the Edgeroi district usings numerical clustering and gamma-ray spectrometry data. In: Minasny B, Malone BP, McBratney A (eds) Digital soil assessments and beyond: proceedings of the 5th global workshop on digital soil mapping. CRC Press, Sydney, pp 187–191
Vašát R, Heuvelink GBM, Borůvka L (2010) Sampling design optimization for multivariate soil mapping. Geoderma 155(3–4):147–153
Viscarra Rossel RA, Adamchuk VI, Sudduth KA et al (2011) Proximal soil sensing: an effective approach for soil measurements in space and time. Adv Agron 113:237–283
Webster R, Oliver MA (2007) Geostatistics for environmental scientists. Wiley, Chichester
Weindorf DC, Bakr N, Zhu Y (2014) Advances in portable X-ray fluorescence (PXRF) for environmental, pedological, and agronomic applications. In: Sparks DL (ed) Advances in agronomy. Academic Press, San Diego, pp 1–45
Wetterlind J (2012) Project final report (in Swedish) to Stiftelsen Svensk Växtnäringsforskning, KSLA, H09-0011-SVX
Wetterlind J, Stenberg B (2010) Near-infrared spectroscopy for within-field soil characterization: small local calibrations compared with national libraries spiked with local samples. Eur J Soil Sci 61:823–843
Witten IH, Frank E, Hall MA et al (2017) Credibility: evaluating what's been learned. In: Data mining, 4th edn. Morgan Kaufmann, pp 161–203

WRB (2006) World Reference Base for Soil Resources. World Soil Resources Reports No. 103. FAO, Rome
Xu D, Chen S, Viscarra Rossel AR et al (2019) Vis NIR and XRF sensor fusion for predicting paddy soil chromium content. Geoderma 352:61–69
Zalik KR (2008) An efficient k-means clustering algorithm. Pattern Recog. Lett 29(9):1385–1391
Zani C, Gowing J, Abbott GD et al (2020) Grazed temporary grass-clover leys in crop rotations can have a positive impact on soil quality under both conventional and organic agricultural system. Euro. J, Soil Sci
Zhang Y, Biswas A, Ji W et al (2017) Depth-specific prediction of soil properties in situ using vis-NIR spectroscopy. Soil Sci Soc Am J 81(5):993–1004

Chapter 5
Sensing with Wireless Sensor Networks

Vasileios Liakos and George Vellidis

Abstract Wireless sensor networks (WSNs) are a group of interconnected autonomous sensor nodes that use wireless communication technology. Their goal is to monitor physical or environmental conditions and to transmit their data cooperatively to a base station. The WSNs are used widely and have proved particularly useful in difficult environments where wired networks cannot be deployed. This feature together with their ability to adapt to changing environments, and the low cost of installing and maintaining them are the main characteristics that make them good sources of data in agricultural settings. The use of WSNs can provide farmers with valuable information on field properties, which can help them understand field variability and make better and quicker decisions on applying agricultural inputs such as pesticides, irrigation and fertilizers. Information on WSNs presented in this chapter is designed to provide potential users with the main concepts of WSNs and the ability to select the best WSN properly for their needs. The chapter is divided into three sections: (5.1) Concepts of Sensing with WSNs; (5.2) Applications of WSNs in Agriculture and (5.3) Conclusions and Future of WSNs. Section 5.1 presents the importance of WSNs. Most of the available wireless communication standards and protocols are described and their performance is compared. Section 5.2 presents the sensors and electronics that are available in the market. Furthermore, it describes and presents examples of applications of WSNs in agriculture such as irrigation, fertilizer application, pest management, crop and weather monitoring. Section 5.3 discusses the importance of WSNs for agricultural applications. Use of WSNs will improve overall agricultural system performance and will propel us closer to precision agriculture and smart farming, by providing the data that will allow farmers to apply the right amount of resources at the right time and in the right place. Future improvements to WSNs will make them more efficient, less complex and self-healing.

V. Liakos (✉)
Crop and Soil Sciences Department, University of Georgia, Athens, GA, USA

PEARL Lab, Department of Agriculture - Agrotechnology, University of Thessaly, Larissa, Greece
e-mail: vliakos@uga.edu

G. Vellidis
Crop and Soil Sciences Department, University of Georgia, Athens, GA, USA
e-mail: yiorgos@uga.edu

R. Kerry, A. Escolà (eds.), *Sensing Approaches for Precision Agriculture*, Progress in Precision Agriculture, https://doi.org/10.1007/978-3-030-78431-7_5

Keywords Smart node · Review · IEEE standards · Precision irrigation

5.1 Concepts of Sensing with Wireless Sensor Networks

5.1.1 Introduction

Rapidly increasing technological developments have resulted in communication technology, nanotechnology and small microchips. Each of these developments has contributed to the establishment and use of wireless sensor networks (WSNs) and smart sensors that are used in a variety of applications. A complete WSN in its simplest form consists of a sensor network made up of sensor nodes to record at least one property and a central gateway known as the base station that collects data from the sensor network. Data transfer among sensor nodes of the same network and between the sensor network and the base station requires the utilization of communication protocols.

A sensor node, sometimes referred to as a mote, consists of three main components: the electronics, the sensor(s) and the antenna. The electronics are small-embedded computers that are responsible for collecting data from sensors, communicating with other sensor nodes and performing some processing, e.g. calibration of sensor readings based on ambient temperature. Sensor response is typically in the form of an electrical property such as voltage, current, or resistance and mathematical formulae are applied to transform the raw data to a measurement unit associated with the property being measured by the sensor. Hence, smart nodes are equipped with embedded computers that run software, which converts the output of the sensors into physical units. For example, some soil moisture sensors respond to changing soil moisture conditions by physically measuring the resistance between two conductors. The change in resistance is measured by the change in the excitation signal which the node captures, compensates for temperature using software and transforms to a measurement in terms of matric potential in units of kPa. In addition, embedded computers run software to establish communication among sensor nodes and the base station. Each sensor node has an antenna to ensure proper connectivity with other sensor nodes and or the base station within the WSNs.

Even though WSNs use embedded computers to accomplish their tasks, the architecture of the network is different from that of other computer networks. A network makes information and services available to anyone on the network regardless of their location. There are many network types such as wide area networks (WAN), local area networks (LAN), personal area networks (PAN) and metropolitan area networks (MAN). The main difference between these types of networks is the arrangement of the elements of the communication network or the topologies and the protocols they use.

5.1.2 Principles of Wireless Sensor Networks in Agriculture

Sensor networks were initially developed some decades ago to monitor information and control systems. Their initial design included hard-wired sensors and a central computer connected through a data bus network (Fig. 5.1). However, the installation cost of hard-wired sensors is large because of the labor, cables, testing and maintenance. In addition, with time wired sensor networks have proved to be unreliable because connectors can become loose, break or are misconnected. These problems are more frequent when sensor networks are installed outdoors. For example, it is cumbersome to use hard-wired sensors in a field for crop monitoring because either regular field activities are impeded by the cables or the cables are damaged by the activities. The WSNs eliminate many of these problems. They enable connectivity of sensors without the need for wires (Gutierrez et al. 2004). This means that this type of network collects data from wireless sensors and sends data to a terminal. The terminal of a WSN can have several forms. For example, it can be an actuator to control a system, a screen to present the information or even a server to store and manipulate the information. Furthermore, the development of micro electro mechanical systems (MEMS) and integrated circuits (IC) contributed to the development of sensor nodes (Schurgers and Srivastava, 2001). Sensor nodes can utilize wireless communication, read sensors and process the sensor readings at the same time because they use a processor. Additionally, sensor nodes are small, low-cost and require a small amount of power for their operation. More advantages and disadvantages of wireless networks are presented in Table 5.1. These characteristics make them ideal for outdoor use compared to hard-wired sensors. A WSN consists of many sensor nodes that can record pressure, humidity, soil properties, temperature, noise and many other variables.

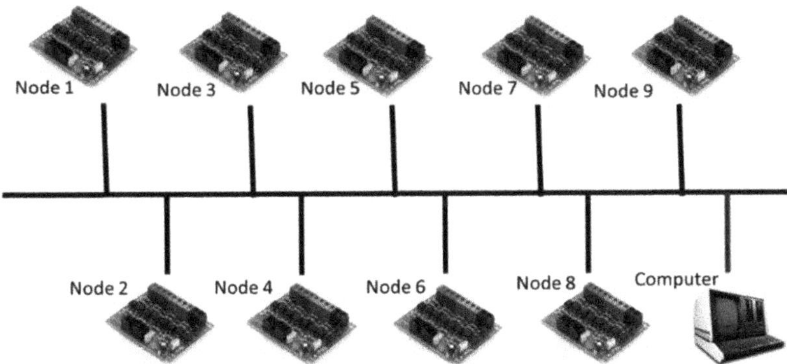

Fig. 5.1 Schematic diagram of a bus network

Table 5.1 Advantages and disadvantages of wireless and wired networks

Activity	Wired network	Wireless network
Use of cables	Yes	No
Connection speed	Faster than wireless	Slower than wired
Security	More secure than wireless.	Less secure than wired
Data transfer	High capacity	Capacity may be limited
Flexibility	Low to moderate	Moderate to high

5.1.2.1 Wireless Communication Standards and Protocols

Sensor nodes within a WSN exchange data with each other and with other devices such as data loggers and laptop computers. However, WSNs when developed must be based on standards to ensure communication among devices and other nodes. These standards are categorized into four types based on their specific application and transmission range. They are:

- The IEEE 802.15 standard that is used for wireless personal area networks (WPAN; Valentini et al. 2010; Mansourkiaie and Ahmed 2016, Luo et al. 2015);
- the IEEE 802.11 standard that is used for wireless local area networks (WLAN; Lin et al. 2007; Shen et al. 2015; Meng et al. 2016, Gupta et al. 2015);
- the IEEE 802.16 standard, known as worldwide interoperability for microwave access networks (WiMAX) and also known as the wireless metropolitan area networks (WMAN; Akashdeep et al., 2014); and
- the IEEE 802.20 standard that is used for wireless wide area networks (WWAN; Zou et al., 2004).

Table 5.2 presents the differences in the four wireless communication standards. Communication protocols are formal descriptions of digital message formats and rules and can include information about authentication, signaling, error detection and correction. In addition, they can describe the syntax and synchronization of analogue and digital communications. There are currently nine wireless communication protocols used for WSNs (Table 5.3). The reason why so many are used is that each WSN application requires a specific combination of data rate, frequency, power consumption and reception range. It is very common in agriculture to use the IEEE 802.15 standard with the IEEE.802.15.4/ZigBee protocol because the power requirements are low, the range of distance for transmitting data is acceptable for agricultural applications and the cost is small. The IEEE.802.15.4 was developed in 2003 for low-power wireless personal area networks (LP-WPAN) to overcome problems related to Wi-Fi and Bluetooth such as power consumption and reception range. Based on the IEEE 802.15.4 standard, it supports two categories of devices. The first is called a reduced function device (RFD) that is responsible for executing simple tasks. This type of device is not standalone, which means that they cannot be used to manage a network. They usually implement a sensing task (e.g. read sensors), report the sensor readings and go to sleep mode to save power. However, RFDs cannot exchange data with other RFDs but they can ‘talk’ to full-function

Table 5.2 Main characteristics of wireless communication standards

Protocol	Standard	Range (m)	Band (GHz)	Data rate (Mbps)	Topology	Power consumption
Wi-fi	IEEE.802.11a	5000	5	54	Star, tree	Very high
Wi-fi 4	IEEE 802.11n	250	2.4	600	Star, tree	Very high
WiMAX	IEEE 802.16	15,000	2.3–5.8	75	Star, tree	High
Bluetooth	IEEE 802.15.1	10	2.4	1–3	Star	High
Bluetooth 4.0 L.E	IEEE 802.15.1	100	2.4	2	Star, mesh	High
Bluetooth 5.0 L.E	IEEE 802.15.1	400	2.4	2	Star, mesh	High
ZigBee	IEEE 802.15.4	100	2.4	0.5	Star, mesh	Low
Z-wave	IEEE 802.15.4	30	0.8–0.9	0.2	Star, mesh	Low
6LoWPAN	IEEE 802.15.4	100	2.4	0.256	Star, mesh	Low
LoRaWAN	201R0	20,000	0.8–0.9	0.05	Star of star	Low

devices (FFD). The FFDs are the second type of device mentioned above and are responsible for tasks that are more complicated. They can run software and also manage networks. Most common devices for outdoor applications are RFDs because they require simple programming code and they are able to set themselves into sleep mode for a certain period to save power before waking up again for the next round of sensing.

The use of the internet of things (IoT) in agriculture increased users' demand for wireless networks that have large coverage, large scale deployment and require low power. To meet these needs, the long range wide area network (LoRaWAN) standard (commonly abbreviated as LoRa) was developed by the LoRaWAN Alliance. It is used for bi-directional communication on wireless sensor networks. The range of the transmitted data rates varies from 0.3 to 50 kbps and depends on the spreading factor (SF) and the adaptive data rate (ADR) algorithm. The aim of the ADR is to select the ideal SF to reduce power consumption and optimize the bandwidth. The LoRa also uses a network server, which manages the communication of nodes and gateways and defines their SF. Nodes transmit data to more than one gateway and then the gateways transmit them to a network server. The network server transmits the data continuously to a specific application server. Communications between nodes and gateways are established in LoRa while communications between gateways and network servers utilize internet protocol (IP) such as WiFi or 5G. The main disadvantages of LoRa are the large number of rejected packet rates (RPR) and the large packet error rates (PER). The RPR is the number of data packages that arrive at the gateway with power that is less than the defined threshold and are considered lost, whereas PER expresses the number of data packages with binary error

Table 5.3 Advantages and disadvantages of wireless communication protocols

Protocols	Advantages	Disadvantages	Applications
3G/4G/5G/ LTE	Fast communication High speed of data sharing Ability to handle big data	Expensive fees for services Require high bandwidth Complicated hardware Power requirements of 4G	Web-based applications Transfer audio and video files
6LoWPAN	Security Simple configuration Connect multiple devices	System issues	Environment monitoring Security Home automation
Bluetooth	Inexpensive - free use Simple installation	Security Very short range Connect only two devices Lose connection	Transfer files Headsets Real-time transition of data
NFC	Easy to use Compatible with other technologies	Short range Low data transfer rate Cost	Share data Phones
RF	Penetration through walls	Security Radiation affect humans and fruits	Telemetry Security systems Access control Remote control
Wi-Fi	High speed High security Inexpensive	Power consumption Network range Configuration	Hotspots Internet access
Wi-Fi Direct (P2P)	Portable Wi-Fi Can be used without internet connection	It is not supported by every device Need to establish connection with other devices	Remote controlled devices Home control application Game controllers
Z-wave	Secure connection Power consumption Cost Simple installation	Limited coverage Security Low communication speed	Home security
ZigBee	Power consumption Less complex than Bluetooth Direct communication	Low speed Cost	Industrial management Consumer electronics Road map products/ tracking

that are received from the gateway. An improved version of LoRa was proposed by Mroue et al. (2020).

5.1.2.2 Network Topologies in WSNs

The topology of a network system is divided into two categories: the physical and the logical. The physical network topology describes how the elements of a network system are connected physically to each other such as wires, wireless connection and much more. On the other hand, the logical network defines the relation among the elements that are included in a network and explains how data are transferred in a network.

Network topologies play a crucial role in network performance and functionality and they help with errors and faults in the network systems. The IEEE 802.15.4 standard, which is widely used in agricultural applications, supports almost all topologies such as star (Fig. 5.2), mesh (Fig. 5.3), tree (Fig. 5.4) and cluster tree networks (Fig. 5.5). These topologies are also used by other networks and they have their advantages and disadvantages as described in Table 5.4.

In a sensor network that uses a star topology, every node is connected directly to a central computer or a central node that controls the network (base station) and is indirectly connected with other nodes through the central node. The benefits of the star topology are that it allows the whole network system to work properly even if there are failures of one or more nodes. It is easy to set up a network system using star topology and it is simple to add and remove nodes from the network. A disadvantage of the star topology is that sensor network performance depends on the central computer (base station) and if it fails, the whole network will be down and will not record data. Another disadvantage, is that this topology typically requires line-of-sight between a sensor node and the base station.

In contrast, in sensor networks that use mesh topology, every node communicates with other nodes in the same network. In other words, it is considered a point-to-point connection where nodes are interconnected (Fig. 5.3). This topology is often used in agriculture, especially when the distance between nodes is large. In this topology data are transmitted by the routing and flooding methods. The routing method uses the logical route that is the shortest distance between the nodes and the central computer or base station, whereas in the flooding method data are sent to every node that belongs to the same network. The advantage of mesh topology is that line-of-sight is not required between all sensor nodes and the base station. Data are transmitted to the base station if at least one node has line-of-sight to the base station. Another advantage is that mesh topology is self-healing. In other words, if one path of transmission is disrupted because of failure of a sensor node, the mesh is re-established using different pathways. However, network systems with mesh topology are complex and require extensive configuration.

Another form of topology is the tree network (Fig. 5.4). The name comes from the shape of the layout which looks like a tree with many branches. The main characteristic of this topology is the hierarchical relationship between the main node (or root node) and the other nodes of the network while there is only one mutual connection between two nodes. The tree topology is ideal for sensor networks where the nodes are located far from each other such as in WANs. This type of topology allows users to add or remove nodes from the system and enables them to find and fix errors in the network. On the other hand, the tree topology has many reliability issues because if the root node fails the whole system fails.

It is also difficult to control sensor networks with tree topology that consists of many nodes. To solve this problem, cluster tree topology is used. The latter is similar to tree topology, but the root node can be connected with many clusters of nodes

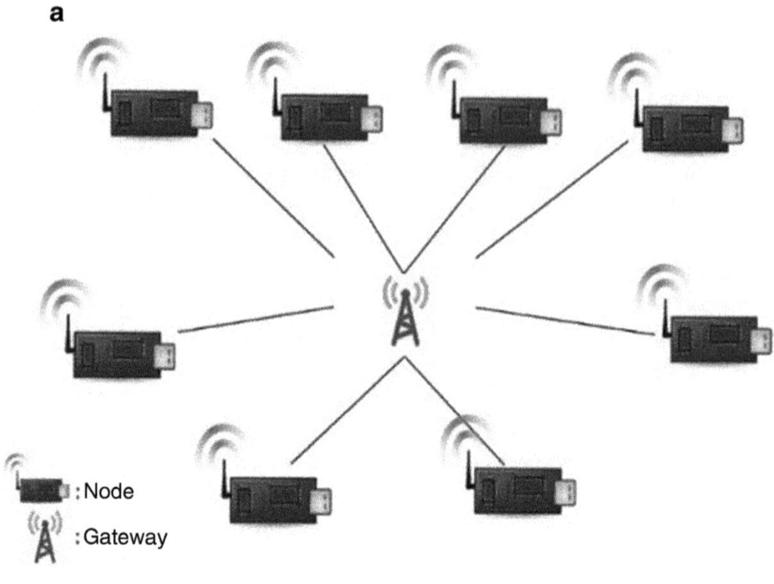

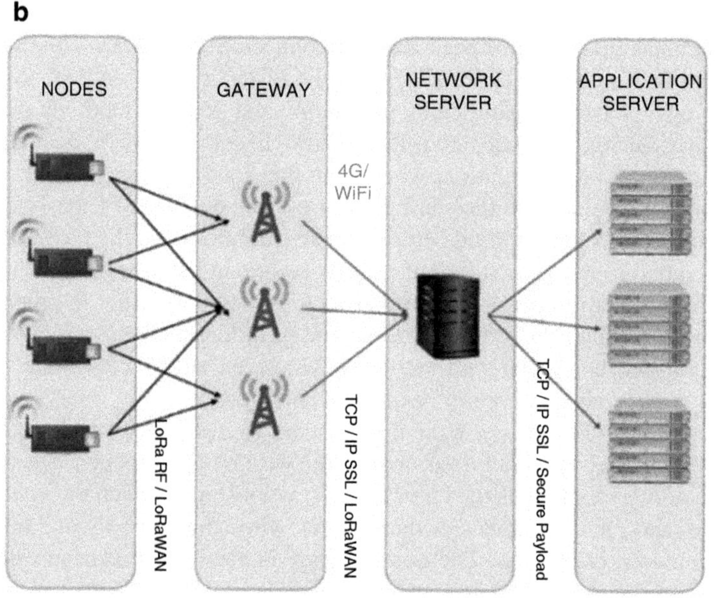

Fig. 5.2 Schematic diagrams of star topologies: (**a**) a typical star topology (**b**) an example of star topology used by LoRa

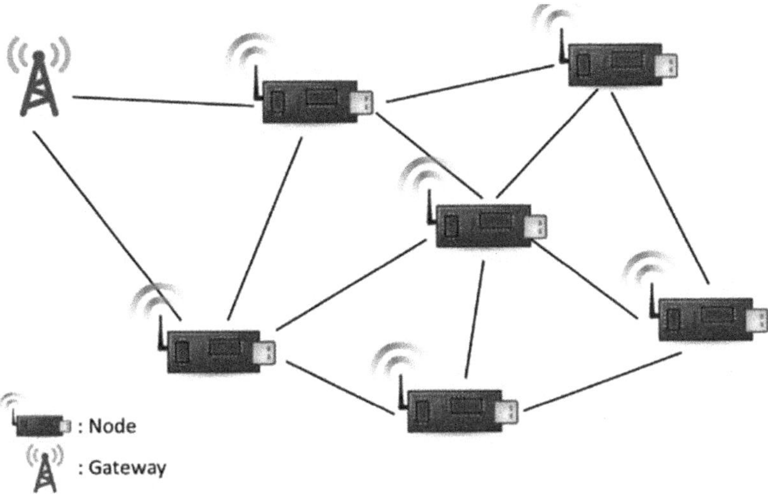

Fig. 5.3 Schematic diagram of mesh network

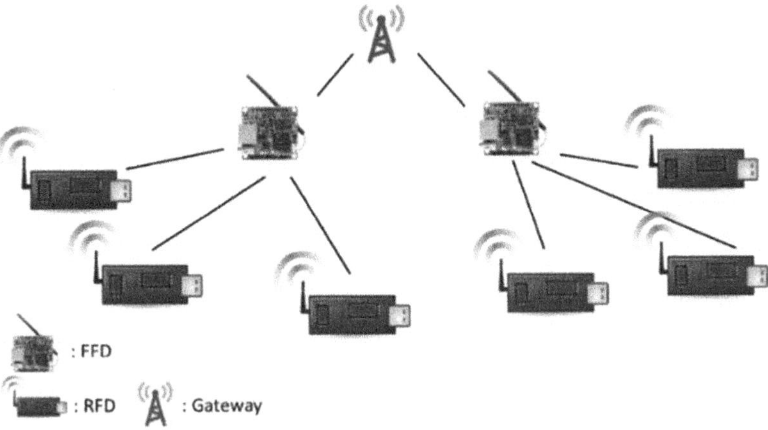

Fig. 5.4 Schematic diagram of a tree network. Reduced function devices (RFDs) devices cannot exchange data with other RFD devices, but they can 'talk' to full function devices (FFDs)

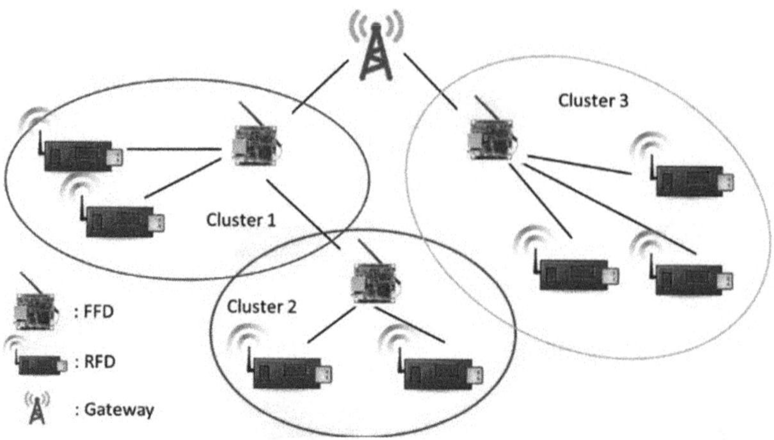

Fig. 5.5 Schematic diagram of cluster tree network

Table 5.4 Advantages and disadvantages of the topologies of WSNs

	Topologies of WSNs				
Activity	Bus	Star	Tree	Cluster Tree	Mesh
Power needs	High	High	High	High	Low
Life time	Low	Low	Low	Low	High
Reliability	Low	Low	Low	Low	High
Path	Single	Single	Single	Single	Multiple
Node failure	High	High	High	High	Low
PRR	Low	Low	Low	Low	High
Load balance	Lowest	Higher than tree	Higher than bus	Higher than bus	Highest

(Fig. 5.5). Unlike the IEEE.802.15.4 protocol, the IEEE.802.15.4/ZigBee supports only the star, tree and mesh topologies.

5.2 Applications of WSNs in Agriculture

Economic pressures for achieving greater production efficiencies in agriculture coupled with global needs for more food production have increased the use of sensors on farms. Tables 5.5, 5.6, 5.7 and 5.8 give examples of the sensors used for agricultural purposes. They are used for sensing weather, soil and plant properties, respectively. Some of these sensors measure two or more properties and can be incorporated into WSNs.

Although sensors play a vital role in a WSN because they produce the required measurements and data, the critical component of a WSN is the micro-controller unit that reads sensor values and translates them into a physical unit. In many agricultural applications, these circuit boards are responsible for the network

Table 5.5 Weather sensors used in agriculture

Sensor Name	Temperature	Wind	Pressure	Humidity	Rainfall	Website
iMETOS					X	http://metos.at
LMPO2					X	https://www.lambrecht.net
IM523					X	http://metos.at
TR-525I					X	https://texaselectronics.com/
Rainew 111					X	www.rainwise.com
MD514D			X			http://metos.at
A660611	X			X		http://metos.at
DS-2		X				www.metergroup.com
05103 L		X				www.youngusa.com
IM512CD		X				http://metos.at
Arable mark	X		X	X	X	www.arable.com

Table 5.6 Soil sensors used in agriculture. EC_a stands for apparent electrical conductivity

Sensor name	Temperature	Moisture	EC_a	Salinity	Website
Tensiometer		X			www.irrometer.com/
MD510SM		X			www.irrometer.com/
SE100S	X	X		X	sentektechnologies.com
IM5041D	X				http://metos.at
ECH-GS3		X	X		www.metergroup.com
ECH-5TE	X	X	X		www.metergroup.com
Hydra probe 2	X	X	X		www.stevenswater.com

communication and data transfer. Table 5.8 presents some commercial circuit boards or motes and their features.

5.3 WSNs for Application of Inputs

5.3.1 Irrigation Applications

Farming is the dominant consumer of freshwater, consuming 70% the world's fresh water withdrawals (Food and Agriculture Organization of the United Nations 2020). Agricultural demands on fresh water supplies are likely to increase as food production increases to feed growing populations and non-farm uses of freshwater continue to grow. During the last decade, groundwater supplies have been depleting at an alarming rate in many agriculture areas, while increasing amounts of industrial activity also demand huge amounts of fresh water. Substantial technological innovation has increased the efficiency of irrigated agriculture over the past several decades, but there is considerable potential for continued improvement. At least half

Table 5.7 Plant sensors used in agriculture

Sensor name	Temperature	Plant growth	Photosynthesis	Moisture status	Website
IM522CD	X				http://metos.at
ECH-LWS				X	www.metergroup.com
IMS521CD				X	http://metos.at
IRTEMP	X				http://metos.at
IM507D			X		http://metos.at
SP Lite2			X		www.kippzonen.com
DN501		X			http://metos.at
Yara-WS				X	www.yara.com
237 LW				X	www.campbellsci.com
CI-340	X		X	X	http://www.solfranc.com/

Table 5.8 Commercial motes and their characteristics. Power requirements are expressed in μA for 'Sleep' mode and in mA for 'Nominal', 'Rx' and 'Tx'. NA stands for not available

	Connectivity					Power requirements			
Name	Bluetooth	Wi-Fi	802.15.4	Zigbee	GPIO	Sleep	Nominal	Rx	Tx
Arduino BT	X	X	X		23	NA	NA	NA	NA
Pan stamp			X		44	1–2	NA	18	36
WiSense			X		24	1	20	NA	NA
BPart	X		X		19	20	1–5	8	8
CoSeN			X	X	34	3	0.7	10	10
FireFly3			X		38	NA	NA	16	18
iSense			X	X	21	3.7	9.4	21	108
Waspmote	X	X	X	X	54	55	15	9	NA
XBee pro			X	X	56	4	NA	62	220
SenseNode			X	X	48	NA	NA	20	17.4
AVRaven			X	X	32	NA	NA	16	17

of the irrigated cropland area across the United States is still irrigated with less efficient, traditional irrigation application systems. If irrigated agriculture is to survive, new irrigation practices and tools should be developed for more efficient water use.

The WSNs are one of the tools that contribute to irrigation management for improving irrigation efficiency worldwide. In Portugal, Morais et al. (1996) used a WSN to record weather data such as solar radiation, soil temperature, air temperature and relative humidity together with soil moisture data. The network was ideal for low data-rate applications and included several solar powered wireless acquisition stations that collected the data. The soil moisture sensors used were dual probe type with a heat pulse. The network protocol used a combination of time division multiple access (TDMA) and carrier sense multiple access (CSMA). The TDMA is a technique that is used in digital mobile radio systems. It allows several users to share the same frequency channel by dividing the timescale into different time slots that are allocated periodically to each user. On the other hand, CSMA is a protocol that checks networks for any other data transmissions that may be in progress prior to transferring data. When data transmission is detected, the network waits for it to end the transferring process. The system worked well; however, there were power supply problems.

In Spain, Damas et al. (2001) developed and evaluated a WSN-based irrigation system for pivots and laterals. The network system was based on the M-Bus standard which used to be called HidroBus. The M-Bus is a European standard for remote reading of water, gas and electricity meters. The WSN is capable of reading meters, water pressure, reservoir levels, and controlling solenoid valves and water pumps. To evaluate the WSN, an area was divided into seven sub-regions and each sub-region was monitored through the WSN. The results showed that utilizing this WSN could reduce irrigation use by 30–60%, increase crop productivity and optimize fertilizer use.

In the USA Evans and Bergman (2003) reported the use of a WSN for irrigation scheduling purposes that used soil moisture sensors, on-site weather data and grower preferences, whereas Kim et al. (2008) developed a closed-loop irrigation system based on a WSN. They used soil water content sensors to monitor soil moisture and IEEE 802.11 and Bluetooth technology to collect weather information, soil moisture data and sprinkler position. The results of the system evaluation showed that such a low-cost wireless system could be useful for remote control of irrigation in precision agriculture (PA) applications. Similar work was done by Kim and Evans (2009) who developed an in-field WSN to implement site-specific sprinkler irrigation using Bluetooth wireless communication. Furthermore, they developed wireless in-field sensing and control software (WISC) which integrates a variable-rate irrigation controller with in-field data feedback. The whole system could be used as a decision support tool for irrigation and real-time monitoring of irrigation operations.

Nam et al. (2016) proposed a WSN for irrigation management that is based on radio frequency identification (RFID) and quick response (QR) technologies. Specifically, they installed water gauges to measure the water levels of irrigation canals and gate regulators to control the water flow of the canals. The WSN was based on RFID and QR codes while a database management system was developed to manage the agricultural water. The sensor network collected data wirelessly and transmitted them to a local server that played the role of a decision support system. The transmission of the data between the sensor nodes and the server was based on the CDMA (Code Division Multiple Access; for long range communication) and Zigbee (for medium-range communication) protocols. The WSNs have also been used to increase irrigation water-use efficiency of drip irrigation (Barkunan et al. 2019). Smartphones were used to take photographs of soil samples to calculate the degree of wetness of the soil. The smartphone transmitted the data to a microcontroller that made irrigation decisions. The global system for mobile (GSM) communication protocol was used for transferring the data from the smartphone to the microcontroller.

Fig. 5.6 The front side of the circuit with the radio frequency transmitter. SM1, SM2 and SM3 are the positive terminals for reading the current of the three soil moisture sensors and GNDs are the negative terminals. T1 and T2 are the terminals for connecting the thermocouple wires. The diode is installed to prevent high currents passing from the antenna to the electronics. This is used for extra protection of the electronics in case of a nearby lightning strike. The number on the sticker 'D60533EB' is the address of the radio transmitter

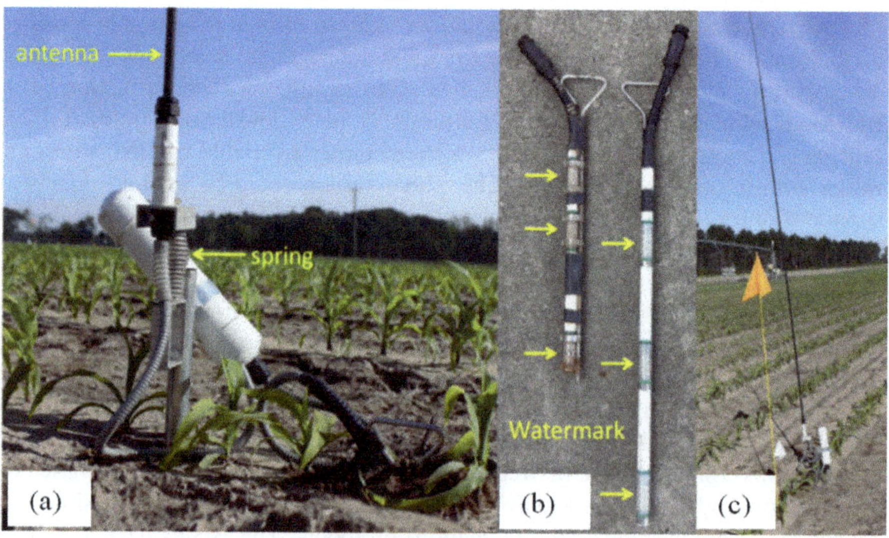

Fig. 5.7 (**a**) A UGA SSA node installed in maize (*Zea mays* L.). The electronics are housed in the white PVC container. The Spring allows the antenna to bend when farm vehicles pass overhead, (**b**) The UGA SSA sensor probe integrates three Watermark sensors and can be customized to any length, (**c**) Antennae can vary in length to accommodate the crop. For cotton (*Gossypium* L.), 2.7 m antennae are used whereas for maize it is 4.2 m

The University of Georgia Smart Sensor Array (UGA SSA) is a system that records soil moisture within fields (Vellidis et al. 2013; Liakos et al. 2017). It consists of a monitoring system, a commercial server that receives soil moisture data wirelessly and a website that presents soil moisture data and recommends irrigation rates (see Figs. 11.3 and 11.4 in Chap. 11). The monitoring system consists of smart sensor nodes and a gateway. Each node has a circuit board, a radio frequency (RF) transmitter (Fig. 5.6), soil moisture sensors, thermocouple wires and an antenna. Each node accommodates two thermocouples for measuring temperature and a probe that consists of up to three Watermark® soil moisture sensors (Fig. 5.7b). The RF transmitter module (Fig. 5.7; RF220SU, Synapse Wireless, Huntsville, Alabama, USA) is a postage stamp-sized intelligent low-cost ($44 in 2020), low-power (2–3.6 V), 2.4 GHz radio module capable of acquiring, analysing and transmitting sensor data up to 2 Mbps. A diode installed close to the antenna plug is used to prevent high current such as that resulting from nearby lightning from passing from the antenna to the electronics.

One characteristic of the soil moisture recording system is that it uses a wireless mesh network to communicate between irrigation sensor nodes. The RF transmitters act as a repeater to pass along data from other nodes to form a meshed network of nodes. If any of the nodes in the network stop transmitting or receiving, or if signal pathways become blocked, the operating software re-configures signal routes to maintain data acquisition from the network. To overcome the attenuating effect of the plant canopy on radio transmissions, the RF transmitter antenna is mounted on spring-loaded, hollow, 6 mm diameter, flexible fiberglass rods approximately 2.5 m above ground level (Fig. 5.7a, c). This design allows field equipment such as tractors and sprayers to pass over the sensors. The published range of the RF transmitter is 500 m although its effective range has been observed to exceed 750 m.

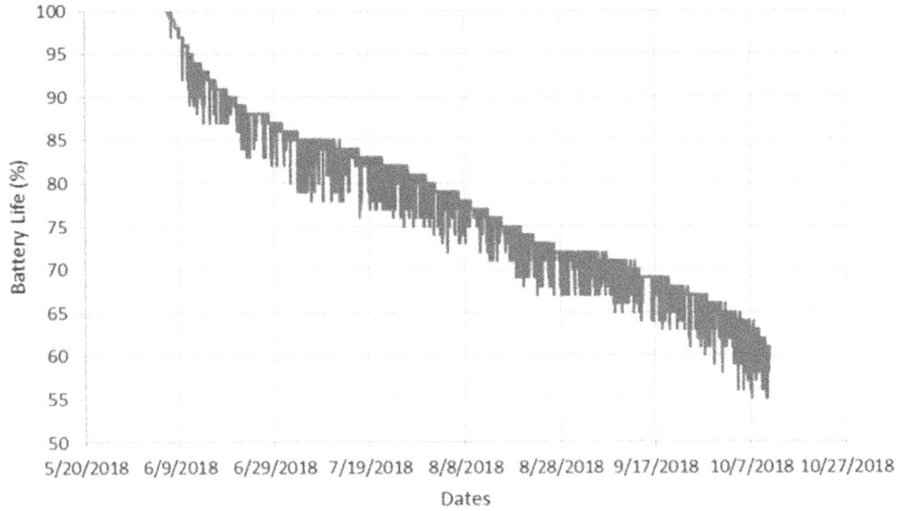

Fig. 5.8 Power consumption graph of one node throughout the growing season of 2018

The smart sensor boards are powered with two 1.5 V alkaline batteries, which last for a growing season (>150 days; Fig. 5.8). Lithium batteries are not a good solution for this system because several field experiments exposed their vulnerability to low voltage, especially below 2 V. Furthermore, to optimize the battery life, the boards are programmed to set themselves in a low-current sleep mode when not transmitting data.

The potential of the UGA SSA is good because significant reductions (up to 50%) in irrigation water and significant increases in irrigation water-use efficiency (up to 40%) have been recorded. The UGA SSA was commercialized by Advanced Ag Systems (Dothan, Alabama, USA). An irrigation case study of the UGA SSA in peanuts is presented as case study 1 in Chap. 11 "Case study from the south-eastern USA: Variable-rate irrigation with centre pivot" (Case Study 11.1). Specifically, the case study gives a detailed description of how prescription maps for irrigation purposes are created and how the UGA SSA web-based user interface helps users to make irrigation decisions. Additionally, it presents performance data of an on-farm evaluation of the UGA SSA in a peanut field.

5.3.2 *Fertilizer Applications*

Fertilizers are used to increase crop yields, but even under the best circumstances, their use efficiency does not exceed 50%. The WSNs have been used to help crop growers fertilize their fields more efficiently with the main aim of increasing crop production while decreasing input costs.

An example of a WSN for fertilizer purposes was developed by Cugati et al. (2003) who created a fertilizer applicator for orchards. The system consisted of input and output modules and a decision support tool (DST). The input modules collected GPS data and real-time sensor data such as tree canopy size, soil electrical conductivity, path planning and weather conditions. The communication between the sensors and the input module was based on a Bluetooth connection and the data were imported into a geographic information system (GIS). The DST consisted of a mathematical model and controllers. The mathematical model was used to estimate optimum fertilizer rates based on the data collected and the controllers to control the actuators on spreaders (output module). The wireless connection between the DST and the actuators was based on the Bluetooth protocol.

Another example of a WSN used for site-specific fertilizer application was described by Elhert et al. (2004) who developed a mechanical sensor that measures crop density. The sensor was mounted in front of a tractor with a spreader. A console received sensor data through a modified controller area network (CAN) bus connection and calculated the optimal amount of fertilizer. The console then controlled the spreader actuators.

He et al. (2011) developed on ASP.NET platform DST for variable-rate application of fertilizers. The system used a LAN type WSN with the IEEE 802.11 protocol to acquire environmental data such as soil moisture, soil electrical conductivity, temperature, pH, air temperature, humidity, CO_2 concentration, and illumination and transmitted them to a server to make them available to participating farmers.

The fertilizer application DST used integrated and optimal fertilization models that utilized these variables to recommend optimal fertilizer rates. Furthermore, the server used a GIS to interpolate data from small experimental plots and extrapolate them to larger plots to reduce the cost of data acquisition.

5.3.3 *Pest Management Applications*

Pest management is the process used to solve problems related to pests. In agriculture, pest management typically involves the application of pesticides to crops either prophylactically or after pests have been detected. Several WSNs have been developed to increase pest management efficiency. For example, Baggio (2005) developed a WSN to protect potato crops from Phytophthora, a fungal disease that causes large economic losses on crops worldwide. The WSN consisted of sensors that measured relative humidity, air temperature, soil moisture and height of the water table. The electronics that read the sensors communicated with each other using the time-out medium access control (T-MAC) protocol. This protocol allows every node to wake up periodically to communicate with its neighbours, and then to sleep again until the next frame. The use of this protocol reduced the power consumption of the nodes and the experiments proved that Phytophthora was significantly reduced in potatoes after using the WSN.

Faical et al. (2014) proposed a more complicated pesticide management system integrated with a WSN. They used an unmanned aerial vehicle (UAV) equipped with a sprayer for pesticide applications and installed a WSN on the ground. The WSN consisted of several nodes installed at different locations within a field. Each node measured the amount of pesticide applied by the UAV. In addition, the authors developed a model that estimated the difference between the prescribed pesticide rate and the applied. When the difference exceeded a specific threshold, the WSN sent messages to the UAV to adjust its route so that each location of the field received the prescribed pesticide rate. The communication between each node and the UAV was achieved using the Xbee–Pro 2 series that supports the IEEE.802.15.4/ZigBee protocol.

5.4 WSNs for Crop Monitoring

Crop monitoring plays an important role in PA because monitoring different crop properties can lead to improvement in agricultural production. Thus, WSNs are often used for crop monitoring. Sanchez et al. (2011) described an integrated WSN solution for PA. The WSN used several nodes to record soil salinity, pH, soil temperature and soil moisture. A gateway was also installed in each field to collect data from each node and send the data to another gateway that was installed at a farmers' cooperative. In addition, a motion detection (PIR, Passive Infrared Sensor) and identification (camera) sensors were used for video-surveillance in case of an intruder, therefore taking care of both crop security and control functions. Data from all sensors were transmitted by the communication modules to a common device denoted as the Gateway that was installed in the field. The gateway was responsible for delivering the information to a farmers' cooperative system that provided farmers with information about their crops in real time such as sensor readings and videos of fields. The communication protocol used for transferring the data from fields to the cooperative was the IEEE 802.15.4.

Liqiang et al. (2011) developed a crop monitoring system using a WSN. The system consisted of two platforms. One platform collected meteorological and soil information such as temperature, humidity, wind, air, rainfall and soil pH while the second platform captured images of the crops to quantify crop growth. The monitoring nodes used general packet radio service (GPRS) and code division multiple access (CDMA) technology for wireless communication.

Jones et al. (2018) developed a wireless sensor network for the study of evapotranspiration in different crops. Infrared temperature sensors and dry reference sensors were used to record canopy temperature and estimate evapotranspiration of cotton. The nodes transmitted the data to a base station located in the field using ultra-high frequency (UHF) radios while the base station sent the data to a server using 3G communication. The server then analyzed the data and visualized visualized them at an online web portal.

Bauer et al. (2019) used a WSN to estimate leaf area index (LAI) of winter wheat. The network consisted of sensors that measured the LAI below and above the canopy. Communication between sensors was achieved using the IEEE 802.15.4/Zigbee protocol.

5.5 The WSNs for Meteorology

Weather data are very important for agriculture because crop yields are strongly related to environmental factors such as temperature, precipitation and evapotranspiration.

The University of Georgia Weather Network (http://weather.uga.edu/) consists of 87 ground weather stations which are equipped with sensors to monitor air temperature, relative humidity, rainfall, solar radiation, wind speed, wind direction, soil temperature at 0.05, 0.1 and 0.2 m depths, atmospheric pressure and soil moisture every second. Data are summarized at 15 minutes intervals while a daily summary is calculated at midnight. At each site, a microcomputer receives data from all sensors and uploads them on a server using a mobile network (GSM). The server stores and analyses the data and delivers them to users. A similar automated wireless weather network is operated by the University of Florida (FAWN; https://fawn.ifas.ufl.edu/). The FAWN consists of 37 ground weather stations and records soil temperature, air temperature, relative humidity, rainfall, barometric pressure, solar radiation, wind speed, wind direction, dew point and evapotranspiration. The nodes at each site communicate with the base station using RF technology and each base station transmits the data to a server using a cellular modem with 2G technology and GPRS network (Table 5.9).

Table 5.9 Applications of WSNs in agriculture

	Application						
	Irrigation	Fertilization	Pesticides	Crop monitoring	Meteorology	Other	Wireless protocol
Morais et al. (1996)					X		RF
Jensen et al. (2000)			X				WiFi
Gomide et al. (2001)	X	X	X	X			WLAN
Hirakawa et al. (2002)						Robotics	WiFi
Lee et al. (2002)						Yield data	Bluetooth
Perkins et al. 2002					X		RF
Charles and Stenz (2003)			X				RF
Flores (2003)				X			WiFi
Liu and Ying (2003)					X		Bluetooth
McKinion et al. (2003, 2004a, b)		X	X				WiFi
Mizunuma et al. (2003)				X			WiFi
Ribeiro et al. (2003)			X				WiFi
Vivoni and Camilli (2003)					X		WiFi
Mahan and Wanjura (2004)				X			RF
Matese et al. (2013)					X		ZigBee
Mendez and Mukhopadhyay (2013)	X				X		WiFi
Tripathy et al. (2013a, b)			X				ZigBee
Zhang et al. (2015)	X						GPRS
Bhanu et al. (2014)					X		ZigBee
Chen et al. (2015)	X				X		Bluetooth
Culibrina and Dadios (2015)	X			X	X		ZigBee
Harun et al. (2015)	X				X		ZigBee
Liu et al. (2015)	X			X	X		ZigBee
Nguyen et al. (2015)					X		ZigBee
Nikolidakis et al. (2015)	X						ZigBee

(continued)

Table 5.9 (continued)

	Application						
	Irrigation	Fertilization	Pesticides	Crop monitoring	Meteorology	Other	Wireless protocol
Sales et al. (2015)	X						ZigBee
Zhao et al. (2015)			X				ZigBee
Aneeth and Jayabarathi (2016)	X						ZigBee
Azaza et al. (2016)					X		ZigBee
Bing (2016)	X				X		Bluetooth
Ferrández-Pastor et al. (2016)	X						ZigBee
Ilie-Ablachim et al. (2016)	X				X		Lora
Jayaraman et al. (2016)	X	X		X	X		WiFi
Li and Li (2016)					X		ZigBee
Martinez (2016)	X	X			X		ZigBee
Mat et al. (2016)	X						ZigBee
Zou et al. (2016)				X	X		ZigBee
Cao-hoang and Duy (2017)					X		WiFi
Lerdsuwan and Phunchongharn (2017)	X				X		WiFi
Liao et al. (2017)	X			X			ZigBee
Mois et al. (2017)	X						WiFi
Patil et al. (2017)	X	X	X		X		WiFi
Reda et al. (2017)					X		LoRa
Vasisht et al. (2017)	X			X			WiFi
Yelamarthi et al. (2017)	X				X		RF
Nurzaman et al. (2018)			X		X		6LoWPAN
Rajakumar et al. (2018)	X			X			WiFi
Sushanth and Sujatha (2018)	X			X	X		WiFi
Patil and Jadhav (2019)	X				X		WiFi

5.6 Conclusions

The use of WSNs is rapidly evolving in the agricultural sector. Smart nodes with embedded computers can read many types of sensors in fields and provide farmers with data that help them make data-driven decisions about agricultural tasks. The WSNs have been used for management of agricultural inputs such as irrigation, fertilizers and pesticides as well as crop and weather monitoring. It is noteworthy to mention that in such studies, WSNs reduced the amount of irrigation and increased the efficiency of fertilizer and pesticide applications. There are many wireless communication standards and protocols because each WSN application requires a specific data rate, frequency, power consumption and reception range. The future of WSNs is expected to be bright because wirelessly they can do more than one task at the same time such as read more than one sensor and transmit data to specific gateways. Nevertheless, WSNs are not yet mainstream in agriculture because each application is highly customized. For this reason, more research and development are required to make the setup and maintenance of WSNs easier. This means that better hardware and more efficient and less complex software must be developed. The WSNs should also be designed to be energy efficient and to use energy sources that do not cause physical impediments in fields. More efficient techniques for fault monitoring and tolerance are necessary. Finally, yet importantly, more research is needed to develop more secure ways to transmit data. If data transmission is not properly secured, packet replay could occur, resulting in potential database corruption and other forms of malice. The WSNs have the potential to trigger the next revolution in computing in agriculture.

References

Akashdeep, Kahlon KS, Kumar H (2014) Survey of scheduling algorithms in IEEE892.16 PMP networks. Egypt Inf J 15(1):25–36

Aneeth TV, Jayabarathi R (2016) Energy-efficient communication in wireless sensor network for precision farming. In: Dash S, Bhaskar M, Panigrahi B, Das S (eds) Artificial intelligence and evolutionary computations in engineering systems, Advances in intelligent systems and computing, vol 394. Springer, New Delhi, pp 417–427

Azaza M, Tanougast C, Fabrizio E et al (2016) Smart greenhouse fuzzy logic based control system enhanced with wireless data monitoring. ISA Trans 61:297–307

Baggio A (2005) Wireless sensor networks in precision agriculture, ACM workshop real-world wireless sensor networks, Stockholm. http://citeseerx.ist.psu.edu/viewdoc/download?doi=10.1.1.120.46&rep=rep1&type=pdf. Accessed 5 Jun 2020

Barkunan SR, Bhanumathi V, Sethuram J (2019) Smart sensor for automatic drip irrigation system for paddy cultivation. Comput Electr Eng 73:180–193

Bauer T, Jarmer S, Schittenhelm S et al (2019) Processing and filtering of leaf area index time series assessed by in-situ wireless sensor networks. Comput Electron Agric 165:104867

Bhanu B, Rao R, Ramesh J et al (2014) Agriculture field monitoring and analysis using wireless sensor networks for improving crop production. In: Proceedings of the 11th international conference on wireless and optical communication networks, Vijayawada, 11–13 September 2014

Bing F (2016) The research of IoT of agriculture based on three layers architecture. In: Proceedings of the 2nd International Conference on Cloud Computing and Internet of Things (CCIOT), Dalian, 22–23 October 2016, pp 162–165
Cao-hoang T, Duy CN (2017) Environment monitoring system for agricultural application based on wireless sensor network. In: Proceedings of the 7th International Conference on Information Science and Technology (ICIST), Da Nang, 16–19 April 2017, pp 99–102
Charles K, Stenz A (2003) Automatic spraying for nurseries. USDA annual report. Project number: 3607-21620-006-03. September 22, 2000–August 31, 2003. USDA, USA
Chen Y, Chanet JP, Hou KM et al (2015) Scalable context-aware objective function (SCAOF) of routing protocol for agricultural low-power and lossy networks (RPAL). Sensors 15(8):19507–19540
Cugati S, Miller W, Schueller J (2003) Automation concepts for the variable rate fertilizer applicator for tree farming. In: Proceedings of the 4th European conference in precision agriculture, Berlin, 14–19 June 2003
Culibrina F, Dadios E (2015) Smart farm using wireless sensor network for data acquisition and power control distribution. In: Proceedings of the 8th IEEE international conference humanoid, nanotechnology, information technology, Cebu, 9–12 December 2015
Damas M, Prados AM, Gomez F et al (2001) HidroBus® system: fieldbus for integrated management of extensive areas of irrigated land. Microprocess Microsyst 25:177–184
Elhert D, Schmerler J, Voelker U (2004) Variable rate nitrogen fertilization of winter wheat based on a crop density sensor. Precis Agric 5(3):263–273
Evans R, Bergman J (2003) Relationships between cropping sequences and irrigation frequency under self-propelled irrigation systems in the Northern Great Plains. USDA annual report. Project number: 5436-13210-003-02. June 11, 2003–December 31, 2007
Faical B, Costa F, Pessin G et al (2014) The use of unmanned aerial vehicles and wireless sensor networks for spraying pesticides. J Syst Archit 60:393–404
Ferrández-Pastor F, García-Chamizo J, Nieto-Hidalgo M et al (2016) Developing ubiquitous sensor network platform using internet of things: application in precision agriculture. Sensors 16:1141
Flores A (2003) Speeding up data delivery for precision agriculture. Agric Res Mag U S Dep Agric (USDA) 51(6):17
Food and Agriculture Organization of the United Nations (2020) Aquastat – FAO's Global information system on water and agriculture. http://www.fao.org/aquastat/en/overview/methodology/water-use. Accessed 2 Feb 2020
Gomide RI, Inamasu RY, Queiroz DM et al (2001) An automatic data acquisition and control mobile laboratory network for crop production systems and spatial variability studies in the Brazilian center-west region, Paper no. 011064. ASAE, St. Joseph
Gupta HP, Rao SV, Yadav AK et al (2015) Geographic routing in clustered wireless sensor networks among obstacles. IEEE Sensors J 15(5):2984–2992
Gutierrez JA, Callaway EH, Barrett RL (2004) Low-rate wireless personal area networks enabling wireless sensors with IEEE 802.15.4. IEEE Press, New York
Harun AN, Kassim MRM, Ma I et al (2015) Precision irrigation using wireless sensor network. In: Proceedings of the international conference on smart sensors and application (ICSSA), Kuala Lumpur, 26–28 May 2015, pp 71–75
He J, Wang J, He D et al (2011) The design and implementation of an integrated optimal fertilization decision support system. Math Comput Model 54:1167–1174
Hirakawa AR, Saraiva AM, Cugnasca CE (2002) Wireless robust robot for agricultural applications. In: Proceedings of the world congress of computers in agriculture and natural resources, Iguacu Falls, 13–15 March 2002, pp 414–420
Ilie-Ablachim D, Pătru GC, Florea IM et al (2016) Monitoring device for culture substrate growth parameters for precision agriculture: acronym: monisen. In: Proceedings of the 15th RoEduNet conference: networking in education and research, Bucharest, 7–9 September 2016, pp 1–7
Jayaraman P, Yavari A, Georgakopoulos D et al (2016) Internet of things platform for smart farming: experiences and lessons learnt. Sensors 16(11):1884

Jensen AL, Boll PS, Thysen I et al (2000) Pl@nteInfo: a web-based system for personalized decision support in crop management. Comput Electron Agric 25(3):271–293
Jones J, Hutchinson P, May T et al (2018) A practical method using a network of fixed infrared sensors for estimating crop canopy conductance and evaporation rate. Biosyst Eng 165:59–69
Kim Y, Evans RG (2009) Software design for wireless sensor-based site-specific irrigation. Comput Electron Agric 66:159–165
Kim Y, Evans RG, Iversen WM (2008) Remote sensing and control of an irrigation system using a distributed wireless sensor network. IEEE Trans Instrum Meas 57(7):1379–1387
Lee WS, Burks TF, Schueller JK (2002) Silage yield monitoring system, ASAE paper no.: 02-1165. The American Society of Agriculture Engineers, St. Joseph
Lerdsuwan P, Phunchongharn P (2017) An energy-efficient transmission framework for IoT monitoring systems in precision agriculture. In: Kim K, Joukov N (eds) Proceedings of information science and applications 2017: Icisa 2017. Springer, Macau, pp 714–721
Li F, Li S (2016) Design and research of intelligent greenhouse monitoring system based on internet of things. In: Proceedings of the international conference on computer science and electronic technology (CSET 2016), Zhengzhou, 13–14 August 2016, pp 76–79
Liakos V, Porter W, Liang X et al (2017) Dynamic variable rate irrigation – a tool for greatly improving water use efficiency. Adv Anim Biosci 8(2):557–563
Liao MS, Chen SF, Chou CY et al (2017) On precisely relating the growth of phalaenopsis leaves to greenhouse environmental factors by using an IoT-based monitoring system. Comput Electron Agric 136:125–139
Lin L, Shroff NB, Srikant R (2007) Asymptotically optimal energy-aware routing for multihop wireless networks with renewable energy sources. IEEE Trans Netw 15(5):1021–1034
Liqiang Z, Shouyi Y, Leibo L et al (2011) A crop monitoring system based on wireless sensor network. Procedia Environ Sci 11:558–565
Liu G, Ying Y (2003) Application of Bluetooth technology in greenhouse environment, monitor and control. Agric Life Sci 29:329–334
Liu D, Cao X, Huang C et al (2015) Intelligent agriculture greenhouse environment monitoring system based on IoT technolog. In: Proceedings of the international conference on intelligent transportation, big data and smart city (ICITBS), Halong Bay, Vietnam, 19–20 December 2015, pp 487–490
Luo J, Hu J, Wu D et al (2015) Opportunistic routing algorithm for relay node selection in wireless sensor networks. IEEE Trans Ind Inf 11(1):112–121
Mahan J, Wanjura D (2004) Upchurch, design and construction of a wireless infrared thermometry system. The USDA annual report. Project Number: 6208-21000-012-03. May 01, 2001–September 30, 2004
Mansourkiaie F, Ahmed MH (2016) Optimal and near-optimal cooperative routing and power allocation for collision minimization in wireless sensor networks. IEEE Sensors J 16(5):1398–1411
Martinez J (2016) Smart viticulture project in Spain uses sensor devices to harvest healthier more abundant grapes for coveted Albario wines. Libelium world. http://www.libelium.com/sensors-mag-smart-viticulture-projectin-spain-uses-sensor-devices-to-harvest-healthier-more-abundant-grapes-for-coveted-albarino-wines/. Accessed 7 May 2020
Mat I, Kassim MRM, Harun AN et al (2016) IoT in precision agriculture applications using wireless moisture sensor network. In: Proceedings of the IEEE conference on open systems (ICOS), Langkawi, 10–12 October 2016, pp 24–29
Matese A, Vaccari FP, Tomasi D et al (2013) Enhancing canopy monitoring management practices in viticulture. Sensors 13:7652–7667
McKinion JM, Jenkins JN, Willers JL et al (2003) Developing a wireless LAN for high-speed transfer of precision agriculture information. In: Proceedings of the 4th European conference on precision agriculture, Berlin, Germany, 15–19 June 2003, pp 399–404
McKinion JM, Turner SB, Willers JL et al (2004a) Wireless technology and satellite internet access for high-speed whole farm connectivity in precision agriculture. Agric Syst 81:201–212

McKinion JM, Willers JL, Jenkins JN (2004b) Wireless local area networking for farm management, ASAE paper: 04-3012. The American Society of Agriculture Engineers, St. Joseph
Mendez GR, Mukhopadhyay SC (2013) A Wi-Fi based smart wireless sensor network for an agricultural environment. In: Mukhopadhyay SC, Jiang JA (eds) Proceedings of wireless sensor networks and ecological monitoring. Springer, Berlin, pp 247–268
Meng T, Wu F, Yang Z et al (2016) Spatial reusability-aware routing in multi-hop wireless networks. IEEE Trans Comput 65(1):244–255
Mizunuma M, Katoh T, Hata S (2003) Applying IT to farm fields—a wireless LAN. NTT Tech Rev 1(2):56–60
Mois G, Folea S, Sanislav T (2017) Analysis of three IoT-based wireless sensors for environmental monitoring. IEEE Trans Instrum Meas 66:2056–2064
Morais R, Cunha JB, Cordeiro M et al (1996) Solar data acquisition wireless network for agricultural applications. In: Proceedings of the 19th IEEE convention of electrical and electronics engineers in Israel, Jerusalem, 5–6 November 1996, pp 527–530
Mroue H, Parrein B, Hamriouri S et al (2020) LoRa+: an extension of LoRaWAN protocol to reduce infrastructure costs by improving the quality of service. Internet Things J 9:100176
Nam W, Hong E, Choi J (2016) Assessment of water delivery efficiency in irrigation canals using performance indicators. Irrig Sci 34:129–143
Nguyen TD, Thanh TT, Nguyen LL et al (2015) On the design of energy efficient environment monitoring station and data collection network based on ubiquitous wireless sensor networks. In: Proceedings of the IEEE RIVF international conference on computing & communication technologies research, innovation, and vision for the future (RIVF), Can Tho, 25–28 January 2015, pp 163–168
Nikolidakis SA, Kandris D, Vergados DD et al (2015) Energy efficient automated control of irrigation in agriculture by using wireless sensor networks. Comput Electron Agric 113:154–163
Nurzaman A, Debashis D, Ifterkhar H (2018) Internet of things (IoT) for smart precision agriculture and farming in rural areas. IEEE Internet Things J 5(6):4890–4899
Patil S, Jadhav M (2019) Smart agriculture monitoring system using IOT. Int J Adv Res Comput Commun Eng 8(4):116–120
Patil G, Gawande P, Bag R (2017) Smart agriculture system based on IoT and its social impact. Int J Comput Appl 176(1):1–4
Perkins M, Correal N, O'Dea B (2002) Emergent wireless sensor network limitations: a plea for advancement in core technologies. In: Proceedings of the 1st IEEE international conference on sensors, Orlando, Florida, USA, 12–14 June 2002, pp 1505–1509
Rajakumar G, Sankari S, Shunmugapriya D, Maheswari U (2018) Iot based smart agricultural monitoring system. Asian J Appl Sci Technol 2(2):474–480
Reda HT, Daely PT, Kharel J et al (2017) On the application of IoT: meteorological information display system based on LoRa wireless communication. IETE Tech Rev 35(3):1–10
Ribeiro A, Garcia–Perez L, Garcia-Alegre et al (2003) A friendly man-machine visualization agent for remote control of an autonomous tractor GPS guided. In: The Proceedings of the 4th European conference in precision agriculture, Berlin, Germany, 14–19 June 2003
Sales N, Remédios O, Arsenio A (2015) Wireless sensor and actuator system for smart irrigation on the cloud. In: Proceedings of the IEEE 2nd World Forum on Internet of Things (WF-IoT), Milan, Italy, 14–16 December 2015, pp 693–698
Sanchez A, Sanchez F, Haro J (2011) Wireless sensor network deployment for integrating video-surveillance and data-monitoring in precision agriculture over distributed crops. Comput Electron Agric 75(2):288–303
Schurgers C, Srivastava MB (2001) Energy efficient routing wireless sensor networks. In: Proceedings of military communications conference on communications for network-centric operations: creating the information force, vol 1. McLean, pp 357–361
Shen H, Li Z, Yu L (2015) A P2P-based market-guided distributed routing mechanism for high-throughput hybrid wireless networks. IEEE Trans Mobile Comput 14(2):245–260

Sushanth G, Sujatha S (2018) IOT based smart agriculture system. In: The proceedings of the international conference on wireless communications, signal processing and networking, Chennai, India, 22–24 March 2018

Tripathy AK, Adinarayana J, Sudharsan D et al (2013a) Data mining and wireless sensor networ for groundnut pest/disease interaction and predictions – a preliminary study. Int J Comput Inf Syst Ind Manag Appl 5:427–436

Tripathy AK, Adinarayana J, Merchant SN et al (2013b) Data mining and wireless sensor network for groundnut pest/disease precision correction. In: Proceedings of National Conference on Parallel Computing Technologies, Bangalore, India, pp 1–8

Valentini G, Abbas CJB, Villalba JGG, Astorga L (2010) Dynamic multi-objective routing algorithm: a multi-objective routing algorithm for the simple hybrid routing protocol on wireless sensor networks. IET Commun 4(14):1732–1741

Vasisht D, Kapetanovic Z, Won J et al (2017) An IoT platform for data-driven agriculture. In: Proceedings of the 14th USENIX symposium on networked systems design and implementation, Boston, 27–29 March 2017, pp 515–529

Vellidis G, Tucker M, Perry C et al (2013) A soil moisture sensor-based variable rate irrigation scheduling system. In: Stafford JV (ed) Proceedings of the 9th European conference on precision agriculture (9ECPA), Lleida, Spain, 7–11 July 2013

Vivoni ER, Camilli R (2003) Real-time streaming of environmental field data. Comput Geosci 29:457–468

Yelamarthi K, Aman MS, Abdelgawad A (2017) An application-driven modular IoT architecture. Wirel Commun Mob Comput 2017(1):16

Zhang X, Zhao Y, Wang S et al (2015) Multiobjective optimization for green network routing in game theoretical perspective. IEEE J Sel Areas Commun 33(12):2801–2814

Zhao G, Guo Y, Sun X et al (2015) A system for pesticide residues detection and agricultural products traceability based on acetylcholinesterase biosensor and internet of things. Int J Electrochem Sci 10:3387–3399

Zou F, Jiang X, Lin Z (2004) IEEE 802.20 based broadband railroad digital network – the infrastructure for m-commerce on the train. In: Proceedings of the 4th international conference electronic business, Beijing, pp 771–776

Zou T, Lin S, Feng Q et al (2016) Energy-efficient control with harvesting predictions for solar-powered wireless sensor networks. Sensors 16(1):53

Chapter 6
Sensing for Health, Vigour and Disease Detection in Row and Grain Crops

David W. Franzen, Yuxin Miao, Newell R. Kitchen, James S. Schepers, and Peter C. Scharf

Abstract Maximum genetically possible crop growth and yield are inhibited by abiotic and or biotic stresses. Biotic stresses may include pests, insects or disease infestations whereas abiotic stress includes nutritional deficiencies. Sensors have been developed and are being developed to detect stresses and to assist in crop stress management. Decision support systems are being developed to help manage structured and unstructured crop stress issues. Weather prediction of disease and insect outbreaks help to assess the risk of these stresses over an area, but do not address the spatial variation of stress outbreak within a field. Most commercially developed sensors are remote sensing sensor technologies that use some form of normalized difference vegetative index (NDVI) to address nitrogen (N) sufficiency or deficiency in crops. Sensors for detection of disease and insect infestations are still under development, and include 'nose' sensors to detect stress-specific volatile organic compounds (VOCs), digital imagery of individual leaves, fluorescence imagery and thermography. Effective remotely sensed strategies need to be easy to implement for general adoption by farmers and crop consultants.

D. W. Franzen (✉)
North Dakota State University, Fargo, ND, USA
e-mail: David.Franzen@ndus.edu

Y. Miao
University of Minnesota, St. Paul, MN, USA
e-mail: ymiao@umn.edu

N. R. Kitchen
USDA-ARS, Columbia, MO, USA
e-mail: Newell.Kitchen@ars.usda.gov

J. S. Schepers
USDA-ARS, Lincoln, NE, USA

P. C. Scharf
University of Missouri, Columbia, MO, USA
e-mail: ScharfP@missouri.edu

R. Kerry, A. Escolà (eds.), *Sensing Approaches for Precision Agriculture*, Progress in Precision Agriculture, https://doi.org/10.1007/978-3-030-78431-7_6

Keywords Remote sensing · Active-optical sensors · Nose sensors · Nitrogen · Sufficiency index · Data normalization · Chlorophyll

6.1 Introduction

A perfectly healthy crop would have adequate nutrients and soil water for maximum growth and grain, seed or forage production (Mosa et al. 2017). The healthy crop would be free of disease and insect pests, and temperature throughout the growing season would be perfect for maximum production. Any difference in an external crop growth factor from its ideal state relative to the crop is considered a stress. Stresses fit into either a biotic or an abiotic stress factor category. Examples of abiotic stressors are unfavourable nutrient availability, soil water excess or deficit, or unfavourable temperature. Examples of biotic stresses include insect feeding, disease infection and parasitic nematode root infection.

In reaction to changes in biotic factors such as pathogens and pests that affect plant health and vigour or abiotic factors in the soil and plant environment such as nutrients, several chemical changes within the plant are initiated. The expression of these chemical changes is first a change in chemicals produced within the plant, which are sometimes emitted as volatile organic compounds (VOCs). Detection of VOCs is the first opportunity for sensor detection of plant stress due to sub-optimal environment, disease infection or pest feeding (Meroni et al. 2010). As the condition continues or worsens, the chemical changes in the plant are expressed as phenotypic symptoms. Stressed plants react with changes, such as in leaf area index (LAI), chlorophyll content and increase in leaf surface temperature, all leading to changes in the spectral signature from the plants (Meroni et al. 2010). Initial changes also include release of fluorescence (West et al. 2003). Changes in spectral signatures from plant tissues are difficult for humans to perceive and identify; however, sensors can detect changes such as differences in healthy and unhealthy plants several days earlier than the human eye can detect the phenotypic changes.

The detection of biotic and abiotic stressors from initial invisible changes to visible phenotypic changes is shown in Fig. 6.1. At the time of sensing, there may be more than one abiotic or biotic factor affecting the chemical and spectral signature. As an initial plant response to stress, VOCs are emitted by the plant (de Moraes et al. 2004). The VOCs emitted are specific to the stress (Baldwin 2010). As the infection spreads to other plants and infection becomes more severe, it can cause significant yield losses. By the time visual symptoms can be detected by the human eye, it is often too late to prevent or treat a stressor. Therefore, there is a need for an early in-field diagnosis to avoid loss of yield. An alternative would be frequent sampling for laboratory analysis such as serological assays or qPCR biotechnological analysis; however, such methods are labour intensive and costly. The use of sensors

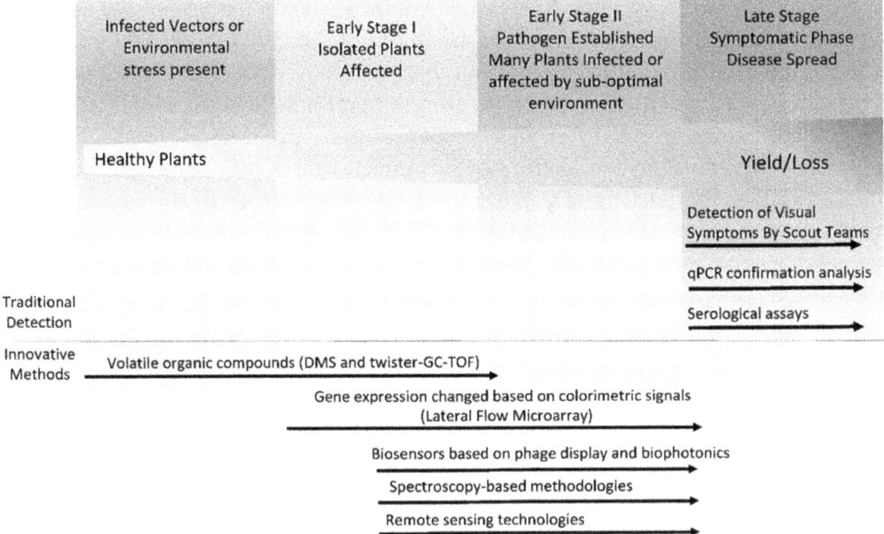

Fig. 6.1 Temporal sequence of plant responses to abiotic and biotic stress. Adapted from Martinelli et al. (2014)

to detect chemical and spectral changes is therefore desirable to reduce cost, yet detect stresses early.

In addition to stress caused by pathogens and pests, detecting stress caused by nutrient deficiency early can be used to synchronize the application of nutrients to the needs of the plants, which can increase nutrient use efficiency and reduction of nutrient losses to the environment.

6.2 Sensors for Decision Support Systems (DSS)

Sensor data used to support DSSs are categorized as remote sensors, proximal sensors, and the imagery developed from the use of the sensors. Remote sensing is the practice of obtaining data from an object without touching it (Lintz and Simonett 1976), while proximal sensing uses sensors that function very close to the sensing object or touch it (Adamchuck and Rossel 2011). A remote sensor is usually placed in the field or in the general area of a field where monitoring some aspect related to crop health is required. Data recorded by the sensor may need to be downloaded in a physical visit to the field by a technician or farmer, or it might deliver the data to a remote computer without a field visit. Remotely-sensed imagery may be generated by a satellite, aeroplane, drone or ground-based mobile sensor. A remotely-sensed image is usually characterized as data from multiple crop plants within an area, although some ground-based systems are exploring plant-to-plant variability.

Decision support systems (DSS) are designed to help the user make decisions when encountering structured and unstructured problems. A structured problem is solved systematically, with information that is known. An example of a structured problem in crop production might be crop mineral nutrition where soil test-based recommendations are based on field response trials. In a region, state or country, agricultural experts have guidelines or recommendations published for soil critical concentrations for nutrients, e.g. phosphorus (P). If the soil analysis result is greater than the 'critical value', no P is required. If the value is less, P addition is recommended at a rate associated with the soil test. The solution of the structured problem of P requirement is systematic.

Some problems in crop production may seem to be structured, e.g. nitrogen (N) nutrition because the recommendations may be systematically based on yield expectation or another 'known' formula. However, crop-available N is hardly structured, as soil N can change continually based on rainfall and temperature, mineralization of N from crop residue and organic matter is subject to seasonal conditions, and soil texture may influence hydrology that increases the risk of nitrate leaching or denitrification. Experts tend to publish systematic solutions to crop nutrition because at the time they might be the best guidelines available, whereas a specific field recommendation might have a very different solution.

An unstructured problem is complex; the data required to solve the problem are unknown or insufficient, or lack spatial pattern (Turban et al. 2005). Nitrogen nutrition, risk of disease and insect pest infection and severity are examples of unstructured problems, where at least part of the uncertainty is related to seasonal weather. Weather sensors have been used for years to predict crop health and yield. Such predictions have been used in decision support systems (DSSs) that help decision makers to solve unstructured problems under complex, uncertain conditions (Shtienberg 2013). However, crop nutrition, disease, insect and weather relations are often solved assuming it is a structured rather than unstructured problem.

Many DSSs for disease forecast are weather-based (Kang et al. 2010). In Korea, Kang et al. (2010) used data from weather stations that were approximately 10 km apart. The weather data were then interpolated to a 250 m × 250 m grid to attempt to predict disease outbreaks at a smaller spatial scale. In North Dakota the NDAWN (North Dakota Agricultural Weather Network) includes the use of weather data to predict the likelihood of *Sclerotinia sclerotiorum (Mont.) de Bary* (white mould) in canola (*Brassica napus*, L.), *Cercospora beticola Sacc., (1876)* in sugar beet (*Beta vulgaris*, L.), potato late blight (*Phytophtora infestans (Mont.) de Bary*) and early blight (*Alternaria solani Sorauer, (1896)*), sugar beet root maggot (*Tetanops myopaeformis*, Roder), orange blossom wheat midge (*Sitodiplosis mosellana (Géhin, 1857)*) infection, and wheat (*Triticum aestivum*, L.) disease (fusarium head blight, *Fusarium graminearum (Gibberella zeae* (Schwein.) Petch, (1936) and tan spot (*Pyrenophortritici-repentis* (Died.) Drechsler, (1923)). The NDAWN includes crop growing degree days for timing of sugar beet herbicide application, and estimates crop water use for irrigation scheduling. In the USA the National Agricultural Statistics Service (NASS) uses precipitation data from the National Climatic Data Center (NCDC) and the Climate Analysis Center (CAC) for input into their regression models for predicting yield.

The spatial density of weather stations from which precipitation data are extracted affects the accuracy of the model to a given farm. Although the weather station disease forecasting in the Korea and North Dakota examples may be useful in forecasting the likelihood of disease infection and severity, the problem within an individual field cannot be solved or predicted using spatial analysis to interpolate possible problems at a small scale. This is due to the lack of observed structure in these diseases at the small spatial scale because of the low spatial density of weather stations (Newlands 2018). The likelihood of incidence and severity of potato late blight in North Dakota within a region can be predicted using the NDAWN disease model, however the actual infection is usually cultivar-based and occurs in specific micro-climates related to field topography (Fig. 6.2). The likelihood of soya bean cyst nematode (SCN) can be predicted using soil sampling the year before soya bean (*Glycine max* L. Merr.) production. However, the actual infection within the field of SCN and severity varies throughout the field. The results from soil sampling might over- or under-estimate the extent of infection and its severity.

Another category of sensors are weather micro-sensors which measure weather data within fields to provide micro-climate information for use in disease and insect pest forecasting models. Weather micro-sensors were placed in peanut (*Arachis hypogaea*, L.) fields in India to develop insect and disease forecasts within fields. These data were analyzed using data-mining techniques to understand the dynamics of barley thrip (*Limothrips denticornis* Haliday) infection and barley mosaic

Fig. 6.2 Within-field variation of potato early blight in a North Dakota field. Brown area (circled) is more severely affected than the surrounding field. Image courtesy of A. Robinson, North Dakota State University

virus (BMV) in peanut (Tripathy et al. 2011). An effective weather micro-sensor would need to be small, inexpensive to use and able to deliver information from the field in a usable format to the user platform.

Proximal imaging is also being investigated for use in DSSs (Moran et al. 1997). Dandawate and Kokare (2015) collected 120 images of soya bean leaves with varying severity of leaf disease from a mobile camera. The images were converted from red, green, blue (RGB) to a hue saturation value (HSV). Variation in the image was separated into classes based on a threshold value (Otsu 1979) after separation of the leaf image into the 'background' and 'leaf'. The resulting categories were used in a DSS to estimate extent of leaf disease using a camera (Fig. 6.3).

A more site-specific approach to plant health assessment is possible with the use of so-called 'electric nose-sensors' (Holopainen and Blande 2012; Sankaran et al. 2010). These sensors measure VOCs that plants emit when specific herbivores eat from the plant tissues. Plants also respond chemically to environmental conditions, i.e. nose sensors may also be used to measure abiotic stress.

Nose sensors are specialized sensors produced from metal oxides with selectivity for different classes of VOCs. To date, nose sensors have been used to detect changes in plant health through changes in crop VOC signature (Laothawomkitkul et al. 2008). The use of nose sensors has been successfully investigated under controlled conditions in laboratories or in growth-chambers; however, it is expected that progress in analytic chemistry and technology will make it feasible to use this type of sensor in fields in the near future (Wilson 2013). Nose sensors specific to an anticipated pest or condition are placed in the field or on/in the plant tissue.

Fig. 6.3 Severe damage from the soya bean cyst nematode in a field with generally less severe infection. (Image courtesy of S. Markell, North Dakota State University)

Promising results have been achieved, although more research needs to be conducted before the technology is ready for general farm use (Cellini et al. 2017; Kwak et al. 2017).

A more mature area of sensor application to crop health is the use of remotely sensed imagery. The relations between crop leaf pigments, light absorption and emission from leaves with crop health is explained in Hatfield et al. (2008). Remotely sensed imagery data from satellites and aerial platforms have been successfully investigated to detect plant health (Mahlein et al. 2013) and specific factors that affect it such as frost and water stress (Hatfield and Pinter 1993). Mahlein (2016) explains that challenges to disease detection from the use of remote sensing and remotely-sensed imagery include calibration of the instrument to detect disease at an early point in time, the ability to differentiate between diseases and other biotic and abiotic stresses, and quantification of disease severity.

Thermography from remote sensing detects differences in leaf surface temperature from emitted infrared electromagnetic spectrum waves. Leaf temperature is affected by transpiration and cooling. One early response from the reduction of plant health is stomatal closure, which results in a warmer leaf surface. Drought stress also results in a warmer leaf surface because of reduced transpiration (Chaerle and Van der Straeten 2000; Costa et al. 2013). However, water stress cannot be estimated by this technique alone because water stress in crops is also related to air temperature, humidity, vapour pressure index, wind speed and incident radiation (Gerhards et al. 2019).

Thermal infrared imagery from ground-based remote sensing was used by Taghvaeian et al. (2013) to develop a crop water stress index (CWSI). Standardization of the canopy temperature using the maximum to minimum temperature difference between the plant canopy and air temperature was used to improve the relationship of the imagery to crop water stress.

The MODIS satellite remotely-sensed imagery was used by Gao (1996) to detect crop water stress using a normalized difference water index (NDWI). Two water absorption bands from MODIS, 860 nm and 1640 nm, were used to develop the index. A shortwave infrared water stress index (SIWI) was developed using the two water absorption bands (Fensholt and Sandholt 2003) to assess water stress in rice (*Oryza sativa* L.) in China. A relative reflectance index (RRI) was successfully investigated for use by Mogensen et al. (1996) to assess water status of oilseed rape (*Brassica napa*, L.). Remote imagery was also used to assess relative crop water stress using canopy equivalent water thickness (EWT) (Champagne et al. 2003) and canopy water content (CWC), which is foliage water per unit ground area (Hunt et al. 2013).

Plant stress leads to the accumulation of metabolic plant compounds and the breakdown of photosynthetic pigments (Chaerle and Van der Straeten 2000). Use of near-infrared imagery or multi-spectrum near-infrared and visible ratios, such as the normalized difference vegetation index (NDVI), have been used to detect plant health and vigour of crops within fields and between fields.

Chlorophyll fluorescent imagery uses the release of fluorescence as a result of light energy absorption by chlorophyll in the leaves. Under stress, chlorophyll is affected and the ability to absorb light is changed i.e. fluorescence changes. The stress may cause the accumulation of metabolites and thus change the VOC composition, which might result in a change in fluorescence (Meroni et al. 2010).

6.3 Crop Health and Vigour Related to Insect Damage and Disease

Plants react chemically with very specific responses to varying abiotic stressors like water and temperature, and biotic stressors like disease attack and insect predation. Plants emit different VOCs specific to the herbivore that is eating its tissues. Saliva from insects also induces a similar plant response. Analysis of the volatiles can be a diagnostic of the specific invasive vector (de Moraes et al. 2004).

Additional responses to pest infestation and damage are 'colour' changes in leaves and other phenotypic structures (Hatfield and Pinter 1993). Remote sensing of 'colour' changes can be used to detect pest damage and other conditions that limit crop productivity including nutrition (Wojtowicz et al. 2016). A hand-held radiometer was used to assess Sunn pest (*Eurygaster integriceps*) damage in wheat (a stink bug family pest) (Genc et al. 2008). Remote imagery was used by Ranjitha et al. (2014) to detect differences in health status and thrip (*Thrips tabaci* Lind) damage in cotton (*Gossypium hirsutum*, L.). Aphids in mustard (*Sinapis alba*, L.) were detected using remote imagery by Kumar et al. (2010). Yang et al. (2005) used remote imagery to detect greenbug (*Schizaphis graminum* Rondani) stress in wheat. Russian wheat aphid (*Diuraphis noxia*, Mordvilko) damage was detected by remote imagery by Riedell and Blackmer (1999) and Mirik et al. (2007). Yang et al. (2007) used the NIR/Red ratio to detect rice (*Oryza sativa* L.) plants infected with leaf folder (*Cnaphalocrocis medinalis* Cabi) and brown planthopper (*Nilaparvata lugens* (Stål, 1854)), using different wavelengths depending on insect species and growth stage. Corn rootworm (*Diabrotica virgifera* LeConte, 1868) infection was detected and the severity was classified using remote imagery (Glaser et al. 2009) with an accuracy of 99 %.

Singh and Misra (2017) used a digital camera to classify disease infection in common bean and tree fruit. The steps were:

1. Image acquisition with a digital camera.
2. Pre-process image by clipping the portion of the image of most interest and then smoothing the remaining image. The image contrast was enhanced.
3. Green-coloured pixels were masked. Green-coloured pixels were selected by choosing a threshold value for green in the RGB colour makeup of the pixel. Pixels with green values below the threshold received a 0 value.
4. In infected clusters, masks were removed.
5. Obtain useful segments to classify disease using a genetic algorithm.

The accuracy of this method to the degree of leaf disease infection, which was always less than 100 % of leaf tissue affected, was about 90 %.

In a more automated approach, Bravo et al. (2003) mounted a spectrograph, which received reflections at 460–900 nm from every point on a line above the crop canopy measuring 0.5 m by 4 mm, at 70 cm above the canopy. The imagery was used to detect early season yellow rust (*Puccinia striiformis* Westend., (1854)) in infected wheat plants. Readings were filtered to eliminate interference from ambient light, and normalized using the 50 % grey technique available as a PhotoShop™ (Adobe Systems, San Jose, CA, USA) tool. Using this method, which included collection of field imagery and laboratory imagery analysis, accuracy of infection detection was about 98 %.

In a sugar beet field study, Mahlein et al. (2013) developed spectral indices using the spectrograph ImSpector V10E (Spectral Imaging Ltd., Oulu, Finland), combined with a mirror scanner (Spectral Imaging Ltd.) for detection of Cercospora leaf spot, sugar beet rust (*Puccinia subnitens* Dietel –syn. *P. aristidae* Tracy) and powdery mildew (*Erysiphe betae*- syn. *E. polygoni*) at early stages. (6 days after inoculation (dai) powdery mildew, 5 dai rust, 8 dai Cercospora). The configuration of the instruments for development of the indices is described in the paper. Specifics for the sugar beet spectral indices follow:

Healthy Plant Index: ((R534–R698)/(R534 + R698)) – ½(R704).
Cercospora Leaf Spot Index: ((R698–R570)/(R698 + R570)) – R734.
Sugar Beet Rust Index: ((R570–R513)/(R570 + R513)) + ½(R704).
Powdery Mildew Index: ((R520–R584)/(R520 + 584)) + R724.

The Healthy Plant Index was also related to Cercospora leaf spot, but the specific index developed for Cercospora leaf spot was more related to infection and damage than health.

Franke and Menz (2007) explored disease detection and severity in winter wheat at several stages of powdery mildew (*Blumeria graminis* (DC.) Speer (1975)) and development of leaf rust (*Puccinia recondite* Dietel & Holw. (1857)). High resolution remotely-sensed imagery was used for spatio-temporal analysis. A decision tree mixture tuned matched filtering (MTMF) process and NDVI were used to classify images into levels of disease severity. At the early onset of disease, the method was only 56.8 % accurate, but improved to 65.9 % and 88.6 % accuracy in later stage images.

Additional studies have been conducted on disease detection using multispectral remote sensing, wheat leaf rust (Ashourloo et al. 2014), take-all disease (*Gaeumannomyces graminis* (Sacc.) Arx & D.L. Olivier, (1952)) in wheat using Landsat multispectral imagery (Chen et al. 2007), brown rust (*Puccinia recondite* f. sp. *tritici*) in wheat using aerial imagery (Mewes 2010), and QuickBird aerial imagery for detection of powdery mildew (*Blumeria graminis*) and leaf rust (*Puccinia recondita*) in wheat at later growth stages. QuickBird imagery, however, was not useful in distinguishing early infection (Franke and Menz 2007). *Alternaria alternate* stress was detected in oilseed rape (*Brassica napus* L.) using multispectral imagery (Baranowski et al. 2015).

Mahlein (2016) conducted a limited review of the use of remote imagery for disease detection and classified remote imagery into red–green–blue (RGB true colour imagery), spectral sensors (multispectral), thermal sensors and fluorescence imagery. The use of sensors to determine the severity of various diseases from the review is given in Table 6.1.

Khaled et al. (2018) state that the main limitation with the use of passive spectroscopy for disease detection is variability of light in the environment. They suggested that researchers find wavelengths sensitive to plant disease that are least likely to be affected by the environment. When using fluorescence, 'photo-bleaching'

Table 6.1 Remote sensors used to detect certain diseases

Sensor	Crop	Disease	Author
RGB	Cotton	Bacterial angular (*Zanhomonas compestris*) (Pammel 1895) Dowson 1939	Camargo and Smith (2009)
	Sugar beet	Leaf spot (*Cercospora beticola*) Sacc., (1876)	Neumann et al. (2014)
		Beet rust (*Uromyces betae*) (Bellynck) Boerema (1987)	Neumann et al. (2014)
		Ramularia leaf spot (*Ramularia beticola*) Fautrey & F. Lamb., (1897)	Neumann et al. (2014)
		Phoma leaf spot (*Phoma betae* syn. *Neocamarosporium beatae*)	Neumann et al. (2014)
		Bacterial leaf spot (*Pseudomonas syringae* pv. Aptata)	Neumann et al. (2014)
Spectral	Barley	Barley net blotch (*Pyrenophora teres*) Drechs. (1923)	Kuska et al. (2015), Wahabzada et al. (2015)
		Brown rust (*Puccinia hordei*)	Kuska et al. (2015), Wahabzada et al. (2015)
		Powdery mildew (*Blumeria graminis hordei*)	Kuska et al. (2015), Wahabzada et al. (2015)
	Wheat	Head blight (*Fusarium graminearem*)	Bauriegel et al. (2011)
		Yellow rust (*Puccinia hordei*) G.H. Otth (1871)	Huang et al. (2007), Moshou et al. (2004)
	Sugar beet	Cercospora leaf spot (*C. beticola*)	Bergsträsser et al. (2015), Steddom et al. (2005), Mahlein et al. (2010, 2013)
		Rust (*Puccinia subnitens* Dietel syn. P. *aristidae* Tracy)	Mahlein et al. (2010, 2013)
		Rhizoctonia root rot (*Rhizoctonia solani*) J.G. Kühn (1858)	Hilnhütter and Mahlein (2008)
		Rhizomania (beet necrotic yellow vein virus)	Steddom et al. (2003)
		Powdery mildew (*Erysiphe betae* syn. *E. polygoni*)	Mahlein et al. (2010, 2013)
Thermal	Sugar beet	Cercospora leaf spot (*C. beticola*)	Chaerle et al. (2004)
Fluorescence	Wheat	Leaf rust (*Puccinia triticina*) Erikss. (1899)	Bürling et al. (2011a, b)
		Powdery mildew (*Blumeria graminis* f. sp. *tritici*)	Bürling et al. (2011a, 2012)
	Sugar beet	Cercospora leaf spot (*C. beticola*)	Chaerle et al. (2004)
	Common bean	Bacterial blight (*Xanthomonas*)	Rousseau et al. (2013)

or over-exposure can be a problem, but it can be solved by reducing light intensity on leaves. Another limitation for the use of remotely sensed imagery in disease detection is interference with nutritional stress in the same field. Although a crop scout in the field may be able to differentiate between a nutritional and disease stress (Figs. 6.4 and 6.5), it is more complicated for a sensor to detect the difference based on reflectance alone.

Because of the phenotypic differences between nutrient deficiencies and disease or insect feeding symptoms, more sophisticated and complex methods of identifying a specific pathogen or insect are required compared to rating nutritional status alone. A method to differentiate between disease and nutritional stress was examined by Bürling et al. (2011a). Blue–green and chlorophyll fluorescence was used

Fig. 6.4 Nitrogen deficiency symptoms in an individual corn plant at V6 (top) and in blocks of V10 corn. (Franzen images)

Fig. 6.5 Cercospora leaf spot in sugar beet (top), individual plant and larger field with area of more severe infection (circled). (Images courtesy of M. Khan, North Dakota State University)

on wheat plants with varying N status with and without inoculation with leaf rust or powdery mildew fungi. The fluorescence peaks were 451 nm (blue, B), 522 nm (green, G), 689 nm (red, R) and 737 nm (far-red, FR). The B/G and R/Fr ratios of the fluorescence differentiated N nutrition status and infection by the pathogens. This method of fluorescence spectroscopy was used to detect powdery mildew successfully on wheat leaves 1 day after inoculation (Bürling et al. 2012).

Leufen et al. (2014) used fluorescence lifetime (combination of 3-, 6-, 9-day measurements) using a laser spectrophotometer, fluorescence imagery with a camera and a portable multiparametric fluorescence sensor. The portable sensor was most practical for field assessment and it could detect early infection by powdery mildew and leaf rust independently, since each exhibit unique reflectance. Although most studies have been conducted through field-based platforms, airborne systems for use in solar-induced fluorescence have also been evaluated (Meroni et al. 2009).

Rumpf et al. (2010) used support vector machines (SVMs) in sugar beet disease management with classification algorithms to identify patterns in data. The data input was the ASD FieldSpec Pro FR spectrophotometer (Malvern Panalytical Ltd., Malvern, UK) with a plant probe fore-optic and leaf clip holder. The instrument output data were also combined with soil–plant analysis development (SPAD) data. The resulting class model was used to predict unlabelled data. For *Cercospora*, rust and powdery mildew, the SVM approach produced the smallest error in accuracy, detection of healthy leaves and diseased leaves compared with use of decision trees and artificial neural networks.

6.4 Nanosensors

Nanosensors allow exploration and detection of compounds within cells and within cell organelles (Kwak et al. 2017). Nanosensors have been used to detect differences in plant health from disease, nutrition and other environmental stresses (Kwak et al. 2017). Phyto-oestrogens are released as part of a plant defense system. Using sophisticated techniques, these volatiles can be detected using nanosensors. Nanoparticle-conjugated fluorescence reflectance energy transfer (FRET) probes have been used to detect some plant volatiles. Researchers used corona phase molecular recognition (CoPhMoRe)-based Bombolitin II nanosensors implanted within wild-type spinach (*Spinacia oleracea* L.) leaf tissues. This converted the tissues that functioned as autosamplers and preconcentrators of analytes from groundwater and could send the information to a smartphone. This is done with a pair of NIR fluorescent nanosensors embedded within the leaf tissues. One nanosensor used single-wall nanotubes (SWNT) attached to the peptide Bombolitin II to recognize nitroaromatics through IR fluorescence emission at 1100 nm after 5–15 min following addition of picric acid to the roots. The second IR channel is a poly vinyl acetate functionalized SWNT that acts as a standard reference signal. When excited by laser light, the difference in fluorescence between the standard and the SWNT that recognises nitroaromatics is distinguished and communicated electronically to a remote receiver (Kwak et al., 2017). Some disease detection instruments are used for quick recognition of chemical differences caused by disease onset. These ‘nose-device’ sensors (Cellini et al. 2017) do not allow for analytical determination of VOCs, but provide a quick-responding tool for disease recognition that can be quantified using other means.

6.5 Sensors for Crop Nutrient Management and Improvement of Crop Health and Vigour from Nutrient Deficiency

Sensors used for crop nutrient management detect phenotypic differences in the plants. Managing agricultural crops involves making many spatial and temporal observations. However, time, resources and ready access to all parts of fields frequently leads to compromises. One approach calls for crop managers to adopt proactive nutrient and pest management strategies, thereby reducing the need for more extensive crop monitoring.

When farmers and consultants monitor their agricultural crops, they usually have preconceived ideas about the kinds of problems to expect based on weather, landscape position and growth stage of the crop. Sometimes the physical effect of nutrient deficiencies is expressed in the form of leaf discolouration. When N is deficient, chlorophyll production will be reduced and yellow pigments will prevail (Fig. 6.4), resulting in pale green or light yellow leaves (Havlin et al. 2014). Because N is a mobile nutrient, chlorosis will appear in older leaves first. In maize, N deficiency starts at leaf tips and develops in a V-shaped pattern towards to the base of the leaf (Mulla and Miao 2016). Nitrogen stress can induce the production of polyphenolics (e.g. flavonoids), which can also be an indicator of N deficiency (Tremblay et al. 2012). These visible symptoms require time for the deficiency to be expressed. During the time lag between the onset of N deficiency and phenotypic expression severe enough to be noticed by the human eye, photosynthesis has been operating at sub-optimal levels and may possibly limit yield potential if the deficiency occurs at a critical growth stage. Nitrogen stress can limit crop growth, resulting in reduced plant height, leaf area index, biomass and grain yield. For these reasons, it is important to have technologies that provide early detection of nutrient stresses before yield potential is reduced.

Plant chlorophyll status is closely linked to photosynthesis and biomass production. As such, technologies to monitor and quantify crop chlorophyll status are at the forefront of optimizing nutrient management practices. Linking nutrient deficiencies to plant chlorophyll content is more of a challenge for some nutrients than others; however, those used in the greatest amounts, such as N, will have the greatest correlation with plant chlorophyll.

Crop leaf and canopy chlorophyll concentration or content have been used as indicators of crop N status. The SPAD chlorophyll meter introduced by Minolta in 1990 offered the first portable instrument used predominantly to quantify the chlorophyll content in leaves of crops such as maize (*Zea mays* L.), sorghum (*Sorghum bicolor* (L.) Moench ssp. *bicolor*), wheat and rice. The SPAD 502 (Spectrum Technologies, Plainfield, IL) chlorophyll meter (CM) is currently the most widely used leaf sensor. It is based on transmittance in the red (640 nm) and infrared (940 nm) spectral regions to estimate crop chlorophyll content, and indirectly estimates crop N status (Muñoz-Huerta et al. 2013). The challenge was that SPAD meter readings are affected by growth stage and cultivar, leaf position and previous crop. Standardized sampling protocols have considerably improved the quality of SPAD meter data (Shapiro et al. 2013).

Standardizing SPAD meter readings compared to an adequately fertilized crop has made it possible to compare data across hybrids and growth stages (Schepers et al. 1992). Rather than reporting the values as SPAD readings, the normalized data may be expressed as a sufficiency index (SI), defined as:

[SPAD reading in general crop area] / [SPAD value for adequately fertilized area]) (Peterson et al. 1993; Schepers et al. 1998).

Prior to the introduction of data standardization, agricultural literature lacked indexed yield, sensor and nutrition data. Current literature commonly reports standardized data for various kinds of plant and soil data.

The most common way to standardize data is to establish or define an N-sufficient reference area within the cultivar and field. Relative or SI values will generally be <1 and represent the percent of 'ideal'. An alternate approach is to invert the SI value resulting in values that are >1, indicating the percentage improvement possible. Mathematically, the SI convention is easier to interpret because the denominator remains constant rather than floating. Yield data are frequently standardized to the mean and then expressed as a percentile of the entire data set. This provides a statistical designation of a reference, e.g. two standard deviations above the mean, which approximates the 95-percentile value for normally distributed data. The N sufficiency index (NSI) is calculated by dividing field CM readings by the average CM value of the N-rich standard. In addition to a reference area, researchers used the mobility principle to calculate ratios of CM readings from leaves in different positions to diagnose crop N status (Lin et al. 2010; Yu et al. 2012).

Farm users observe differences in colour and vigour, but blending of colours between areas of N-sufficiency and the adjacent areas sometimes make it difficult to differentiate between some N rates. Further, the most appropriate N rate might change during the growing season and between field positions within a field. A larger scale approach was to generate a ramped N-rate strip with rectangular blocks (perhaps 5 × 15 m) receiving increasingly larger or smaller N rates (Raun et al. 2008). This approach enabled vehicle-mounted sensing and machine harvesting, but interpretation was frequently confounded by changes in soil type over the several hundred metres needed to accommodate a reasonable number of N rates (Fig. 6.6). Solari et al. (2008) working with maize increased the size of square blocks to the width of the planter to minimize the effect of wheel traffic. The 3 × 3 design accommodated a control plot and eight N rates within an area having the same soil type. These plots were large enough to facilitate active sensor and yield measurements without the influence of border effects. Normalized data were used to generate an in-season N rate algorithm (Solari et al. 2008). A farmer would use a similar width utilizing a small fertilizer application early in the season to compare field N availability to a sufficient N rate in the ramp.

The SPAD meters were designed to measure red waveband transmittance through leaf tissue. An abundance of red light is directed onto the surface of a leaf. Leaves with greater chlorophyll content can capture more red light and so less is

Fig. 6.6 Nitrogen application ramp used to aid in-season N application to winter wheat. (Image courtesy of B. Arnall, Oklahoma State University Extension)

transmitted through the leaf. This is in contrast to absorbance, cameras, spectroradiometers and hyperspectral scanners which capture reflectance of incident light. The SPAD meters generate light that is calibrated internally each time the device is used, whereas those that quantify reflectance are subject to changing light intensity (time of day, cloud cover and shadows). The devices that rely on light reflection from the sun are classified as 'passive' sensors in contrast to 'active' sensors that generate their own specific types and amounts of radiant energy. Spectroradiometers and hyperspectral scanners are used predominately in research to identify and characterize relations between reflectance in specific or combinations of wave bands and related plant or soil properties. One major outcome of experimentation has been the development of active sensors that measure reflectance in specific wavebands that are correlated with plant and soil properties of interest.

The first active sensors were the outgrowth of experiments with mini-computers in the 1970s. These systems were bulky and crude compared to modern devices that feature sophisticated electronics and precision optical systems. Early active sensors used mechanical choppers to segment a beam of light and photo-multiplier tubes to record reflectance. A tractor carried the mini-computer and an external generator needed to provide the high voltage required by the photo-multiplier circuitry. Development of transistors and integrated circuits for military applications has led to stable and high-speed circuitry, light-emitting diode (LED) sources of wave-band specific radiation and very sensitive photo-detectors. These advances have been complimented by the development of global navigation satellite system (GNSS) technologies again for military applications that have made site-specific management a reality.

Crop canopy sensors are more efficient than leaf sensors and have been used to estimate crop N status. The sensing difference is that the leaf sensors are specific to a plant, while output from a crop canopy sensor is an integration of multiple leaves from multiple plants. Previous research indicates that the green (530–590 nm) and red edge (around 710 nm) wavebands are much more sensitive to the variation in chlorophyll than blue, red or near-infrared wavebands at canopy closure (Gitelson and Merzlyak 1996; Hatfield 2015). Several vegetation indices have been developed to use proximal and remote sensing technologies to estimate crop canopy chlorophyll content, including normalized pigment chlorophyll index (NPCI) (Fillela et al. 1995), modified chlorophyll absorption in reflectance index (MCARI) (Daughtry et al. 2000), chlorophyll index using green band (CI_{green}) or red edge band (CI_{RE}) (Gitelson et al. 2005). An integrated vegetation index, transformed chlorophyll absorption in reflectance index/optimized soil-adjusted vegetation index (TCARI/OSAVI), was developed to improve sensitivity to canopy chlorophyll content and reduce the influence of leaf area index (LAI) and solar zenith angle (Haboudane et al. 2002). The CI_{green} can be calculated using active sensor Crop Circle ACS 210 (Holland Scientific, Inc., Lincoln, NE, USA) (590± 5.5 nm and or 880 ± 10 nm) or the Crop Circle ACS 470 (Holland Scientific, Inc., Lincoln, NE, USA) with a choice of six bands, while CI_{RE} can be calculated using active sensors like Crop Circle ACS 430 (670 nm, 730 nm, and 780 nm) and RapidSCAN CS-45 (Holland Scientific, Inc., Lincoln, NE, USA) sensor (670 nm, 730 nm and 780 nm) (Cao et al. 2013; Li et al. 2014). The Crop Circle ACS 430 sensor performed more consistently than the ACS470 sensor because of its independence of operational measurement height over the range of 0.3–2.0 m (Cao et al. 2018).

Sensing for crop N status is complex because anything short of tissue analysis involves one or more proxies that can be measured or monitored more easily. In most cases, the independent variable is a property related to N status and the dependent variable is usually crop yield. Photosynthesis is the process that links N status and yield, with chlorophyll being the component that captures the solar energy. Many factors and considerations affect photosynthesis with N status being the nutrient required in the largest amount. As such, measuring plant chlorophyll status is the most appropriate proxy to assess crop N status. Blue and red light are the primary wavebands captured in photosynthesis, but red light is more abundant and so it is more commonly measured to assess chlorophyll content. Living vegetation is especially effective at reflecting near-infrared (NIR) radiation. In combination, as plants grow, red reflectance decreases while NIR reflectance increases. These opposing trends are the basis for the normalized difference vegetation index (NDVI) and account for strong sensitivity to quantify properties such as leaf area index (LAI). The general formula for NDVI is:

$$\text{RNDVI}(\text{red-based}) = (\text{NIR} - \text{Red}) / (\text{NIR} + \text{Red}), \tag{6.1}$$

where NIR is near-infrared.

The NDVI measurements represent a point-in-time assessment, however, the value represents an integration of variables and conditions that influence photosynthesis earlier in the growing season. Plant N status, together with water, is a key requirement for photosynthesis. As such, the red-NDVI (RNDVI) of cereal crops is considered an acceptable proxy for plant N status, assuming other plant growth considerations are adequate and the leaves do not cover the row, which leads to saturation (Sharma et al. 2015). The RNDVI expression becomes progressively more insensitive as the amount of vegetation increases. This is because as LAI increases, red reflectance approaches zero (perhaps 2–5 %) and the response of RNDVI to the amount of living biomass decreases significantly (Gitelson and Merzlyak 1996, 1997). Green (550 nm) and amber (590 nm) reflectance also decreases as LAI increases, but not as rapidly as red reflectance. As such, specialized versions of the NDVI expression for these wavebands extend the sensitivity of measurements over a wider range of LAI situations.

The RNDVI also does not account for the three-dimensional development of plants. It is sensitive mainly to LAI, it does not respond to differences in biomass once leaf surfaces cover the soil nadir (Sharma et al. 2016). For example, a plant can be yellow and stunted, yet the leaves cover the row. As long as the leaves cover the row, the crop will be perceived by RNDVI to be as healthy and vigorous as a plant that is green from top to bottom with leaves also covering the row. Early work with RNDVI in active optical sensors was conducted on early spring growth of Bermuda grass (*Cynodon dactylon* L.) and winter wheat in Oklahoma (Mosali et al. 2007; Raun et al. 2001). Since the plants were small and the leaves did not completely cover the row, performance of RNDVI was not a deterrent to its ability to relate sensor readings to yield. However, the poor performance of RNDVI in more mature spring wheat, maize and sugar beet due to greater LAI at row closure is a problem for its use to direct mid- to late-season supplemental N to these crops were a problem (Bu et al. 2017).

The red edge waveband (~730 nm) has no practical bearing on photosynthesis and its reflectance from plant tissue is approximately the same as that of bare soil. It is commonly quoted or perceived that the red edge version of NDVI (NDRE) is more sensitive to plant chlorophyll content than RNDVI once the crop canopy closes. However, red edge reflectance changes very little throughout the growing season, be it from soil or plant tissue. The red edge version of NDVI is relatively more sensitive to LAI once the canopy closes because the numerator increases by the value of red edge reflectance and the denominator increases accordingly. The result is that the NDRE detects differences in 'tint' of leaves, light to dark, corresponding in N nutrition with yellow to green, meaning deficient to non-deficient healthy leaves (Horler et al. 1983; Sharma et al. 2015; Bu et al. 2017).

Algorithms that relate active optical sensor readings to crop yield have been developed for several crops, including winter wheat, spring wheat and maize (Franzen et al. 2014, 2016). An important part of most algorithms is to standardize the readings over a series of crop growth stages because it would be impractical to use the sensor at exactly the same growth stage over many fields in a commercial farming setting. The standardization calculation is called 'INSEY' (in-season estimate of yield) (Raun et al. 2002). Use of INSEY allows the algorithms to be used over at least a growth stage ±1 growth stage, allowing practical use in the field for in-season N fertilizer application.

Assessing late-season N status of maize supports the identification and correction of N deficiency, which is most severe when N has been lost from soil. But it can also occur when the initial N application was not sufficient for seasonal crop needs. Nitrogen loss is most commonly due to excess rainfall, leading to either leaching (on well-drained soils) or denitrification (on warm, poorly-drained soils). Only the nitrate form of N is subject to these loss processes, but conversion of fertilizer and soil-derived N to nitrate is a continuous (but temperature-and pH-dependent) soil microbial process.

Plant height is also related to crop yield and vigour (Bu et al. 2016; Sharma et al. 2016; Varela et al. 2017). Bu et al. (2016) and Sharma et al. (2016) used a ground-based platform with an acoustic crop height sensor for relating sensor height to sugar beet and maize yields, respectively. Varela et al. (2017) used a UAV approach subtracting a digital terrain model (DTM) of the study area from GNSS-related crop surface model (CSM). The model was developed mathematically from the triangulation of points of known distance. Another method using a modified difference between CSM and DTM is the structure from Motion (SfM) model (Holman et al. 2016). Holman used the SfM model in determining wheat plant height at two growth stages and monitoring differences in the rate of plant growth related to wheat height.

Multispectral imagery has been related to crop yield, which is the integrated result of crop health and vigour through a growing season (Shanahan et al. 2001). Estimates of maize mid-grain filling yield were best when made using green-based NDVI (GNDVI) compared with red-based NDVI and transformed soil adjusted vegetative index (TSAVI). Landsat multispectral imagery was combined with regression trees to compare the prediction of wheat yield in Mexico to determine variation in production between years. In some years, timing of the first irrigation was important and in another year fertilizer application was important.

In-season N losses and subsequent N deficiency can occur on any N-fertilized crop, but maize is the most vulnerable for several reasons. A large majority of the United States maize crop receives all N fertilizer before planting; however, rapid N uptake does not begin until 6–7 weeks after planting, and continues for another 4–6 weeks. This is longer than the time between N fertilizer application and N uptake for most other crops (e.g. winter wheat, cotton and forage grasses), meaning that there is more time for loss processes to occur in maize. In-season fertilizer application is a common nutrient management tool for winter wheat, cotton and forage grasses, partly because these crops are shorter and ground-based fertilizer application equipment can usually be driven through them. Because maize can be much taller and grows quickly, there is a risk of not being able to drive through it with the relevant equipment, which motivates farmers to fertilize earlier.

Although both soil-derived and fertilizer N can be lost, little can be done from a management perspective to limit loss of soil-derived N if seasonal rainfall is excessive. Applying N fertilizer later in the growing season reduces the risk of N loss before uptake; however, it increases the risk of the crop growing past critical stages of growth that are linked to important crop yield component decisions before N is applied. Later applications of N fertilizer to maize reduce but do not eliminate the risk of N deficiency. Under some conditions, fertilizer N can be lost in large amounts in a short time, especially in poorly-drained soils. Even if no fertilizer N is lost, loss of soil-derived N prior to fertilizer application can lead to N deficiency (Fig. 6.4), though it is unlikely to be severe. Nitrogen deficiency in maize is easily seen from the air (Fig. 6.4). Widespread N deficiency along the Iowa–Minnesota border in 2018 was due to excessive spring rainfall in this area (Fig. 6.7).

Evidence indicates that N loss from excessive rainfall has increased in the Midwestern USA by as much as 50 % from 1900 to 2002 (Groisman et al. 2004). Streamflow in the USA, especially the eastern USA, began to increase around 1970 (McCabe and Wolock 2002).

Given the short amount of time available for farmers or crop advisors to diagnose and treat N deficiency of maize, speed of diagnosis and application of supplemental N is important. The taller the maize, the fewer machines can physically apply fertilizer to alleviate N deficiency without damaging the crop. Remote-sensing imagery may be the fastest way to assess N sufficiency or deficiency over large areas, especially since the launch of 88 small satellites by Planet Labs in February, 2017. However, in growing seasons most susceptible to in-season N losses from excessive

Fig. 6.7 Aerial scene during a wet growing season near Clarion, IA in 2014. Dark green areas indicate corn that is sufficient in N nutrition, while yellow areas covering large parts of fields indicate N deficient corn. (Scharf image)

rainfall, the number of days where fields are not obscured by cloud cover is small (Bu et al. 2017).

Scharf and Hubbard (2017) showed that it is possible to estimate how much maize grain yield will be lost to N deficiency with aerial remotely-sensed images, and also how much fertilizer will be needed to correct the problem. Although most farmers can probably look at an image like Fig. 6.6 and identify correctly which fields have no problem, which have a small N deficiency problem, and which have a major problem, it may still be difficult for them to decide what N rate to use and whether it would be profitable to apply it. For fields with a major problem, applying more N fertilizer may be the correct decision. For fields with no visible problem, applying more N is unlikely to increase yield. But for fields with a 'small' problem, a quantitative estimate in dollars of the actual size of the problem would help farmers make a better decision regarding the economic return from applying supplemental N.

Examination of Fig. 6.7 reveals that most fields with lighter-coloured corn have large differences in plant colour, corresponding to wide differences in N deficiency. Nitrogen loss is nearly always spatially variable, with greater losses where soils accumulate water or where areas are sandier in texture. Greater deficiency requires greater supplemental rates of N. Remotely-sensed imagery and proximal remote sensing using passive or active reflectance to form the basis for automated supplemental N application. Digital data behind the digital images can be converted into rate control files to apply a greater rate of N where N deficiency is more severe (Scharf and Hubbard 2017). Proximal sensing can also help to guide variable N fertilizer applications in fields where N loss has occurred. However, those sensors are not a feasible solution for identifying which fields would benefit from additional N and which do not. Agricultural fertilizer supply businesses prefer to know how much fertilizer will be applied in a field before the fertilizer is applied. Therefore, proximal sensors can be used as a logistical prediction tool to project N fertilizer use. To correct these logistical issues, a remotely-sensed image of fields, each containing an N-sufficient area might be used to predict fields with N deficiency problems and estimate the total N fertilizer that would be required to correct the deficiency (Holland and Schepers 2013; Bu et al. 2016).

To have value, sensing of N deficiency and application of additional N fertilizer must occur at a time when the crop can provide a profitable response. Several studies have established that yield improvement in N deficient maize is possible in late vegetative stages until at least the silking stage (Miller et al. 1975; Binder et al. 2000; Scharf et al. 2002; Nelson et al. 2011; Mueller and Vyn 2018).

While NDVI is the most commonly used vegetation index, many other indices have been proposed depending on the plant variable in question. Hatfield and Prueger (2010) evaluated various vegetation indices and suggested appropriate ones for selected properties such as canopy greenness, LAI, vegetative cover and chlorophyll status. Active sensors are generally specific in terms of the featured wavebands (e.g. red, red edge and NIR). The target or scale of sensing is generally at the canopy level whereas many nutritional deficiencies are expressed uniquely at the leaf level. This is because some nutrients are mobile within plants and others are not. In general, canopy level sensors mask these colour patterns within leaves that are visible with individual plant observations. In the case of N deficiency, the visible symptoms can be observed first on lower leaves that are masked by upper leaves, so the only hope for imagery to detect this stress is through colour and density of the upper canopy. It is even more difficult to detect deficiencies of other nutrients that are required in smaller quantities than N. All things considered, sensors that assess the amount of living biomass and plant chlorophyll content have the greatest potential to improve crop management in general.

6.6 Indices Used to Identify Nutrient Deficiencies

Plant N concentration (PNC) and plant N uptake (PNU) have been used as indicators for assessing plant N status. Threshold values have been established for different crops (Fageria 2009). Researchers have used proximal and remote sensing technologies to estimate PNC and PNU directly to guide in-season N management (Li et al. 2010, 2014; Cao et al. 2015). Cao et al. (2015) found that the green difference vegetation index (GDVI) based on the Holland Scientific Crop Circle ACS 470 sensor (Holland Scientific, Lincoln, NE, USA) was strongly related to winter wheat PNC ($R^2 = 0.60$). Canopy chlorophyll content index (CCCI) is based on the theory of a two-dimensional planar domain illustrated by Clarke et al. (2001) and has been found to be a superior method to estimate N-related indicators (El-Shikha et al. 2008; Rodriguez et al. 2006; Fitzgerald et al. 2010; Li et al. 2014). In the calculation of CCCI, NDVI is used as a surrogate for ground cover and red edge NDVI (NDRE) is used as a measure of canopy N status to separate soil signal from plant signal, permitting a relative measure of N under different ground cover conditions (Fitzgerald et al. 2010). Li et al. (2014) found CCCI performed the best among the indices evaluated for estimating maize PNC at the V6–V7 ($R^2 = 0.65–0.68$) and at V10–V12 growth stages ($R^2 = 0.33–0.34$) followed by the MERIS terrestrial chlorophyll index (MTCI). Four red edge-based indices, CCCI, MTCI, NDRE and CIred edge, performed similarly well for estimating maize PNU ($R^2 = 0.76–0.91$) at the V10–V12 and V6–V12 growth stages. Chen et al. (2010) developed the double-peak canopy nitrogen index (DCNI), which was found to be the best spectral index among those evaluated for predicting PNC ($R^2 = 0.72$ for corn, 0.44 for wheat and 0.64 for both species combined).

Optimum PNC depends on above-ground biomass and decreases during the growth cycle in dense canopies (Greenwood et al. 1986). This relation to the above-ground biomass is independent of the climatic conditions of the year, or the species and genotype (Lemaire et al. 2005). Plant N uptake also varies within a single year, between years, sites and crops, even under sufficient N supplies (Gastal and Lemaire 2002). The critical N dilution curve describes the relation between the above-ground biomass and critical N concentration, which is defined as the minimum PNC necessary to achieve maximum above-ground biomass production (Greenwood et al. 1986). The N nutrition index (NNI) is a more reliable indicator to access the relative plant N status and can be calculated as the ratio of the measured N concentration over the critical N concentration (Lemaire and Gastal 1997). An NNI value of one indicates optimum N status, while a value greater or less than one indicates N surplus or deficient conditions. Other threshold values have been proposed to guide practice applications, such as 0.95–1.05 to indicate optimum N status (Huang et al. 2015; Xia et al. 2016).

There are several different approaches to the use of proximal and remote sensing technologies to estimate crop NNI non-destructively. One mechanistic approach is to use spectral vegetation indices to estimate crop above-ground biomass and PNC. The critical PNC at a known above-ground location can be calculated using the established critical N dilution curves for most crops (Bélanger et al. 2001; Huang et al. 2018; Lemaire et al. 2008; Plenet and Lemaire 1999). The NNI can then be calculated accordingly. This approach was termed as the NNI–PNC approach (Xia et al. 2016; Lu et al. 2017). Research indicates that crop above-ground biomass can be predicted quite reliably using passive or active canopy sensors, UAV, aerial or satellite remote sensing (Cao et al. 2015; Gnyp et al. 2014; Huang et al. 2015; Lu et al. 2017; Xia et al. 2016). However, the prediction of PNC early in the growing season to guide side- or top-dressing of N has been very challenging because of the influence of soil and water, plant structure, and LAI or biomass differences (Cao et al. 2015; Li et al. 2010, 2014; Yu et al. 2012; Huang et al. 2015). Later in the growing season, PNC can be predicted well using proximal and remote sensing technologies, but that would be too late to guide in-season N management for most crops.

A second mechanistic approach is to estimate PNU instead of PNC. The critical PNU can be calculated by multiplying above-ground biomass and critical PNC, and NNI can be calculated using the ratio of predicted PNU over critical PNU. This approach is termed NNI–PNU (Lu et al. 2017). Research indicates that PNU can be predicted better with proximal or remote sensing technologies than PNC, and this approach is more promising (Cao et al. 2015; Huang et al. 2015).

A third approach is NNI–VI, which is a semi-empirical approach using VIs to estimate NNI directly. This approach is influenced by crop phenological changes across growth stages and growth stage-specific models are needed (Chen 2015; Xia et al. 2016).

A fourth approach is a semi-empirical approach termed NNI–NSI–VI or NNI–RI–VI. This approach uses vegetation indices to calculate NSI or the response index (RI), which will then be used to estimate NNI (Lu et al. 2017; Xia et al. 2016). The RI is the inverse of SI. Nitrogen-rich reference plots or strips are used to normalize the sensor readings to reduce the influence of many different factors other than N (e.g. varieties, seasonal variation, plant water status, diseases and pests, plant population and growth stages) (Samborski et al. 2009). Another normalization approach is to calculate RI as an inverse of NSI (Mullen et al. 2003). The NSIs or RIs of CM readings or vegetation indices were closely related to NNIs across years, cultivars and growth stages and performed more consistently than the original CM readings or vegetation indices (Ziadi et al. 2008; Xia et al. 2016; Lu et al. 2017).

A fifth approach is based on terrestrial light detection and ranging (LiDAR) scanning (TLS) to quantify directly physical above-ground biomass proxies (crop height or volume) and derive PNC from green (532 nm) TLS point return intensity (Eitel et al. 2014). Results with winter wheat indicated that the above-ground biomass could be predicted well using TLS-derived vegetation volume across growth stages and seasons ($R^2 > = 0.72$), and the relation between PNC and green laser return intensity ($R^2 = 0.10–0.75$). The NNI could be predicted moderately well ($R^2 = 0.45–0.54$) based on the TLS-predicted above-ground biomass and PNC (Eitel et al. 2014).

Fluorescence sensors have also shown promise in the detection of crop nutrient deficiency. The Dualex sensor (ForceA, Orsay, France) is a fluorescence leaf sensor for estimating polyphenolics. Dualex 4 Scientific (D × 4) is a portable leaf-clip fluorescence sensor that measures leaf epidermal flavonoids at 375 nm based on the chlorophyll fluorescence screening method and leaf chlorophyll content based on the ratio of transmittance at 710 and 850 nm (Cerovic et al. 2012). The N balance index (NBI) can be calculated using the ratio of chlorophyll / flavonoid content. When N is deficient, leaf chlorophyll content will be decreased, but flavonoid content will be increased, consequently NBI was more sensitive to crop N status than chlorophyll or flavonoid alone (Tremblay et al. 2012). It was also more responsive to N fertilizer application in maize and less sensitive to the variation in soil-water content (Zhu et al. 2011).

The Multiplex (ForceA, Orsay, France) is a handheld canopy fluorescence sensor that can produce 66 fluorescence ratios that can be used to estimate canopy or leaf chlorophyll, flavonoids, anthocyanin and NBIs. This sensor can detect variation in maize as early as V5, showing great potential for the early detection of crop N stress (Longchamps and Khosla 2014). There were strong relations between Multiplex fluorescence indices and rice N indicators (leaf N concentration, PNC and NNI, with $R^2 = 0.40–0.78$) (Huang et al. 2019). One limitation is that the measurement can be made at a distance of only 10 cm from the light source and the measurement surface is a circle of 8 cm in diameter (Tremblay et al. 2012).

6.7 Use of UAV Remote Sensing Platforms

The exploration of unmanned aerial vehicles (UAVs) is a relatively recent development in remote sensing for nutrient management. Zermas et al. (2015) developed a methodology using high resolution RGB images collected with UAV flying at a low altitude and a computer vision program to identify maize leaves with V-shape chlorosis characteristics of N deficiency. They achieved 84.2 and 72.9 % accuracy under heavily and lightly deficient conditions, respectively. This research indicates that UAV imagery may have the potential to identify other stress factors when combined with computer vision technologies. Chen et al. (2019) used fixed-wing UAV remote sensing to estimate winter wheat NNI of smallholder farmer fields across a village and demonstrated the potential of UAV remote sensing for monitoring crop N status in small-scale farming systems.

6.8 Plant Health Assessment Through Soil Sensing

Another approach to understanding crop health is by sensing of the soil supporting the crop. While crop sensing technologies are performed to assess health of the above-ground plant while it is growing, soil sensing is generally done prior to crop establishment to gauge the soil's suitability to support plant growth. Among many functions, soil provides the medium giving physical support, water and nutrients for arable crops. Quantifying soil physical, chemical and biological properties that directly affect germination, emergence, growth and yield expression has been a major focus of agronomic research for centuries. However, soil sampling is labour intensive, requires multiple cores from each location with adequate mixing to produce a homogenous sample not only to represent area, but also depth for laboratory analysis to provide good representation of the root-zone for the location. Further, soils within cropped fields are often very variable spatially. This variation is the expression of many processes of soil formation working in different ways across soil landscapes. In addition to this natural variation is variation from historical crop and soil management within fields that has not always been uniform (e.g. tillage, manuring, fertilizer application). Consequently, arable fields are generally a composite of widely varying soil properties. This variability produces many unique environments at different scales, and therefore many phenotypic expressions of crop growth and yield.

Realization of soil variability within fields, and the need to develop economic methods and technologies to quantify soil properties, is the basis on which precision agriculture (PA) has evolved over the last three decades. That evolution has taken what was laboratory assessment of soil samples, to in-field stationary measurements, to on-the-go soil sensing systems with multiple properties being measured simultaneously. Interest in these in-situ and on-the-go soil sensing systems is large because it addresses variation that is intrinsic to soils. A more comprehensive description of the physics, engineering and performance of different soil sensing systems is described by Adamchuk et al. (2018) and is given in Chap. 4 of this book.

An example of how soil sensing has been used to understand and manage crop health is within-field bulk soil electrical conductivity (most often designated as apparent electrical conductivity, EC_a). Interest in soil EC_a began with assessing soil salinity, and in particular where salinity from irrigation caused crop stress (Rhoades and van Schilfgaarde 1976). Different commercial sensing technologies have emerged over recent decades to measure soil EC_a, including electromagnetic induction (Doolittle and Brevik 2014) and galvanic coupled resistivity that uses electrodes in direct contact with soil to both inject and measure a low-frequency current (Allred et al. 2008). In general, such technologies measure the soil's resistivity or conductivity, and this has been found to be affected by several soil properties that govern crop growth, such as salinity, clay content and cation exchange capacity (CEC), clay mineralogy, soil pore size and distribution, soil moisture content, bulk density and temperature (Rhoades et al. 1999).

The most successful applications of soil EC_a sensing in PA occur when a specific soil property has a dominant influence on the within-field variation of EC_a and that also has a direct impact on crop heath. For example, because soil EC_a integrates texture and moisture availability, two characteristics that both vary over the landscape and also affect crop growth and yield, EC_a sensing shows promise in interpreting the variation in crop yield in response to water stress. Certain soils especially express this relationship (Kitchen et al. 1999). A notable example of this relation is for soils with pronounced argillic horizons. Often the depth from the soil surface to the argillic horizon varies, either from natural landscape variation or because of topsoil erosion. The contrast in topsoil and the argillic subsoil conductivities using EC_a can be calibrated directly to the thickness of soil layers. Examples include EC_a regression for the depth of flood-induced sand deposition (Kitchen et al. 1996) and for topsoil depth above a subsoil argillic horizon (Doolittle et al. 1994; Sudduth et al. 2010) An example of a field mapped with soil EC_a converted to topsoil thickness above an argillic horizon is shown together with an aerial image and yield map (Fig. 6.8).

6.9 Conclusions for the Chapter

There are a variety of sensing tools that have been studied to identify crop stresses, both biotic and abiotic. Of the biotic stressors, the development of nose sensors may hold the best promise for advancement in insect feeding detection and injury severity. There have been many studies that have related remote sensors and remotely-sensed imagery with disease infection and severity. However, the studies that have demonstrated success indicate that disease identification and differentiation with other stressors is complicated and may require multiple steps, tools and algorithms before farmers can take advantage of the science. Despite the weight of research that has been devoted to disease and remote sensing, effective use of the science in commercial agriculture is still in the development stage.

Of all the stressors, use of remote sensing and remotely-sensed imagery is at the stage of development that they could be used by farmers today for N management stress. Even though research into the use of active sensors and remotely-sensed imagery shows that it could be used immediately by practitioners to direct in-season N application, its actual use by farmers is very low. Oklahoma leads the USA in the use of active-optical sensors for in-season N application in its winter wheat crop, with about 10 % of hectares fertilized with the system most recently (personal communication, Brian Arnall). The success in Oklahoma is due to a strong, consistent Extension program and demonstrations on farm fields over the past 11 years. The percentage of wheat hectares in Kansas where the system is used is estimated at about 2 %. In Kansas, a directed Extension program promoting and demonstrating

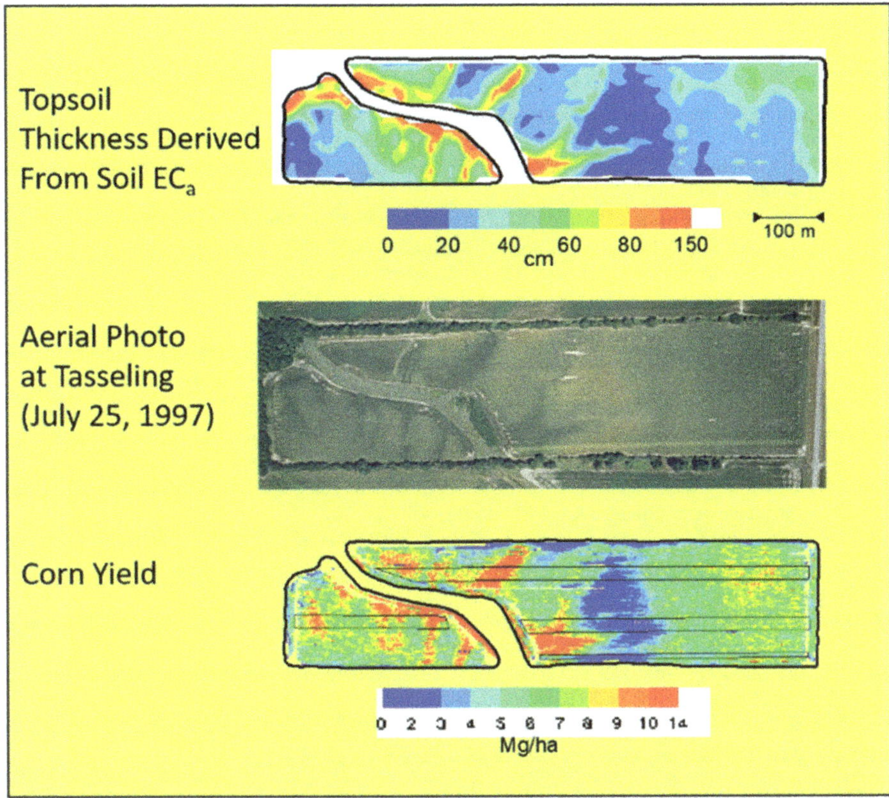

Fig. 6.8 A field in Missouri, USA showing soil EC_a converted to topsoil depth, aerial image, and yield map of maize. Smaller EC_a indicted deeper topsoil and thus greater potential water holding capacity, which translated into larger yields. (N, Kitchen images, University of Missouri)

the system has been in place for only about five years. In North Dakota, where algorithms for maize have been published for about three years, only a couple farmers use the system, although substantially more use the N-sufficient area to determine the need for supplemental N on a field basis. Feedback from North Dakota farmers indicates that although they are comfortable letting a tractor steer itself, they are far less comfortable allowing an automatic device to vary the rate of N on-the-go (Franzen et al. 2019).

The continued development of sensors for crop stress management should include consideration of the practical ease of its application. Use of the technology should not have to require undue effort by the user to reprogram present equipment, require purchase of expensive peripherals, or additional tools and connectors that might be a barrier to implementation. In addition, and maybe most importantly, there needs to be support from equipment suppliers or consulting services that understand and service the system and can guide individual farmers in their initial use, and would be readily available to answer questions and troubleshoot any issues that arise with the technologies.

References

Adamchuk V, Rossel W (2011) Precision agriculture: proximal soil sensing. In: Glinski J, Horabik J, Lipiec (eds) Encyclopedia of agrophysics. Springer, New York, pp 650–656

Adamchuk V, Rossel W, Gebbers RV et al (2018) Proximal soil and plant sensing. In: Shannon DK, Clay DE, Kitchen NR (eds) Precision agriculture basics. ASA, CSSA, and SSSA, Madison, pp 119–140

Allred BJ, Groom D, Ehsani MR et al (2008) Chapter 5: resistivity methods. In: Allred BJ, Daniels JJ, Ehsani MR (eds) Handbook of agricultural geophysics. CRC Press, Boca Raton, pp 85–108

Ashourloo D, Mobasheri MR, Huete A (2014) Developing two spectral disease indices for detection of wheat leaf rust (*Pucciniatriticina*). Remote Sens 6:4723–4740

Baldwin IT (2010) Plant volatiles. Curr Biol 20:R392–R397

Baranowski P, Jedryczka M, Mazurek W et al (2015) Hyperspectral and thermal imaging of oilseed rape (*Brassica napus*) response to fungal species of the genus *Alternaria*. PLoS One 10:e0122913

Bauriegel E, Giebel A, Geyer M et al (2011) Early detection of *Fusarium* infection in wheat using hyper-spectral imaging. Comp Electron Agric 75:304–312

Bélanger G, Walsh JR, Richards JE et al (2001) Critical nitrogen curve and nitrogen nutrition index for potato in eastern Canada. Am J Potato Res 78:355–364

Bergsträsser S, Fanourakis D, Schmittgen S et al (2015) HyperART: non-invasive quantification of leaf traits using hyperspectral absorption-reflectance-transmittance imaging. Plant Methods 11:1

Binder DL, Sander DH, Walters DT (2000) Maize response to time of nitrogen application as affected by level of nitrogen deficiency. Agron J 92:1228–1236

Bravo C, Moshou D, West J et al (2003) Early disease detection in wheat fields using spectral reflectance. Biosyst Eng 84:137–145

Bu H, Sharma LK, Denton A et al (2016) Sugar beet yield and quality prediction at multiple harvest dates using active-optical sensors. Agron J 108:273–284

Bu H, Sharma LK, Denton A et al (2017) Comparison of satellite imagery and ground-based active optical sensors as yield predictors in sugar beet, spring wheat, corn and sunflower. Agron J 109:1–10

Bürling K, Hunsche M, Noga G (2011a) Use of blue-green and chlorophyll fluorescence measurements for differentiation between nitrogen deficiency and pathogen infection in wheat. J Plant Physiol 168:1641–1648

Bürling K, Hunsche M, Noga G et al (2011b) UV-induced fluorescence spectra and lifetime determination for detection of leaf rust (*Puccinia triticina*) in susceptible and resistant wheat (*Triticum aestivum*) cultivars. Funct Plant Biol 38(4):337–345

Bürling K, Hunsche M, Noga G (2012) Presymptomatic detection of powdery mildew infection in winter wheat cultivars by laser-induced fluorescence. Appl Spectrosc 66:1411–1419

Camargo A, Smith JS (2009) Image pattern classification for the identification of disease causing agents in plants. Comp Elec Agric 66:121–125

Cao Q, Miao Y, Wang H et al (2013) Non-destructive estimation of rice plant nitrogen status with crop circle multispectral active canopy sensor. Field Crops Res 154:133–144

Cao Q, Miao Y, Feng G et al (2015) Active canopy sensing of winter wheat nitrogen status: an evaluation of two sensor systems. Comp Electron Agric 112:54–67

Cao Q, Miao Y, Shen J et al (2018) Evaluating two crop circle active canopy sensors for in-season diagnosis of winter wheat nitrogen status. Agronomy 8:201

Cellini A, Blasioli S, Biondi E et al (2017) Potential applications and limitations of electronic nose devices for plant disease diagnosis. Sensors 17:2596

Cerovic ZG, Masdoumier G, Ghozlen NB et al (2012) A new optical leaf-clip meter for simultaneous non-destructive assessment of leaf chlorophyll and epidermal flavonoids. Physiol Plant 146:251–260

Chaerle L, Van der Straeten D (2000) Imaging techniques and the early detection of plant stress. Trends Plant Sci 5:495–501. http://www.cell.com/trends/plant-science/pdf/S1360-1385(00)01781-7.pdf

Chaerle L, Hagenbeek D, De Bruyne E et al (2004) Thermal and chlorophyll-fluorescence imaging distinguish plant-pathogen interactions at an early stage. Plant Cell Physiol 45:887–896

Champagne CM, Staenz K, Bannari A et al (2003) Validation of a hyperspectral curve-fitting model for the estimation of plant water content of agricultural canopies. Remote Sens Environ 87:295–309

Chen P (2015) A comparison of two approaches for estimating the wheat nitrogen nutrition index using remote sensing. Remote Sens 7:4527–4548

Chen X, Ma J, Qiao H et al (2007) Detecting infestation of take-all disease in wheat using Landsat thematic mapper imagery. Int J Remote Sens 28:5183–5189

Chen P, Haboudane D, Tremblay N et al (2010) New spectral indicator assessing the efficiency of crop nitrogen treatment in corn and wheat. Remote Sens Environ 114:1987–1997

Chen Z, Miao Y, Lu J et al (2019) In-season diagnosis of winter wheat nitrogen status in smallholder farmer fields across a village using unmanned aerial vehicle-based remote sensing. Agronomy 9:619

Clarke TR, Moran MS, Barnes EM et al (2001) Planar domain indices: a method for measuring a quality of a single component in two-component pixels. In: Proceedings of IEEE international geoscience and remote sensing symposium (CD ROM) Sydney, Australia, 9–13 July 2001

Costa JM, Grant OM, Chaves MM (2013) Thermography to explore plant-environment interactions. J Exp Bot 64:3937–3949

Dandawate Y, Kokare R (2015) An automated approach for classification of plant diseases towards development of futuristic decision support system in Indian perspective. In: Proceedings of the 2015 International Conference on Advances in Computing, Communications and Informatics (ICACCI) August 10–13, 2015, Keralaq, India, pp 794–799. https://ieeexplore.ieee.org/stamp/stamp.jsp?tp=andarnumber=6141424

Daughtry CST, Walthall CL, Kim MS et al (2000) Estimating corn leaf chlorophyll concentration from leaf and canopy reflectance. Remote Sens Environ 74:229–239

de Moraes CM, Schultz JC, Mescher MC et al (2004) Induced plant signaling and its implications for environmental sensing. J Toxicol Environ Health A 67(8–10):819–834

Doolittle JA, Brevik EC (2014) The use of electromagnetic induction techniques in soils studies. Geoderma 223–225:33–45

Doolittle JA, Sudduth KA, Kitchen NR et al (1994) Estimating depths to claypans using electromagnetic induction methods. J Soil Water Conserv 49:572–575

Eitel JUH, Magney TS, Vierling LA et al (2014) LiDAR based biomass and crop nitrogen estimates for rapid, non-destructive assessment of wheat nitrogen status. Field Crops Res 159:21–32

El-Shikha DM, Barnes EM, Clarke TR et al (2008) Remote sensing of cotton nitrogen status using the canopy chlorophyll content index (CCCI). Trans ASAE 51:73–82

Fageria NK (2009) The use of nutrients in crop plants. CRC Press/Taylor and Francis Group, Boca Raton

Fensholt R, Sandholt I (2003) Derivation of a shortwave infrared water stress index from MODIS near- and shortwave infrared data in a semiarid environment. Remote Sens Environ 87:111–121

Fillela I, Serrana L, Sevra J, Peruelas J (1995) Evaluating wheat nitrogen status with canopy reflectance indices and discriminant analysis. Crop Sci 35:1400–1405

Fitzgerald G, Rodriguez D, O'Leary G (2010) Measuring and predicting canopy nitrogen nutrition in wheat using a spectral index – the canopy chlorophyll content index (CCCI). Field Crops Res 116:318–324

Franke J, Menz G (2007) Multi-temporal wheat disease detection by multi-spectral remote sensing. Precis Agric 8:161–172

Franzen D, Sharma LK, Bu H (2014) Active optical sensor algorithms for corn yield prediction and a corn side-dress nitrogen rate aid. North Dakota St. Univ. Ext. Cir. SF1176–5. https://www.ag.ndsu.edu/publications/crops/site-specific-farming-5-active-optical-sensor-algorithms-

for-corn-yield-prediction-and-a-corn-side-dress-nitrogen-rate-aid/sf1176-5.pdf. Accessed Jan 2019
Franzen D, Kitchen N, Holland K et al (2016) Algorithms for in-season nutrient management in cereals. Agron J 108:1775–1781
Franzen DW, Sharma LK, Schultz EC et al (2019) Integrated approach for site-specific nitrogen management in North Dakota, USA. In: Stafford JV (ed) Precision agriculture '19. Wageningen Academic, Wageningen, pp 525–530
Gao BC (1996) NDWI – a normalized difference water index for remote sensing of vegetation liquid water from space. Remote Sens Environ 58:257–266
Gastal F, Lemaire G (2002) N uptake and distribution in crops: an agronomical and ecophysiological perspective. J Exp Bot 53:789–799
Genc H, Genc L, Turhan H et al (2008) Vegetation indices as indicators of damage by the sunn pest (*Hemiptera:Scutelleridae*) to field grown wheat. Afr J Biotechnol 7(2):173–180
Gerhards M, Schlerf M, Mallick K et al (2019) Thermal infrared remote sensing for crop water-stress detection: a review. Remote Sens 11:1240
Gitelson AA, Merzlyak MN (1996) Signature analysis of leaf reflectance spectra: algorithm development for remote sensing of chlorophyll. J Plant Phys 148:494–500
Gitelson AA, Merzlyak MN (1997) Remote estimation of chlorophyll content in higher plant leaves. Int J Remote Sens 18:2691–2697
Gitelson AA, Viña A, Rundquist DC et al (2005) Remote estimation of canopy chlorophyll content in crops. Geophys Res Lett 32:L08403
Glaser J, Cases J, Copenhaver K et al (2009) Development of a broad landscape monitoring system using hyperspectral imagery to detect pest infestation. In: Proceedings of the First Workshop on Hyperspectral Image and Signal Processing- Evolution in Remote Sensing (WHISPERS'09). Grenoble, France, pp 1–4
Gnyp ML, Miao Y, Yuan F et al (2014) Hyperspectral canopy sensing of paddy rice above-ground biomass at different growth stages. Field Crops Res 155:42–55
Greenwood D, Neeteson J, Draycott A (1986) Quantitative relationships for the dependence of growth rate of arable crops on their nitrogen content, dry weight and aerial environment. Plant Soil 91:281–301
Groisman PY, Knight RW, Karl TR et al (2004) Contemporary changes of the hydrological cycle over the contiguous United States: trends derived from *in situ* observations. J Hydrometeorol 5:64–85
Haboudane D, Miller JR, Tremblay N et al (2002) Integrated narrow-band vegetation indices for prediction of crop chlorophyll content for application to precision agriculture. Remote Sens Environ 81:416–426
Hatfield JL (2015) Precision nutrient management and crop sensing. In: Kumar J, Pratap A, Kumar S (eds) Phenomics in crop plants: trends, options and limitations, Springer, New Delhi, pp 207–221
Hatfield JL, Pinter PJ Jr (1993) Remote sensing for crop protection. Crop Prot 12:403–413
Hatfield JL, Prueger JH (2010) Value of using different vegetative indices to quantify agricultural crop characteristics at different growth stages under varying management practices. Remote Sens 2:562–578
Hatfield JL, Gitelson AA, Schepers JS et al (2008) Application of spectral remote sensing for agronomic decisions. Agron J 100:117–131
Havlin JL, Tisdale SL, Nelson WL et al (2014) Soil fertility and fertilizers: an introduction to nutrient management, 8th edn. Pearson, Upper Saddle River, p 309
Hillnhütter C, Mahlein A-K (2008) Neue Ansätze zur frühzeitigen Erkennung und lokalisierung von zuckerrübenkrankheiten. (Early detection and localization of sugar beet disease: new approaches- in German). Gesunde Pflanz 60:143–149
Holland KH, Schepers JS (2013) Use of a virtual-reference concept to interpret active crop canopy sensor data. Precis Agric 14:71–85

Holman FH, Riche AB, Michalski A et al (2016) High throughput field phenotyping of wheat plant height and growth rate in field plot trials using UAV based remote sensing. Remote Sens 8:1031

Holopainen JK, Blande JD (2012) Molecular plant volatile communication. In: Lopez-Larrea C (ed) Sensing in nature. Springer, New York, pp 17–31

Horler DNH, Dockray M, Barber J (1983) The red edge of plant leaf reflectance. Int J Remote Sens 4:273–288

Huang W, Lamb DW, Niu Z et al (2007) Identification of yellow rust in wheat using in-situ spectral reflectance measurements and airborne hyperspectral imaging. Precis Agric 8:187–197

Huang S, Miao Y, Zhao G et al (2015) Satellite remote sensing-based in-season diagnosis of rice nitrogen status in Northeast China. Remote Sens 7:10646–10667

Huang S, Miao Y, Cao Q et al (2018) A new critical nitrogen dilution curve for rice nitrogen status diagnosis in Northeast China. Pedosphere 28:814–822

Huang S, Miao Y, Yuan F et al (2019) In-season diagnosis of rice nitrogen status using proximal fluorescence canopy sensor at different growth stages. Remote Sens 11:1847

Hunt ER, Ustin SL, Riano D (2013) Remote sensing of leaf, canopy and vegetation water contents for satellite environmental data records. In: Qu J, Powell A, Sivakumar MVK (eds) Satellite-based applications on climate change. Springer, Dordrecht, pp 335–357

Kang WS, Hong SS, Han YK et al (2010) A web-based information system for plant disease forecast based on weather data at high spatial resolution. Plant Pathol J 26:37–48

Khaled AY, Aziz SA, Bejo SK et al (2018) Early detection of diseases in plant tissue using spectroscopy-applications and limitations. Appl Spectrosc Rev 53:36–64

Kitchen NR, Sudduth KA, Drummond ST (1996) Mapping of sand deposition from 1993 Midwest floods with electromagnetic induction measurements. J Soil Water Conserv 51:336–340

Kitchen NR, Sudduth KA, Drummond ST (1999) Soil electrical conductivity as a crop productivity measure for claypan soils. J Prod Agric 12:607–617

Kumar J, Vashisth A, Sehgal VK et al (2010) Identification of aphid infestation in mustard by hyperspectral remote sensing. J Agric Phys 10:53–60

Kuska M, Wahabzada M, Leucker M et al (2015) Hyperspectral phenotyping on microscopic scale towards automated characterization of plant-pathogen interactions. Plant Methods 11:28

Kwak S-Y, Wong MH, Lew TTS et al (2017) Nanosensor technology applied to living plant systems. Annu Rev Anal Chem 10:113–140

Laothawomkitkul J, Moore JP, Taylor JE et al (2008) Discrimination of plant volatile signatures by an electronic nose: a potential technology for plant pest and disease monitoring. Environ Sci Technol 42:8433–8439

Lemaire G, Gastal F (1997) N uptake and distribution in plant canopies. In: Lemaire G (ed) Diagnosis of the nitrogen status in crops. Springer, Berlin, pp 3–43

Lemaire G, Avice JC, Kim TH et al (2005) Developmental changes in shoot N dynamics of Lucerne in relation to leaf growth dynamics as a function of plant density and hierarchical position within the canopy. J Exp Bot 56:935–943

Lemaire G, Jeuffroy MH, Gastal F (2008) Diagnosis tool for plant and crop N status in vegetative stage: theory and practices for crop N management. Eur J Agron 28:614–624

Leufen G, Noga G, Hunsche M (2014) Proximal sensing of plant-pathogen interactions in spring barley with three fluorescence techniques. Sensors 14:11135–11152

Li F, Miao Y, Hennig SD et al (2010) Evaluating hyperspectral vegetation indices for estimating nitrogen concentration of winter wheat at different growth stages. Precis Agric 11:335–357

Li F, Miao Y, Feng G et al (2014) Improving estimation of summer maize nitrogen status with red edge-based spectral vegetation indices. Field Crops Res 157:111–123

Lin F, Qiu L, Deng J et al (2010) Investigation of SPAD meter-based indices for estimating rice nitrogen status. Comput Electron Agric 71:S60–S65

Lintz J, Simonett DS (1976) Remote sensing environment. Addison-Wesley, Reading, p 1, 694 pp

Longchamps L, Khosla R (2014) Early detection of nitrogen variability in maize using fluorescence. Agron J 106:511–518

Lu J, Miao Y, Shi W et al (2017) Evaluating different approaches to non-destructive nitrogen status diagnosis of rice using portable RapidSCAN active canopy sensor. Sci Rep 7:14073

Mahlein A-K (2016) Plant disease detection by imaging sensors-parallels and specific demands for precision agriculture and plant phenotyping. Plant Dis 100:241–251

Mahlein A-K, Steiner U, Hillnhütter C et al (2010) Spectral signatures of sugar beet leaves for the detection and differentiation of diseases. Precis Agric 11:413–431

Mahlein A-K, Rumpf T, Welke P et al (2013) Development of spectral indices for detecting and identifying plant diseases. Remote Sens Environ 128:21–30

Martinelli F, Scalenghe R, Davino S et al (2014) Advanced methods of plant disease detection. A review. Agron Sustain Dev

McCabe GJ, Wolock DM (2002) A step increase in streamflow in the conterminous United States. Geophys Res Lett 29:2185

Meroni M, Rossini M, Guanter L et al (2009) Remote sensing of solar-induced chlorophyll fluorescence: review of methods and applications. Remote Sens Environ 113:2037–2051

Meroni M, Rossini M, Colombo R (2010) Characterization of leaf physiology using reflectance and fluorescence hyperspectral measurements. In: Maselli F, Menenti M, Brivio PA (eds) Optical observation of vegetation properties and characteristics. Research signpost. Trivandrum. Thycaud, Kerala, pp 165–187

Mewes T (2010) The impact of the spectral dimension of hyperspectral datasets on plant disease detection. PhD dissertation, University of Bonn, Bonn, Germany. http://hss.ulb.uni-bonn.de/2011/2475/2475.htm. Accessed October 2019

Miller HF, Kavanaugh J, Thomas GW (1975) Time of N application and yields of corn in wet, alluvial soils. Agron J 67:401–404

Mirik M, Michels GJ, Kassymzhanova-Mirik S et al (2007) Reflectance characteristics of Russian wheat aphid (*Hemiptera: Aphididae*) stress and abundance in winter wheat. Comput Electron Agric 57:123–134

Mogensen VO, Jensen CR, Mortensen G et al (1996) Spectral reflectance index as an indicator of drought of field grown oilseed rape (*Brassica napus* L.). Eur J Agron 5:125–135

Moran MS, Inoue Y, Barnes EM (1997) Opportunities and limitations for image-base remote sensing in precision crop management. Rem Sens Env 61:319–346

Mosa KA, Ismail A, Helmy M (2017) Introduction to plant stresses. In: Plant stress tolerance, Springer briefs in systems biology. Springer, Cham

Mosali J, Girma K, Teal RK et al (2007) Use of in-season reflectance for predicting yield potential in Bermudagrass. Commun Soil Sci Plant Anal 38:1519–1531

Moshou D, Bravo C, West J et al (2004) Automatic detection of 'yellow rust' in wheat using reflectance measurements and neural networks. Comput Elecron Agric 44:173–188

Mueller SM, Vyn TJ (2018) Physiological constraints to realizing maize grain yield recovery with silking-stage nitrogen fertilizer applications. Field Crops Res

Mulla DJ, Miao Y (2016) Precision farming. In: Thenkabail PS (ed) Land resources monitoring, modeling, and mapping with remote sensing. CRC Press, Boca Raton, pp 161–178

Mullen RW, Freeman KW, Raun WR et al (2003) Identifying an in-season response index and then potential to increase wheat yield with nitrogen. Agron J 95:347–351

Muñoz-Huerta RF, Guevara-Gonzalez RG, Contreras-Medina LM et al (2013) A review of methods for sensing the nitrogen status in plants: advantages, disadvantages and recent advances. Sensors 13:10823–10843

Nelson KA, Scharf PC, Stevens WE et al (2011) Rescue nitrogen applications for corn. Soil Sci Soc Am J 75:143–151

Neumann M, Hallau L, Klatt B et al (2014) Erosion band features for cell phone image based plant disease classification. In: Proceedings of the 22nd International Conference on Pattern Recognition (ICPR 2014). Los Alamitos, CA, USA: IEEE 3315–3320

Newlands NK (2018) Model-based forecasting of agricultural crop disease risk at the regional scale, integrating airborne inoculum, environmental, and satellite-based monitoring data. Front Environ Sci 6(63):1–16

North Dakota Agricultural Weather Network (NDAWN) (2010) A threshold selection method from gray-level histograms. IEEE Trans Syst 9: 62–66. https://ndawn.ndsu.nodak.edu. Accessed Oct 2019
Otsu N (1979) A threshold selection method from gray level histograms. IEEE Trans. Sys Man Cyber 9:62–66. https://doi.org/10.1109/TSMC.1979.4310076
Peterson TA, Blackmer TM, Francis DD et al (1993) Using a chlorophyll meter to improve N management. Coop. Ext. Service, Univ. Nebraska, NebGuide G93-1171A
Plenet D, Lemaire G (1999) Relationship between dynamics of nitrogen uptake and dry matter accumulation in maize crops. Determination and critical nitrogen concentration. Plant Soil 2116:65–82
Ranjitha G, Srinivasan MR, Rajesh A (2014) Detection and estimation of damage caused by thrips (*Thrips tabaci*, Lind) of cotton using hyperspectral radiometer. Agrotech 3(1):1–5
Raun WR, Johnson GV, Stone ML et al (2001) In-season prediction of potential grain yield in winter wheat using canopy reflectance. Agron J 93:131–138
Raun WR, Solie JB, Johnson GV et al (2002) Improving nitrogen use efficiency in cereal grain production with optical sensing and variable rate application. Agron J 94:815–820
Raun WR, Solie JB, Taylor RK et al (2008) Ramp calibration strip technology for determining midseason nitrogen rates in corn and wheat. Agron J 100:1088–1093
Rhoades JD, van Schilfgaarde J (1976) An electrical conductivity probe for determining soil salinity. Soil Sci Soc J 40:647–651
Rhoades JD, Corwin DL, Lesch SM (1999) Geospatial measurements of soil electrical conductivity to assess soil salinity and diffuse salt loading from irrigation. In: Corwin DK (ed) Assessment of non-point source pollution in the vadose zone, Geophysical monograph, vol 108. American Geophysical Union, Washington, DC, pp 197–215
Riedell WE, Blackmer TM (1999) Leaf reflectance spectra of cereal aphid-damaged wheat. Crop Sci 39:1835–1840
Rodriguez D, Fitzgerald GJ, Belford R et al (2006) Detection of nitrogen deficiency in wheat from spectral reflectance indices and basic crop eco-physiological concepts. Aust J Agric Res 57:781–789
Rousseau C, Belin E, Bove E et al (2013) High throughput quantitative phenotyping of plant resistance using chlorophyll fluorescence image analysis. Plant Methods 9:17
Rumpf T, Mahlein A-K, Steiner U et al (2010) Early detection and classification of plant diseases with support vector machines based on hyperspectral reflectance. Comput Electron Agric 74:91–99
Samborski SM, Tremblay N, Fallon E (2009) Strategies to make use of plant sensors-based diagnostic information for nitrogen recommendations. Agron J 101:800–816
Sankaran S, Mishra A, Ehsani R, Davis C (2010) A review of advanced techniques for detecting plant diseases. Comp Elec Agric 72:1–13
Scharf PC, Hubbard VC (2017). Method of predicting crop yield loss due to nitrogen deficiency: U.S. Patent No. 9,652,691. http://patft.uspto.gov/netacgi/nph-Parser?Sect1=PTO1andSect2=HITOFFandd=PALLandp=1andu= %2Fnetahtml %2FPTO %2Fsrchnum.htmandr=1andf=Gandl=50ands1=9,652,691.PN.andOS=PN/9,652,691andRS=PN/9,652,691. Accessed Oct 2019
Scharf PC, Wiebold WJ, Lory JA (2002) Corn yield response to nitrogen fertilizer timing and deficiency level. Agron J 94:435–441
Schepers JS, Francis DD, Vigil M et al (1992) Comparison of corn leaf nitrogen and chlorophyll meter readings. Commun Soil Sci Plant Anal 23:2173–2187
Schepers JS, Blackmer TM, Francis DD (1998) Chlorophyll meter method for estimating nitrogen content in plant tissue. In: Kalra YP (ed) Handbook on reference methods for plant analysis. CRC Press, Baton Rouge, pp 129–135
Shanahan JF, Schepers JS, Francis DD et al (2001) Use of remote-sensing imagery to estimate corn grain yield. Agron J 93:583–589

Shapiro CA, Francis DD, Ferguson RB et al (2013) Using a chlorophyll meter to improve N management. Univ. Nebraska Ext. Pub. G1632. https://www.specmeters.com/assets/1/22/SPAD_Using_a_chlorophyll_meter.pdf. Accessed Oct 2019
Sharma LK, Bu H, Denton A et al (2015) Active-optical sensors using red NDVI compared to red edge NDVI for prediction of corn grain yield in North Dakota, U.S.a. Sensors 15:27832–27853
Sharma LK, Bu H, Franzen DW et al (2016) Use of corn height measured with an acoustic sensor improves yield estimation with ground based active optical sensors. Comput Electron Agric 124:254–262
Shtienberg D (2013) Will decision-support systems be widely used for the management of plant diseases? Annu Rev Phytopathol 51:1–16
Singh V, Misra AK (2017) Detection of plant leaf diseases using image segmentation and soft computing techniques. Inf Process Agric 4:41–49
Solari F, Shanahan J, Ferguson R et al (2008) Active sensor reflectance measurements of corn nitrogen status and yield potential. Agron J 100:571–579
Steddom K, Heidel G, Jones D et al (2003) Remote detection of Rhizomania in sugar beet. Phytopathology 93:720–726
Steddom K, Bredehoeft MW, Khan M et al (2005) Comparison of visual and multispectral radiometric disease evaluations of *Cercospora* leaf spot of sugar beet. Plant Dis 89:153–158
Sudduth KA, Kitchen NR, Myers DB et al (2010) Mapping depth to argillic soil horizons using apparent electrical conductivity. J Environ Eng Geophys 15:135–146
Taghvaeian S, Chavez JL, Altenhofen J et al (2013) Remote sensing for evaluating crop water stress at field scale using infra-red thermography: potential and limitations. In: Proceedings of the 2013 Hydrology Days Conference. Colorado State University, Fort Collins, Colorado, pp 73–83. https://pdfs.semanticscholar.org/38cf/7b65d11b67ce9241ed7f99aec516506e357c.pdf. Accessed October 2019
Tremblay N, Wang Z, Cerovic ZG (2012) Sensing crop nitrogen status with fluorescence indicators. A review. Agron Sustain Dev 32:451–464
Tripathy AK, Adinarayana J, Sudharsan D et al (2011) Data mining and wireless sensor network for agriculture pest/disease predictions. In: Proceedings of the 2011 World Congress on Information and Communication Technologies. Dec 11–14, 2011, Mumbai, India, pp 1229–1234
Turban E, Aronson JE, Liang TP (2005) Decision support systems and intelligent systems, 7th edn. Pearson Prentice Hall, Upper Saddle River
Varela S, Assefa Y, Vara Prasad PV et al (2017) Spatio-temporal evaluation of plant height in corn via unmanned aerial systems. J Appl Remote Sens 11:036013
Wahabzada M, Mahlein A-K, Bauckhage C et al (2015) Metro maps of plant disease dynamics – automated mining of differences using hyperspectral images. PLoS One
West JS, Bravo C, Oberti R et al (2003) The potential of optical canopy measurement for targeted control of field crop diseases. Annu Rev Phytopathol 41:593–614
Wilson AD (2013) Diverse applications of electronic-nose technologies in agriculture and forestry. Sensors 13:2295–2348
Wojtowicz M, Wojtowicz A, Piekarczyk J (2016) Application of remote sensing methods in agriculture. Commun Biometry Crop Sci 11:31–50
Xia T, Miao Y, Wu D et al (2016) Active optical sensing of spring maize for in-season diagnosis of nitrogen status based on nitrogen nutrition index. Remote Sens 8:605
Yang Z, Rao MN, Elliott NC et al (2005) Using ground based multispectral radiometry to detect stress in wheat caused by greenbug (*Homoptera:Aphididae*) infestation. Comput Electron Agric 47:121–135
Yang C-M, Cheng C-H, Chen R-K (2007) Changes in spectral characteristics of rice canopy infested with brown planthopper and leaffolder. Crop Sci 47:329–335
Yu W, Miao Y, Feng G et al (2012) Evaluating different methods of using chlorophyll meter for diagnosing nitrogen status of summer maize. In: Proceedings of the First International Conference on Agro-Geoinformatics (Agro-Geoinformatics 2012), Shanghai, China, August 2–4, 2012

Zermas D, Teng D, Stanitsas P et al (2015) Automation solutions for the evaluation of plant health in corn fields. In: Proceedings of 2015 IEEE/RSJ International Conference on Intelligent Robots and Systems (IROS), Sept. 28–Oct. 2, 2015, Hamburg, Germany, pp 6521–6527

Zhu J, Tremblay N, Liang Y (2011) A corn nitrogen status indicator less affected by soil water content. Agron J 103:890–898

Ziadi N, Brassard M, Bélanger G et al (2008) Chlorophyll measurements and nitrogen nutrition index for the evaluation of corn nitrogen status. Agron J 100:1264–1273

Chapter 7
On-Combine Sensing Techniques in Arable Crops

Dan S. Long and John D. McCallum

Abstract A combine harvester provides unique capabilities as a mobile sensing platform. This chapter aims to contribute to the advancement of on-combine sensor use for obtaining site-specific crop data by trying to convince potential users in the agricultural community of its value and accessibility. Today, mass/volume flow and electrical capacitance sensors are widely used for measuring grain yield and moisture. A variety of other sensors have been used in crop analysis and process control that include photoelectronic spectrometers for analysis of crop quality attributes as well as ultrasonic and laser sensors for quantifying aboveground biomass. Applications of this information include precision N management, post-harvest assessment of crop stress, grain segregation by protein concentration and mapping of late-season weed infestations. Barriers challenging wider adoption of on-combine sensing techniques include the need for (*i*) software for exploring multi-year yield data and constructing profit zones, (*ii*) inexpensive spectrometers for grain quality measurement and mapping, (*iii*) commercial firms offering services in spectroscopy, custom mapping and data fusion, (*iv*) stand-alone units with user interface and firmware for multi-sensor data collection, and (*v*) field studies demonstrating economic benefits of various applications of information from on-combine sensing.

Keywords Yield mapping · Near-infrared spectroscopy · Optical sensing · Multi-sensor data · Grain segregation

7.1 Introduction

Advances in electronic and information technologies have led to the development of various sensing systems for yield mapping and prediction (Arslan and Colvin 2002), soil sensing (Kitchen et al. 2003), nutrient (Raun et al. 2002) and herbicide

D. S. Long (✉) · J. D. McCallum
USDA-Agricultural Research Service, Adams, Oregon, United States
e-mail: junco.hyemalis@protonmail.com

R. Kerry, A. Escolà (eds.), *Sensing Approaches for Precision Agriculture*, Progress in Precision Agriculture, https://doi.org/10.1007/978-3-030-78431-7_7

(Christensen et al. 2009) applications, and irrigation control (Sadler et al. 2005). Precision agriculture (PA) is an information intensive discipline that relies upon highly complex geo-referenced data to address the site-specific needs of spatially variable application (Blackmore et al. 2003). Data with high spatial resolution, wide thematic range and high thematic resolution are desperately needed for characterizing environmental heterogeneity (Lee et al. 2010). By serving as a ground-based, mobile platform for sensors to collect and record detailed information about the crop and its grain at harvest, combine harvesters can help fill that gap.

Grain yield data derived from yield monitors have been a primary source of information for making crop management recommendations and decisions for many years (Schuster et al. 2017). More recently, spectrophotometers have appeared that are capable of measuring concentrations of nitrogen (N), protein, oil, starch and water of grain during harvest. Optical sensors that operate on the principle of light detection and ranging (LiDAR) have been used that are capable of quantifying aboveground biomass and straw yield. Various combinations of these data permit the computation of secondary crop attributes such as grain N removal and harvest index that in turn are useful post-harvest bio-indicators of environmental stress, and when and where it occurred in the growing season.

Various sensors mounted on a combine can operate at large rates of measurement and these measurements can be geo-spatially positioned autonomously with satellite navigation systems that provide global coverage. Therefore, because combines cover every hectare of land at relatively low ground speeds, the sensors on them can record information on different yield components of the crop at a spatial resolution comparable to aerial and satellite systems in the direction of travel (Table 7.1). Aerospace systems can provide in-season measures of canopy greenness, surface temperature, and biomass that indirectly indicate soil water content and plant N status. Combines equipped with yield monitors and grain spectrometers offer the possibility of site-specific scale estimates of crop biomass, grain yield and grain protein. However, data are retrospective in that they provide a post-harvest look at

Table 7.1 Comparison of sensing platforms used in precision agriculture

	Satellite[a]	Aircraft[b]	Tractor[c]	Combine[d]
Spatial resolution (m)	1.84 × 1.84	0.05 × 0.05	1 × 30	1 × 10
Temporal resolution (days)	1.1	<1	<14	365
Perspective	in-season	in-season	in-season	post-harvest
Information	canopy greenness, temperature, cover, density	canopy greenness, temperature, cover, density	canopy greenness, temperature, cover, density	grain quality and yield

[a]Digital Globe Worldview-2 providing eight-band multispectral imagery
[b]Unmanned aerial system providing four-band digital imagery
[c]High wheeled sprayer with single optical crop canopy sensor and 30 m boom
[d]Commercial combine equipped with mass flow yield monitor and 10 m header

previous growing season conditions that might not be applicable in the following season.

This chapter aims to contribute to the advancement of on-combine sensor use for obtaining site-specific crop data by trying to convince potential users in the agricultural community of its value and accessibility. After a general review of on-combine sensing systems that are presently used to collect information, there follows several examples of how various sensing technologies can be used, either individually or in combination, to provide sub-metre analysis of crop conditions within farm fields. The final section describes further developmental needs.

7.2 On-Combine Sensing Systems

Today, the grain yield monitor and grain moisture sensor are the most widespread sensing techniques used on combine harvesters, however, photoelectric, laser and ultrasonic sensors have also been deployed. Photoelectric sensors emit light from a light emitting element and receive a portion of the light that was emitted at an electro-optical detector. LiDAR sensors emit and receive a laser beam to measure time-of-flight or phase shift and precisely measure distance to objects at different angles. Ultrasonic sensors emit ultrasonic wave trains and measure distance by timing the reception of the wave reflected from the target. Detection methods used by these sensors and applications in crop measurement and process control are listed in Table 7.2, which are referenced to the most recent reports in the literature.

7.2.1 Sensing Grain Yield

Grain yield monitors are commercially available for a variety of arable grain crops including rice, wheat, maize, barley and grain sorghum. Using data from the USDA Agricultural Resource Management Survey, Schimmelpfennig (2016) reported that roughly 34% of maize and 26% of soybean planted hectares in the United States were harvested with a combine equipped with a yield monitor in 2010. Yield monitoring has had a modest rate of adoption, reaching 30% during the 20 years since commercialization of the first yield monitor in 1992.

In general, two types of systems are used that operate based on sensing of either mass flow or volumetric flow. In each case, a harvester is equipped with a global navigation satellite system (GNSS) receiver for geo-registering yield monitor values with locations in the landscape. Mass flow yield monitoring systems consist of a compression load cell that creates an electrical signal in proportion to the force of grain impacting a plate and a computer that converts that signal to mass flow rate in units of grain yield (Risius 2014). A typical example of this type is the impact-type mass flow grain yield monitor developed by Myers (1994). In contrast, volumetric systems use an optical array of photodetectors to measure the amount of light

Table 7.2 Various detection methods used by sensors on combine harvesters and associated uses in crop analysis and process control during harvest.

Detection method	Use	References
Crop Analysis		
Mass/volume flow	Grain yield	Chung et al. (2016)
Electrical capacitance	Grain moisture	Risius et al. (2017)
Photoelectric (spectrophotometer)	Grain protein	Long et al. (2008)
Photoelectric (spectrophotometer)	Test weight	Bonfil et al. (2008)
Photoelectric (spectrophotometer)	Falling number	Risius et al. (2015)
Ultrasonic	Crop height/straw yield	Maertens et al. (2004)
Laser (LiDAR)	Crop height/straw yield	Long and McCallum (2013)
Mass flow + laser (LiDAR)	Harvest index	Long and McCallum (2015)
Photoelectric (spectrophotometer)	Weed index	Barroso et al. (2017)
Process Control		
Laser (LiDAR)	Feed-rate and speed control	Saeys et al. (2008)
Photoelectric (spectrophotometer)	Grain segregation by protein	Long et al. (2013)

blocked by the grain pile on upward moving paddles of the combine's clean grain elevator (Schuster et al. 2017). As a mature technology, the measurement principles of yield monitors have been described in many articles and will not be repeated here. Interested readers may wish to consult the review by Chung et al. (2016).

Yield monitors output grain yield data continuously at measurement rates of up to 1 Hz. Because yield is expressed as grain weight per area, a monitor calculates yield based on inputs from different components that individually sense grain flow (mass or volume basis), grain moisture and harvested area (Chung et al., 2016). Area represented by each flow rate data point is calculated based on cutting width and travel speed. Locational information from a GNSS receiver is added to each yield record calculated. The measured wet rate of flow is corrected to dry yield by means of a grain moisture sensor (Reyns et al. 2002). Grain moisture is predicted by sensing the change in electrical charge (capacitance) stored between two insulated conductors as determined by the dielectric properties of grain. Risius et al. (2017) reported total measurement error to be within 2% compared with direct methods (oven drying), which is accurate enough for characterizing spatial variability in grain moisture within fields.

Calibration of flow rate, grain moisture and area is essential to achieve the high degree of accuracy of yield measurement needed by growers. Error of a typical yield monitor is <3%, but accuracy might deviate with use over time due to buildup of material on the impact plate, chain tension of the clean grain elevator, elevator paddle wear and other factors. Calibration is needed each year to maintain measurement quality. The process of calibrating a yield monitor involves comparing measurements of known accuracy, such as those provided by either a weigh cart in the field or farm truck scale at the elevator. For example, Nelson et al. (2015) harvested a series of strip trials with maize that provided a total of 195 yield comparisons across six South Dakota environments and showed an R^2 of 0.967 between measurements obtained by weigh carts and combine-mounted impact-type yield monitors. They concluded that mass flow yield monitors can be used effectively in on-farm maize hybrid strip trials provided the yield monitor and accompanying moisture sensor are properly calibrated, and large areas are harvested to improve accuracy.

Schuster et al. (2017) used a combine test stand to vary the rate of grain flow when comparing the accumulated grain weights derived from mass- and volumetric-flow commercial yield monitors. Yield readings from the volumetric flow sensor were highly sensitive to clean grain elevator configuration and machine orientation, whereas those from the mass flow sensor were more sensitive to variation in rate of flow across all flow ranges during testing. Beam-based volumetric yield sensors were more accurate than impact-based mass yield sensors at flow rates typical of small grains and conditions not covered by calibration. Error primarily became a problem with mass flow sensors at low flow rates below 5 kg s^{-1}, typical of small grains in dryland environments.

Kettle and Peterson (1998) determined that mass flow yield monitors are influenced by harvesting direction either upslope or downslope in the hilly Palouse region of northern Idaho. Errors were 18.2% in the uphill versus 60.7% in the downhill direction when harvesting on 6–9% slopes with a GreenStar™ yield mapping system. Fulton et al. (2009) examined the influence of varying roll and pitch on the accuracy of a mass flow sensor mounted to a laboratory test stand. Errors observed during roll tests were small (–3.64 to 3.65%) and within the manufacturer reported accuracy. In contrast, errors during pitch tests were significantly higher (–6.4 to 5.5%) resulting from overestimation of mass flow when the monitor was pitched forward and underestimation when pitched backward. Overestimation was the result of gravity aided impact forces during downhill operation. Sensor response is reduced in the uphill direction because of the opposite effect of gravity. The linear relationship between error and slope could be used to model pitch corrections and reduce mass flow error rates.

Blackmore and Marshall (1996) identified six main groups of errors affecting the production of yield maps: (*i*) unknown crop width entering the header, (*ii*) time lag of grain through the threshing mechanism, (*iii*) wandering GPS error, (*iv*) surging grain through the combine transport system, (*v*) grain losses from the combine and (*vi*) sensor accuracy and calibration. Assessment of the delay time between cutting of the crop in front and sensing in the clean grain elevator was one of the most significant errors for map creation. Incorrect delay time causes an offset of position of yield measurements in a field and may be identified by visually inspecting where grain yield is the same for two different directions across a known zero-yield point in a field (Stott et al. 1993). It is especially problematic when it varies across a field, and Chung et al. (2002) successfully developed geostatistical and data segmentation methods to estimate delay time based on objective criteria. More recently, engineers have focused on data-driven grain flow models to estimate grain throughput on the combine in real-time (Hermann et al. 2016). Sudduth et al. (2012) developed the Yield Editor 2.0 software that provides automated removal of a variety of errors from raw yield monitor data, including outliers and variable delay times, to produce accurate and reliable yield maps. In addition, density-based spatial clustering has been applied to filter out outliers that occur alone in low density regions and other defective observations likely to be present in yield datasets (Leroux et al. 2018a, b).

Crop yield is a complex variable that is influenced by soil fertility, plant available water and other underlying environmental factors during the growing season. Therefore, measurement and mapping of within-field variation in crop yield is essential to developing a PA system. At the end of the season, a yield map is useful for evaluating the results of precision farming decisions that are made before and during the growing season. However, spatial yield patterns in a given field may change from year to year because of interaction of soil factors influencing yield with climate variability (Machado et al. 2000). For example, there may be years of below- or above-average precipitation and spatial variation of plant available water across growing seasons (Schepers et al. 2004). This temporal variability detracts from the ability to apply a single season yield map for identifying management zones in variable-rate application.

A series of yield maps over several years can provide sufficient information for evaluating long-term trends in crop productivity and dividing fields into generalized zones of low, medium and high yield (Stafford et al. 1998). However, to enable comparison of multi-year yield maps, data for each year must be brought to a common grid having the same point of origin and cell size. Grid harmonization typically involves kriging, inverse distance weighting, or other interpolation techniques to aggregate data to a regular grid with the result that the spatial resolution may be less than that of the original raw data. In practice, yield data are converted numerically to the same measurement scale to normalize for the variation in yield potential among different crops comprising a crop rotation over a series of years. This practice is also assumed to account for variability in rainfall and other environmental factors across years. Normalization rescales yield values into a quotient by dividing cell yield values by a divisor such as the mean, maximum, median or range in yield for all cells in a field. Standardization typically rescales yield data to have a mean of zero and standard deviation of unity, but also involves rescaling to a median of zero and interquartile range of unity.

After rescaling of data, yield maps from multiple years can then be combined mathematically by several techniques. Many researchers have simply calculated the mean yield and standard deviation at each point on the regular grid over the years of interest (i.e. averaging) and produced a single map showing at least three yield zones: high yielding and stable, low yielding and stable, and unstable (Blackmore 2000; Taylor et al. 2001; Ping and Dobermann 2005; Flowers et al. 2005; Maestrini and Basso 2018) where stability is indicated by the amount of variance in yield within a zone across years. Kleinjan et al. (2006) divided average maize yield of four years into classes using arbitrary cut-off values based on either the standard deviation or coefficient of variation. A summary of selected methods for scaling yield data and combining them across years is shown in Table 7.3.

Clustering and 'region growing' are sophisticated mathematical methods of combining multi-year yield data. Jaynes et al. (2003, 2005) used cluster analysis to partition the population into groups such that differences among groups are minimized while differences between groups are maximized. Recently, Leroux et al. (2018a) derived yield zones from time series yield data through the method of region growing

Table 7.3 Comparison of methods for scaling yield data and combining multiple yields across years

Method of scaling	Method of combining	Crops	Reference
Normalized to a range between 0 and1	Averaging	Wheat, rape	Blackmore (2000)
Normalized to a range between 0 and 1	Averaging	Maize	Taylor et al. (2001)
None	Averaging	Barley, wheat,	Blackmore et al. (2003)
Normalized to a range between 0 and 1	Averaging	Maize	Ping and Dobermann (2005)
Normalized to a range between 0 and 1	Averaging	Soybean, wheat, maize	Flowers et al. (2005)
Normalized to a range between 0 and 1	Averaging	Maize	Kleinjan et al. (2006)
Normalized to a range between 0 and 255	Clustering	Cotton, maize	McKinion et al. (2010)
Standardized to zero mean and unit variance	Averaging	Maize	Diker et al. (2003)
Standardized to zero mean and unit variance	Averaging	Maize, cotton, soybean, wheat	Maestrini and Basso (2018)
Standardized to zero mean and unit variance	Region growing	Wheat, canola	Leroux et al. (2018a, b)
Standardized to zero median and unit interquartile range	Clustering	Maize	Jaynes et al. (2003)
Standardized to zero median and unit interquartile range	Clustering	Soybean	Jaynes et al. (2005)
None	Fuzzy clustering	Wheat	Milne et al. (2012)
Reclassified to binary values	GIS overlay	Maize	Diker et al. (2004)

in which a seeded region-growing algorithm operates in a multivariate attribute space. Multi-temporal yield data are partitioned into multiple segments that assign a label to every pixel such that pixels with the same label share certain characteristics. When image segmentation is applied to multiyear yield data, neighbouring pixels of initial seed points are examined to determine whether pixel neighbours should be added to the region. The process is iterated until no further pixels can be found.

7.2.2 *Near-Infrared Spectroscopy of Grain Quality*

Near-infrared (NIR) spectroscopy is widely used in the grain industry for the laboratory analysis of concentrations of protein, starch, oil and moisture in whole grain. In 2000, the first spectrometer appeared that was designed for installation on combine harvesters to measure and map these attributes across farm fields (von Rosenberg Jr.

et al. 2000). Today, the instruments that are advertized as designed specifically for combines are the CropScan 3300H On-Combine Analyzer (Next Instruments, Condell Park, NSW, Australia) and AccuHarvest On-Combine Grain Analyzer (Zeltex, Inc., Hagerstown, MD, USA). These products come with GNSS interface and software for field mapping. Other spectrometers include the PSS-1721 (Polytec GmbH, Waldbronn, Germany) and Corona extreme (Carl Zeiss Jena GmbH, Jena, Germany) that were designed for process control in harsh environments and could be adapted to operate on a combine. An excellent review of the initial history of the development of this technology can be found in Taylor and Whelan (2007).

Spectroscopy quantifies how much a chemical substance absorbs light by measuring the amount of light that either passes through the substance (transmittance) or reflects from the substance (reflectance). Different on-combine spectrometers collect raw spectra in either transmittance or reflectance mode, which differs in the geometry between the light source and detector. Both methods have similar accuracy and reproducibility (Williams et al. 1998). Transmittance instruments have longer detection times because a sampling mechanism is required to route the grain from the clean grain elevator through a holding chamber inside the instrument to isolate the grain temporarily for transmittance measurements. Reflectance spectroscopy needs no sampling mechanism and can run continuously, rapidly sensing the grain as it flows freely past the sensor head. Based on the authors' experience, reflectance instruments are easier to install to the combine's grain bin filling auger and have fewer moving parts.

A basic instrument consists of a light source, monochromator and detector. The light source is often an incandescent halogen lamp consisting of a tungsten element sealed into a compact transparent envelope that is filled with a halogen gas. The monochromator disperses light into multiple wavebands of the electromagnetic spectrum. The dispersed light is incident upon the detector array that provides sensitivity in multiple channels. The detector is made of semiconductor material such as silicon or germanium. Silicon photovoltaic detectors are widely available and absorb radiation in the ultraviolet, visible and NIR wavelength regions between 300 and 1100 nm. Germanium diodes reach further into the NIR region, but have been supplanted by an alloy of gallium arsenide and indium arsenide (InGaAs): a compound semiconductor obtained by combining group III elements (In and Ga) with the group V element (As). Standard InGaAs detectors have a greater responsivity, and cover the NIR and short-wavelength NIR spectral regions between 800 and 1700 nm. Extended range InGaAs detectors are also available that extend the observable region to 2600 nm.

Incoming light interacts with whole grain as absorption, reflection and transmission. Molecules absorb NIR light when the bond between atoms in the molecule vibrate at the same frequency as the frequency of incident NIR light. The resulting spectrum yields information about four major functional molecular groups ubiquitous in biological molecules: O–H, C–H, N–H, and double bond C=O. Many of the stretching and bending vibrational modes have been assigned to frequencies in the infrared, with a series of overtones and combination bands with decreasing intensity running into the NIR region. Since N occurs primarily in proteins and not in starches and oils, absorbances in the NIR associated with N bonds can be useful in determining protein concentration.

The lower cost, silicon-based charge-coupled device (CCD) detector instruments lose sensitivity at 1000 nm and have limited wavelength range (Fig. 7.1). Back illuminated CCD detectors can extend this range to include the O–H and N–H bands near 1050 nm. InGaAs-based detector instruments that extend out to 1650 nm include the second overtone regions, which have stronger absorbances. The broad peak at 1000 nm is due primarily to overlapping N–H and O–H stretches, whereas the strong peak around 1900 nm has been assigned to a carbonyl stretch (Shenk et al. 2001). Proteins, carbohydrates, oils and moisture contain these types of bond. Spectra shown in Fig. 7.1 are whole wheat kernels presenting a combination of these biomolecules and functional groups varying in vibrational frequency with environment.

The development of NIR spectroscopy as an analytical tool depended on the development of computer systems to extract information (Blanco and Villarroya 2002). Protein is usually determined through modelling the spectra of calibration samples of grain with known protein concentration as determined by the Kjeldahl and Dumas reference laboratory methods. Both these methods have been automated in recent years and both have been determined to have similar precision (Williams et al. 1998). A 'model' is a vector which, when multiplied by the spectrum, generates the protein concentration. Modeling has been traditionally done with partial least squares analysis (PLS, also referred to as 'projection onto latent structures'),

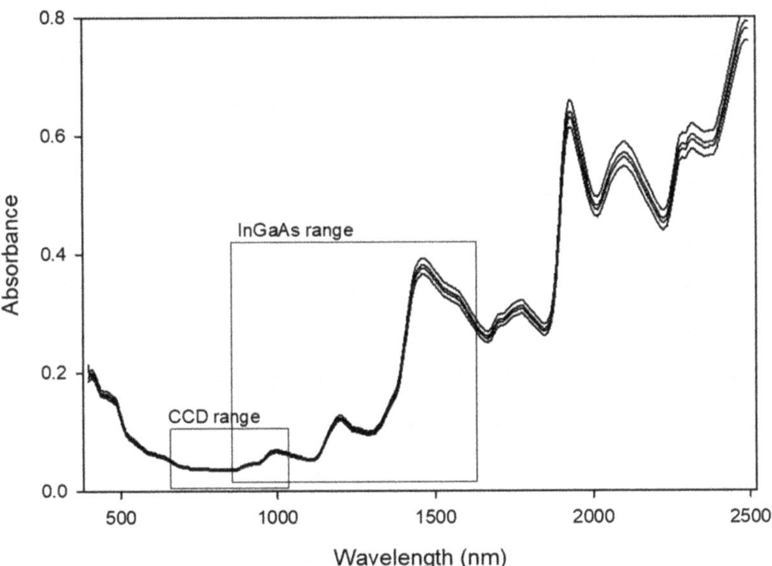

Fig. 7.1 Spectra of whole grain with wavelength ranges in sensitivity of charge-coupled device (CCD) and indium–gallium–arsenide (InGaAs) detectors. Data provided by Bradford Seabourn, USDA-Agricultural Research Service, Lincoln, NE

but is also beginning to use artificial neural networks and binomial machines (Büchmann et al. 2001). Since, the dominant feature between spectra is a baseline shift, models are usually developed from transformed spectra that eliminate the baseline using a standard normal variate or a derivative spectrum. Depending on instrument geometry or resolution, other pre-processing steps such as smoothing can be used to enhance the signal. Chemical knowledge should be used to help select the optimal wavelengths to use because certain regions of the full grain spectrum may contain information that is unrelated to protein (e.g. grain colour, water content, etc.).

Prediction error of NIR spectroscopy is reported to be within 5 g protein kg^{-1} grain when evaluated on the combine as the grain is conveyed through the threshing system into the bulk tank (Long et al. 2008). As an aside, other grain quality attributes have been predicted, due to correlation with grain protein concentration, using bench-type NIR whole grain analyzers. Dowell et al. (2006) found test weight to be strongly correlated with NIR spectra of wheat, which probably arises from to a direct relationship between starch and test weight. As a major constituent of grain, starch has strong absorption bands throughout the NIR region (Williams 2001). Forming in the later stages of kernel development, starch gives plumpness to the kernel (Shollenberger and Kyle 1927). Drought conditions that arrest development of the kernel and shorten the grain filling period will produce kernels that have low test weight and high protein. In contrast, grain grown under abundant moisture will have high test weight and low protein. Based on these relationships, Bonfil et al. (2008) in Israel was able to predict test weight from the transmission spectra of grain acquired with an on-combine spectrometer.

The falling number method is a laboratory test for measuring the viscosity of a water–flour mixture as an indicator of the alpha-amylase activity of wheat and the bread making qualities in flour. Studies of laboratory NIR spectroscopy of falling number have had limited success with large prediction errors (Osborn 1984). Risius et al. (2015) evaluated NIR spectroscopy on a combine and determined the standard error of prediction to be 37 s, which was good enough for use as a screening tool and to characterize falling number within fields. Unfortunately, the study presented no spectral models to evaluate which wavebands in the NIR spectrum corresponded with alpha-amylase and other known chemical entities in the grain. Recently, Delwiche et al. (2018) found that NIR spectroscopy was imprecise for alpha-amylase and recommended that it should not be used for making decisions on segregating wheat lots according to alpha-amylase activity.

7.2.3 *Optical Sensing of Crop Density, Cover or Height*

Sensing the torque required to move the straw from the header to the feeder unit of a combine harvester has been used to measure straw yield directly (Schueller et al. 1985, Missotten et al. 1997). Separator load, grain loss and engine load are used today for automatic speed control of the harvester to maintain optimal crop flow through the machine (Taylor et al. 2005). By accommodating spatial variability in biomass, variable speed control maximizes harvest efficiency and thus is a form of precision conservation. Sensors have also been investigated that can characterize the density, cover and height of the crop ahead of the combine. For example, Taylor et al. (1986) placed a laser beam in front of the header at the height of the cutter bar. Maertens et al. (2003) experimentally tested an ultrasonic sensor for measuring the crop density between two points by sending an acoustic wave with known energy level and measuring the transmitted energy at the other side of the crop. They found that the energy received was closely related with the volumetric density of the wheat crop.

LiDAR scanners compute distance based on the constancy of the speed of light and time-of-flight or phase shift between the emission and detection of a laser beam that is reflected from objects in front of the scanner. A rotating mirror within the instrument is used to project a laser beam in a single (2D) or multiple (3D) scanning planes. Precise direction is measured by an encoder on the mirror. LiDAR sensors are described in detail in Chap. 3. Relatively inexpensive LiDAR scanners have been deployed on agricultural equipment to obtain field measurements of crop height, crop volume and biomass density (Polo et al. 2009; Ehlert et al. 2009). Saeys et al. (2008) evaluated the feasibility of LiDAR sensors for online measurement of crop density in front of the combine and applying this information to speed and feed-rate control. Crop volume could be determined accurately ($R^2 \leq 0.96$) from the ground and crop profiles by fitting thin plate splines through the crop height values obtained with a high frequency LiDAR scanner. By accounting for lateral inclination and the unevenness of the ground, the LiDAR-based method was an improvement over measurements of crop density based on ultrasonic and laser sensing.

Biomass is a production factor that also determines the amount of straw that is available for soil conservation purposes versus its harvest and export from the field as bedding material or biofuel feedstock. Long and McCallum (2013) positioned a LiDAR sensor on the roof of the cab of a combine to clear the top of the header when lowered to cut. The laser beam was programmed to scan the standing wheat crop through a total angle of 45° resulting in 91 points taken in 0.5° increments over a scanning path of 6.1 m. Based on the principal of similar triangles, crop canopy height was calculated from LiDAR distance between the scanner and the crop, sensor height above the ground surface and sensor inclination angle. LiDAR-measured crop height was a good predictor of straw yield when tested in three wheat fields. Straw yield maps were generated during harvest showing where excess crop residue could be removed for commercial purposes, with allowance for straw yield required to maintain soil carbon levels. Further research is needed to address the problem of floating chaff and flying insects in front of the combine that caused aberrations in crop height measurements.

7.3 Applications of On-Combine Sensor Information

7.3.1 *Precision Nitrogen Management*

On-combine protein sensing technology is a potentially useful tool to characterize spatial variation in the adequacy of N nutrition across a wheat field. The rationale for this approach is that crops are excellent bio-indicators of conditions in the root zone, and grain protein is an important post-harvest indicator of soil N fertility. For example, in a wheat field in northern Montana, lower, wetter areas with large production potential had large yield (blue colour) and small protein (cyan color) because they were deficient in soil N (see Fig. 7.2). In contrast, higher, drier areas had small yield (blue color) and large protein content (red color) because they were deficient in plant available soil water. These grain yield and protein maps revealed the location and extent of N-deficient areas more effectively than a soil N map derived from limited test data.

The protein concentration of grain can be used as a post-harvest diagnostic tool in evaluating whether N was sufficient or deficient for grain yield once a critical protein level has been identified for a region (Goos et al. 1982). Critical protein levels have been found to vary by wheat class, cultivar and region. Critical protein concentrations can be useful for evaluating the potential yield loss to N deficiency. For example, Glenn et al. (1985) found a 9.8% reduction in yield of 'Stephens' soft white winter wheat for each grain protein percentage below a critical level of 85 g^{-1} kg^{-1}. Selles and Zentner (2001) concluded that the critical protein level as an indicator of N deficiency works well when water is not limiting, but is unreliable when grain yields are adversely affected by water stress because a large protein content can also occur under drought. Long et al. (2017) determined that spring wheat under severe water stress with protein content below the critical level did not guarantee that yields were compromized by a lack of N. However, N was sufficient for yield when protein was above this level.

Site-specific yield and protein data can be used to calculate the N application required to replace that removed from the soil by the crop (N removed, NR) during the previous crop phase, as: $NR = (GY \times GPC)/0.17$ where GY is wheat yield in kg of grain ha^{-1}, GPC is grain protein concentration expressed as a fraction and 0.17 is the fractional part of protein that is N (Engel et al. 1999; Long et al. 2000). Conceptually, applying fertilizer with such a map would maintain plant available N for the following crop by replacing only what was removed by the crop in the previous season. If the yield of the harvested crop was already limited by poor nutrition, as indicated by protein below the critical level, additional N can be applied in deficient areas of fields to raise protein to the critical level with local knowledge of the amount of fertilizer N equivalent (FNE) to a 10 g kg^{-1} change in protein (Engel et al. 1999).

Current estimates of FNE are reported to be 5.9–8.2 kg N ha^{-1} for hard red spring wheat compared with 5.4–6.7 kg N ha^{-1} for soft white spring wheat in eastern Oregon (Long et al. 2017). In northern Montana, the FNE is reported to be 1.4–2.2 kg N ha^{-1}

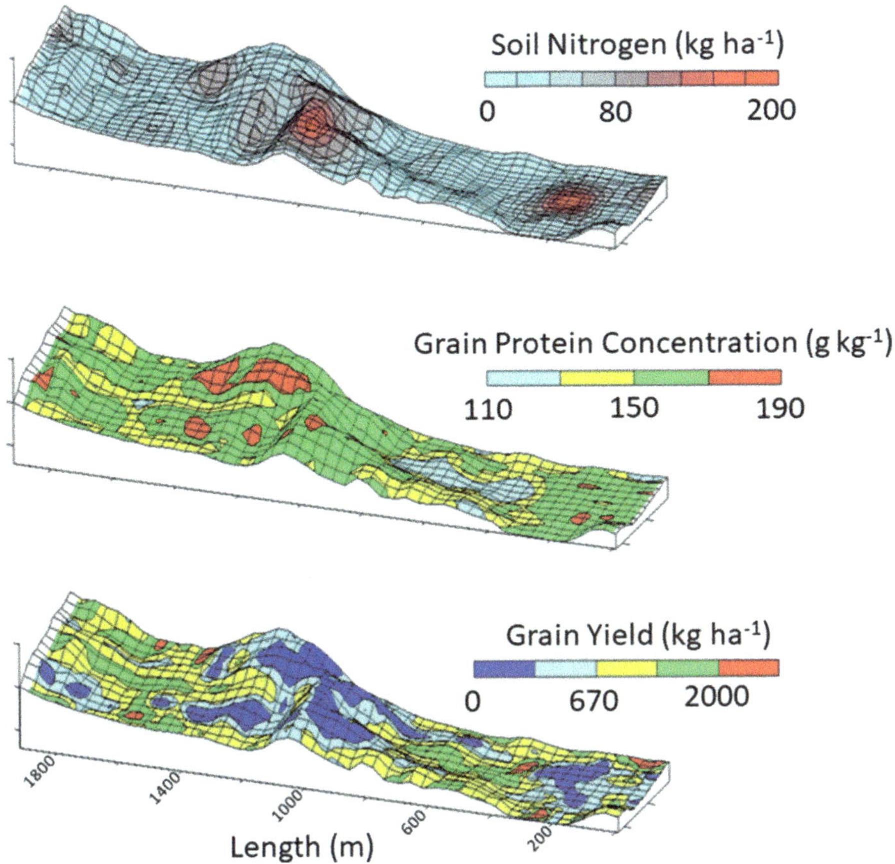

Fig. 7.2 Block diagrams showing the spatial distributions of soil profile nitrogen, grain protein concentration and grain yield within a dryland wheat field in northern Montana. Maps created from interpolation of point soil, protein, and yield data

for hard red spring wheat yields below 3000 kg ha^{-1} (Engel et al. 1999). Additional N could be applied based on a mid-range FNE value of 1.8 if more protein was desired than that obtained in the wheat crop of the previous year. For example, if wheat is harvested at 120 g protein kg^{-1}, but 140 was desired to capture price premiums, then 36 kg N ha^{-1} would be needed to overcome an 'N deficit' and reach 140 g protein [36 = (140 − 120) × 1.8]. Summation of N removed and N deficit equal the total N application to the subsequent crop. Management zones can be easily constructed to guide N fertilization of the following crop (Fig. 7.3).

A field experiment was conducted in the Montana wheat field to determine if the N replacement approach could improve grain yields and protein levels (Long et al. 2000). Variable- or uniform-rate N application was randomly assigned to each member of numerous pairs of treatment strips oriented across the field. Both treatments received nearly the same amount of fertilizer, but N in the variable-rate

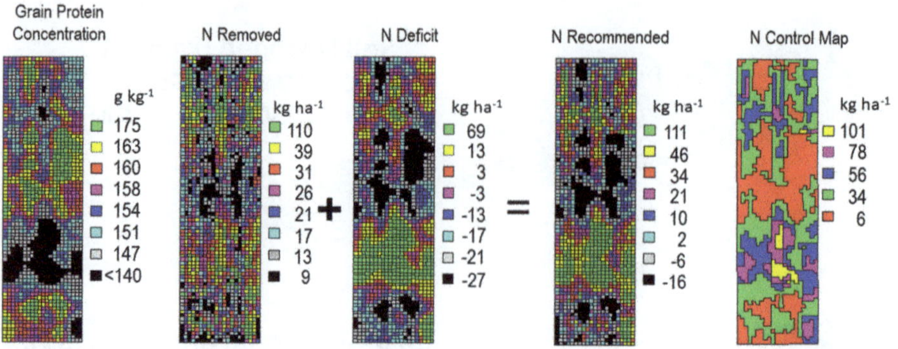

Fig. 7.3 Flow diagram of the process of deriving a nitrogen recommendation map from maps of nitrogen removed and nitrogen deficit. A map for controlling the nitrogen application is constructed from the nitrogen recommendation map. Map for grain yield not shown. All maps created from interpolation of point protein and yield data except N control map, which was derived by aggregating N recommended into five classes

treatment was varied according to N removed and N deficit that were computed from the previous year's yield and protein levels. Yields did not differ significantly between fertilizer management systems, but protein contents improved considerably and variability reduced, by variable-rate N application. The overall improvement in grain quality could be attributed to placement of more N in areas where protein had been below the critical level and less in areas where protein had exceeded this level in the prior season.

7.3.2 Grain Segregation by Protein on the Combine

Certain cereal crops qualify for price premiums that provide an incentive for farmers to produce and deliver grain of higher than average quality. Quality payment systems exist in many grain producing countries. In the U.S.A., premiums are often available for malting barley with grain protein concentration <120 g kg^{-1}, soft white wheat with protein concentration <95 g kg^{-1} and hard red spring wheat >140 g kg^{-1}. For example, on 5 October 2018, Portland market prices for the dark northern subclass of hard red spring wheat were US$244 vs. US$252 Mg^{-1} at protein concentrations of 130 and 150 g kg^{-1} (Pacific Northwest Grain Market News, www.ams.usda.gov/mnreports/jo_gr110.txt).

Protein concentration of mature grain varies within farm fields because of spatial variation in soil fertility (Delin 2004), topography (Fiez et al. 1994), plant available water (Stewart et al. 2002) and other crop determining factors. Segregating the high quality grain from certain field areas and binning it separately may offer profit opportunities in markets that offer protein premiums. Stewart et al. (2002) found that segregating durum wheat into two batches would have increased profits by

AU$34 ha^{-1} over conventional harvesting. Meyer-Aurich et al. (2008) investigated site-specific N fertilization and grain segregation for different price structures that varied as a function of GPC. Grain segregation, which isolated and sold valuable grain at higher prices, increased marginal returns by €50 ha^{-1} (US$67 ha^{-1}) over that of conventional harvesting.

Long et al. (2013) developed an optical-mechanical system for automatically segregating wheat by protein concentration on a Case-International Harvester 1470 combine. The on-combine segregator consisted of a reflectance-type NIR spectrometer, hydraulic-mechanical diverter valve, and electrical control unit (Fig. 7.4). The Textron ProSpectra™ spectrometer continually measured the protein content of the grain stream in the grain bin filling auger and sensed whether protein was above or below a certain cut-off value. Mounted to the end of the filling auger, the diverter valve routed the grain into one bin or the other depending on the protein content. The control unit, consisting of a notebook computer, input/output interface and relay box, acquired data from the spectrometer, determined if grain was above or below the cut-off and instructed the diverter valve by means of the input/output interface. The combine's bulk tank was divided into two bins: one for high quality grain and a second for common quality.

Sivaraman et al. (2002) found that the economic benefit of grain segregation may be achieved with only two bins. Segregation was profitable if the slope of a price function increased above a cut-off point with further increases in protein such that the value gained exceeded the value lost. These findings were based on continuous mathematical functions even though real price schedules are discontinuous, step-wise functions. Martin et al. (2013) showed that it was possible to exploit the stepped nature of the price schedules and profit either by removing an amount of lower protein wheat thereby increasing the value of the remaining bulk or by removing an amount of higher protein wheat that can be sold at a premium price while maintaining the price of the remaining bulk. They developed the Grain Segregation Profit Calculator (available online at http://agweb.montana.edu/

Fig. 7.4 Mechanical on-combine segregation system consisting of near-infrared reflectance spectrometer, hydraulic-mechanical diverter valve, electrical control unit, lift auger, and front and rear bins

grainsegregationprofitcalculator/Default.aspx, accessed 22 Jan. 2020) for computing the cut-off value for segregating wheat into two bins such that prices received for average protein maximize gross returns. Output requires the user to enter the grain price schedule and mean and standard deviation of the grain protein concentration within the field. The latter values can be derived from a single pass across the field provided the pass is representative of the entire field.

Long et al. (2015) quantified the net dollar returns per hectare for the segregation of hard red spring wheat grown in a wheat–fallow rotation in northern Montana. Site-specific protein and yield data were obtained from 21 fields over an 11-year period. Segregation increased the dollar value of each Mg of grain and showed potential for revenue gain in 17 of these fields. However, an economic analysis based on the segregator designed by Long et al. (2013) and long-term grain prices showed that the increased income would not offset the increased equipment costs such that net returns were smaller than that received from conventional harvesting. Sensitivity analyses showed that grain segregation would be profitable only in years of large differences between price steps, above average yields and if costs could be spread over more land in production.

To avoid high equipment costs, growers could use grain quality spectrometers to map fields and identify the areas that consistently yield grain with low or high quality and selectively harvest them once this information is known. An advantage of this approach is that the combine does not need to be fitted with a mechanical segregator. Segregation is profitable if the slope of a price function increases above a cut-off value with further increases in protein such that the value gained exceeds the value lost. In the U.S.A., this condition is true at each point of the price function for hard red spring wheat, which is priced relative to 2.5 g kg^{-1} incremental changes in protein between 120 and 150 g kg^{-1}. Current optical sensors designed for on-combine use measure protein to within 5.0 g kg^{-1} and thus grain segregation may be better suited for crops with price schedules that pay to the nearest 5.0 g kg^{-1} protein concentration. Soft white winter wheat is an example of such a crop, but premiums are not regularly available every year.

7.3.3 *Post-Harvest Evaluation of Crop Stress*

On-combine yield monitors are widely used in PA for locating areas within fields where yields are reduced. Yet, large unexplained yield differences within fields are common. Variation in the local relationship between grain yield and grain protein can be used to evaluate whether a wheat crop underwent drought or N stress. In Australia, Whelan et al. (2009) used on-combine sensors to map wheat yields and proteins across numerous fields. The local correlation coefficient was computed between grain yield and protein using the closest 100 neighbouring pairs to each protein sample location. Significant negative correlations indicated zones of water stress, whereas positive correlations indicated N stress. These results suggested that on-combine sensors for yield and protein are potentially useful for improving N fertilizer management and environmental auditing within fields.

In addition, the ratio of grain yield to total aboveground biomass, or harvest index (HI), is a simple post-harvest indicator of the influence of environment on a crop's ability to partition and accumulate dry matter in mature grain (Snyder and Carlson 1984). Under ideal growing conditions, the maximum HI of maize is 0.6 indicating that the crop is capable of partitioning 60% of dry matter to grain (Prihar and Stewart 1990). Deviations in HI below the maximum level indicate that environmental stress negatively affected the partitioning and accumulation of dry matter. Copeland et al. (1996) applied the HI concept using data from N fertility trials with maize on each of four 5 ha field sites in Minnesota. They split a scatter plot of grain yield vs. straw yield into four environmental stress categories as delineated by the midline of the lines of maximum and minimum grain to stover ratio and by the most efficient stover weight defined by highest grain yield for least stover. These stress categories corresponded with field areas of (*i*) low HI and low stover associated with stress during vegetative growth and grain filling, (*ii*) high HI and low stover indicating stress during vegetative growth, (*iii*) low HI and high stover denoting stress during grain filling, and (*iv*) high HI and high stover with absence of stress. A map of HI was useful for identifying where different environmental factors occurred and indicating how those areas might be managed differently.

Missotten et al. (1997) obtained on-combine grain and straw data from selected fields in France, the Netherlands and Belgium. Grain yield was measured with a mass flow sensor on the clean grain elevator and straw yield was measured by sensing the torque needed to drive the auger in the combine's header. Cutting width of the combine, necessary for computing straw yield, was determined by means of two ultrasonic distance sensors on each side of the cutter bar of the header. Today, yield monitors are available that use information from the GNSS receiver to determine position overlap and automatically adjust the crop width setting (Hawkins et al. 2017). By showing areas that differed in soil water content as determined by spatial variation in soil texture, these researchers suggested that maps of grain and straw yield in ratio gave more information that a yield map alone. Furthering this work, Reyniers et al. (2006) sensed and mapped grain and straw yield within a 7.2 ha winter wheat field in Belgium. Both variables were interpolated to a common estimation grid to enable their arithmetic combination in a geographic information system (GIS). Fuzzy clustering was applied to both grain and straw yield maps and used to divide the field into five classes representing zones with different grain and straw yield potential. Areas of the field with low grain and high straw yield (low HI) and both low grain and straw yield were associated with slumping of the soil surface during wet periods, soil crusting during dry periods and soil compaction by equipment. Areas with high grain yield and low straw yield (high HI) on steep slopes had great potential for erosion. Flat places in the field had both high grain and straw levels that had experienced minimal erosion.

In eastern Oregon, Long and McCallum (2015) developed an on-combine, multi-sensing system consisting of grain yield monitor, NIR spectrometer to measure grain protein, and LiDAR to measure straw yield. Multiple data streams were recorded and registered jointly to the same locations over a farm field such that the local correlation between yield and protein could be computed without need for interpolation to a common estimation grid. It was not necessary to interpolate irregularly spaced data to a common grid as needed for fusion of grain yield, protein and straw yield. Strong negative correlations occurred in overwatered areas of high grain yield and low protein. Weak correlations were associated with relatively uniform field areas where yield and protein showed little variability. Class maps of environmental stress based on scatter plots of grain and straw yield effectively identified four specific regions where grain yield was affected by early-season water stress, late-season water stress, both early and late stress, or none. This information would facilitate an improved understanding of how to increase production efficiency with precision application of irrigation water and other crop inputs.

7.3.4 Detecting and Mapping Weeds

Kochia, Russian thistle and prickly lettuce are weeds that are green and growing at the time of harvest when a dryland wheat crop has matured and senesced. Green foreign material enters the grain stream when these late-maturing weeds enter the combine and are ground up in the cylinder. In eastern Oregon, Barroso et al. (2017) showed that a relatively low-cost spectrometer (AvaSpec UL2048RS, Avantes, Apeldoorn, The Netherlands) with sensitivity over the 638–1163 nm wavelength range could be used to map the presence of these weeds across two wheat fields by detecting the presence of chlorophyll in the grain stream during harvest. Foreign weed matter deepened the absorbance feature in the red waveband such that a 'chlorophyll peak' appeared in the grain spectrum. Weedy areas could be identified spectroscopically by a weed index defined by a ratio of chlorophyll peak at 670 nm to baseline area under that peak, or a normalized difference index. Grain spectra influenced by green weed material are shown in Fig. 7.5.

Reference maps were constructed by partitioning each field into 7 m × 7 m cells, visually estimating the weed density in each cell, and coding them into categories of none (0), low (1), moderate (2) and high (3) density. In contrast, the irregularly spaced weed index took on values between –1 and 10 from low to high weed infestation. Map data from the reference method and spectrometer were interpolated to a common grid to enable their visual comparison. Cut-off values were then used to divide the distribution of the interpolated map values into two map classes: weeds absent and weeds present. Observing the coincidence between the histograms of the accumulated percentage of cells in the gridded data of each method assisted with this process. On average, moderately to highly infested areas could be correctly identified 72% of the time provided the user could tolerate a small amount of economic loss by letting some lightly infested areas go untreated. A key conclusion from this study was that mapping with a relatively inexpensive spectrometer might ease the high cost of commercial spot spray systems and thus help accelerate adoption of site-specific weed management.

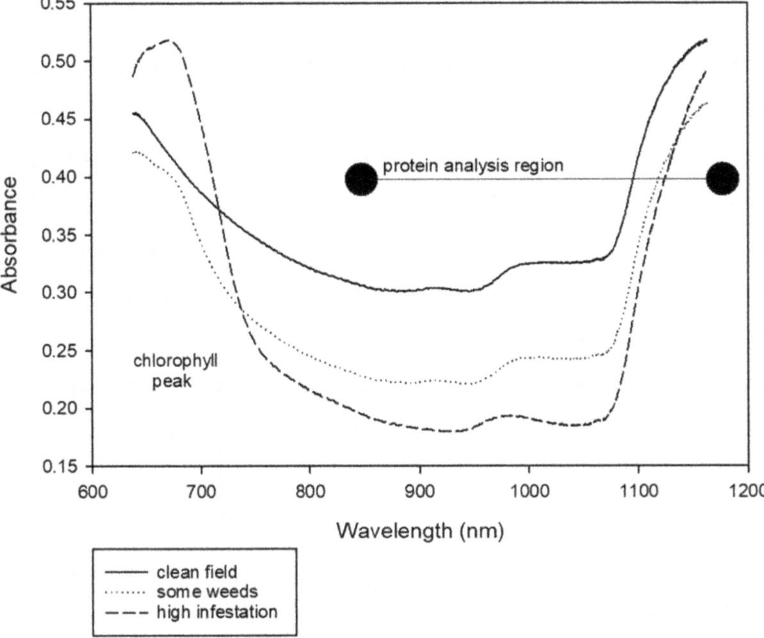

Fig. 7.5 Spectral absorbance versus wavelength for grain differing in amount of green material from weeds. The separate grain protein analysis region is shown from 850 nm

7.3.5 Conclusions for the Chapter

Sensing the mature crop from a combine at harvest is an excellent approach that provides growers with an ability to generate data to consider site-specific management and to evaluate and monitor the results of their precision practices within fields. For example, growers in the U.S. Midwest now rely upon grain moisture readings from yield monitors to adjust grain dryers for optimum drying performance during winter. In addition, yield maps have been useful for assessing yield losses in double planted areas, wet field spots and compacted soils; and for comparing the performance of different crop varieties, tillage treatments and other cultural practices. However, the accuracy of yield and protein maps is influenced by the transportation delay time of grain through the combine's threshing mechanism. Dynamic grain flow models show promise in generating delay free material flow estimates of grain throughput and should be pursued. Yield monitors are a mature technology but practitioners struggle with the collection, analysis and interpretation of data (Luck et al. 2015). Training is needed in the understanding and use of yield maps.

Growers intimately familiar with their fields are often able to identify potential sources of the variation they see in yield maps. However, the accumulation of yield data with each year has led to a large volume of information that may be difficult to study because of the presence of time. The next step associated with yield monitoring may be the development of software for the exploration of multi-year yield data. Analysis techniques would not only include the traditional approach of combining yield data into one multi-year layer from which are derived high, medium and low yield zones, but also focus on extracting information about the crop patterns that is valuable for decision making in crop management. Data mining techniques have been found to be useful in finding hidden information in large amounts of small-scale data (Ruß 2009). Analysing change for anomaly detection in multi-year yield data can be useful for uncovering spatial patterns that are not consistently significant over time. Eigenvector spatial filtering techniques promoted by Griffith and Chun (2012) that accommodate spatial structure over time may also have an important role.

Provided information is available on production costs and expected crop prices, some agricultural economists have suggested that yield monitors should also display profit on the screen as needed to identify unprofitable areas in fields that should not be planted next season if crop prices do not improve (Doster 2018). Use of multi-year yield data have been encouraged to represent the range of possible yield outcomes (Lamb et al. 1997, Dobermann et al. 2003, Schepers et al. 2004), especially since the outcome depends on the interaction of genetics, environment and management (Hatfield and Walthall 2015). By allowing the aggregation and comparison of yield data across crops and years in \$ ha^{-1}, profit maps enable decision makers to see what areas of fields were above or below economic benchmarks across years and assess how specific areas of a field differ in profitability by crops over years (Massey et al. 2008). Profitability zones, defined by cost categories, can be specified to identify breakeven points where producers might have the incentive to make changes. Yield monitors and associated mapping software should be expanded to enable growers to consider profit mapping.

Many farmers did not realize why they needed a yield monitor until after they started using one. We expect that growers will also find value in the information from on-combine spectrometers. It is now possible to map grain protein levels in fields at the same spatial resolution as yield, therefore, studying the local relationship between yield and protein offers a way to interpret yield maps better. Furthermore, grain yield and protein maps can reveal the location and extent of areas within fields where more (or less) N was needed for yield more effectively than soil N maps derived from a limited number of soil samples. High cost (>\$15,000) has been a barrier preventing wider adoption of on-combine protein sensing technology. In Oregon, we have tested a relatively low-cost (<\$5000) laboratory-grade, reflectance NIR instrument in the field on a combine harvester. Preliminary results are sufficiently promising to suggest that wheat protein can be measured and mapped with a relatively inexpensive spectrometer on a combine. Instrument makers should develop and commercialize lower cost spectrometers for this purpose.

Growers are unwilling to learn and adopt new technologies unless they have been proved, profitable, complete, ready to be used and easy to operate. In addition, because federal minimum wages are too costly for many U.S.A. farmers and prevent them from hiring extra labor, they are already over worked and may not have time to learn new tasks. Therefore, adoption of grain protein sensing technology will require industry support and scientific explanation. New businesses in the agricultural market will be needed that offer services in spectroscopy including instrument repair and maintenance, calibration, and chemometric modelling. These firms will also need to know how to produce robust chemometric models that can predict grain protein concentration over a diversity of cultivars and growing environments. Numerous calibration samples of known protein concentration are needed for this purpose that are likely to be obtained from grain crops grown under varying N and water regimes.

The number of custom mapping and GIS consulting services will need to increase in proportion to expanded adoption of on-combine sensing technologies. Specialist software will be needed for processing and removing errors from yield and protein data and turn on-combine readings into task control maps for variable-rate N application. Software tools already exist for editing of yield monitor data to correct for flow delay and erroneous values, but will need to be extended to include site-specific protein data. Data fusion algorithms and analysis techniques will be needed for computation of secondary attributes such as grain N removal and HI. The derivation of variable-rate N recommendations based on N removal, grain yield and protein will require knowledge of the FNE specific to different crop cultivars and growing regions. This information can be acquired either from the scientific literature, extension bulletins, or actual field experiments.

On-combine multi-sensor systems have been described for simultaneous measurement of grain yield, grain protein concentration and straw yield that can be combined to identify areas of environmental stress in fields (Long and McCallum 2013). The fusion of multiple sensors promises to form an integral component of PA, but data acquisition will depend on the architecture of each sensor system. Various sensors differ in processing ability, memory and power requirements. Data streams will be fed to a central processing unit acting as an automated storage and retrieval system that in turn accommodates heterogeneity in sensors' signal types, data rates and packet sizes. Ideally, each sensor should have an export serial port for connectivity with the central processing unit. Such a framework improves data acquisition and supports the preparation of data for their later processing. In turn, efficient data analysis methods will be needed to process the sheer volume of data that is generated (Mahmood et al. 2012).

Yield monitors and protein analysers provide the information necessary to replace N at the rate of removal, but instrument error may weaken the ability to capture the full economic potential of this approach. For example, the typical precision of an NIR spectrometer is ±5 g of protein kg^{-1} of grain for an error rate of ±3.6% at a protein concentration of 140 g kg^{-1} ($3.6 = 5/140 \times 100$). Therefore, if protein was determined as 125 g kg^{-1} indicating that N had been deficient for yield, but was overestimated by 5 g kg^{-1}, then soil N would be increased by 37.5 kg ha^{-1} instead of 50 kg ha^{-1} in the following season to reach a critical protein level of 140 g kg^{-1}, given that the FNE is 50 kg ha^{-1} for hard red spring wheat in eastern Oregon. In addition, a measurement error of ±5 g of protein kg^{-1} exceeds the interval of 2.5 g of protein kg^{-1} in the incremented price schedule for this grain, which would make it difficult to segregate by protein to capture price premiums. Grain segregation by protein apparently may be better suited for cereal crops having price schedules with wider price steps. Regarding instrument error, further work is needed to determine whether on-combine sensing exhibits robustness or gives good performance for precision management of N and grain quality. As always, field studies of different cropping systems are necessary to demonstrate the economic benefits of variable-rate N application, grain segregation and other aspects of PA before wider adoption of on-combine sensing systems can be expected.

Crop sensing from the combine has tremendous potential to increase the amount of information available to growers for site-specific field management. This chapter was limited to reports in the literature of a few aspects of crop management that have benefited from applications of on-combine sensing. Additional opportunities will probably present themselves in the future that may have consequences for the grain processing industry. For example, variation in grain composition that is affected by genetic and environmental factors is important when the grain is grown for its protein, starch, or oil for human or animal consumption. This variation can cause havoc in crushing, milling, baking, fractionation and other processes that are becoming increasingly automated in food processing plants. By providing within-field information about the crop at harvest, on-combine sensing technologies may help growers manage to obtain uniformity in grain quality that is desired by end-users.

References

Arslan S, Colvin TS (2002) An evaluation of the response of yield monitors and combines to varying yields. Prec Agric 3:107–122

Barroso J, McCallum J, Long D (2017) Optical sensing of weed infestations at harvest. Sensors 17:2381

Blackmore S (2000) The interpretation of trends from multiple yield maps. Comput Electron Agric 26:37–51

Blackmore BS, Marshall CJ (1996) Yield mapping: errors and algorithms. In: Proceedings of the 3rd International Conference on Precision Agriculture. Minneapolis, p 403

Blackmore S, Godwin RJ, Fountas S (2003) The analysis of spatial and temporal trends in yield map data over six years. Biosyst Eng 84:455–466

Blanco M, Villarroya I (2002) NIR spectroscopy: a rapid-response analytical tool. Trends Anal Chem 21:240–250

Bonfil DJ, Mufradi I, Asido S, Long DS (2008) On-combine near infrared spectroscopy applied to prediction of grain test weight. [CD-ROM computer file]. In: Khosla R (ed) Proceedings of the 9th International Conference on Precision Agriculture, 20–23 July 2008. Denver, CO

Büchmann NB, Josefsson H, Cowe IA (2001) Performance of European Artificial Neural Network (ANN) calibrations for moisture and protein in cereals using the Danish Near-Infrared Transmission (NIT) network. Cereal Chem 78:572–577

Christensen S, Sogaard HT, Kudsk P et al (2009) Site-specific weed control technologies. Weed Res. 49:233–241

Chung SO, Sudduth KA, Drummond ST (2002) Determining yield monitoring system delay time with geostatistical and data segmentation approaches. Trans ASAE 45:915–926
Chung S, Moon-Chan C, Kyu-Ho L, Yong-Joo K, Soon-Jung H, Minzan L (2016) Sensing technologies for grain crop yield monitoring systems: A review. J Biosyst Eng 41:408–417
Copeland PJ, Malzer GI, Davis JG, Lamb JA, Robert PC, Bruulsema TW (1996) Using harvest index to locate environmental stress. In: Robert PC et al (eds) Proceedings of the Third International Conference on Precision Agriculture. ASA, CSSA, SSSA, Madison, p 531
Delin S (2004) Within-field variations in grain protein content – relationships to yield and soil nitrogen and consistency in maps between years. Prec Agric 5:565–577
Delwiche SR, Higginbotham RW, Steber CM (2018) Falling number of soft white wheat by near-infrared spectroscopy: a challenge revisited. Cereal Chem 95:469–477
Diker K, Heerman DF, Buchleiter GW (2003) Analysis of multiyear yield data for delineating yield response zones. American Society of Agricultural Engineering Annual Meeting, ser. 031086. https://elibrary.asabe.org/pdfviewer.asp?param1=s:/8y9u8/q8qu/tq9q/5tv/L/13BIGGJ/GJHGOM.5tv¶m2=HI/HG/IGHP¶m3=HPP.HJJ.LH.PL¶m4=13974 Accessed 10 Dec 2019
Diker K, Heerman DF, Brodahl MK (2004) Frequency analysis of yield for delineating yield response zones. Prec Agric 5:435–444
Dobermann A, Ping JL, Adamchuk VI, Simbahan GC, Ferguson RB (2003) Classification of crop yield variability in irrigated production fields. Agron J 95:1105–1120
Doster H (2018) Why not have a profit map on yield monitor? Indiana Prairie Farmer. https://www.indianaprairiefarmer.com/management/why-not-have-profit-map-yield-monitor. Accessed 22 Oct 2018
Dowell FE, Maghirang EB, Xie F, Lookhart GL, Pierce RO, Seabourn BW, Bean SR, Wilson JD, Chung OK (2006) Predicting wheat quality characteristics and functionality using near-infrared spectroscopy. Cereal Chem 83:529–536
Ehlert D, Adamek R, Horn HJ (2009) Laser rangefinder-based measuring of crop biomass under field conditions. Prec Agric 10:395–408
Engel RE, Long DS, Carlson GR, Meier C (1999) Method for precision nitrogen management in spring wheat: I Fundamental relationships. Prec Agric 1:327–338
Fiez TE, Miller BC, Pan W (1994) Winter wheat yield and grain protein across varied landscape postitions. Agron J 86:1026–1032
Flowers M, Weisz R, White J (2005) Yield-based management zones and grid sampling strategies: describing soil test and nutrient variability. Agron J 97:968–982
Fulton JP, Sobolik CJ, Shearer SA, Higgins SF, Burks TF (2009) Grain yield monitor flow sensor accuracy for simulated varying field slopes. Appl Eng Agric 25:15–21
Glen DM, Carey A, Bolton FE, Vavra M (1985) Effect of N fertilizer on protein content of grain, straw, and chaff tissues in soft white winter wheat. Agron J 77:229–232
Godwin RJ et al (1999) Cumulative mass determination for yield maps of non-grain crops. Comput Electron Agic 23:85–101
Goos RJ, Westfall DG, Ludwick AE, Goris JE (1982) Grain protein content as an indicator of N sufficiency for wheat. Agron J 74:130–133
Griffith DA, Chun Y (2012) Spatial autocorrelation and eigenvector spatial filtering. In: Fischer M, Nijkamp P (eds) Handbook of regional science. Springer, Berlin
Hatfield JL, Walthall CL (2015) Meeting global food needs: Realizing the potential via genetics × environment × management interactions. Agron J 107:1215–1226
Hawkins E, Fulton J, Port K (2017) Tips for calibrating grain yield monitors – maximizing value of your yield data. Fact Sheet ANR-8 Ohio State University Extension. https://ohioline.osu.edu/factsheet/anr-8. Accessed 26 July 2019
Hermann D, Bilde ML, Andersen NA, Ravn O (2016) A framework for semi-automated generation of a virtual combine harvester. IFAC-PapersOnLine 49-16:55–60
Jaynes DB, Kaspar TC, Colvin TS, James DE (2003) Cluster analysis of spatiotemporal corn yield patterns in an Iowa field. Agron J 95:574–586

Jaynes DB, Colvin TS, Kaspar TC (2005) Identifying potential soybean management zones from multi-year yield data. Comput Electron Agric 46:309–327

Kettle LY, Peterson FL (1998) An evaluation of yield monitors and GPS systems on hillside combines operating on the steep slopes in the Palouse. ASAE Paper No. 981046. ASAE, St, Joseph

Kitchen NR, Drummond ST, Lund ED et al (2003) Soil electrical conductivity and topography related to yield for three contrasting soil-crop systems. Agron J 95:483–495

Kleinjan J, Clay DE, Carlson CG, Clay SA (2006) Developing productivity zones form multiple years of yield monitor data. Site specific management guidelines SSMG-45 PPI-FAR, Norcross, GA. Accessed 31 Oct 2018

Lamb JA, Dowdy RH, Anderson JL, Rehm GW (1997) Spatial and temporal stability of corn grain yields. J Prod Agric 10:410–414

Lee WS, Alchanatis V, Yang C, Hirafuji M, Moshou D, Li C (2010) Sensing technologies for precision specialty crop production. Comput Electron Agric 74:2–33

Leroux C, Jones H, Taylor J, Clenet A, Tisseyre B (2018a) A zone-based approach for processing and interpreting variability in multi-temporal yield data sets. Comput Electron Agric 148:299–308

Leroux C, Jones H, Clenet A, Dreux B, Becu M, Tisseyre B (2018b) A general method to filter out defective spatial observations from yield mapping datasets. Prec Agric 19:789–808

Long DS, McCallum JD (2013) Mapping straw yield using on-combine light detection and ranging (lidar). Int J Remote Sens 34:6121–6134

Long DS, McCallum JD (2015) On-combine, multi-sensor data collection for post-harvest assessment of environmental stress in wheat. Prec Agric 16:492–504

Long DS, Engel RE, Carlson GR (2000) Method for precision nitrogen management in spring wheat: II. Implementation. Prec Agric 2:25–38

Long DS, Engel RE, Siemens MC (2008) Measuring grain protein concentration with in-line near infrared reflectance spectroscopy. Agon J 100:247–252

Long DS, McCallum JD, Scharf PA (2013) Optical-mechanical system for on-combine segregation of wheat by grain protein concentration. Agron J 105:1529–1535

Long DS, Whitmus JD, Engel RE, Brester GW (2015) Net returns from terrain-based variable-rate nitrogen management on dryland spring wheat in northern Montana. Agron J 107:1055–1067

Long DS, McCallum JD, Reardon CL, Engel RE (2017) Nitrogen requirement to change protein concentration of spring wheat in semiarid Pacific Northwest. Agron J 109:675–683

Luck JD, Fulton JP, Rees J (2015) Hands-on precision agriculture data management workshops for producers and industry professionals: development and assessment. J Extension 53(4). https://www.joe.org/joe/2015august/pdf/JOE_v53_4tt10.pdf. Assessed 26 July 2019

Machado S, Bynum ED, Archer TL, Lascano RJ, Wilson LT, Bordovsky J, Segarra E, Bronson K, Nesmith DM, Xu W (2000) Spatial and temporal variability of corn grain yield: Site-specific relationships of biotic and abiotic factors. Prec Agric 2:359–376

Maertens K, Reyns P, de Clippel J, Baerdemaeker D (2003) First experiments on ultrasonic crop density measurement. J Sound Vibration 266:655–665

Maertens K, Reyns P, De Baerdemaeker J (2004) On-line measurement of grain quality with NIR technology. Trans. ASAE 47:1135–1140

Maestrini B, Bassso B (2018) Predicting spatial patterns of within-field crop yield variability. Field Crops Res 219:106–112

Mahmood H, Hoogmoed W, van Henten E (2012) Sensor data fusion to predict multiple soil properties. Precis Agric 13:628–645

Martin CT, McCallum JD, Long DS (2013) A web-based calculator for estimating the profit potential of grain segregation by protein concentration. Agron J 105:721–726

Massey RE, Myers DB, Kitchen NR, Sudduth KA (2008) Profitability maps as an input for site-specific management decision making. Agron J 100:52–59

McKinion JM, Willers JL, Jenkins JN (2010) Spatial analyses to evaluate multi-crop yield stability for a field. Comput Electron Agric 70:187–198

Meyer-Aurich A, Gandorfer M, Weersink A, Wagner P (2008) Economic analysis of site-specific wheat management with respect to grain quality and separation of the different quality fractions. In: European Association of Agricultural Economists, 2008 International Congress, Ghent, Belgium. 26–29 August 2008. AgEcon Search, Department of Applied Economics, University of Minnesota. http://ageconsearch.umn.edu/bitstream/43649/2/012.pdf. Accessed 31 Oct 2018

Milne AE, Webster R, Ginsburg D, Kindred D (2012) Spatial multivariate classification of an arable field into compact management zones based on past crop yields. Comput Electron Agric 80:17–30

Missotten B, Strubbe G, De Baerdemaeker J (1997) Straw yield mapping: A tool for interpretation of grain yield differences within a field. In: Stafford JV (ed) Proceedings of the First European Precision Agriculture Conference. BIOS Scientific Publishers, Oxford, pp 735–742

Myers A (1994) Method and apparatus for measuring grain mass flow rate in harvesters. U.S. Patent 5343761, filed 17 June 1991 and issued 6 September 1994

Nelson BP, Elmore RW, Lenssen AW (2015) Comparing yield montors with weigh wagons for on-farm corn hybrid evaluation. Crop Forage Turfgrass Manag

Osborn BG (1984) Investigations into the use of near infrared reflectance spectroscopy for the quality assessment of wheat with response to its potential for bread baking. J Sci Food Agric 35:106–110

Ping JL, Dobermann A (2005) Processing of yield map data. Prec Agric 6:193–212

Polo JRR, Sanz R, Llorens J, Arno J, Escolà A, Ribes-Dasi M, Masip J, Camp F, Gracia F, Solanelles F, Palleja T, Val L, Planas S, Gil E, Palacin J (2009) A tractor-mounted scanning LIDAR for the non-destructive measurement of vegetative volume and surface area of tree-row plantations: a comparison with conventional destructive methods. Biosyst Eng 102:128–134

Prihar SS, Stewart BA (1990) Using upper-bound slope through origin to estimate genetic harvest index. Agron J 82:1160–1165

Raun WR, Solie JB, Johnson GV et al (2002) Improving nitrogen use efficiency in cereal grain production with optical sensing and variable rate application. Agron J 94:815–820

Reyniers M, Maertens K, Vrindts E, De Baerdemaeker J (2006) Yield variability related to landscape properties of a loamy soil in central Belgium. Soil Tillage Res 88:262–273

Reyns P, Missotten B, Ramon H, De Baerdemaeker J (2002) A review of combine sensors for precision farming. Prec Agric 3:169–182

Risius NW (2014) Analysis of a combine grain yield monitoring system. Graduate M.S. thesis, Iowa State University, Ames. https://lib.dr.iastate.edu/cgi/viewcontent.cgi?article=4806&context=etd Accessed 1 Nov 2018

Risius H, Hahn J, Huth M, Tölle R, Korte H (2015) In-line estimation of falling number using near-infrared diffuse reflectance spectroscopy on a combine harvester. Prec Agric 16:261–274

Risius H, Prochnow A, Ammon C, Mellmann J, Hoffmann T (2017) Appropriateness of on-combine moisture measurement for the management of harvesting and postharvest operations and capacity planning in grain harvest. Biosystems Eng 156:120–135

Ruß G (2009) Data mining of agricultural yield data: a comparison of regression models. In: Perner P (ed) Advances in data mining. Applications and theoretical aspects. ICDM 2009. Lecture notes in computer science, vol 5633. Springer, Heidelberg

Sadler EJ, Evans RG, Stone KC, Camp CR (2005) Opportunities for conservation with precision irrigation. J Soil Water Cons. 60:371–379

Saeys W, Lenaerts B, Craessaerts G, De Baerdemaeker J (2008) Estimation of the crop density of small grains using LiDar sensors. Biosystems Eng 102:22–30

Schepers AR, Shanahan JF, Liebig MA, Schepers JS, Johnson SH, Luchiari A Jr (2004) Appropriateness of management zones for characterizing spatial variability of soil properties and irrigated corn yields across years. Agron J 96:195–203

Schimmelpfennig D (2016) Farm profits and adoption of precision agriculture. ERR-217, U.S. Department of Agriculture, Economic Research Service

Schueller J, Mailander M, Krutz G (1985) Combine federate sensors. Trans ASAE 28:2–6

Schuster JN, Darr MJ, McNaull RP (2017) Performance benchmark of yield monitors for mechanical and environmental influences. ASABE Paper No. 1700881. St. Joseph

Selles F, Zentner RP (2001) Grain protein as a postharvest index of N sufficiency for hard red spring wheat in the semiarid prairies. Can J Plant Sci 81:631–636

Shenk JS, Workman JJ, Westerhaus MO (2001) Application of NIR spectroscopy to agricultural products. In: Burns DA, Ciurczak EW (eds) Handbook of near-infrared analysis, 3rd edn. CRC Press, New York, p 419

Shollenberger JH, Kyle CF (1927) Correlation of kernel texture, test weight per bushel, and protein content of hard red spring wheat. J Agric Res 35:1137–1151

Sivaraman E, Lyford CP, Brorsen RW (2002) A general framework for grain blending and segregation. J Agribusiness 20:155–161

Snyder FW, Carlson GE (1984) Selecting for partitioning of photosynthetic products in crops. Adv Agron 37:47–72

Stafford JV, Ambler B, Lark RM, Catt J (1998) Mapping and interpreting the yield variation in cereal crops. Comput Electron Agric 14:101–119

Stewart CM, McBratney AB, Skerritt JH (2002) Site-specific durum wheat quality and its relationship to soil properties in a single field in northern New South Wales. Prec Agric 3:155–168

Stott BL, Borgelt SC, Sudduth KA (1993) Yield determination using an instrumented Claas combine. ASAE Paper No. 93-1507

Sudduth KA, Drummond ST, Myers DB (2012) Yield Editor 2.0: software for automated removal of yield map errors. Paper No. 12-1338243, ASABE, St. Joseph

Taylor JA, Whelan BM (2007) On-the-go grain quality monitoring: a review. In: Proceedings of the 4th International Symposium of Precision Agriculture (SIAP07). Viscosa, Brasil

Taylor RK, Stone ML, Downs HW (1986) Laser based crop density detection. ASAE Paper No. 86-1619

Taylor RK, Kluitenberg GJ, Schrock MD, Zhang N, Schmidt JP, Havlin JL (2001) Using yield monitor data to determine spatial crop production potential. Trans ASAE 44:1409–1414

Taylor RK, Hobby M, Schrock MD (2005) Evaluation of an automatic feedrate control system for a grain combine. In: ASAE Annual International Meeting. ASAE, Tampa, pp 1–12

Von Rosenberg Jr CW, Abbate A, Drake J (2000) A rugged near-infrared spectrometer for real-time measurement of grains during harvest. Spectroscopy 15:34–38

Whelan BM, Taylor JA, Hassall JA (2009) Site-specific variation in wheat grain protein concentration and wheat grain yield measured on an Australian farm using harvester-mounted on the-go sensors. Crop Pasture Sci 60:808–817

Williams PC (2001) Implementation of near-infrared technology. In: Williams PC, Norris K (eds) Near-infrared technology in the agricultural and food industries, 2nd edn. American Association of Cereal Chemists, St. Paul

Williams P, Sobering D, Antoniszyn J (1998) Protein testing methods. In: Fowler DB et al (eds) Wheat protein production and marketing. Proceedings of the wheat protein symposium, Saskatoon, 9–10 March 1998, University of Saskatchewan Extension Press, p 37

Chapter 8
Sensing in Precision Horticulture

Manuela Zude-Sasse, Elnaz Akbari, Nikos Tsoulias, Vasilis Psiroukis, Spyros Fountas, and Reza Ehsani

Abstract Information technology is playing an increasingly important role in today's agricultural production systems, regardless of operation size, commodity or management approach. Precision horticulture is an information-based management strategy that relies on collecting site-specific or plant-specific data. These data can be converted to useful information that helps growers make informed management decisions. Precision horticulture can benefit growers because of the high value of their products and the large amounts of crop inputs used in producing horticultural crops. Any improvement in reducing production costs can greatly increase profit for producers. Also, the optimal use of crop inputs in precision horticulture can potentially reduce the environmental impact of horticultural crop production. Implementation of precision horticulture relies heavily on sensors and systems that can collect weather, soil and plant-specific data cost-effectively. Plant data, in particular, allow a direct feedback for production and harvest management. Examples of data that need to be recorded by the plant sensors include biotic and abiotic stress detection at asymptomatic or early stages, canopy size and density, yield estimation and crop quality. For example, LiDAR-based or computer vision-based sensors are being used for measuring tree canopy size and density. The quantifying of variation in canopy size of orchards is needed for variable-rate crop inputs. With advances in sensing technology, various types of sensors have been developed commercially and are becoming available for precision horticulture. Optical sensors are most commonly used and several techniques have shown the potential for efficient, rapid, non-invasive field detection of plant diseases and yield estimation. This chapter reviews the current applications of sensor technologies being used in horticultural production systems.

M. Zude-Sasse (✉) · N. Tsoulias
Leibniz Institute for Agricultural Engineering and Bioeconomy (ATB), Potsdam, Germany
e-mail: mzude@atb-potsdam.de

E. Akbari · R. Ehsani
University of California, Merced, CA, USA
e-mail: rehsani@ucmerced.edu

V. Psiroukis · S. Fountas
Agricultural University of Athens, Athens, Greece
e-mail: sfountas@aua.gr

R. Kerry, A. Escolà (eds.), *Sensing Approaches for Precision Agriculture*,
Progress in Precision Agriculture, https://doi.org/10.1007/978-3-030-78431-7_8

Keywords Abiotic stress · Fruit quality · Precision horticulture · Proximal sensing · Remote sensing · Spectral-optical · Yield prediction

8.1 The Situation in Horticulture

In the production and harvest of fruit and vegetables (horticultural crops), large investment costs and expected high productivity of the land provide strong incentives for implementing technology-supported production methods because the application of more technology possibly results in more precise, optimized management of production. The horticultural product is highly perishable, of high value and has heterogeneous quality. This heterogeneity of horticultural produce often requires the handling of individual plants (Schouten et al. 2007).

With increasing demands regarding the yield, sustainable use of land and water resources, and requests for good crop quality production methods have been changing. Conventional citrus orchards with fixed tree spacing have been replaced by systems with greater tree density supported by water-efficient drip irrigation. Intensive production systems using adapted cultivars and pruning techniques show densities up to 3200 trees per ha, which is in the range that we find in apple production with small spindle tree forms enabling planting density of 2600–3700 trees per ha. In stone fruit, such as mango (*Mangifera indica*), sweet cherry and olive production similar trends can be seen after rootstocks became available that reduce vegetative growth of the cultivar. In addition, sophisticated tree training has been practised for over 30 years (Fig. 8.1). In olive production, for example, the system is called hedgerow or super high density allowing 600–1250 trees per ha. Mechanization of spraying, foliar application of fertilizers and thinning agents, pruning, and

Fig. 8.1 Tree forms in olive (*Olea europaea* L.) (**a**), sweet cherry (*Prunus avium* L.) as slender spindle (**b**), and apple (*Malus* x *domestica* Borkh) in Y-shaped trellis system (**c**). Photographs courtesy of: Zude / Zude / Penzel

harvesting fruit (Fig. 8.2) has been introduced because machinery ensure the safety of application and labour costs have also become a crucial factor in fruit and vegetable production.

Applications are usually carried out uniformly over the orchard blocks because information on the individual plant has either not been available or utilized. A huge step forward in precision horticulture was achieved when the variable-rate application of chemical inputs was introduced, allowing treatment to be based on the requirement of the crop. In the first approach, recognition of plant biomass, tree height, gaps due to missing plants as this occurs regularly in perennial orchards represent the crop's variability and, therefore, the 'variable', i.e. adaptive component of the precise production. Efforts are also being made to integrate variable-rate application legally into the pest management of tree fruit production. In Europe according to common agricultural policy (CAP), the amount of foliage needs to be known instead of a simple area-dependant calculation. The actual plant information on the foliage has been used as the basis for calculating the expenditure since 2020. Companies are now marketing variable-rate equipment and software so that innovation is entering horticultural practises.

Furthermore, horticulturists seek new or renewed tree training such as planar, cordon and y-shaped trellis systems that enhance light interception, which is closely related to the yield and fruit quality that the plants can achieve. Coincidently, such light-effective orchards (Fig. 8.1) might also favour the application of robots for harvesting and other tasks. Mechanical harvesting of table fruit is a benchmark (Fig. 8.2c), which still needs to be achieved because harvest costs amount to >50 % of all production costs in many countries.

The advantage of an intensive production system is to grow fruit more economically and in a way that allows variable-rate application in plant protection, foliar application of fertilizers, flower and fruit thinning, and harvest. The drawback is the increased susceptibility of the system to unfavourable conditions because the soil's capacity to act as a buffer is diminishing. Furthermore, the tree is tied in the production system possibly reducing the tree's adaptation capacity. Taking into account the

Fig. 8.2 Machine pruning in citrus (**a**) and apple (**b**) production, and harvester prototype (**c**). Photographs courtesy of: Zude / Betz / Karkee

enhanced investment costs of intensive production systems, the introduction of sensors for *in situ* analysis of specific traits of the plant or even the plant's physiological responses appears increasingly reasonable with feedback from the plant to reduce errors in the management of production and harvest.

In situ data acquisition can now be applied to reflect the plant status in digital format. At present, we see close cooperation between industry and research partners to develop digital twins of orchards and vegetable greenhouse production. The digital twins represent a model tree or single fruit in digital format that can be used to follow or simulate a process with an underlying physiological model. Here tree or fruit sensors provide information, preferably in real time, on the plant or fruit to define the digital twin, y. The environmental and production system variables form the X matrix to simulate the development of tree or fruit. The new information might provide valuable knowledge for sustainable fruit and vegetable production because, for the first time, agronomic models based on sensor data can be applied easily with cloud computing. This approach can possibly lead to new services and business models based on selling information rather than sensor devices.

From a fruit grower's perspective 'farming with sensors is so much easier' because knowledge of the crop in real-time assists decision-making for precise management. To meet this goal, sensors should collect data *in situ* during production and postharvest. Information and communication technology tools, such as satellites, drones, autonomous platforms, wireless networks and data management techniques are available for all scales to support data acquisition by remote and proximal sensors directly. On the other hand, the translation of sensor data into information on the crop and knowledge of the process is still challenging.

8.2 Biotic and Abiotic Plant Stresses

8.2.1 Background

Pests and diseases can cause major economic loss in horticulture, and managing pests and diseases accounts for a significant cost of production. Consequently, one of the major components of precision horticulture is the detection of pests and disease at an early stage. Detection of anomalies at an early stage, in most cases, could allow the growers to manage the stress cost-effectively and prevent the adverse effects of that stress on yield and profit. Plant stresses can be divided into biotic and abiotic stresses. Abiotic stress is defined as the stress caused by non-living factors on living organisms under a specific environment (Husain et al. 2017). Abiotic stress can be caused by high irradiation, salinity, heat, frost, drought, flooding or undesirable conditions such as nutritional availability, nutrient imbalance or contamination by toxic chemical compounds. When the environmental conditions change and are beyond the normal range, abiotic stress can occur and adversely affect the performance of the plants. Abiotic stress is one of the major harmful factors affecting the growth and productivity of crops worldwide (Gao et al. 2007).

Biotic stress occurs as a result of plant interactions with other living organisms, such as arthropods, bacteria, viruses, fungi, parasites, insects, weeds, and cultivated or native plants (Gupta and Senthil-Kumar 2017; Shabala 2017; Atkinson et al. 2015). Herbivores can be harmful to the plants causing mechanical damage and competition for growth factors, resulting in reduced growth rate. On the other hand, plant pathogens can cause different types of diseases. Once an attack is perceived, plant metabolism must balance the demands for resources to support defence mechanisms against the requirements for cellular maintenance, growth and reproduction (Berger et al. 2007). Hence defence processes can be costly in terms of plant growth and health. Plants, herbivores and their natural enemies have coexisted for at least 100 million years, evolving a variety of beneficial and deleterious interactions (Karimzadeh and Wright 2008).

Currently, stress detection mainly relies on manual scouting, which is labour intensive and costly. Furthermore, human scouts rely on their senses, which are often not sufficiently quantitative and often fail to detect early stress. For example, certain plant stresses can be detected easily in the near-infrared range, but are not detectable by the human eye, which can only see in the visible bands. Much effort has been applied by engineers and scientists to develop sensor systems for detecting plant stress. Sensors for detecting plant stress in the field need to be cost-effective, easy to use, non-invasive, real-time or near real-time, and robust. They can be very valuable if they can detect the stress at an asymptomatic stage. The majority of the sensors that have these capabilities are based on spectroscopy or imaging techniques in the electromagnetic wavelength range from UV to far-infrared (Table 8.1). Other emerging techniques are volatile profiling-based and polymerase chain reaction-based sensors. These are new and growing areas of research, and more commercially available sensors that use these principles are anticipated. The advantage of these types of sensors is that they can be very accurate, and crop and disease-specific. The material in this chapter describes the optical techniques, which are currently used for biotic and abiotic stress detection. Sensors can be operated at the production site with tractors, autonomous platforms or robots, cranes, unmanned aerial vehicles (UAV), wireless sensor networks and manual readings (Fig. 8.3). Earlier approaches used cranes and sensors mounted on tractors (Zude-Sasse et al. 2016), whereas now trees can be monitored on all scales (Figs. 8.3b and 8.4).

8.2.2 *Remote Sensing for Stress Detection*

One of the significant applications of remote sensing in agriculture has been stress and disease detection. Numerous different types of platforms can be used for remote sensing in precision agriculture (PA), such as manned aircraft (Hunt and Daughty 2018; Kamal et al. 2020), satellite (Hegarty-Craver et al. 2020; Zhou et al. 2017), and UAVs (Xiang and Tian 2011). Satellite-based remote sensing technologies are used more often for general stress detection over a large area because they have the capacity for continuously monitoring the earth's surface. However, the measurement uncertainty is increased due to mixed pixels capturing the plant rows and the soil or ground cover in between the rows.

Table 8.1 Examples of techniques reported in the literature for biotic stress detection in different crops

Type of sensing	Crop species	Name of disease	Classification or numerical analysis method	References
Hyper- and multispectral	Vineyards	Phylloxera	Various	Vanegas et al. (2018)
	Lettuce	Anomalities	Various	Ren et al. (2017)
RGB imaging	Strawberry leaf	Powdery mildew	ANN, SVM	Shin (2020)
	apple	apple scab (*Venturia inaequalis*), apple rot, apple blotch	Multi-class SVM	
	Mango leaf	Red rust, bacterial canker, gall flies, powdery mildew,	SVM, PCA	Padhye et al. (2014)
	Cacao	Cacao black pod rot (BPR)	SVM	Tan et al. (2018)
Thermal sensors	Cucumber	Downy mildew (Pseudoperonospora cubensis); powdery mildew (Podosphaera xanthii)	Correlation analysis	Berdugo et al. (2014) and Mahlein et al. (2013)
	Apple	Defect	*K*-nearest neighbours, Otsu method	Yogeshi et al. (2018)
Florescence	Grapefruit	Anomalities	PCA, PLSR	Saleem et al. (2020)
	Bean	Common bacterial blight (Xanthomonas fuscans)	–	Rousseau et al. (2013)
	Lettuce	Downy mildew (Bremia lactucae)	Generalised linear mixed model with binomial distribution	Brabandt et al. (2014)
Spectral sensors	Banana	Fusarium wilt	SVM, ANN	Ye et al. (2020)
	apple	Blotch, rot, and scab	Multi-class SVM	Dubey and Jalal (2016)
	Tomato	Anomalities	Histogram	Zhu et al. (2018)

Recent advancements in UAV technologies resulted in the production of more affordable UAVs; lower costs and ease of use made them the platform of choice for stress detection when a high-resolution image is needed. Different types of sensor such as thermal cameras, multispectral cameras, NIR cameras, and regular digital cameras have been used on-board UAVs to collect data or monitor plant health (Martinelli et al., 2015). Computer vision and image processing have also been applied to remotely sensed images to make decisions in agricultural applications. These decisions could be carried out after collecting and processing the data or on-board while the UAV is flying (Maes and Steppe 2019).

Fig. 8.3 Sensor platforms supporting optical sensors: (**a**) an agricultural vehicle and (**b**) unmanned aerial vehicle. Photographs courtesy of: Ehsani / Zude

Extensive reviews on the application of remote sensing for the detection of plant diseases were published by, e.g., Jackson (Jackson 1986), and West and co-workers (West et al. 2003). The topic was also covered in a review of video remote sensing systems for agricultural assessment (Abd El-Ghany et al. 2020). Remote sensing has been a useful tool in detecting foliar and soil-borne plant diseases (Oerke 2020). Assessments of plant disease infestations are typically carried out visually by workers. However, individual differences in perception of colour and light, as well as lack of concentration or fatigue, can significantly reduce the accuracy of visual disease assessments. Here as well as in instrumental remote sensing, other factors, such as cloud cover causing varying levels of light on different parts of the plant canopy, also limit accuracy of visual and instrumental disease assessments.

Instrumental remote sensing can be applied to plant disease detection because of pathogen-induced colour changes from chlorosis or necrosis. Furthermore, water-stressed crop canopies consistently have higher temperatures than non-stressed crop canopies because of reduced transpiration; thermal IF emissivity may be measured to detect these types of changes (Martinelli et al. 2015). Remote sensing may provide earlier detection of plant stress, even prior to the appearance of visual symptoms (Matese et al. 2018). Utilizing remote sensing instruments capable of detecting non-visible wavelengths could allow these reduced photosynthesis rates to be detected earlier than by visible disease symptoms (Reddy 2018; James et al. 2020; Vitrack-Tamam et al. 2020). A wide variety of remotely sensed data can be collected, and these data can help plan and schedule irrigation, water management (Gonzalez-Dugo et al. 2020), detect insect infestation problems and weed infiltration, and determine plant stress by carrying out small scale photogrammetric surveys using RGB and/or 4-band multispectral imaging and LiDAR laser scanning.

Plant health is detectable by airborne sensors because of the plant's reflectivity and absorption of electromagnetic radiation. The pigmentation of the plants controls this reflectivity and absorption, creating incident radiation depending on the plant size, orientation, and colour. Plant pigment heavily relies on the amount of

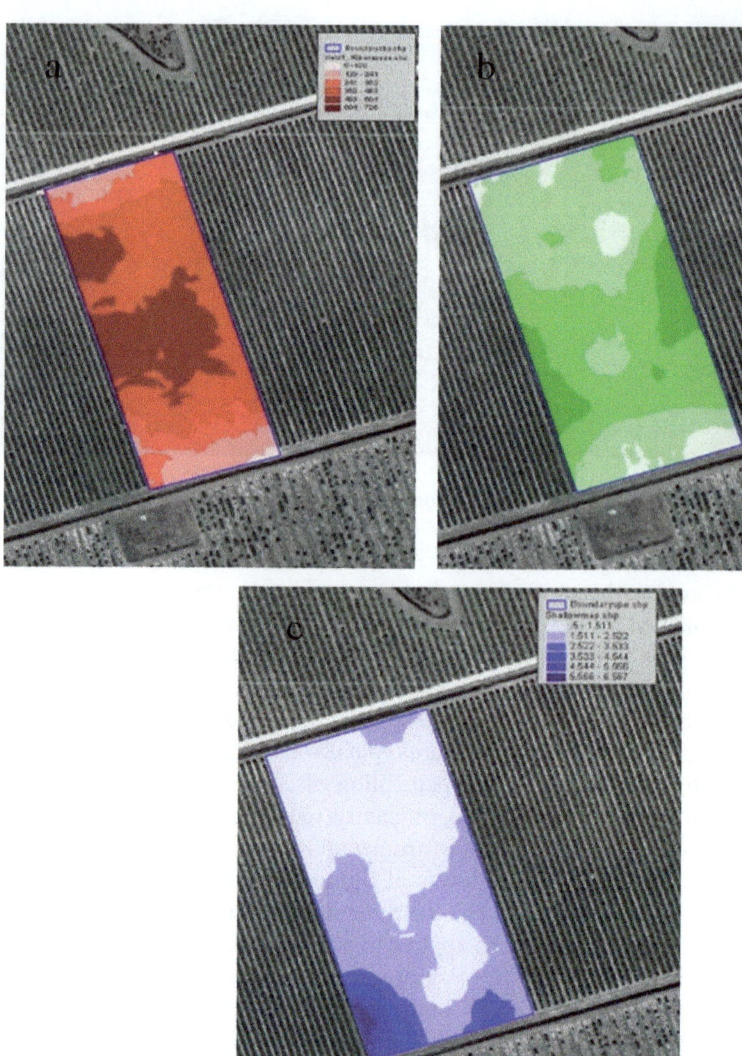

Fig. 8.4 (**a**) Contour yield map of a citrus orchard, (**b**) contour map of tree volume (in cubic metres), (**c**) map of apparent soil electrical conductivity. Photographs courtesy of: Ehsani

chlorophyll (Saglam et al. 2019), which intensely absorbs radiation within the visible spectrum. When a plant is stressed, chlorophyll production declines, increasing the reflectance of wavelengths in the visible spectrum, including those in the red bands.

Plant health and heterogeneity can be quantitatively and qualitatively measured by means of calculating a series of remotely sensed vegetation indices (VIs) (Zare et al. 2020). This is done through a series of image band calculations (Rasmussen et al. 2016). The most common calculation performed is related to crop status such as leaf area index (LAI), canopy cover, biomass, vegetation wellbeing, or 'greenness'. Most estimations are based on the chlorophyll content, which is frequently calculated by the normalized difference vegetation index (NDVI), and is calculated using the sensor's red band (Rouse et al. 1973), which registers the absorption of red wavelengths due to chlorophyll content (Rouse 1973; c.f. Walsh et al. 2020). The NIR channel can be useful in determining plant stress due to higher reflection in plants containing more chlorophyll and vice versa. Meggio et al. (2010) used hyperspectral imagery via manned aerial vehicles to calculate a series of VI for detecting iron chlorosis in a vineyard. It was concluded that this type of imagery is useful in determining plants with iron chlorosis, which is a huge problem in vineyards and citrus plantations. Most sensing principles of remotes sensing for stress detection have been focused on optical techniques so far, but in recent years more complex approaches are emerging using multiple sensors and sensor fusion techniques. For example, it is common to see more cost-effective sensors that combine multispectral images and thermal images. Multi-sensor remote sensing could potentially enhance the possibility of early stress detection and has a lot of potential for large scale adaptation by growers.

8.2.3 *Visible and NIR Bands*

Visible and NIR bands are ideal for developing sensors for disease detection because the detectors used in this part of the electromagnetic spectrum are relatively inexpensive (Table 8.1). Sankaran and Ehsani (2013) used a portable spectrophotometer (SVC HR-1024, SpectraVista, USA) in the 350–2500 nm range for detection and classification of citrus greening or Huanglongbing (HLB) and citrus canker from healthy trees. They used quadratic discriminant analysis and a *K*-nearest neighbour classifier; both techniques provided large detection accuracy. Similar studies were performed with different classification techniques. Sankaran and co-workers (Sankaran et al. 2012) used a hand-held spectrophotometer to collect data from asymptomatic, symptomatic, freeze-damaged and healthy plants. Linear discriminant analysis, quadratic discriminant analysis, Naïve-Bayes and bagged decision trees were used as classifiers with 77, 92, 84 and 99 % accuracy, respectively. All classifiers were able to discriminate symptomatic-infected leaves from freeze-damaged leaves, but some asymptomatic leaves were incorrectly detected as healthy. Low-cost, rugged disease-specific sensors can be built that use only a few selected bands rather than continuous spectral data in the whole range of the visible and NIR bands. Active multispectral sensors using high-density illuminators at a specific narrow band for detecting disease and stress might be even better suited to field application. These bands are usually tailored for a specific diseases rather than for general stress detection. They use two bands in the red and NIR range to calculate the NDVI based on changes in chlorophylls (Olsen et al. 1969; Zude 2003) and flavonoids (Moran and Moran 1998) that provide data on the plant and its stress response. Previous experiments indicated that these approaches can distinguish between adaptation and irreversible damage to plants. Consequently, reflectance measurements based on remote sensing may be used to estimate integrated stress responses of the canopies (e.g., Peñuelas and Filella 1998). Mishra et al. (Mishra et al. 2011) used a four-band active optical sensor for detecting Huanglongbing (HLB) disease in citrus, two bands of which were in the visible region (570 and 670 nm), whereas the other two were in the NIR region (750 and 870 nm). They used *K*-nearest neighbour, support vector machines (Rançon et al. 2019) and decision trees for classification and achieved 96, 97 and 95.5 % accuracy, respectively. However, it can be assumed that distinguishing diseases by means of chlorophyll, water and microstructural changes is almost impossible because they are almost universal in all fruit and leaf responses. An additional trait, but still unspecific, is the intensity difference of reflected light between 450 and 570 nm, which is caused by changes in the xanthophyll cycle. The xanthophyll cycle describes variation in the actual composition of carotenoids. Changes in the related carotenoids (zeaxanthin, violaxanthin and antheraxanthin) appeared measurable with reflectance spectra and were recorded in remote- or proximal-sensing with the leaf under various stress conditions (Peñuelas and Filella 1998). They have been little studied on fruit because reproductive plant organs tend to accumulate carotenoids during seasonal development to enhance attraction to vectors. However, wavelength-specific indices or multivariate methods for processing the spectra of harvest products or leaves might provide relevant data for detecting anomalies.

8.2.4 Mid-Infrared Bands

Mid-infrared (MIR) spectra range from 2500–50,000 nm. The MIR spectroscopy usually requires sample preparation, which is a disadvantage compared to NIR spectroscopy. However, MIR has some advantages over NIR spectroscopy. One of the major advantages of MIR spectroscopy is that many chemicals have a unique peak or signature in this region which makes spectral interpretation easy compared to the NIR region where the spectra comprise overlapping overtones of absorption bands of many interfering functional groups of molecules. The application of MIR for measuring the nitrogen content in impregnated tomato leaf has been indicated. Sankaran et al. (2010) used a portable MIR spectrometer (InfraSpec VFAIR, Spectro Scientific, USA) in the range of 5.15–10.72 μm to detect healthy leaves as well as those infected with HLB and canker. The leaves were ground into a fine powder and were placed on the crystal window of a MIR spectrometer. The scan time was 1 minute. Quadratic discriminant analysis and *K*-nearest neighbour classifiers were used, and resulted in large accuracies of 98 ± 0.9 % and 99 ± 0.9 %, respectively.

8.2.5 Fluorescence Spectroscopy

Fluorescence spectroscopy is an optical sensing technique that takes advantages of the re-emission of light from a sample. In this process, excitation light is absorbed by the material of interest and then emits this light at longer wavelengths. There are commercial sensors that use this technique. A hand-held multiparameter optical sensor (Multiplex_3, Force A, France) was used to detect HLB disease in leaves of two different sweet orange cultivars, Hamlin and Valencia. Four excitation wavelengths are employed in the instrument: UV, blue, green, and red (c.f. Zude 2009). For each excitation wavelength, yellow, red and far-red fluorescence can be measured. The classifiers used were Naïve-Bayes and bagged decision tree with accuracies of 85 % and more than 94 %, respectively. The bagged decision tree classifier performed better than Naïve-Bayes; however, it needed more time for the computation process, at least 10 times more than the Naïve-Bayes classifier (Sankaran et al. 2012).

8.2.6 Laser-Induced Breakdown Spectroscopy

Laser-induced breakdown spectroscopy (LIBS) is destructive, but it is an *in situ* spectroscopic technique that can be used for qualitative and quantitative analysis of elemental composition in solids, liquids and gases. This technique uses a high-power laser to generate plasma around the target and emissions from the atoms in the plasma are analyzed by a detector. The advantages of this technique include little to no sample preparation, analysis is done on very small samples and there is simultaneous measurement of multiple elements. The technique can also be rapid and cost-effective. The LIBS has been used in several applications, many of which include elemental analysis in different media (Yamamoto et al. 1996; Xu et al. 1997; Hussain and Gondal 2008). In addition to these studies, LIBS has also been used to measure nutrients in plant materials (Santos Jr. et al. 2012). Trevizan et al. (2009) used LIBS for analyzing microelements such as B, Cu, Fe, Mn and Zn in plant materials. They compared the laser-induced breakdown spectrometer results with those of conventional acid digestion. The detection limit of the LIBS method using their analysis protocol was 2.2 mg kg^{-1} for B, 3.0 mg kg^{-1} for Cu, 3.6 mg kg^{-1} for Fe, 1.8 mg kg^{-1} for Mn and 1.2 mg kg^{-1} for Zn. Similarly, Yao et al. (2010) used the LIBS technique to identify nutrients in orange leaves. The authors found that the spectral peaks indicate the specific elements and that their intensity is proportional to the elemental concentration. Thus, they concluded that the nutrient status can be evaluated from the spectral characteristics of the orange leaves using a LIBS system. Despite being able to demonstrate the potential of LIBS for nutrient analysis in citrus leaves, the study did not evaluate different citrus anomalies with the LIBS system. Sankaran et al. (2015) studied the variation in LIBS spectra obtained from healthy citrus leaves and leaves with anomalies such as diseases (HLB, canker) and nutrient deficiencies (Zn, Fe, Mg and Mn). Pattern recognition algorithms, support vector machine (SVM) and quadratic discriminant analysis (QDA) were applied successfully to classify the healthy leaves from leaves with anomalies.

8.2.7 Thermal Bands

The water status of fruit and vegetable plants as well as the detection of fruit have been analyzed frequently with thermal imaging. Thermal data can be obtained from satellites, airplanes and drones or by proximal sensing over the entire thermal range or with filters at a specific wavelength. The crop water stress index (CWSI) (Jones and Corlett 1992) and adapted indices have been employed for water stress detection. The CWSI is a surface temperature-based index between 1 and 0, with 1 representing the temperature of non-transpiring dry leaves and 0 representing fully transpiring wet leaves. Thermal imaging of canopies has been carried out by cranes, drones, terrestrial robots and tractors (c.f. González-Dugo et al. 2013). The CWSI has been used to guide irrigation protocols, for example in olives in arid climates (Ben-Gal et al. 2009) and in peach orchards considering the fruit development stages in the semi-arid environment (Bellvert et al. 2016). Interpretation of the data differs considerably in semi-arid conditions from that in arid sub-tropics in relation to emissivity itself and even more considering the thresholds indicating water stress (Jones and Corlett 1992). In differently irrigated apple trees under a hail net, CWSI values ranged between 0.08 and 0.55 with values >0.3 defined as stressed trees under the given conditions (Nagy 2015).

8.2.8 Ranging Sensors

Ranging sensors, particularly those based on light detection and ranging (LiDAR) and the ultrasonic sensors, gained importance in fruit tree management for variable-rate application in plant protection and foliar fertilizer application. These types of sensors are mainly utilized to estimate the geometric (height, width, volume) and structural (leaf area, leaf density, stem diameter, etc.) characteristics of fruit trees. The initial applications of ranging sensors in horticulture have been deployed with ultrasonic sensors (McConnell et al. 1984; Giles et al. 1987). The sensors were arranged at different heights along a vertical pole, facing the side of the tree row. Each ultrasonic unit measured its distance to the canopy as the system moved along the row at constant speed and aimed to estimate height, width and tree-row volume (TRV). This system was used to control variable-rate application in real-time, resulting in spray volume savings up to 52 % in apples and 28 % in peach (Giles et al. 1987) and 30 %–37 % in citrus (Moltó et al. 2000). In a similar study, the spray application based on the actual tree width measured by the ultrasonic sensor reduced spray deposits by 70, 28 and 39 % in olive, pear and apple orchards, respectively (Solanelles et al. 2006). Moreover, Zaman et al. (2005) created a real-time variable application of nitrogen fertilizer in citrus, considering the TRV and nitrogen content of leaves; it decreased the costs by 40 %. However, in comparison with LiDAR sensors, the ultrasonic sensors had a lower resolution and poorer accuracies because of sound divergence or distance within the sensors and slower sampling rate (Tumbo et al. 2002).

The implementation of LiDAR sensors enabled the detection of geometric and structural properties of fruit trees with enhanced accuracy, but the sensor devices were more costly. The LiDAR sensors followed the arrangement of the ultrasonic sensor and were mounted on a vehicle and moved along the tree rows to generate 3D point clouds. In earlier studies, the structural properties of apple and citrus trees have been described by simplified geometric equations (Walklate et al. 2002; Lee and Ehsani 2009; Tsoulias et al. 2019a). Polo et al. (2009), after analysing the TRV in 3D, obtained a strong correlation and high coefficients of determination with leaf area in pear trees ($R^2 = 0.85$), apple trees ($R^2 = 0.81$) and a vineyard ($R^2 = 0.80$). In a similar experiment, a logarithmic equation was used to describe the relation between TRV and leaf area density ($R^2 = 0.89$) in pear and apple trees, and vineyards denoting the reciprocal relation between both properties (Sanz et al. 2018). The LAI, which is one of the most widely used indices to characterize grapevine vigour, was well estimated with TAI (tree area index) from LiDAR scanning with $R^2 = 0.91$ (Polo et al. 2009).

Escolà et al. (2017) developed tools for point cloud data analyzes from the LiDAR-based system to extract further geometric and structural information. Meanwhile the estimation of the leaf area (Fig. 8.5a) is state of the art in research (Tsoulias et al. 2019b). Fruit detection and fruit size estimation are currently approached based on geometric and full waveform (intensity) information derived from LiDAR systems (Tsoulias et al. 2020). Particularly LiDAR-based laser scanning seems to be promising for *in situ* fruit size analysis due to high density 3D point clouds (Fig. 8.5b), which are not affected by varying lighting conditions making segmentations easier. Irrigation treatments showed a positive effect on canopy growth in grapevines measured by LiDAR (Chakraborty et al. 2019). Spatial dependence between soil electrical conductivity and leaf area has been observed in apple production with a terrestrial 2D LiDAR-based laser scanner. The spatial information of LiDAR-estimated leaf area was implemented in a water balance model, revealing variation in water needs in an apple orchard (Tsoulias et al. 2019a) providing a good example of how plant sensor data can be integrated in existing agronomic models, aimed at more precise management.

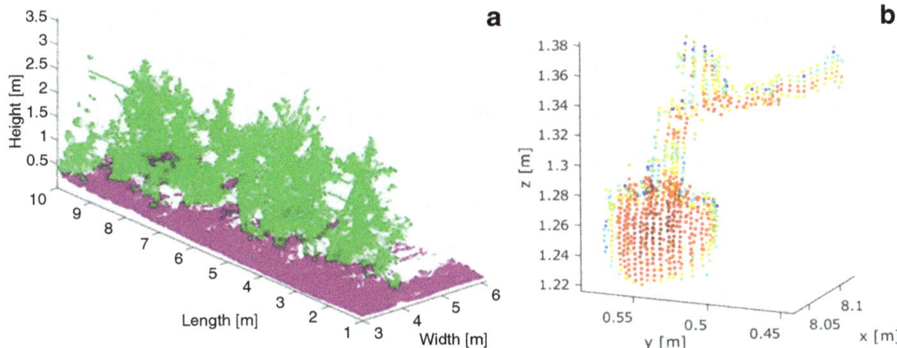

Fig. 8.5 3D point cloud of apple trees (green) and ground (purple) obtained by means of a light detection and ranging (LiDAR) –based system (**a**); apple fruit segmented from the tree 3D point cloud 120 days after full bloom, fruit size is 69.7 mm (**b**). Photo courtesy of: Tsoulias

8.3 Proximal fruit Sensing

8.3.1 *Background*

Proximal sensing in fruit production can be approached with gas exchange analysis, which provides information on the growth efficiency, but this has not been automated for monitoring the crop throughout the season. Dendrometers can be employed for measuring growth and shrinkage versus swelling of stem, branch or fruit. Dendrometers can be placed in orchards and data can be continuously recorded with a wireless network (Vougioukas 2013). For real-time analysis of produce after harvest and for grading and monitoring during processing, machine vision systems became commercially available in 1980. Research groups in cooperation with industry developed new sorting lines using spectroscopic methods that became commercially available e.g. 2002 by Greefa, Netherlands. Meanwhile it is possible to classify fruit and vegetables according to their soluble solids content (Walsh et al., 2020), various internal defects (e.g. internal browning, glassiness, stone cracking, bruising, bitter pit), storage reserve level (soluble sugar, starch or oil) and pigment contents (Merzlyak et al., 2003), but attempts have been made to analyze the fruit flesh for firmness only (Lu et al. 2020; Zude-Sasse et al. 2019). Desktop modules and portable instruments for individual product testing became available at the same time based on the same technology, in 2001 by Fantec, Japan, 2002 by CP, Germany and 2004 by Integrated Spectronics, Australia. With the new sensors, the quality of fruit can be assessed *in situ* at the production site and subsequently followed postharvest. It is precisely this repeated analysis along the supply chain that is essential to optimize the processes for variable-rate application in production and at postharvest.

8.3.2 *Maturity*

Horticultural maturity or commercial maturity is the stage of development when the produce develops several attributes and characteristics that make them desirable to consumers. Physiological maturity refers to the stage that fruits or vegetables reach the characteristic, fully ripe eating sensation. For certain crops, maturity can be defined chronologically considering the amount of heat energy an organism accumulates over a period of time. Based on the principle that plant organs grow in proportion with ambient temperature, a certain amount of heat energy is required over a period of time for crops to reach maturity. This amount is expressed as growing degree-days (GDD) and is frequently used in practise as the optimal harvest time indicator (Holmes and Robertson 1959). Researchers have experimented with models to estimate crop maturity using meteorological data. Jenni et al. (1998) used heat unit formulae to forecast cantaloupe melon (*Cucumis melo* var. *cantalupensis*) yield. The relation between heat energy and time that iceberg lettuce needs to reach maturity has been modelled by Wurr et al. (1988). In fruit trees many models exist based on the GDD.

The stage of maturity at the time of picking influences the storage life and quality of the crop (Rahman et al. 2016). If harvesting is done too early, fruits often fail to ripen on reaching the market, whereas on the other hand harvesting too late will result in fruits softening too early in the supply chain. Consumers demand high quality food and are likely to refuse to buy unripe or overripe fruit. Harvesting time is one of the most important variables that affects final economic benefit. However, deciding on the optimal harvest time is a major problem because specific indices that determine maturity have been developed for only a few fruits and vegetables, while for the vast majority such indices do not exist.

Several studies demonstrate a connection between vegetation indices and fruit maturity, suggesting fruit sensing is a reliable method for maturity assessment. Reduced vegetation growth and shading can lead to changes in sugar content in grapes, suggesting that NDVI or other vegetation indices would be a good tool for quality estimation (Bonilla et al. 2015). Guthrie and Walsh (1997) used temporal satellite data to determine the optimum harvest date for pineapple fields and mango orchards. Early remote sensing applications to identify the optimal date of harvesting of vegetable crops were also reported by Nageswara Rao et al. (2004), who used aerial imagery to quantify colour changes of the visible spectrum related to several horticultural vegetables and fruit ripening processes in the canopy. Mango maturity was analyzed using spectral data recorded with a ground-based vehicle (Wendel et al. 2018). However, as experienced recently in PA applications, the chlorophyll-based NDVI can be influenced by several factors that may limit the robustness of the harvest date model; for example, in citrus production, physiological iron deficiency can cause yellowing of leaves. Furthermore, large within-field variation of fruit maturity was described for most fruit and vegetables, which can indicate the need for selective harvesting in some high value climacteric crops needing to be maintained at high quality in long-term storage. Consequently, the analysis of maturity and quality is requested at the individual fruit level. Mobile applications have been investigated already to provide decision support systems for fruit maturity monitoring (King 2017).

8.3.3 Visible and NIR Bands

With reflectance readings in the visible to infrared wavelength range, changes in pigment profiles, overtones of a molecule's vibration and rotation, temperature and emission coefficients can be measured. Most commonly silicon-based detectors in the 400–1100 nm range are applied because they are inexpensive and allow easy handling in outdoor conditions. With a light source and a detector, attenuation of the radiation by the fruit tissue can be recorded. Colour data are related to the red (700.0 nm), green (546.1 nm) and blue (435.8 nm) bands, which can be transferred in colour spaces such as the L*a*b* space described in the CIE standards. However, to approach the pigment contents in the visible and functional groups such as –OH in the SWNIR range (700–1100 nm), an enhanced spectral resolution is required. The coinciding absorption of red pigments and chlorophylls can be addressed now with a spectral resolution of 10 nm provided that calibration and interpolation are carried out. For example, the NDVI calculated on two wavelengths or, with enhanced sensitivity, the red edge (nm) calculated by means of the second and third derivative of the full spectrum considering the wavelength at zero in the range 660–690 nm (Zude-Sasse et al. 2002) (Fig. 8.6) can be employed to calculate the chlorophyll at its absorption peak even if visually masked by red pigments. It is even possible to detect the specific chlorophylls out of the total pool of chlorophylls (Seifert et al. 2015). Consequently, the possible perturbation by means of coinciding absorption spectra from the complex matrix of fruit tissue can be better addressed with high spectral resolution.

The wavelength of the molecule under question can be approached by the relevant spectrum of the molecule in solvent, but the binding conditions may change the actual absorption peak *in vivo*. For chlorophyll, the peak of dissolved chlorophyll, an absorption, appears at 660 nm, while *in vivo* the peak changes during fruit development, ranging between 682 nm and 686 nm (Seifert et al. 2015). This discrepancy is usually disregarded, but it provides considerable potential for more indicative plant sensing. In summary, changes in the chlorophylls can be detected non-destructively *in situ* assuming rather isotropic pip fruit such as apples.

In more anisotropic fruit, the changes in scattering coefficients of the sample can perturbate the measurement (Cubeddu et al. 2001; Seifert et al. 2015). Here,

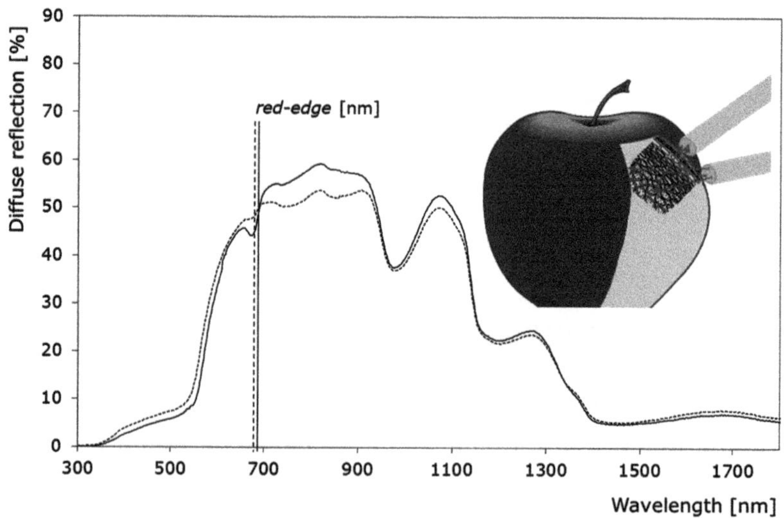

Fig. 8.6 Spectrum of unripe (solid line) and ripe (dotted line) apple measured in optical geometry of diffuse reflection with a halogen lamp, integrating sphere, silicon-based and InGAs detectors. Photograph courtesy of: Zude

fluorescence analysis might be a better option to detect chlorophyll even if the measurement appears less robust than visible spectroscopy (Kuckenberg et al. 2008; Baluja et al. 2012).

An interesting development is that radiometric information can be obtained with ranging sensors that have little information on surface colour. Such data might provide a reasonable approach to measure fruit location and size in the tree (Wang et al. 2017; Tsoulias et al. 2020) (Fig. 8.6).

8.4 Data Analyses

Sensors usually provide raw data that cannot be used directly for data analysis. Figure 8.7 shows the typical steps involved in processing data obtained from optical sensors. It includes baseline correction and background removal, dimension reduction and feature extraction, and then estimation by numerical analysis or a classification algorithm. In the estimation of certain plant properties, the adjusted coefficient

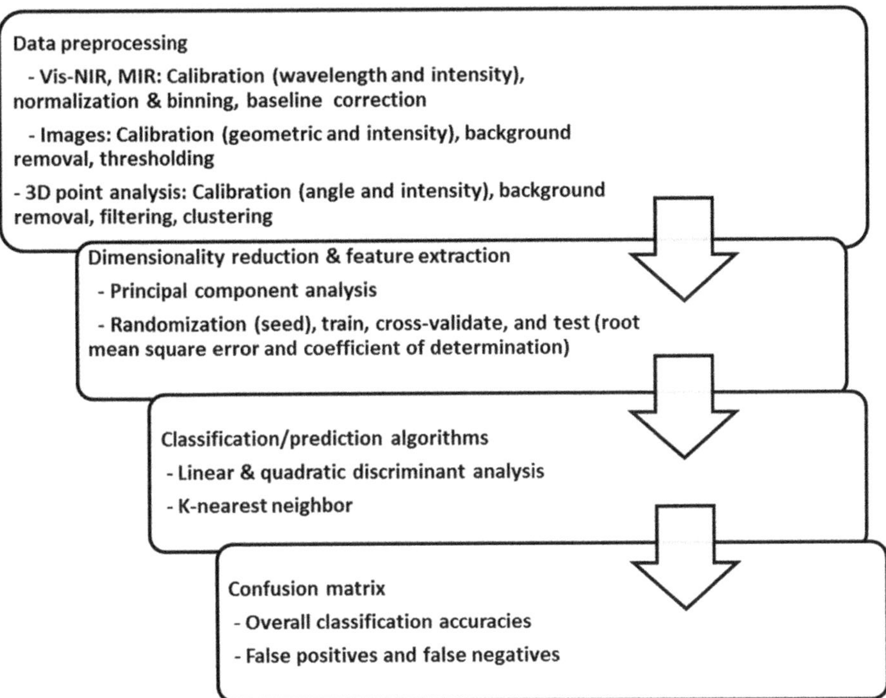

Fig. 8.7 Typical steps involved in data processing of plant stress detector sensors with examples of methodology and necessary reporting of results

of determination (R^2_{adj}) in calibration and cross-validation, e.g. using the leave-one-out method, needs to be reported. Obviously better than a cross-validation on the same data set coming from the same population of samples, an independent test-set validation on samples of different origin should be carried out. This will result in robust calibrations. In addition to the coefficient of determination, the root mean square error (RMSE) of calibration, cross-validation, and test-set validation are requested (Walsh et al. 2020). Recently, machine learning algorithms have been used extensively for classification for grading fruit inline, and subsequently for spatial or tree-wise information of the planting system. Classification data measured *in situ* in the planting system provides a 3D reconstruction for detection and monitoring (Zha et al. 2020). The results usually report both false positives and false negatives. In most cases it is desirable to have the minimum number of false negatives because they mean that there are stressed trees that were not detected which could lead to more problems in the future. The work flow can also be visualized considering the ground truth data (Fig. 8.8).

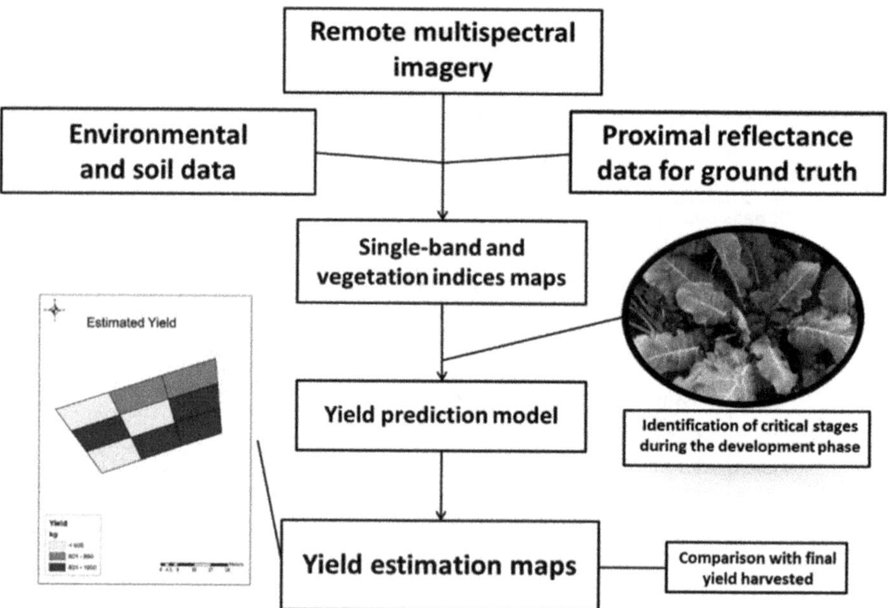

Fig. 8.8 Workflow to develop a calibration model and achieve yield prediction mapping by integrating reference or frequently named ground truth data, critical interfering or beneficial stages, and final validation

8.5 Yield Estimation

8.5.1 Background

Crop yield is the most important information for crop management in PA. Yield monitors are commercially available for many crops and are used increasingly, while the estimation of yield before harvest is implemented far less in practise. Yield monitor data can be obtained after harvest, whereas some problems such as nutrient deficiencies, water stress or disease occurrence that affect yield, should be managed during the growing season (Usha and Singh 2013). These problems can be solved with continuous data sources that can provide valuable information on crop health and stress, nutrient requirements and, ultimately, yield estimates. The need for accurate yield predictions has become evident since precision viticulture was first adopted, because vineyards have considerable variability in both yield and quality (Bramley 2005). Yield estimation can help the growers to optimize the timing of harvest operations, as well as storage and shipping of their products. However, the difficulty of sampling and lack of efficient methods are obstacles that greatly limit the growth and development of the sector. The traditional method used by wine growers for predicting yield is based on the weight of bunches (Wolpert and Villas 1992), an inefficient and time-consuming operation, which also fails to provide accurate estimates (Clingeleffer et al. 2001).

8.5.2 Weather Data

Quality and yield estimation of various crops is also a function of weather data (Frioni et al. 2017). An early assessment of yield reduction could help to avert a disastrous situation and help in strategic planning to meet market demands. Lobell et al. (2006) developed yield estimation models for 12 crops, including horticultural crops, cultivated in California using data from proximal meteorological stations as inputs. The models showed high accuracy in cross-validation testing for a period of 23 years (1980–2003). Laxmi and Kumar (2011) developed a neural network-based yield forecasting model for various crops that used several meteorological data sources as inputs. Similarly, yield forecasting models based on applications of artificial intelligence with open access to weather data have been developed for various other horticultural crops recently (Kartika et al. 2016). McKeown et al. (2005) found that yield of cool season vegetable crops was inversely proportional to the number of hot days with temperatures >30 °C. In the same year, Koller and Upadhyaya (2005) predicted tomato yield with a more complex model based on soil, crop and environmental variables. While there was no clear correlation with yield, similar yield patterns were observed in the maps. Solar radiation and temperature data were applied by Higashide (2009) for estimating the yield of greenhouse tomato at different growth stages. The results showed a strong correlation between fruit yield and solar radiation. Radiation, in particular the light interception, has been identified as the driving force for the yield in apple production leading to the development of cordon and V-shaped trellis systems (Fig. 8.1).

Analysis of trellis tension continuously measures the tension in the horizontal (cordon) support wire of the trellis and is indirectly related to plant data (Tarara et al., 2004). As a field-based method, it allows the collection of temporal information on the growing crop, but also provides information on changes during the growing stages and in different growing seasons (Blom and Tarara, 2009).

8.5.3 *Remote Sensing*

Remote sensing applications can provide data throughout the season, and also present an alternative when yield monitor data are not available (Li et al., 2010). Yang and Liu (2008) evaluated aerial photogrammetry and field reflectance data for estimating the physical properties of cabbage, showing that both air-borne images and reflectance spectra can be used to extract reliable information on cabbage plant growth and yield. Applying machine vision to detect and count flowers was a crucial step in the estimation of yield in apple (Aggelopoulou et al., 2011) and mango (Koirala et al., 2020). Smart et al. (1990) described the relation between canopy management and yield in vineyards. Vegetation indices are robust and feasible predictors of vegetative growth (Bonilla et al., 2015) and are strongly related to yield and quality of grapes (Fiorillo et al., 2012) provided that no unfavourable event takes place affecting the generative growth. High correlation can be found between proximal NDVI measurement, yield, and fruit quality in apple (Aggelopoulou et al. 2010; Liakos et al. 2017) and pear orchards (Vatsanidou et al. 2017), if the optimum (Penzel et al. 2020) leaf:fruit ratio is adjusted in crop management. The variable-rate thinning (Penzel et al. 2021) would be an economically interesting approach for including plant sensor data in agronomic models that can be used for precise crop load management.

Much research has been done on the application of satellite data. Nageswara Rao et al. (2004) estimated yield of potato fields as well as other crops with an accuracy exceeding 90 % with satellite imagery-derived NVDI data. O'Connell and Goodwin (2005) used canopy coverage data derived from high resolution aerial imagery of a peach orchard and estimated yield. Shrivastava and Gebelein (2007) found a significant correlation between citrus yields with remotely sensed Landsat images of canopy coverage. Anastasiou (2018) examined several vegetation indices derived from both proximal and satellite sensing to predict yield and quality in table grapes. Although satellite data that can be converted into vegetation indices show a strong correlation to yield, the number of factors affecting the values of indices is believed to be too large to be considered a reliable stand-alone data source by many researchers. Best et al. (2005) combined soil electrical conductivity and NDVI values in multifactorial spatial regression models to generate the yield map of a vineyard. In the same study, selective harvesting of grapes was introduced as a method to improve grape quality, a method that is well-known in tree fruit production. Tagarakis and Käthner also combined soil variables (elevation, electrical conductivity, soil texture) and plant data to delineate management zones based on the correlations with yield and quality in vineyards and plum orchards, respectively (Tagarakis et al. 2013, Käthner et al. 2017). In plum orchards, the ECa was combined with thermal imaging (Käthner et al., 2017).

Considering that the thermodynamic properties of fruit and vegetables differ from those of the surrounding objects, yield forecasts can also be made with thermal imagery data, as demonstrated by Stajnko et al. (2004) and Bulanon et al. (2008). The latter, as well as Wachs et al. (2009), combined thermal and RGB images and managed to increase the accuracy of estimates. Grossetete et al. (2012) introduced an easy-to-use application that allows producers to create yield estimation maps using only their cell phones. Serrano et al. (2005) introduced a handheld system equipped with a camera and a GPS receiver that could make estimates based on image data from the fields. Mango fruits were detected based on colour images, and particularly testing colour spaces, with a high coefficient of determination (Payne et al. 2013). Various data processing techniques were tested on green citrus fruits (Maldonado and Barbosa 2016). In vineyards, on-the-go yield estimation based on photogrammetry has been advanced (Millan et al. 2018) by the reconstruction of fruit bunches (Herrero-Huerta et al. 2015) by structure from motion 3D point clouds.

The spatio-temporal monitoring of structural and geometric plant properties has become more reliable in outdoor applications with the implementation of ranging sensor technology. However, their relation with yield is complex and needs to be investigated further. The use of ultrasonic sensors, again revealed a correlation between yield and canopy volume similar to the vegetation indices studied earlier (Mann et al. 2011). Whitney et al. (2002) and Zaman et al. (2006) used ultrasonic sensors to create vegetation maps and accurate yield estimates in citrus groves. Zaman et al. (2005) and Schumann et al. (2006) found high coefficients of determination ($R^2 = 0.80$ and 0.64, respectively) between canopy volume and fruit yield in commercial citrus plots.

The spatial dependence of height and volume with yield was investigated in three different citrus groves (Colaço et al. 2019). The canopy geometry showed consistent spatial coregionalization with historical yield in two of the three groves, yet the delineation of management zones based on the canopy volume has been proposed. A similar relation has also been identified between canopy volume and yield ($R^2 = 0.77$) in almond trees (Underwood et al. 2016). However, this is not always the case because the fruit load can vary considerably in trees with similar canopy volume, for example because of pollination problems, pruning measures, frost events and alternate bearing (Uribeetxebarria et al. 2019; Penzel et al. 2020). Yield prediction has been attempted from canopy volume and trunk circumference clustering to minimize sampling effort in mango trees (Anderson et al., 2019). However, the frequently reported low coefficient of determination between trunk circumference and yield ($R^2 = 0.17$) suggests that the leaf area represented in canopy volume or vegetation indices provides better predictions than trunk measurements in pruned and well managed production. Furthermore, the tree structure is a useful indicator for identifying the fruiting potential of orchard crops.

Tagarakis et al. (2018), found that the pruning weight is strongly related with yield, suggesting the potential of LiDAR-based systems to assess the spatial variation in vine vigour to regulate site-specific application of inputs (irrigation, fertilization) and to adapt vineyard management operations (i.e. pruning) to improve vine balance and grape quality. Despite the use of LiDAR and photogrammetric data collection methods, it has been suggested that in dwarf apple trees the yield prediction results appear highly variable and difficult to predict simply from structural assessment (Murray et al. 2020). Fruit detection has been reviewed recently (Fu et al. 2020). Stein et al. (2016) developed a multi-view approach with a LiDAR mask to detect, track, count and locate fruit within a mango orchard. Gené-Mola et al. (2019), detected and measured apples in the trees with different LiDAR systems using the ranging and radiometric (full wave form) information. The systems revealed a < 5 % error compared to ground truth, while the authors note that denser canopies could prevent a view of all fruit on the tree (Gené-Mola et al. 2019; Tsoulias et al. 2020). A similar drawback was pointed out in photogrammetry earlier and attempts have been undertaken to estimate weighting factors to link the number of fruit in the visible first layer with the crop load of the entire tree.

Holistic models such as the one proposed by Martínez-Casanovas et al. (2012) or Bonilla et al. (2015) can integrate spectral reflectance data and various other field variables and historical patterns, which might increase the accuracy of predictions in the horticultural sector.

8.6 Harvesting

Selective fruit picking is a financially viable strategy considering the need to maintain quality along the supply chain. However, the harvesting of fruit and vegetables is a time-consuming operation. Mechanization of the process has been desired for decades, but several aspects such as the strong susceptibility of fresh produce to mechanical stress and the inability of robots to separate mature from immature fruits create very challenging conditions. Van Henten et al. (2003) developed an autonomous harvesting robot for cucumbers that could assess maturity in real time and harvest only the cucumbers that were mature enough based on two RGB images. The images were also used to calculate the exact position of the cucumbers and to harvest with considerable accuracy. A robotic system was reported that could instantly detect and assess the maturity of strawberries in real time while harvesting (Hayashi et al., 2005). The robot determined the maturity of the fruit based on ripening index data extracted from RGB images. The group also created a similar system for tomato harvesting, which demonstrated high accuracy of 85 % yield success. A sweet pepper-harvesting robot achieved success rates between 26 and 33 % and a cycle time of 94 seconds for a full harvesting operation in a greenhouse (Bac et al., 2017). A 76.5 % success rate, validated on 68 fruit, was reported recently (Lehnert et al. 2020). Si et al. (2015) developed a robot with a stereo camera system to recognize, locate, and harvest mature apples in tree canopies. The robot successfully recognized over 89.5 % of the crop, while the errors were less than 20 mm when the measuring distances were between 400 and 1500 mm. An apple harvesting robot used a spoon-shaped end-effector to grip and an electric blade to cut the stem of each fruit, reporting an average harvesting time of 15.4 seconds per apple (Zhao et al. 2011). Silwal et al. (2017) designed and evaluated an apple robotic harvester for a V-shaped trellis orchard system with an overall success rate of 84 % and an average picking time of 6.0 seconds per fruit. Still, 13 % of the fruit was missed because the end-effector finger made unplanned contact with obstacles during grasping. Even cherry harvesting has been approached (Amatya et al. 2016). However, the cooperation between engineering and research is ongoing (Fig. 8.3c) and it seems that we are rather close to obtaining a harvest robot. The harvest accounts for roughly 50 % of the entire production costs in horticulture and, due to the high costs and decreased availability of workers, more developments are needed.

8.7 Conclusions for the Chapter

At present, proximal and remote sensors can be used in precision horticulture. However, the implementation of sensor data in agronomic models to control processes directly would be the next logical step. Here we speak about digital twins representing the crop information obtained by employing the crop data in physiological models. Using the digital twins for simulating the process can support the optimization of this process.

However, there is a further need to develop cost-effective sensors that can detect stress of the whole plant and harvest products at an early stage. Currently, optical-based sensors are the most widely used tools for stress detection. In general, optical sensors based on emission, reflectance, and fluorescence are feasible for detecting anomalies in orchards, vegetables production, and berry fields (pre-screening) because they can detect stressed plants and are very cost effective. Some of these sensors can also be used to detect a specific stressor. However, most methods need a re-calibration for a given site or given variety with appropriate reference data that need to be produced by extension services or technical service providers. Another business model would be that the manufacturer remains the owner of the sensor system and sells the service. This would result in easy to access information for the fruit grower as well as big data sets that are interesting for further deep learning analysis. Recent developments in information and communication technologies provide some incentive for implementation in, e.g., yield prediction and harvest. Mobile applications on the smart phone for precise horticulture have been developed based on fruit sensor data (King 2017) to inform stakeholders in real time, but the economic advantage needs to be calculated.

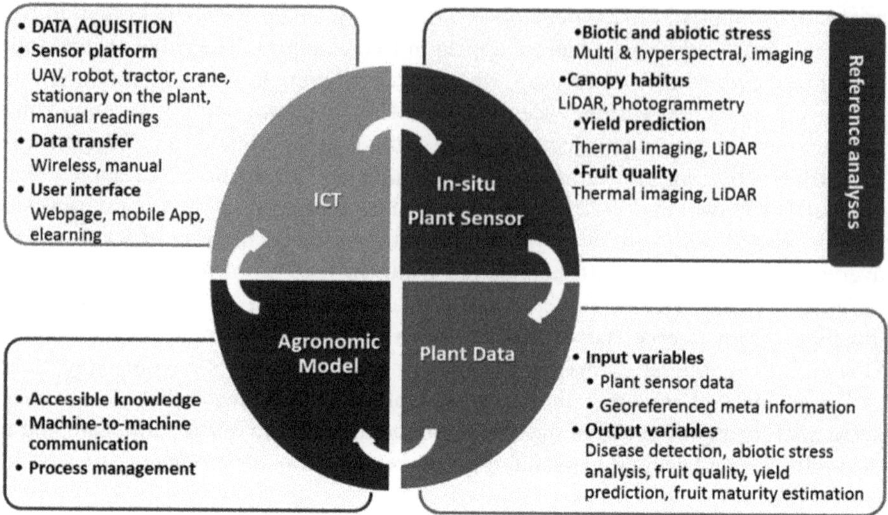

Fig. 8.9 Current situation of the steps of data acquisition by sensors (measuring), deriving plant information with physiological model (deriving information) and agronomic modelling of the production process (application) supported by information and communication technology, when approaching precision horticulture

The use of the newly available plant data in agronomic models such as precise yield prediction, variable rate thinning, and precise harvest, represent the challenge for the next step of automation (Fig. 8.9).

References

Abd El-Ghany NM, Abd El-Aziz SE et al (2020) A review: application of remote sensing as a promising strategy for insect pests and diseases management. Environ Sci Pollut Res:1–13

Aggelopoulou KD, Wulfsohn D et al (2010) Spatial variation in yield and quality in a small apple orchard. Precis Agric 11(5):538–556

Aggelopoulou AD, Bochtis D et al (2011) Yield prediction in apple orchards based on image processing. Precis Agric 12(3):448–456

Amatya S, Karkee M et al (2016) Detection of cherry tree branches with full foliage in planar architecture for automated sweet-cherry harvesting. Biosyst Eng 146:3–15

Anastasiou E, Balafoutis A et al (2018) Satellite and proximal sensing to estimate the yield and quality of table grapes. Agriculture 8(7):94

Anderson NT, Underwood JP et al (2019) Estimation of fruit load in mango orchards: tree sampling considerations and use of machine vision and satellite imagery. Precis Agric 20(4):823–839

Atkinson NJ, Jain R et al (2015) The response of plants to simultaneous biotic and abiotic stress. In: Combined stresses in plants. Springer, Cham, pp 181–201

Bac CW, Hemming J et al (2017) Performance evaluation of a harvesting robot for sweet pepper. J Field Robot 34(6):1123–1139

Baluja J, Diago MP et al (2012) Assessment of the spatial variability of anthocyanins in grapes using a fluorescence sensor: relationships with vine vigour and yield. Precis Agric 13(4):457–472

Bellvert J, Zarco-Tejada PJ et al (2016) Vineyard irrigation scheduling based on airborne thermal imagery and water potential thresholds. Aust J Grape Wine Res 22:307–315
Ben-Gal A, Agam N et al (2009) Evaluating water stress in irrigated olives: correlation of soil water status, tree water status, and thermal imagery. Irrig Sci 27:367–376
Berdugo CA, Zito R et al (2014) Fusion of sensor data for the detection and differentiation of plant diseases in cucumber. Plant Pathol 63(6):1344–1356
Berger S, Sinha AK et al (2007) Plant physiology meets phytopathology: plant primary metabolism and plant-pathogen interactions. J Exp Bot 58(15–16):4019–4026
Best S, León L et al. (2005) Use of precision viticulture tools to optimize the harvest of high quality grapes. Proceedings of the fruits and nuts and vegetable production engineering TIC (Frutic05), Montpellier France, 12–16 September 2005
Blom PE, Tarara JM (2009) Trellis tension monitoring improves yield estimation in vineyards. HortScience 44(3):678–685
Bonilla I, De Toda FM et al (2015) Vine vigor, yield and grape quality assessment by airborne remote sensing over three years: analysis of unexpected relationships in cv. Tempranillo. Span J Agric Res 13(2):8, e0903
Brabandt H, Bauriegel E et al (2014) ΦPSII and NPQ to evaluate Bremia lactucae-infection in susceptible and resistant lettuce cultivars. Sci Hortic 180:123–129
Bramley RGV (2005) Understanding variability in winegrape production systems −2. Within vineyard variation in quality over several vintages. Aust J Grape Wine Res 11(1):33–42
Bulanon DM, Burks TF et al (2008) Study on temporal variation in citrus canopy using thermal imaging for citrus fruit detection. Biosyst Eng 101(2):161–171
Chakraborty M, Khot LR et al (2019) Evaluation of mobile 3D light detection and ranging based canopy mapping system for tree fruit crops. Comput Electron Agr 158:284–293
Clingeleffer PR, Martin SR et al. (2001) Crop development, crop estimation and crop control to secure quality and production of major wine grape varieties: a national approach: final report to Grape and Wine Research & Development Corporation/principal investigator, Peter Clingeleffer; (prepared and edited by Steve Martin and Gregory Dunn). Adelaide: Grape and Wine Research & Development Corporation; 2001. http://hdl.handle.net/102.100.100/201731?index=1
Colaço AF, Molin JP et al (2019) Spatial variability in commercial orange groves. Part 2: relating canopy geometry to soil attributes and historical yield. Precis Agric 20(4):805–822
Cubeddu R, D'Andrea C et al (2001) Time-resolved reflectance spectroscopy applied to the non-destructive monitoring of the internal optical properties in apples. Appl Spectrosc 55:1368–1374
Dubey SR, Jalal AS (2016) Apple disease classification using color, texture and shape features from images. Signal Image Video Processing 10:819–826
Escolà A, Martínez-Casasnovas JA et al (2017) Mobile terrestrial laser scanner applications in precision fruticulture/horticulture and tools to extract information from canopy point clouds. Precis Agric 18(1):111–132
Fiorillo E, Crisci A et al (2012) Airborne high-resolution images for grape classification: changes in correlation between technological and late maturity in a Sangiovese vineyard in Central Italy. Aust J Grape Wine R 18(1):80–90
Frioni T, Green A et al (2017) Impact of spring freeze on yield, vine performance and fruit quality of Vitis interspecific hybrid Marquette. Sci Hortic 219:302–309
Fu L, Gao F et al (2020) Application of consumer RGB-D cameras for fruit detection and localization in field: a critical review. Comput Electron Agric 177:105687
Gao JP, Chao DY et al (2007) Understanding abiotic stress tolerance mechanisms: recent studies on stress response in rice. J Integr Plant Biol 49(6):742–750
Gené-Mola J, Gregorio E et al (2019) Fruit detection in an apple orchard using a mobile terrestrial laser scanner. Biosyst Eng 187:171–184
Giles DK, Delwiche MJ et al (1987) Control of orchard spraying based on electronic sensing of target characteristics. T ASAE 30(6):1624–1636

Gonzalez-Dugo V, Zarco-Tejada P et al (2013) Using high resolution UAV thermal imagery to assess the variability in the water status of five fruit tree species within a commercial orchard. Precis Agric 14:660–678

Gonzalez-Dugo V, Testi L et al (2020) Empirical validation of the relationship between the crop water stress index and relative transpiration in almond trees. Agric For Meteorol 292:108128

Grossetete M, Berthoumieu Y et al. (2012) Early estimation of vineyard yield: site specific counting of berries by using a smartphone. Paper presented at the International Conference of Agricultural Engineering—CIGR-AgEng

Gupta A, Senthil-Kumar M (2017) Concurrent stresses are perceived as new state of stress by the plants: overview of impact of abiotic and biotic stress combinations. In: Plant tolerance to individual and concurrent stresses. Springer, New Delhi, pp 1–15

Guthrie J, Walsh K (1997) Non-invasive assessment of pineapple and mango fruit quality using near infra-red spectroscopy. Aust J Exp Agr 37(2):253–263

Hayashi S, Ota T et al. (2005) Robotic harvesting technology for fruit vegetables in protected horticultural production. Information and Technology for Sustainable Fruit and Vegetable Production (Frutic05), Montpellier France, 12–16 September 2005, p 227–236

Hegarty-Craver M, Polly J et al (2020) Remote crop mapping at scale: using satellite imagery and UAV-acquired data as ground truth. Remote Sens 12:1984

Herrero-Huerta M, González-Aguilera D et al (2015) Vineyard yield estimation by automatic 3D bunch modelling in field conditions. Comput Electron Agr 110:17–26

Higashide T (2009) Prediction of tomato yield on the basis of solar radiation before Anthesis under warm greenhouse conditions. HortScience 44(7):1874–1878

Holmes RM, Robertson GW (1959) A modulated soil moisture budget. Monthly Weather Rev 87:101–106

Hunt ER Jr, Daughtry CS (2018) What good are unmanned aircraft systems for agricultural remote sensing and precision agriculture? Int JRemote Sens 39:5345–5376

Husain M, Rathore JP et al (2017) Abiotic stress management in herbaceous crops using breeding and biotechnology approaches. J Pharm Innov 6(10):269–276

Hussain T, Gondal MA (2008) Monitoring and assessment of toxic metals in gulf war oil spill contaminated soil using laser-induced breakdown spectroscopy. Environ Monit Assess 136(1–3):391–399

Jackson RD (1986) Remote sensing of biotic and abiotic plant stress. Annu Rev Phytopathol 24(1):265–287

James K, Nichol CJ et al (2020) Thermal and multispectral remote sensing for the detection and analysis of archaeologically induced crop stress at a UK site. Drones 4(4):61

Jenni S, Stewart KA et al (1998) Predicting yield and time to maturity of muskmelons from weather and crop observations. J Am Soc Hortic Sci 123(2):195–201

Jones HG, Corlett EJ (1992) Current topics in drought physiology. J Agric Sci 119(3):291–296

Kamal M, Schulthess U et al (2020) Identification of mung bean in a smallholder farming setting of coastal South Asia using manned aircraft photography and Sentinel-2 images. Remote Sens 12(22):3688

Karimzadeh J, Wright DJ (2008) Bottom-up cascading effects in a tritrophic system: interactions between plant quality and host-parasitoid immune responses. Ecol Entomol 33(1):45–52

Kartika ND, Astika IW et al (2016) Oil palm yield forecasting based on weather variables using artificial neural network. Indones J ElectrEng Comp Sci 3(3):626–633

Käthner J, Ben-Gal A et al (2017) Evaluating spatially resolved influence of soil and tree water status on quality of European plum grown in semi-humid climate. Front Plant Sci 8:1053

King A (2017) Technology: the future of agriculture. Nature 544:S21–S23

Koirala A, Walsh KB et al (2020) Deep learning for mango (Mangifera indica) panicle stage classification. Agronomy 10(1):143

Koller M, Upadhyaya SK (2005) Prediction of processing tomato yield using a crop growth model and remotely sensed aerial images. T ASAE 48(6):2335–2341

Kuckenberg J, Tartachnyk I et al (2008) Evaluation of fluorescence and remission techniques for monitoring changes in peel chlorophyll and internal fruit characteristics in sunlit and shaded sides of apple fruit during shelf-life. Postharvest Biol Technol 48(2):231–241

Laxmi RR, Kumar A (2011) Weather based forecasting model for crops yield using neural network approach. Statistics Appl 9(1&2):55–69. New Series

Lee KH, Ehsani R (2009) A laser scanner based measurement system for quantification of Citrus tree geometric characteristics. Appl Eng Agric 25(5):777–788

Lehnert C, Mccool C et al (2020) Performance improvements of a sweet pepper harvesting robot in protected cropping environments. J Field Robot

Li G, Wan S et al (2010) Leaf chlorophyll fluorescence, hyperspectral reflectance, pigments content, malondialdehyde and proline accumulation responses of castor bean (Ricinus communis L.) seedlings to salt stress levels. Ind Crop Prod 31(1):13–19

Liakos V, Tagarakis A et al (2017) In-season prediction of yield variability in an apple orchard. Eur J Hortic Sci 82(5):251–259

Lobell D, Cahill K et al (2006) Weather-based yield forecasts developed for 12 California crops. Calif Agr 60(4):211–215

Lu R, Van Beers R et al (2020) Measurement of optical properties of fruits and vegetables: a review. Postharvest Biol Technol 159:111003

Maes WH, Steppe K (2019) Perspectives for remote sensing with unmanned aerial vehicles in precision agriculture. Trends Plant Sci 24(2):152–164

Mahlein AK, Rumpf T et al (2013) Development of spectral indices for detecting and identifying plant diseases. Remote Sens Environ 128:21–30

Maldonado W Jr, Barbosa JC (2016) Automatic green fruit counting in orange trees using digital images. Comput Electron Agric 127:572–581

Mann KK, Schumann AW et al (2011) Delineating productivity zones in a citrus grove using citrus production, tree growth and temporally stable soil data. Precis Agric 12(4):457–472

Martinelli F, Scalenghe R et al (2015) Advanced methods of plant disease detection. A review. Agron Sustain Dev 35(1):1–25

Martínez-Casasnovas JA, Agelet-Fernandez J et al (2012) Analysis of vineyard differential management zones and relation to vine development, grape maturity and quality. Span J Agric Res 10(2):326–337

Matese A, Baraldi R et al (2018) Estimation of water stress in grapevines using proximal and remote sensing methods. Remote Sens 10(1):114

McConnell RL, Elliot KC et al. (1984) Electronic measurement of tree-row-volume. Paper presented at the National Conference on Agricultural Electronics Applications, Hyatt Regency Illinois Center, Chicago, Ill.(USA), 11–13 Dec 1983. American Society of Agricultural Engineers

McKeown A, Warland J et al (2005) Long-term marketable yields of horticultural crops in southern Ontario in relation to seasonal climate. Can J Plant Sci 85(2):431–438

Meggio F, Zarco-Tejada PJ et al (2010) Grape quality assessment in vineyards affected by iron deficiency chlorosis using narrow-band physiological remote sensing indices. Remote Sens Environ 114(9):1968–1986

Merzlyak MN, Solovchenko AE et al (2003) Reflectance spectral features and non-destructive estimation of chlorophyll, carotenoid and anthocyanin content in apple fruit. Postharvest Biol Technol 27:197–211

Millan B, Velasco-Forero S et al (2018) On-the-go grapevine yield estimation using image analysis and Boolean model. J Sens 2018

Mishra A, Karimi D et al (2011) Evaluation of an active optical sensor for detection of Huanglongbing (HLB) disease. Biosyst Eng 110(3):302–309

Moltó E, Martín B et al (2000) Pm—power and machinery: design and testing of an automatic machine for spraying at a constant distance from the tree canopy. J Agric Eng Res 77(4):379–384

Moran JA, Moran AJ (1998) Foliar reflectance and vector analysis reveal nutrient stress in prey-deprived pitcher plants (Nepenthes rafflesiana). Int J Plant Sci 159:996–1001

Murray J, Fennell JT et al (2020) The novel use of proximal photogrammetry and terrestrial LiDAR to quantify the structural complexity of orchard trees. Precis Agric 21(3):473–483

Nageswara Rao PP, Ravishankar HM et al (2004) Production estimation of horticultural crops using irs-1d liss-iii data. J Indian Soc Remote Sens 32(4):393–398

Nagy A (2015) Thermographic evaluation of water stress in an apple orchard. J Multidisc Eng Sci Technol 2:2210–2215

O'Connell MG, Goodwin I (2005) Spatial variation of tree cover in peach orchards. Paper presented at the International Symposium on Harnessing the Potential of Horticulture in the Asian-Pacific Region. Acta Hortic. ISHS 694:203–205

Oerke E-C (2020) Remote sensing of diseases. Annu Rev Phytopathol 58:225–252

Olsen KL, Schomer HA et al (1969) Segregation of 'Golden delicious' apples for quality by light transmission. Amer Soc Hort Sci:821–828

Padhye P, Rajani K, Shikalgar S, Khot ST (2014) Machine vision guided system for classification and detection of plant diseases using support vectorMachine. Int J Electron Commun Comput Eng 5(4). Technovision-2014, ISSN2249–071X

Payne AB, Walsh KB et al (2013) Estimation of mango crop yield using image analysis – segmentation method. Comput Electron Agric 91:57–64

Peñuelas J, Filella I (1998) Visible and near-infrared reflectance techniques for diagnosing plant physiological status. Trends Plant Sci 3:151–156

Penzel M, Lakso AN et al (2020) Carbon consumption of developing fruit and individual tree's fruit bearing capacity of 'RoHo 3615' and 'Pinova' apple. Int AgroPhys 34:409–423

Penzel M, Pflanz M et al. (2021) Tree adapted mechanical flower thinning prevents yield loss caused by over thinning of trees with low flower set in apple. Eur J Hortic Sci, in press

Polo JRR, Sanz R et al (2009) A tractor-mounted scanning LIDAR for the non-destructive measurement of vegetative volume and surface area of tree-row plantations: a comparison with conventional destructive measurements. Biosyst Eng 102(2):128–134

Rahman MM, Moniruzzaman M et al (2016) Maturity stages affect the postharvest quality and shelf-life of fruits of strawberry genotypes growing in subtropical regions. J Saudi Soc Agric Sc 15(1):28–37

Rançon F, Bombrun L et al (2019) Comparison of SIFT encoded and deep learning features for the classification and detection of esca disease in Bordeaux vineyards. Remote Sens 11(1):1

Rasmussen J, Ntakos G et al (2016) Are vegetation indices derived from consumer-grade cameras mounted on UAVs sufficiently reliable for assessing experimental plots? Eur J Agron 74:75–92

Reddy GO (2018) Satellite remote sensing sensors: Principles and applications. Geospatial Technologies in Land Resources Mapping, Monitoring and Management. Springer, pp 21–43

Ren DD, Tripathi S et al (2017) Low-cost multispectral imaging for remote sensing of lettuce health. J Appl Remote Sens 11(1):016006

Rouse JW et al. (1973) Monitoring vegetation systems in the Great plains with ERTS. Third ERTS Symposium 309–317 NASA SP-351 I

Rousseau C, Belin E, Bove E et al (2013) High throughput quantitative phenotyping of plant resistance using chlorophyll fluorescence image analysis. Plant Methods 9:17

Saglam A, Chaerle L et al. (2019) Promising monitoring techniques for plant science: thermal and chlorophyll fluorescence imaging. Photosynthesis, Productivity and Environmental Stress, 241–66

Saleem M et al (2020) Laser-induced fluorescence spectroscopy for early disease detection in grapefruit plants. Photochem Photobiol Sci 19:713–721

Sankaran S, Ehsani R (2013) Comparison of visible-near infrared and mid-infrared spectroscopy for classification of Huanglongbing and citrus canker infected leaves. Agric Eng Int CIGR J 15(3):75–79

Sankaran S, Ehsani R et al (2010) Mid-infrared spectroscopy for detection of Huanglongbing (greening) in citrus leaves. Talanta 83(2):574–581

Sankaran S, Ehsani R et al (2012) Evaluation of visible-near infrared reflectance spectra of avocado leaves as a non-destructive sensing tool for detection of Laurel wilt. Plant Dis 96(11):1683–1689
Sankaran S, Ehsani R et al (2015) Detection of anomalies in Citrus leaves using laser-induced breakdown spectroscopy (LIBS). Appl Spectrosc 69(8):913–919
Santos D, Nunes LC et al (2012) Laser-induced breakdown spectroscopy for analysis of plant materials: a review. Spectrochim Acta B 71-72:3–13
Sanz R, Llorens J et al (2018) LIDAR and non-LIDAR-based canopy parameters to estimate the leaf area in fruit trees and vineyard. Agric For Meteorol 260:229–239
Schouten RE, Huijben TPM et al (2007) Modelling the acceptance period of truss tomato batches. Postharvest Biol Technol 45(3):307–316
Schumann AW, Hostler K et al (2006) Relating citrus canopy size and yield to precision fertilization. Proc Fla State Hort Soc 119:148–154
Seifert B, Zude M et al (2015) Optical properties of developing pip and stone fruit reveal underlying structural changes. Physiol Plant 153:327–336
Serrano E, Roussel S et al. (2005) Early estimation of vineyard yield: Correlation between the volume of a vitis vinifera bunch during its growth and its weight at harvest. Information and Technology for Sustainable Fruit and Vegetable Production (Frutic05), Montpellier France, 12–16 September 2005 5:311–318
Shabala S (2017) Plant stress physiology, 2nd edn. Cabi Publishing, Boston, pp 1–362
Shin J, Young K et al (2020) Effect of directional augmentation using supervised machine learning technologies: a case study of strawberry powdery mildew detection. Biosyst Eng 194:49–60
Shrivastava RJ, Gebelein JL (2007) Land cover classification and economic assessment of citrus groves using remote sensing. ISPRS J Photogramm 61(5):341–353
Si YS, Gang L et al (2015) Location of apples in trees using stereoscopic vision. Comput Electron Agr 112:68–74
Silwal A, Davidson JR et al (2017) Design, integration, and field evaluation of a robotic apple harvester. J Field Robot 34(6):1140–1159
Smart RE, Dick JK et al (1990) Canopy management to improve grape yield and wine quality-principles and practices. S Afr J Enol Vitic 11(1):3–17
Solanelles F, Escolà A et al (2006) An electronic control system for pesticide application proportional to the canopy width of tree crops. Biosyst Eng 95(4):473–481
Stajnko D, Lakota M et al (2004) Estimation of number and diameter of apple fruits in an orchard during the growing season by thermal imaging. Comput Electron Agr 42(1):31–42
Stein M, Bargoti S et al (2016) Image based mango fruit detection, localisation and yield estimation using multiple view geometry. Sensors 16(11):1915
Tagarakis A, Liakos V et al (2013) Management zones delineation using fuzzy clustering techniques in grapevines. Precis Agric 14(1):18–39
Tagarakis AC, Koundouras S et al (2018) Evaluation of the use of LIDAR laser scanner to map pruning wood in vineyards and its potential for management zones delineation. Precis Agric 19(2):334–347
Tan DS, Leong RN et al (2018) AuToDiDAC: automated tool for disease detection and assessment for cacao black pod rot. Crop Prot 103:98–102
Tarara JM, Ferguson JC et al (2004) Estimation of grapevine crop mass and yield via automated measurements of trellis tension. T ASAE 47(2):647–657
Trevizan LC, Santos D et al (2009) Evaluation of laser induced breakdown spectroscopy for the determination of micronutrients in plant materials. Spectrochim Acta B 64(5):369–377
Tsoulias N, Paraforos DS et al. (2019a) Calculating the water deficit spatially using LiDAR laser scanner in an apple orchard. ECPA, Technical Paper:14762
Tsoulias N, Paraforos DS et al (2019b) Estimating canopy parameters based on the stem position in apple trees using a 2D LiDAR. Agronomy 9:740
Tsoulias N, Paraforos DS et al (2020) Apple shape detection based on geometric and radiometric features using a LiDAR laser scanner. Remote Sens 12:2481

Tumbo SD, Salyani M et al (2002) Investigation of laser and ultrasonic ranging sensors for measurements of citrus canopy volume. Appl Eng Agric 18(3):367–372
Underwood JP, Hung C et al (2016) Mapping almond orchard canopy volume, flowers, fruit and yield using lidar and vision sensors. Comput Electron Agr 130:83–96
Uribeetxebarria A, Martínez-Casasnovas JA et al (2019) Assessing ranked set sampling and ancillary data to improve fruit load estimates in peach orchards. Comput Electron Agr 164:104931
Usha K, Singh B (2013) Potential applications of remote sensing in horticulture-a review. Sci Hortic 153:71–83
Van Henten EJ, Van Tuijl BJ et al (2003) Field test of an autonomous cucumber picking robot. Biosyst Eng 86(3):305–313
Vanegas F, Bratanov D et al (2018) A novel methodology for improving plant Pest surveillance in vineyards and crops using UAV-based hyperspectral and spatial data. Sensors 18(1):260
Vatsanidou A, Nanos GD et al (2017) Nitrogen replenishment using variable rate application technique in a small hand-harvested pear orchard. Span J Agric Res 15(4):e0209
Vitrack-Tamam S, Holtzman L et al (2020) Random forest algorithm improves detection of physiological activity embedded within reflectance spectra using stomatal conductance as a test case. Remote Sens 12(14):2213
Vougioukas S, Anastassiu hat et al. (2013). Influence of foliage on radio path losses (PLs) for wireless sensor network (WSN) planning in orchards. Biosyst Eng 114: 454–465
Wachs J, Stern H, et al. (2009) Multi-modal registration using a combined similarity measure. Paper presented at the applications of Soft Computing, Springer, Berlin, Heidelberg, 52:159–168
Walklate PJ, Cross JV et al (2002) Comparison of different spray volume deposition models using LIDAR measurements of apple orchards. Biosyst Eng 82(3):253–267
Walsh KB, Blasco J et al (2020) Visible-NIR 'point' spectroscopy in postharvest fruit and vegetable assessment: the science behind three decades of commercial use. Postharvest Biol Technol 168:111246
Wang ZL, Walsh KB, Verma B (2017) On-tree mango fruit size estimation using RGB-D images. Sensors 17(12):2738
Wendel A, Underwood J et al (2018) Maturity estimation of mangoes using hyperspectral imaging from a ground based mobile platform. Comput Electron Agric 155:298–313
West JS, Bravo C et al (2003) The potential of optical canopy measurement for targeted control of field crop diseases. Annu Rev Phytopathol 41(1):593–614
Whitney JD, Tumbo SD et al. (2002) Comparison between ultrasonic and manual measurements of citrus tree canopies. Paper presented at the ASAE Annual International Meeting, St. Joseph, MI, USA, Paper:021052, July 2002
Wolpert JA, Vilas EP (1992) Estimating vineyard yields - introduction to a simple, 2-step method. Am J Enol Viticult 43(4):384–388
Wurr DCE, Fellows JR et al (1988) Crop continuity and prediction of maturity in the crisp lettuce variety Saladin. J Agr Sci 111(3):481–486
Xiang H, Tian L (2011) Development of a low-cost agricultural remote sensing system based on an autonomous unmanned aerial vehicle (UAV). Biosyst Eng 108(2):174–190
Xu L, Bulatov V et al (1997) Absolute analysis of particulate materials by laser-induced breakdown spectroscopy. Anal Chem 69:2103–2108
Yamamoto KY, Cremers DA et al (1996) Detection of metals in the environment using a portable laser-induced breakdown spectroscopy instrument. Appl Spectrosc 50(2):222–233
Yang CH, Liu TX (2008) Estimating cabbage physical parameters using remote sensing technology. Crop Prot 27(1):25–35
Yao M, Liu M et al. (2010) Identification of nutrition elements in orange leaves by laser induced breakdown spectroscopy. Paper presented at the 2010 3rd International Symposium on Intelligent Information Technology and Security Informatics, Jinggangshan, China, 2–4 April 2010, IEEE:398–401
Ye H et al (2020) Recognition of Banana Fusarium wilt based on UAV remote sensing. Remote Sens 12(6):938

Yogesh, Dubey AK, Agarwal A, Sarkar A, Arora R (2018) Adaptive thresholding based segmentation of infected portion of pomefruit. J Statistics Manag Syst 20, 2017 - (4): Machine Learning and Software Systems

Zaman QU, Schumann AW et al (2005) Variable rate nitrogen application in Florida citrus based on ultrasonically-sensed tree size. Appl Eng Agric 21(3):331–335

Zaman QU, Schumann AW et al (2006) Estimation of citrus fruit yield using ultrasonically-sensed tree size. Appl Eng Agric 22(1):39–43

Zare M, Drastig K et al (2020) Tree water status in apple orchards measured by means of land surface temperature and vegetation index (LST–NDVI) trapezoidal space derived from Landsat 8 satellite images. Sustainability 12:70

Zha FS, Fu Y et al (2020) Semantic 3D reconstruction for robotic manipulators with an eye-in-hand vision system. Appl Sci 10:1183

Zhao DA, Lv JD et al (2011) Design and control of an apple harvesting robot. Biosyst Eng 110(2):112–122

Zhou QB, Yu QY et al (2017) Perspective of Chinese GF-1 high-resolution satellite data in agricultural remote sensing monitoring. J Integr Agric 16(2):242–251

Zhu WJ et al (2018) Application of infrared thermal imaging for the rapid diagnosis of crop disease. IFAC-Papers 51(17):424–430

Zude M (2003) Comparison of indices and multivariate models to non-destructively predict the fruit chlorophyll by means of visible spectrometry in apples. Anal Chim Acta 481:119–126

Zude M, ed. (2009). Optical monitoring of fresh and processed agricultural crops. ISBN: 9781420054026 , ISBN-10: 1420054023 , CRC Press, 576 pp

Zude-Sasse M, Truppel I et al (2002) An approach to non-destructive apple chlorophyll determination. Postharvest Biol Technol 25:123–133

Zude-Sasse M, Fountas S et al (2016) Applications of precision agriculture in horticultural crops – review. Eur J Hortic Sci 81:78–90

Zude-Sasse M, Hashim N et al (2019) Validation study for measuring absorption and reduced scattering coefficients by means of laser-induced backscattering imaging. Postharvest Biol Technol 153:161–168

Chapter 9
Sensing from Unmanned Aerial Vehicles

Ryan R. Jensen, Perry J. Hardin, Eduardo Galilea, José A. Martínez-Casasnovas, and Austin Hopkins

Abstract Remote sensing data acquired from traditional satellite and aerial platforms have been used for many years to support precision agriculture studies and decisions. More recently, small unmanned aerial vehicles (UAVs) have become ubiquitous in precision agriculture for a variety of applications, such as, crop vigour mapping and disease monitoring and for a variety of reasons, such as, ease of use, small pixel size of images and relatively low cost. This chapter describes some of the basics of using UAVs in precision agriculture, including data acquisition and processing, characteristics of available and commonly used UAVs, example applications, sample imagery, pros and cons of using UAVs, a brief UAV case study, and future needs and considerations.

Keywords Unmanned aerial vehicles · Remote sensing · Orthophotos · Orthomosaics · Image processing · Crop vigour

9.1 Introduction

Remote sensing data and their derivative products from orbital and aerial platforms have a long history in precision agriculture (PA) (Zhang and Kovacs 2012; Deng et al. 2019). However, the spatial and temporal resolutions of these data and products make them less useful to many PA applications, which often require very fine resolutions (Torres-Sánchez et al. 2013; Adao et al. 2018; Huang et al. 2018).

R. R. Jensen (✉) · P. J. Hardin
Department of Geography, Brigham Young University, Provo, UT, USA
e-mail: rjensen@byu.edu

E. Galilea · J. A. Martínez-Casasnovas
Department of Environmental and Soil Sciences, Research Group on AgroICT & Precision Agriculture – GRAP, Universitat de Lleida/Agrotecnio-CERCA Centre, Lleida, Catalonia, Spain
e-mail: joseantonio.martinez@udl.cat

A. Hopkins
Department of Plant and Wildlife Science, Brigham Young University, Provo, UT, USA

R. Kerry, A. Escolà (eds.), *Sensing Approaches for Precision Agriculture*, Progress in Precision Agriculture, https://doi.org/10.1007/978-3-030-78431-7_9

Ground sample distance (GSD) refers to the distance between pixel centres measured on the ground or pixel size of an image, while temporal resolution is related to the frequency that a sensor can acquire data over a specific area or field. Also, cloud cover reduces the temporal resolution that useable satellite data can be acquired for. In contrast, small unmanned aerial vehicles (UAVs) or drones can provide very fine GSD and very frequent image acquisition (Khot and Zhou 2016), and because of these and other reasons, the use of UAVs in agriculture is becoming very common (Jorge et al. 2014; Tripicchio et al. 2015; Perera et al. 2019). Nevertheless, operating drones requires expertise in pre-flight planning, in flying the device, and processing data gathered into useful information (Huuskonen and Oksanen 2018). As a summary, Table 9.1 describes the main characteristics of satellite, manned aircraft, and UAV remote sensing systems.

The research and development of UAVs and their applications in agriculture and other areas have increased markedly in recent years, and it was expected that the UAV market would reach US $200 billion in 2020 (Puri et al. 2017). Some estimates suggest that 80–90% of this growth will be in agriculture (Stehr 2015; Montes De Oca et al. 2018). This estimate is so high because PA has created a need for high accuracy spatial data for regularly observing many field and crop characteristics (Gnadinger and Schmidhalter 2017; Yao et al. 2019). For example, Puri et al. (2017) listed five needs/benefits of using UAVs in PA: farm analysis, time saving, higher agriculture yields, GIS mapping integration, and imaging of crop health status. Further, Rokhmana (2015) described four necessary characteristics of an agricultural UAV: good spatial resolution (<20 cm), portable and easy to operate, able to produce stereo viewing capabilities, and good spatial accuracy. Farmers are typically very knowledgeable about their land and crops, but it is much less common for farmers to possess the specific knowledge necessary to acquire, process and interpret UAV imagery and other geospatial information into management practices (Basso et al. 2017). Further, despite the recent boom of UAVs in PA, the implementation of UAV data and their derivatives have not always been used effectively (Inoue and Yokoyama 2019).

Table 9.1 Basic description of the three main remote sensing platforms

Remote sensing platform	Altitude	Spatial resolution	Temporal resolution	Comments
Satellite	600–800 km	0.31–1000 m/pixel	Sub-daily to every 16 days	Good for different types of studies, from regional to within field-scale studies
Manned aircraft	500–3000 m	0.20–1.50 m/pixel	Dependent on funds	Can be very expensive
UAVs	10–200[a] m	0.01–0.10 m/pixel	Dependent on weather	Capable of rapid, ad hoc flights

Adapted from Huang et al. (2018)
[a]May depend on country regulations

This chapter describes the process of UAV data collection, data processing, common UAV platforms and sensors, and some PA studies that used UAV data. The chapter also describes the pros and cons and some current and future considerations of using UAVs in PA and presents a brief case study about the cost/benefit relationship in extensive crops.

9.2 Platforms and Sensors

Implementation of UAVs in agriculture has been accelerated by the development of platforms and sensors that are both less expensive and smaller, thus making the technology more accessible (Jorge et al. 2014). The main UAV airframe classes are fixed-wing or multi-copter (Jensen 2018). Fixed-wing UAVs usually have one wing with one or more propellers. Fixed-wing UAVs can be launched by manually throwing them into the wind or by accelerating down a runway into the wind. While in flight, fixed-wing UAVs are controlled using aileron and other controls. Piloting a fixed-wing UAV requires much practice – especially when following flight lines at specific heights above ground level. Nevertheless, in this respect, pre-designed flight plans facilitate piloting and data acquisition. Both launching and landing a fixed-wing UAV requires up to 30 m of unobstructed space devoid of trees, shrubs, power lines, buildings, etc. Fixed-wing UAVs are generally able to fly faster and for longer periods of time than multi-copter UAVs (Jensen 2018).

Multi-copter UAVs are usually configured with an even number of rotors (e.g., 4, 6, 8, etc.). Multi-copter UAVs are able to take-off and land in very small areas and controlling them while in flight is generally easier than controlling a fixed-wing UAV while in flight. For example, with no input from the controls, a multi-copter simply maintains its location in the air – even if environmental factors, such as wind speed, would otherwise force it to move. Most multi-copters also have a 'return to home,' or the take-off location, command that allows the remote pilot in command (RPIC) to bring the UAV home at any point during a flight. Further, most multi-copters automatically return to home when communication is lost to the remote controller or when its battery capacity descends below a certain threshold (e.g., <25%; Jensen 2018). However, those options may also be included in fixed-wing UAV and both types of UAV can be operated in autonomous flight.

A large number of companies manufacture UAVs in many different configurations. Apart from multi-copters and fixed-wing UAVs, it is possible to find single-rotor and hybrid solutions. A single-rotor UAV is modelled on a conventional helicopter to be operated by remote control or autonomously. A hybrid UAV mixes the advantages of multi-rotor and fixed-wing UAVs. They are also named vertical take-off and landing UAVs and are able to take off and land vertically, in a multi-rotor configuration, while flying and taking pictures as a fixed-wing UAV. This can be achieved in different ways such as changing the rotation axis of the rotors from vertical to horizontal or including different sets of rotors. Dilep et al. 2020analyse various types of UAV and their applications in agriculture. Nevertheless, one of the most common UAV types currently in use are quadcopters (Jensen 2018), and most of the discussion in this chapter is about these airframes.

There are many commercially available sensors that can be mounted on UAVs (Table 9.2). Most sensors are passive; they measure electromagnetic energy ranging from Red, Green, and Blue cameras to multispectral sensors that measure electromagnetic energy in the visible spectrum and in other parts of the electromagnetic spectrum, such as near-infrared and mid-infrared. Both traditional RGB and multispectral sensors typically measure electromagnetic energy in very wide bands – e.g., 60–90 nm width. Also, while traditional RGB and multispectral data have proven to be useful in many applications, they lack the spectral information needed to create detailed spectral profiles of plants and vegetation (Adao et al. 2018). Conversely, hyperspectral sensors are capable of measuring electromagnetic energy in tens to hundreds of very narrow (e.g., 2–5 nm) spectral bands. Hyperspectral data provide much more spectral information about objects. This information helps model biophysical parameters. Because of the large number of bands, hyperspectral sensors do not usually acquire images in an array (e.g., 4000 × 3000 pixels). Rather, these sensors rely on the forward motion of the aircraft while they scan one image line at a time. This may result in final images that require much more complex geometric corrections – especially if the aircraft exhibited much airframe movement, such as roll and pitch.

Each of these sensors has advantages and disadvantages that end users should be aware of before deciding on a sensor for a project. Table 9.2 gives a list of some common cameras and/or sensors and their characteristics.

Thermal infrared sensors can also be mounted on UAVs. Thermal sensors measure emitted energy from objects that is converted into surface temperature. Temperature conversion is possible because all objects that have a temperature above absolute zero (0 K) emit electromagnetic energy at a temperature dependent wavelength (Jensen 2007). Thermal sensors have potential for studies in

Table 9.2 Multispectral sensors suitable for hosting on UAVs

Model	RGB resolution (megapixels)	Other bands	Weight (g)
Sentera quad sensor	No	RGB + 655 nm, 725 nm, 800 nm at 1.2 MP	170
Sentera NDVI single sensor	No	RG+NIR NDVI capable at 1.2 MP	30
Parrot sequoia	16	Green, red, red edge, near-infrared at 1.2 MP	135
MicaSense RedEdge-M	No	Blue, green, red, red edge, near-infrared at 1.2 MP	173
Tetracam ADC lite	No	Red, green, near-infrared at 3.2 MP	200
Sequoia (MS)	No	Green, red, red edge, NIR at 1.2 MP	72
senseFly eBee MultiSPEC 4C	No	Green, red, red edge, near-infrared at 1.2 MP	160

Adapted from Hardin et al. (2018)

PA. According to Messina and Modica (2020) plant temperature changes rapidly under stress conditions, thus making thermal UAV data valuable for real-time detection of crop stress. Raeva et al. (2019) used thermal imagery acquired from a UAV to study the relationship between vegetation and soil temperatures. Thermal sensors generally have much smaller sensor arrays – and therefore coarser spatial resolution – because the amount of emitted thermal electromagnetic energy is less (at a longer wavelength) than that of reflected electromagnetic energy. The sensor arrays may be 640 × 480 pixels instead of 4000 × 3000 pixels which is common for sensors measuring reflected electromagnetic energy. To demonstrate this, Fig. 9.1 contains two images acquired at exactly the same time on a DJI Mavic Enterprise Dual that has two sensors – an RGB camera and thermal sensor.

A burgeoning sensor class available for use on UAVs is light detection and ranging (LiDAR; Hardin et al. 2018). LiDAR sensors are active sensors that measure the returns from an electromagnetic pulse that the sensor emits. Each pulse may have multiple returns (e.g., top-canopy return, mid-canopy return, and bare-earth return) that allow researchers to estimate canopy height and structure accurately. These samples are then used to create an interpolated surface that estimates surface values between the sample points. LiDAR cost has decreased significantly for use on UAVs because of the growing market using LiDAR sensors on automobiles. Table 9.3 lists several LiDAR sensors designed for UAV LiDAR scanning as opposed to being designed for transportation. LiDAR sensors mounted on UAVs have found application in many fields, including PA. For example, Wallace et al. (2012) found a UAV LiDAR system competent to measure tree location, height, and crown width in an experiment conducted at a university farm in Tasmania. Shendryk et al. (2020) compared sugarcane biomass predictions derived from LiDAR and multispectral data measured from a UAV. They found that multispectral data estimated sugarcane biomass more accurately early in the season while LiDAR estimated biomass more accurately later in the season.

The number of UAV platforms and sensors is substantial and constantly changing. Those interested in learning more about UAV platforms and sensors may consult Puri et al. (2017) and Mogili and Deepak (2018) who describe additional UAV platforms and sensors used in PA.

9.3 UAV Data Collection and Data Processing Process

9.3.1 Data Collection

Recent advances in mission planning, data acquisition, and data processing have greatly simplified UAV data collection and processing. These advances, probably more than other factors, have enabled the widespread use of UAVs in PA. For example, flight planning applications such as, DroneDeploy, Pix4D capture, AirMap, and others allow users to digitize a specific field where data will be collected. This is

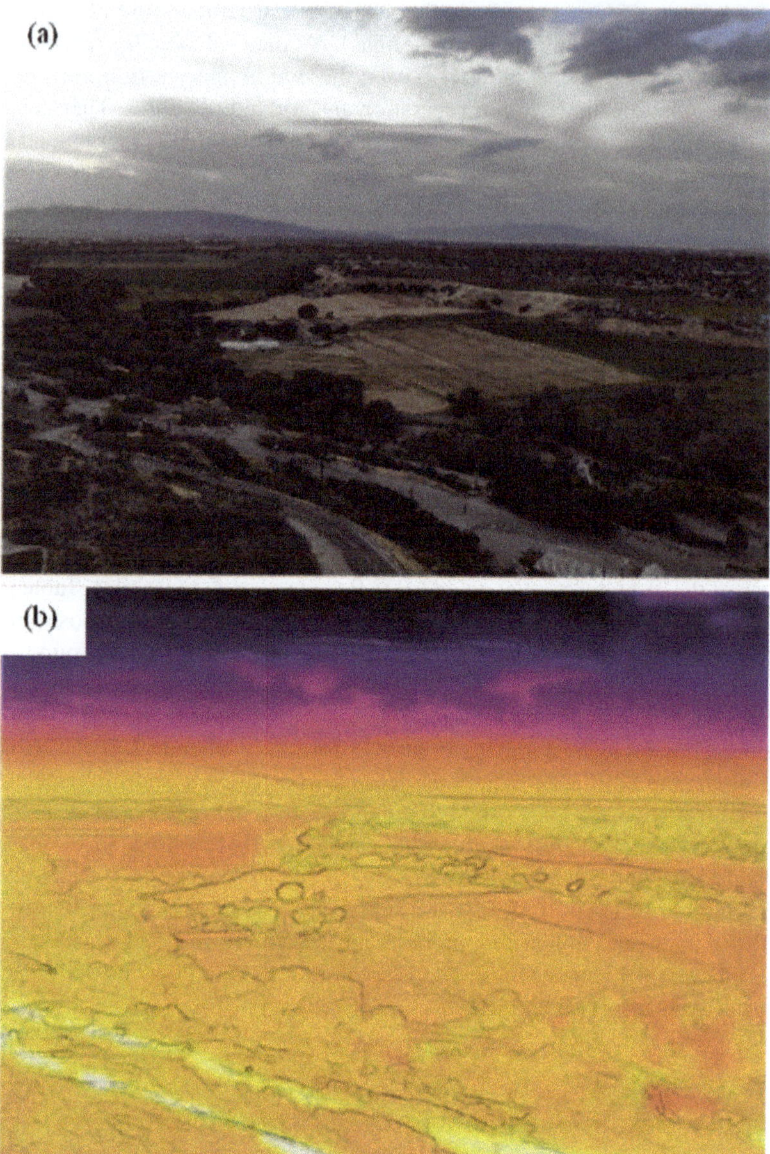

Fig. 9.1 An oblique RGB image (**a**) and an oblique thermal IR image (**b**) acquired at exactly the same time (September 2019) over a recently harvested maize field in Spanish Fork, Utah. In (**b**), brighter or whiter pixels denote higher surface temperature. Note the greater level of detail in the RGB image. Locate the road in both images and note how the road has a higher surface temperature than the surrounding areas. The RGB image (**a**) has 4000 × 3000 pixels; the thermal image (**b**) has 640 × 480 pixels

Table 9.3 LiDAR sensors available for use on UAVs

Model	Range (m)	Accuracy (+/− cm)	Thousands of PPS[a]	Weight (kg)	Power consumption (watts)
Riegl VUX-1UAV	55–350[b]	1	50–550[b]	3.7	60
Revolution 120	100–120	3.8	300	1.7	25
Revolution HD	50–80	3.1	700	2.5	25
RouteScene	100	2	700	2.5	11[c]
Velodyne HDL-32E	100	2	695–1400	1.0	12
Velodyne Puck Hi-Res	100	3	300–600	0.8	8
Velodyne Puck Lite	100	3	300–600	0.6	8
YellowScan Mapper II	150[d]	5	18.5–40	2.1	10

Adapted from Hardin et al. (2018)
[a]Points Per Second
[b]Airborne applications. Depends on selected laser pulse repetition rate. Lower altitudes permit higher rates. At 100 m above ground level, maximum rate is approximately 550,000 PPS
[c]Estimated
[d]Absolute. 75–100 m typical

done by drawing a polygon around the field where the data are to be acquired. The user is able to determine spatial resolution via aircraft height, image endlap and sidelap, aircraft speed, and many other parameters (Fig. 9.2). Image endlap (sometimes called forward overlap) refers to the overlap between images collected along the same flightline. High image endlap does not require extra flightlines; rather, endlap can be managed by varying the interval between images being acquired. Image sidelap (sometimes called lateral overlap) refers to the amount of overlap between images acquired on adjacent flightlines (Seifert et al. 2019). The greater the image overlap (both endlap and sidelap) the more common points can be found between adjacent images to create the orthomosaic and digital surface model. The trade-off is additional processing time and image storage space.

After determining the flight area and image acquisition details, the UAV pilot RPIC simply initiates the flight and the application uploads all flight and image acquisition parameters to the aircraft. The aircraft then takes off and flies the preplanned route to acquire the images. After data acquisition, the UAV lands in the same location as take-off. The RPIC should be ready to take control of the aircraft throughout the entire automated flight.

Another consideration is image GSD, sometimes referred to as spatial resolution or pixel size. Optimum GSD is generally determined by considering the smallest dimension (length or width) of the smallest object needed to be identified on the resulting images. Then, making half of that the optimum GSD (Jensen 2007). For example, if a mature leaf is the smallest element to be identified on an image and its smallest dimension is 4 cm, then the ideal GSD would be 2 cm. The GSD is determined by a variety of factors using the following equation (Seifert et al. 2019):

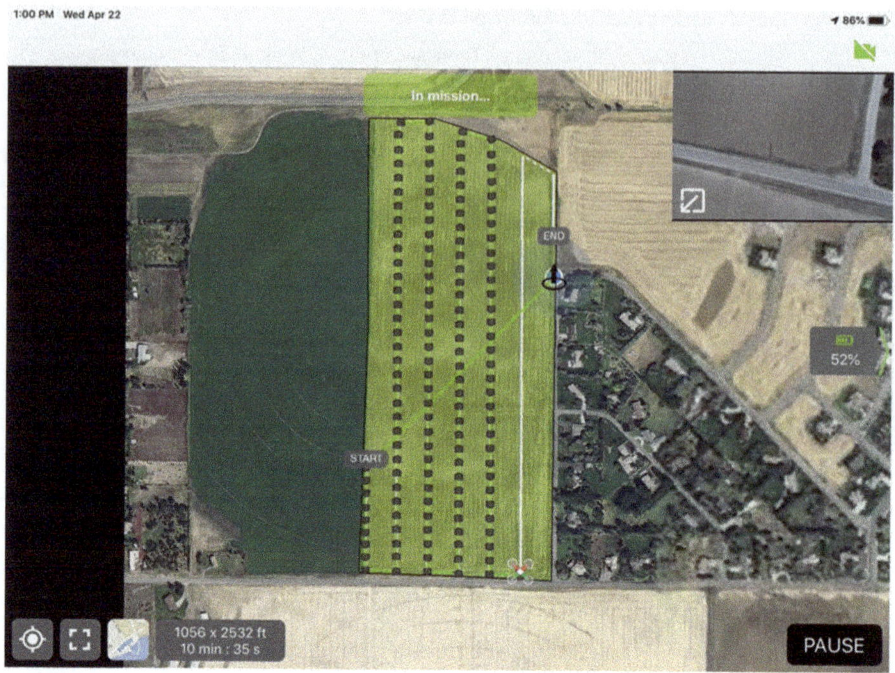

Fig. 9.2 Screen capture of a flight plan of a wheat field in Rexburg, Idaho, USA. The flight plan is uploaded to the UAV, and the UAV will take off and acquire the data. During flight, an icon showing where the UAV is located (bottom right of field), what the camera is imaging (upper right of screen), and where images have been captured (camera icons on field) are all displayed on the screen (Pix4D XE "Pix4D" 2020)

$$\text{GSD} = \frac{p}{f} H \tag{9.1}$$

where p is the size of the pixel on the sensor (mm), f is the focal length (mm), and H is the distance of the camera to the ground (cm; also called flight height). For example, the DJI Phantom 4 camera has a 25.4 mm complementary metal oxide semiconductor (CMOS). Its individual pixels are 2.4 μm (0.0024 mm), and the camera has a focal length of 24 mm. Assuming a flight height of 100 m, the GSD would be 1 cm. The higher the flight height, the larger the GSD and image footprint are. Therefore, a higher flight height results in fewer images that need to be processed. However, a lower flight height will have a smaller GSD that may result in a higher probability of detecting more detail (Seifert et al. 2019).

One disadvantage of the UAV flights is the large number of images acquired in the surveys. It depends on the size of the area being measured, the aircraft height, end lap and side lap settings. For example, when collecting data over a 20 ha field with 80% end lap and 80% side lap, a quad copter at 120 m above ground level collected 155 georeferenced images. When the same quadcopter acquired data at 60 m above ground level, 210 images were collected.

9.3.2 Image Processing

The ability to process and analyse UAV imagery rapidly is necessary to inform prompt field decision making. Recent advances in image processing/photogrammetric applications allow users to process and analyse data rapidly with minimal user input. After the data acquisition flight is complete, the user downloads the images from the aircraft—usually via a microSD card. Then, using photogrammetric techniques, point clouds, digital elevation/surface models, orthophotos are derived from the images. To create digital surface models, the same objects (or pixels) are found in many overlapping photographs, which allows the stereoscopic parallax of each pixel to be determined. If the imaged area is completely flat, all pixels in all images will move consistently from photo to photo (Jensen 2018). However, most areas are not completely flat and/or they consist of trees or buildings that are higher than the terrain. These objects will have a greater amount of parallax that is then able to be mathematically modelled to determine the relationship between the amount of parallax associated with each pixel and an elevation or height value. These data can then be displayed as a point cloud and used to create a digital surface model and it is now possible to move each pixel geometrically to its correct location. This process creates digital orthophotography (Jensen 2018). Orthophotos are generated based on orthorectification where image distortions are removed by registering it to a coordinate system and map projection and removing the effects of elevation (Ngadiman et al. 2018). Provided enough matching points (e.g., >1000) are found in the images, orthophotos are able to correct camera perspective and image scale (Pix4D 2020). Therefore, an orthophoto has the geometric accuracy of a map with the visual characteristics of an aerial image (Liu et al. 2018). Individual orthophotos are typically mosaicked together to create orthophotomosaics (sometimes referred to as orthomosaics) of an entire field or other study area. Just as with an orthophoto, an orthomosaic can be used as if it were a traditional map to measure distance, area, length, and direction (Jensen 2018). A number of other data products, such as visible vegetation indices and near-infrared based vegetation indices (if near-infrared data were collected), can also be computed at this time.

At a minimum, orthomosaics and digital elevation models (DEM) are usually generated from UAV imagery (Rokhmana 2015). Both orthomosaics and DEMs can be created using either desktop or cloud-based applications. In general, desktop applications require more user-interaction and selection and they allow for greater control of the end products. Further, desktop applications use a large amount of computer memory (RAM) and require multiple processing cores. As a result, desktop image processing applications consume much of a computer's resources, thus making any other computer work difficult. Depending on the specifications of individual computers, processing even small areas (e.g., < 10 ha) with fewer than 100 images can take hours to complete. Cloud-based applications simply require users to upload their images, determine what kinds of end products they want, and then wait for the application to complete the process. In general, cloud-based processing takes less time than desktop processing. However, most cloud-based processing applications limit the number of projects and the total number of images a user may process during a specific time – e.g., per month or per annum.

Popular desktop and cloud-based (or both) applications include DroneDeploy, Pix4D, Web Open Drone Map (WebODM), and Drone2Map. Both DroneDeploy and Pix4D have applications to both plan missions, collect the data, and then process the data. These applications are available for purchase by subscription in increments of 1 month or 1 year. DroneDeploy processes all images in the cloud while Pix4D allows both cloud-based and desktop processing. Drone2Map is available as a supplement to ESRI's ArcGIS Pro, and this enables near seamless integration of end products into ArcGIS Pro. Conversely, WebODM is an open-source application that allows users to download and use for free. Although WebODM is not as intuitive to install and use as the other applications, it affords much flexibility with data processing. Further, because it is open-source, WebODM relies on the drone community for improvements in the application.

Transforming the individual images acquired in UAV surveys to orthomosaics can be a very time consuming process. It may take a significant amount of time (e.g., > 5 h) to create an orthomosaic, point cloud, digital surface model, and other data products. Further, a robust desktop or laptop computer is essential for timely image processing. We have found that current 'gaming' computers are very good for processing UAV image data. These computers usually have at least 16 GB of RAM, solid state hard drives, and the fastest available processors and video cards.

9.3.3 UAV Example Imagery

A wheat field in Rexburg, Idaho, USA was imaged on 25 June 2019 with the Phantom 4 RGB camera and the Sentera NDVI Single Sensor. The images were acquired as part of an ongoing project to study variable-rate irrigation in different parts of the field. The RGB images (Fig. 9.3) and NDVI (normalized difference vegetation index) images (Fig. 9.4) were acquired at exactly the same time as both sensors were on board a DJI Phantom 4 flown at 91.4 m above ground level. The flight was planned and implemented using DroneDeploy installed on a sixth generation iPad. The flight resulted in 210 RGB images and 129 NIR (near infra-red) images with 80% overlap and 80% sidelap that were used to create orthomosaics. Figures 9.3 and 9.4 show the two orthomosaics derived from the two datasets. The data were processed in Web OpenDroneMap.

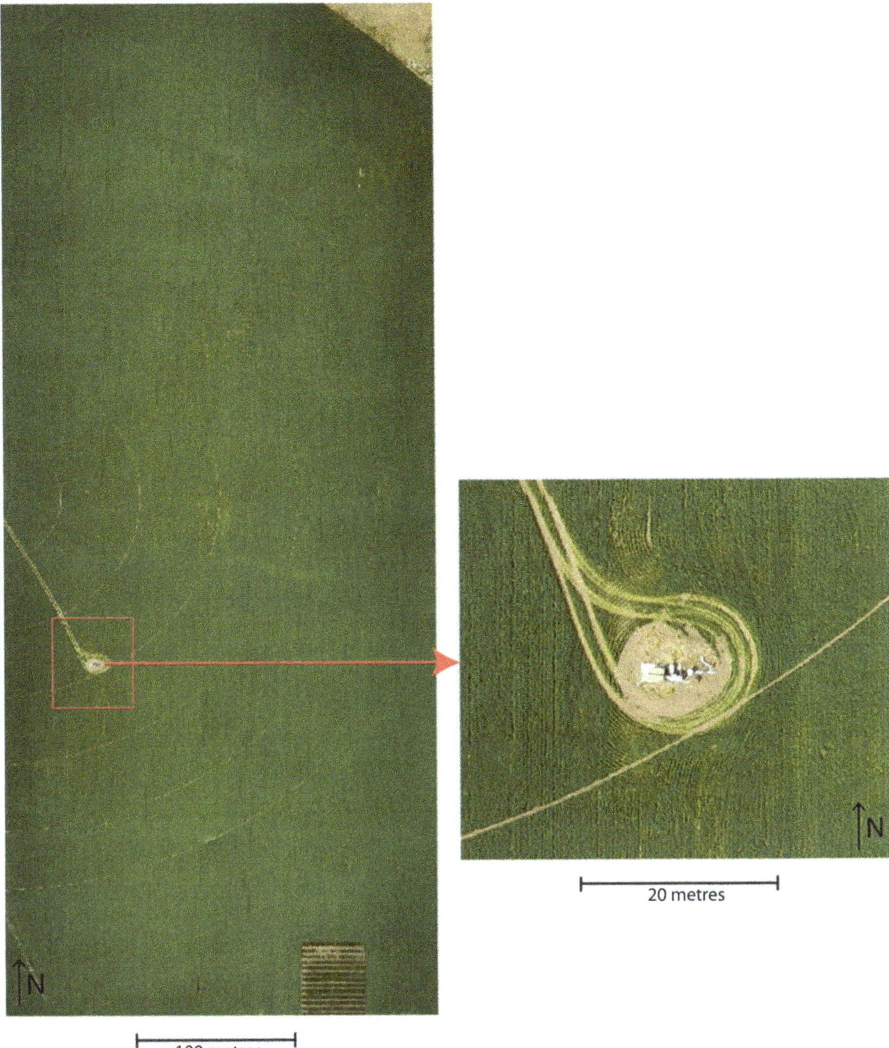

Fig. 9.3 Traditional RGB orthomosaic generated from images acquired over a winter wheat field 25 June 2019. The images were acquired with a DJI Phantom 4 flying at 91.4 m above ground level and processed with Web OpenDroneMap

9.4 Sample Applications of UAVs in Agriculture

The UAVs have been used in a large variety of ways in PA. Gnadinger and Schmidhalter (2017) used an octocopter and a 10 megapixel RGB camera to count maize plants. The authors found high agreement with maize counted by both UAV images and *in situ* counts. Also, in maize, Maresma et al. (2016) analysed different

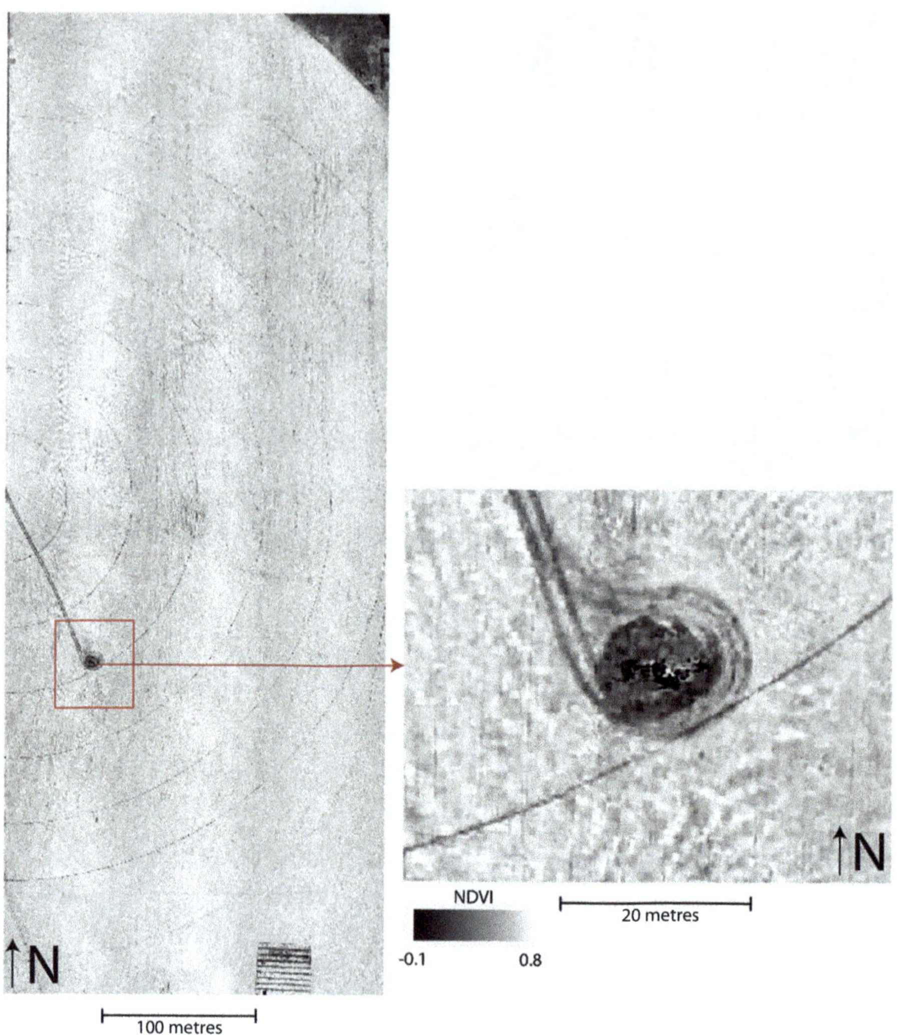

Fig. 9.4 The NDVI orthomosaic acquired over a winter wheat field 25 June 2019. The images were acquired using a Sentera NDVI Single Sensor mounted on a DJI Phantom 4 flying at 91.4 m above ground level. The initial images were converted to an orthomosaic using WebODM and then converted to NDVI in ArcGIS Pro. In this image the brighter (whiter) pixels represent healthier or greener vegetation – e.g., higher NDVI values

multispectral vegetation indices and crop height derived from a standard UAV commercial service to determine the required amount of side-dress N fertilizer just before flowering. The study analysed the spectral response of seven different inorganic N rates, two different pig slurry manure rates and four inorganic–organic N combinations in experimental plots (10 × 15 m). The image was acquired by an Atmos-6 UAV (CATUAV, Moià, Catalonia, Spain), a fixed-wing UAV with a wingspan of 1.80 m, a length of 1.29 m and a payload of 500 g. The camera was a

VEGCAM-Pro with a 14 Mp Foveon X3 image sensor, acquiring data in three wide spectral bands (green, red and near-infrared). The spectral indices WDRVI, NDVI and the crop height showed no significant response to extra N application at the economic optimum rate of fertilization (239.8 kg N ha^{-1}). This demonstrates the potential of UAV multispectral images to determine and adjust the side-dress N fertilization rates.

De Castro et al. (2018) used a DJI Inspire 2 and two remote sensing cameras—both with red, green, and near-infrared bands—to detect two levels of water stress in six ornamental plants. The authors suggest good potential for identification of water stressed plants, but more analysis, including using vegetation indices, is needed. Montes De Oca et al. (2018) developed and tested a low-cost multispectral imaging system with a quadcopter and two digital cameras to monitor crops. The system also included in-flight image acquisition and processing of the acquired images, and it was deemed a low-cost and reliable system compared to existing market options. Michez et al. (2019) compared pasture height derived from an RGB camera mounted on an octocopter UAV to a terrestrial light detection and ranging (LiDAR) dataset. They found good agreement ($R^2 = 0.62$), and they calculated a biomass map using the UAV data. The methods were determined to be a potential alternative to more labour and time intensive field methods for estimating pasture height and biomass. Candiago et al. (2015) used a hexacopter and a multispectral camera to calculate several vegetation indices of vineyards and tomatoes. Normalized difference vegetation index (NDVI), green normalized difference vegetation index (GNDVI), and soil adjusted vegetation index (SAVI) were all calculated for each crop. The study demonstrated that the UAV data and their derivatives, such as vegetation indices, are a fast, reliable, and economic resource for assessing crops. Honkavaara et al. (2013) developed a workflow using a Fabry-Perot interferometer-based camera on an experimental wheat field. The authors found promising results for wheat biomass estimation. Torres-Sánchez et al. (2019) demonstrated the feasibility of using UAV photogrammetry to detect leaf trimming and leaf removal in a vineyard.

In a very practical sense, UAVs have been used to spray pesticides and fertilizers over agricultural fields (Faical et al. 2014), allowing UAVs to be able to improve crop productivity through the mitigation of diseases (Mogili and Deepak 2018). Jorge et al. (2014) showed it is possible to use UAVs for crop disease detection in citrus fruit. According to Krishna (2020), UAVs are expected to have an even larger impact on crop production techniques and on the input of fertilizers, fungicides, pesticides, herbicides and irrigation in the future.

The UAVs have been used to map soil characteristics and water content in plants. For example, Huuskonen and Oksanen (2018) used UAV data and augmented reality to create a top soil map with management zones for precision farming. The authors used a DJI Phantom 4 Pro to help complete the project. Izzo et al. (2019) used hyperspectral UAV data and partial least squares regression to model water content in grapevines. Their best model produced a coefficient of determination of 0.68. The UAVs have also been used to support satellite imagery analysis in PA. Guo et al. (2019) used UAV data to develop three-dimensional ground reference information and training data for rice mapping and monitoring in Australia. In a different approach, Martínez-Casasnovas et al. (2013) applied multi-temporal UAV images to quantify the effects of ephemeral gully erosion and to determine the decrease of plant vigour in vineyards. For that, an Atmos-6 fixed-wing UAV (above mentioned) equipped with a Sony NEX-C3-na camera was used. The 9 cm resolution pictures allowed the quantification of ephemeral gully volumes with high precision.

9.5 Pros and Cons of UAVs in Agriculture

There are several pros and cons to using UAVs in PA. The prices of current UAVs and their sensors continue to decrease —even as the quality and specifications of both UAV platforms and sensors get better. These factors will enable many farmers, researchers, and farm managers to be able to rapidly analyse fields and make more informed decisions. Further, given the constraints of traditional aerial photography and satellite remote sensing images (temporal and spatial resolution, cost, etc.), the flexibility of obtaining UAV imagery on an *ad hoc* basis is a distinct advantage. For example, most satellites, such as Landsat 8, are only capable of imaging individual fields on very specific days and times of day. This is especially a problem if the scheduled day for data acquisition is overcast. Conversely, UAV data can be obtained over a field on any given day provided an acceptable time window is available, and image data collection is largely independent of time and atmospheric conditions, such as overcast skies (Gnadinger and Schmidhalter 2017). Further, UAVs provide a much finer GSD that provides pure pixels of objects/crops being observed (Guo et al. 2019). That is of special interest when dealing with row crops, and ground pixels need to be removed.

Despite the many benefits of using UAVs in PA, many challenges remain. For example, interpretation of UAV data and their end products should be done with care, and McKee et al. (2018) demonstrated how uncertainty in orthorectification can negatively impact agricultural end products. Their results suggest the need for caution when comparing remote sensing products across sensor types and through time. Further, some researchers feel that the hype surrounding UAVs may lead to disillusionment with using them in agriculture (Freeman and Freeland 2015). Other disadvantages of using UAVs for PA include the short duration of flights, lack of engine power (thrust), inability to fly in certain conditions (too windy, foggy, etc.), payload weight, sun illumination, and others (Hardin et al. 2018; Mateo-Aroca et al. 2019). Further, operation of UAVs in general has been hampered by the lack of certification standards and regulations (Hayhurst et al. 2016; Hardin et al. 2018). While the United States and many other countries have regulations and certifications for both UAV pilots and UAV platforms, regulatory differences continue to confound UAV applications in many locations. Reger et al. (2018) compared UAV laws in Germany, the EU, the USA, and Japan. The authors found that all these countries are currently in the process of updating existing laws or drafting new laws governing UAV use. Further, while there are some similarities in each country's laws, many differences exist—e.g., maximum flight altitude, proximity to restricted areas such as airports, etc. The UAV users must be aware of all laws and regulations—locally and nationally—before conducting any UAV flight.

9.5.1 Changing Technologies and Software

The UAV aircraft and sensors are constantly being improved. These improvements are essential as UAVs continue to be integrated into PA and many other applications. However, the speed at which new UAV aircraft and sensors become available may be intimidating to many practitioners as there always seems to be something else to learn and/or purchase. Further, the ever-changing and improving nature of flight and processing software, rapid advancements in UAVs and advancements in other high-tech farming procedures may be daunting (Basso et al. 2017). So, while a practitioner may be quite comfortable with UAVs and their derivative data products today, being unable to keep up with advances in flight, data processing, etc. may seem overwhelming.

9.5.2 Imaging Large Fields

Currently, UAV pilots in the United States are restricted to visual line-of-sight flight (VLOS) where the remote pilot in command must maintain unaided vision (e.g., without the use of binoculars or other vision aids) of the UAVs while in flight. Other licensed pilots assisting the remote pilot in command may be strategically placed—and be in contact with the pilot (e.g., with a two-way radio or cell phone)—to maintain sight of the UAV. However, even with a team of people to assist in the flight, it is still difficult map large fields because of limited flight time. This may limit the usefulness of the data as only a subset of a field may be captured. Even if a fixed-wing aircraft is being flown to image a larger field, the visual line-of-sight regulation is still enforced. Primarily because of these constraints, Reger et al. (2018) estimated that UAVs have a cost advantage relative to satellite and traditional airborne remote sensing in fields up to 20 hectares; in larger fields, satellite and airborne remote sensing becomes more economical.

9.6 Case Study: Application of UAV Images in Variable Side-Dress Fertilization of Winter Cereals

One of the most usual applications of PA techniques in extensive crops is variable fertilization. To know the intra-field variability of the crop vigour to determine differential management zones, farmers face the dilemma of what remote sensing data to use, either open access multispectral images, at 10 m resolution (Sentinel-2) or very detailed images acquired by means of UAVs (< 0.10 m resolution). Of course it depends on the size of the fields, but the availability of commercial services based on UAV images makes farmers think about which is the better option. In this section, a case study is presented in which variable-rate side-dress nitrogen fertilization in barley was determined on the basis of a UAV image in a 75 ha field located in NE Spain (72.1 ha effective crop surface) (Fig. 9.5).

The study plot was sown on January fourth 2019 with barley (*Hordeum vulgare* L., var. Gustav, Svalöf Weibull AB, Sweden), which is a two-row spring barley. The seeding dose was uniform at a rate of 200 kg ha^{-1}. The plot was irrigated by means of a central pivot and sprinklers were used in the zones outside the areas covered by the pivot. Soils of the area were classified as Typic Xerorthents and Typic Xerofluvents (Soil Survey Staff 2014). Prior to sowing, the plot was classified into three classes of relative potential productivity on the basis of the yield map for previous years (barley) (Fig. 9.5). These zones were used to apply a variable base dressing fertilization: 25-37-37 NPK units in zone 1 (low potential), 30-44-44 NPK

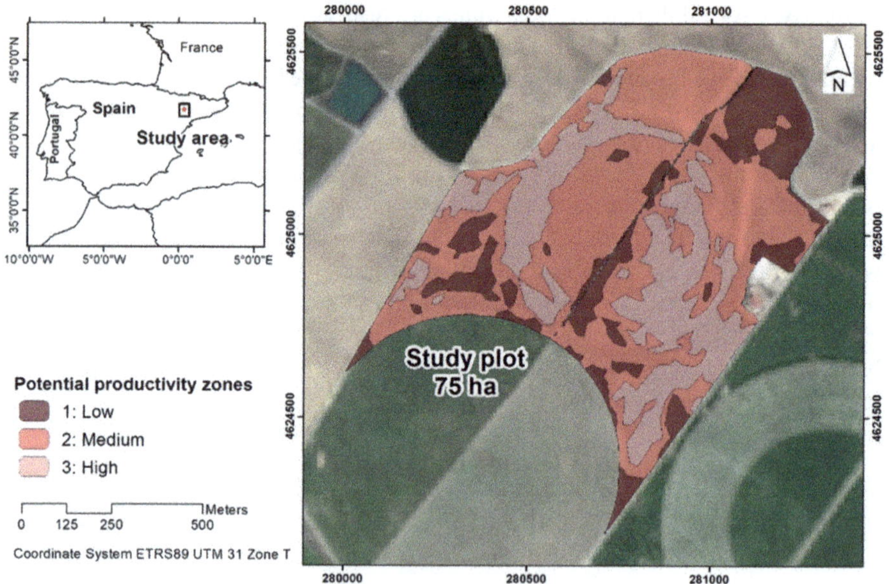

Fig. 9.5 Location of the case study area and potential productivity zones on the basis of a previous barley yield map

units in zone 2 (medium potential) and 35-52-52 NPK units in zone 3 (high potential). The side-dress N fertilization was decided on the basis of the crop vigour at half tillering (Decimal Code 25-27 of the scale of Zadoks, Zadoks et al. 1974). For that, a multispectral image was acquired on March 26th, 2019 using a Sequoia camera (Parrot SA, Paris, France) mounted in a DJI Inspire 1-T600 UAV (DJI Sciences and Technologies Ltd., Shenzhen, China) (Fig. 9.6).

The pixel size of the acquired images was about 0.07 m. The study analysed the cost of data acquisition and processing to establish commercial advice service for farmers. The UAV flight was configured with the aid of Pix4DCapture software (Pix4D S.A., Prilly, Switzerland) (Fig. 9.6). The main characteristics were: flight height 70 m above ground, speed 8.5 m s^{-1}, 2 shots s^{-1}, total survey time 2 h, including the time to change UAV batteries (6 battery changes).

As shown in Table 9.2, the Sequoia acquires four multispectral bands centred on 5500 nm (green), 660 nm (red), 735 nm (red edge) and 790 nm (near-infrared). A downwelling light sensor (DLS) was connected to the camera to measure the ambient light during the flight for each band to correct for lighting changes. In addition, a calibrated reflectance panel was used to convert the pixel values to reflectance values.

A total of 2850 photos were acquired and a calibrated mosaic for the four spectral bands was created with the aid of Pixel4D software. From here, the NDVI was calculated and classified in 3 classes according to the ISODATA unsupervised classification algorithm (Jensen 2007) implemented in ArcGIS Desktop 10.7 (ESRI, Redlands, USA). After that, the NDVI classified layer was resampled to pixels of 0.3 m to avoid a very large number of regions in the prescription map. Then, the layer was converted to a polygon layer and polygons less than 200 m^2 were integrated in the neighbouring polygon with larger areas according to the width of the fertilizer application machinery.

The service and the cost/benefit for the farmer were assessed taking into account a cost of 32 € h^{-1} for data acquisition and processing, a price of 200 € Mg^{-1} for the fertilizer and a price obtained for the barley of 180 € Mg^{-1}. The average yield in

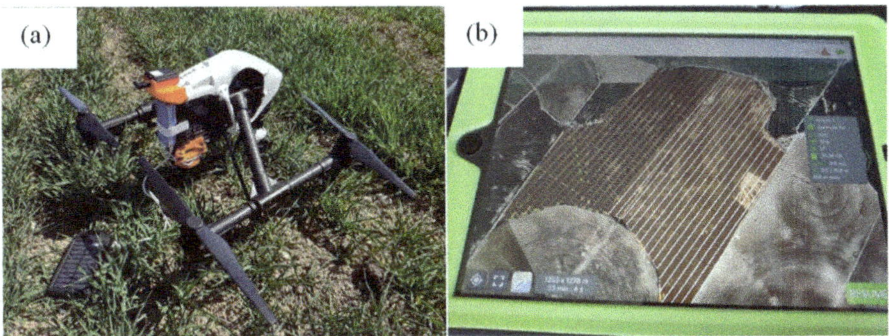

Fig. 9.6 DJI Inspire 1-T600 UAV with the Sequoia multispectral camera used to acquire the image in the study plot (**a**) and flight plan with the rows across the plot to acquire the images (**b**)

each zone was: 4300 kg ha^{-1} in the low vigour zone, 6600 kg ha^{-1} in the medium vigour zone and 8900 kg ha^{-1} in the high vigour zone.

Figure 9.7 shows the results of the NDVI map derived from the multispectral image acquired on March 26th, 2019 to assess the vigour of the barley during tillering. In addition, the variable-rate prescription map with the zones for differential side-dress fertilization is shown.

In total, the acquisition and processing time to elaborate the prescription map was 12.5 h, with a total cost of 400 € (unitary cost: 5.5 € ha^{-1}). According to the fertilization plan in each zone, the total cost of the side-dress fertilization is shown in Table 9.4.

According to Table 9.4, the average cost of the side-dress fertilization based on the UAV image was 88.6 € ha^{-1}. This cost was added to the cost of the advice service (5.5 € ha^{-1}) and compared to the total incomes generated by the barley harvest (1242 € ha^{-1}). In summary, the gross benefit based on the UAV service was 1147.9 € ha^{-1}.

If compared with the hypothetical gross benefit obtained with the uniform side-dress fertilization based on the average yield (1105.1 € ha^{-1}), the differential fertilization based on the UAV service generated higher gross benefits of about 42.8 € ha^{-1}.

Although not described in the methodology, in the same study the gross benefit of the advice service was assessed on the basis of an NDVI map of the same date derived from a Sentinel-2 image. In comparison, the gross benefits obtained with

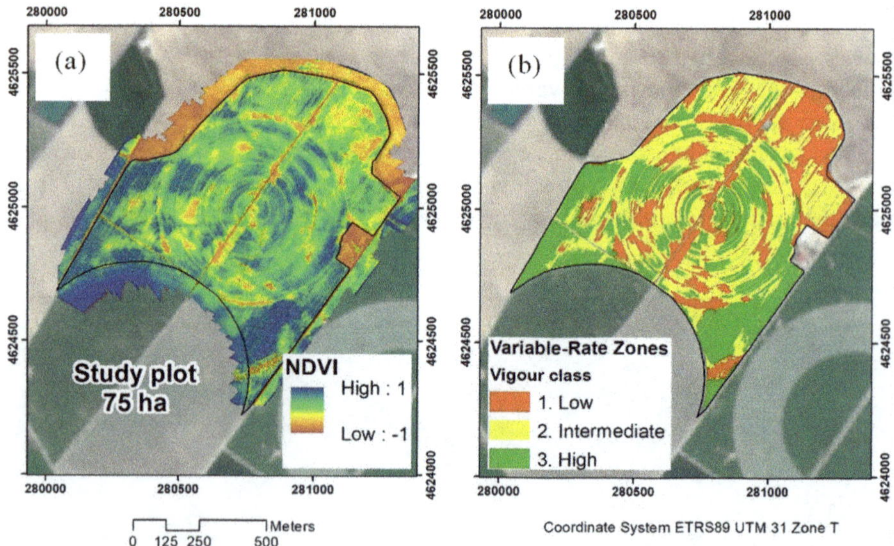

Fig. 9.7 The NDVI derived from the Sequoia multispectral image acquired on March 26th, 2019 to assess the vigour of the barley to establish the variable-rate side-dress fertilization (**a**) and variable-rate zones for side-dress fertilization (**b**). The low vigour class areas were fertilized with less nitrogen (N) due to soil salinity which limits the uptake of N by the crop. High vigour class areas were fertilized with higher N doses

Table 9.4 Side-dress fertilization plan on the basis of the NDVI zones and cost of the fertilizer

Zone	Surface (ha)	N32 (kg ha^{-1})	N32 fertilizer applied (kg)	Cost (€)
Low vigour	21.1	265.6	5612.1	1122.4
Medium vigour	20.1	421.8	8461.3	1692.3
High vigour	30.9	578.1	17,880.6	3576.1
Total	72.1		31,954.0	6390.8

either advice service was slightly higher in the case of the UAV-based prescription (6.4 € ha^{-1}): 1147.9 € ha^{-1} based on the UAV image and 1141.5 € ha^{-1} based on the Sentinel-2 image, with both significantly higher when compared with uniform management.

The results obtained are in the range of the maximum economic advantages of variable-rate nitrogen application in grain crops analysed by Colaço and Bramley (2018). These authors reviewed different works related to the use of crop sensors in N management in grain crops, reporting fertilizer savings of 5–45%, with little effect on grain yield. Reported impacts on profit usually ranged between losses of US$ 30 ha^{-1} (about 24.6 € ha^{-1}) and profits of US$ 70 ha^{-1} (about 57.4 € ha^{-1}), with an overall average profit of US$ 30 ha^{-1} (about 24.6 € ha^{-1}). Nevertheless, these studies were mainly based on hand held or on-board machine sensors such as Crop Circle™ or GreenSeeker™, and not on multispectral images as in the present case.

In conclusion, the analysis showed that the average price of UAV-acquired NDVI images was 5.5 € ha^{-1}, based on a labour cost of 32 € h^{-1}. This could be considered a low cost in relation to other services of about 80 € h^{-1} reported in the literature. The average economic benefit provided by variable nitrogen application was 42.8 € ha^{-1} compared to uniform fertilization. Therefore, one can conclude that the UAV-based service was affordable for variable-rate nitrogen application. Nevertheless, the Sentinel-2 based service produced similar results, avoiding data acquisition in the field and the complexity of data processing. However, despite the widespread use of UAV services in the last few years, and because of the variability in the price of data acquisition and processing services, UAV-based services may not be as profitable as they seem for variable-rate nitrogen fertilization, in particular in large plots that require high acquisition and processing times.

9.7 Conclusions for the Chapter

We look forward to further development and advances in UAVs for applications in PA including: advances in aircraft; greater availability of multispectral, hyperspectral, thermal-infrared, and LiDAR-based sensor systems; and longer-life intelligent battery technology. We believe that the future of UAVs in PA is very bright—especially as the quantity and quality of UAV platforms and sensors continues to increase and improve, respectively. We project that the two most significant needs in the future that will encourage greater adoption in PA are increased flight times and increased payload weight capability. Increased flight time will allow for larger areas to be imaged at one time—thus simplifying flight planning and some post-processing procedures. Increased payload ability would provide for additional cameras and sensors to be placed on UAVs for simultaneous data capture from multiple sensors.

Nevertheless, and according to the results of the case study presented in this chapter, further studies are needed to quote the profitability of UA platforms for varied tasks in different types of crop and size of field. In this respect, remote sensing data acquired from satellite platforms, such as Sentinel-2 could be enough to advise zonal management based on crop vigour in large fields. In these cases, advice services based on UAV images are more expensive due to the necessary fieldwork and the greater complexity of image processing.

Finally, as noted above, the speed at which new UAV platforms, sensors, and applications are being developed can be daunting. We encourage the development of robust UAV PA methods that are significant and repeatable in many different agricultural areas with various environmental conditions.

References

Adao T, Hruska J, Padua L et al (2018) Hyperspectral imaging: a review on UAV-based sensors, data processing and applications for agriculture and forestry. Remote Sens 9:1110

Basso B, Dobrowolski J, McKay C (2017) From the dust bowl to drones to big data. Georget J Int Aff 18:158–165

Candiago S, Remondino F, De Giglio M et al (2015) Evaluating multispectral images and vegetation indices for precision farming applications from UAV images. Remote Sens 7:4026–4047

Colaço AF, Bramley RGV (2018) Do crop sensors promote improved nitrogen management in grain crops? Field Crop Res 218:126–140

De Castro AI, Maja JM, Owen J et al (2018) Experimental approach to detect water stress in ornamental plants using UAVs-imagery. In: Proceedings SPIE 10664, autonomous air and ground sensing systems for agricultural optimization and phenotyping III, 106640N (21 May 2018)

Deng L, Mao A, Li X et al (2019) UAV-based multispectral remote sensing for precision agriculture: a comparison of different cameras. ISPRS J Photogramm Remote Sens 146:124–136

Dilep MP, Navaneeth AV, Ullagaddi S et al (2020) A study and analysis on various types of agricultural drones and its applications. In: Proceedings of the 2020 fifth international conference on research in computational intelligence and communication networks (ICRCICN). IEEE, Bangalore, pp 181–185

Faical BS, Costa FG, Pessin G et al (2014) The use of unmanned aerial vehicles and wireless sensor networks for spraying pesticides. J Syst Archit 60:393–204

Freeman PK, Freeland RS (2015) Agricultural UAVs in the U.S.: potential, policy, and hype. Remote Sens Appl Soc Environ 2:35–43

Gnadinger F, Schmidhalter U (2017) Digital counts of maize plants by unmanned aerial vehicles (UAVs). Remote Sens 9:544

Guo Y, Jia X, Paull D et al (2019) A drone-based sensing system to support satellite image analysis for rice farm mapping. IGARSS 2019–2019 IEEE international geoscience and remote sensing symposium, 28 July – 2 August

Hardin PJ, Lulla V, Jensen RR et al (2018) Small unmanned aerial systems (UAVs) for environmental remote sensing: challenges and opportunities revisited. GISci Remote Sens 56(2):309–322

Hayhurst KJ, Maddalon JM, Neogi NA et al (2016) Safety and certification considerations for expanding the use of UAS in precision agriculture. In: Proceedings of the 13th international conference on precision agriculture (unpaginated, online). International Society of Precision Agriculture, Monticello

Honkavaara E, Saari H, Kaivosoja J et al (2013) Processing and assessment of spectrometric, stereoscopic imagery collected using a lightweight spectral camera for precision agriculture. Remote Sens 5:5006–5039

Huang Y, Reddy KN, Fletcher RS et al (2018) UAV low-altitude remote sensing for precision weed management. Weed Technol 32:2–6

Huuskonen J, Oksanen T (2018) Soil sampling with drones and augmented reality in precision agriculture. Comput Electon Agric 154:25–35

Inoue Y, Yokoyama M (2019) Drone-based optical, thermal, and 3d sensing for diagnostic information in smart farming – systems and algorithms. In: IGARSS 2019 – IEEE international geoscience and remote sensing symposium

Izzo RR, Lakso AN, Marcellu ED et al (2019) An initial analysis of real-time UAVs-based detection of grapevine water status in the finger lakes wine country of upstate New York. In: Thomasson JA, McKee M, Moorhead RJ (eds) Autonomoous air and ground sensing systems for agriculture optimization and phenotyping IV, proceedings of SPIE volume 11008

Jensen J (2007) Remote sensing of the environment – an earth resource Perspectiv, 2nd edn. Pearson – Prentice Hall, Upper Saddle River

Jensen J (2018) Drone aerial photography and videography – data collection and image interpretation. https://www.amazon.com/Drone-Aerial-Photography-Videography-Interpretation-ebook/dp/B07929YHLS/ref=sr_1_4?dchild=1&keywords=drone+aerial+photography+and+videography&qid=1606962327&sr=8-4

Jorge L, Brandao A, Inamsu R (2014) Insights and recommendations of use of UAV platforms in precision agriculture in Brazil. In Proceedings SPIE 9239, remote sensing for agriculture, ecosystems, and hydrology XVI, 923911 (21 October 2014)

Khot L, Zhou J (2016) Precision agriculture: beyond the domain of small UAS. Resour Mag 23(3):20–21

Krishna K (2020) Unmanned aerial vehicle Systems in crop production: a compendium. Apple Academic Press, Palm Bay

Liu Y, Ai G, Zhang Y et al (2018) Generating a high-precision true digial orthophoto map base don UAV images. ISPRS Int J Geo-Inf 7:333

Maresma A, Ariza M, Martínez E et al (2016) Analysis of vegetation indices to determine nitrogen application and yield prediction in maize (*Zea* mays L.) from a standard UAV service. Remote Sens 6:973

Martínez-Casasnovas J, Ramos M, Balasch C (2013) Precision analysis of the effect of ephymeral gully erosion on vine vigour using NDVI images. In: Stafford J (ed) Precision agriculture 13, ISBN 978-90-8686-224-5, Wageningen Academic Publishers (The Netherlands), pp 777–783

Mateo-Aroca A, Garcia-Mateos G, Ruiz-Canales A et al (2019) Remote image capture system to improve aerial supervision for precision irrigation in agriculture. Water 2019(11):255

McKee M, Nassar A, Torres-Rua A et al (2018) Implications of sensor inconsistencies and remote sensing error in the use of small unmanned aerial systems for generation of information products for agricultural management. In: Proceedings SPIR 10664, autonomous air and ground sensing systems for agricultural optimization and phenotyping III, 1066402 (21 May 2018)

Messina G, Modica G (2020) Applications of UAV thermal imagery in precision agriculture: state of the art and future research outlook. Remote Sens 12:1491

Michez A, Lejeune P, Bauwens S et al (2019) Mapping and monitoring of biomass and grazing in pasture with an unmanned aerial system. Remote Sens 11:473

Mogili U, Deepak B (2018) Review on application of drone systems in precision agriculture. Proc Comput Sci 133:502–509

Montes De Oca A, Arreola L, Flores A et al (2018) Low-cost multispectral imaging systems for crop monitoring. In: 2018 International Conference on Unmanned Aircraft Systems (ICUAS), 12–15 June 2018

Ngadiman, N., Kaamin M, Sahat S et al (2018) Production of orthophoto map using UAV photogrammetry: a case study in UTHM campus. In: Proceedings of the 3rd International Conference on Applied Science and Technology (ICAST'18) AIP Conf. Proc. 2016, 020112-1–020112-5
Perera TANT, Priyankara ACP, Jayasinghe GY (2019) Unmanned aerial vehicles (UAV) in smart agriculture: trends, benefits and future perspectives. In: International Research conference of UWU, 2019
Pix4d (2020) Photo stitching vs orthomosaic generation. https://support.pix4d.com/hc/en-us/articles/202558869-Photo-stitching-vs-orthomosaic-generation. Last accessed June 2020
Puri V, Nayyar A, Raja L (2017) Agricultural drones: a modern breakthrough in precision agriculture. J Stat Manag Syst 20(4):504–518
Raeva PL, Sedina J, Dlesk A (2019) Monitoring of crop fields using multispectral and thermal imagery from UAV. Eur J Remote Sens 52:192–201
Reger M, Bauerdick J, Bernhardt H (2018) Drones in agriculture: current and future legal status in Germany, the EU, the USA and Japan. Landtechnik 73(3):62–79
Rokhmana CA (2015) The potential of UAV-based remote sensing for supporting agriculture in Indonesia. Procedia Environ Sci 24:245–253
Seifert E, Seifert S, Vogt H et al (2019) Influence of drone altitude, image overlap, and optical sensor resolution on multi-view reconstruction of forest images. Remote Sens 11:1252
Shendryk Y, Sofonia J, Garrard R et al (2020) Fine-scale prediction biomass of leaf nitrogen content in sugarcane using UAV LiDAR and multispectral imaging. Int J Appl Earth Obs Geoinf 92:102177
Soil Survey Staff (2014) Keys to soil taxonomy, 12th edn. USDA-Natural Resources Conservation Service, Washington, DC
Stehr NJ (2015) Drones: the newest technology for precision agriculture. Nat Sci Edu 44:15-04-0772
Torres-Sánchez JF, López-Granados, De Castro AI et al (2013) Configuration and specifications of an unmanned aerial vehicle (UAV) for early site specific weed management. PLoS One 8:e58210
Torres-Sánchez J, Marín D, De Castro AI et al (2019) Assessment of vineyard trimming and leaf removal using UAV photogrammetry. In: Stafford JV (ed) Precision Agriculture'19, p 1030. eISBN: 978-90-8686-888-9 | ISBN: 978-90-8686-337-2
Tripicchio P, Satler M, Dabisias G et al (2015) Towards smart farming and sustainable agriculture with drones. In: Proceedings of the 2015 international conference on intelligent environments IE 2015, Prague, Czech Republic, 15–17 July 2015, pp 140–143
Wallace L, Lucieer A, Watson C et al (2012) Development of a UAV-LiDAR system with application to forest inventory. Remote Sens 4:1519–1543
Yao H, Qin R, Chen X (2019) Unmanned aerial vehicle for remote sensing applications – a review. Remote Sens 11:1443
Zadoks JC, Change TT, Konzak CR (1974) A decimal code for the growth stages of cereals. Weed Res 14:415–421
Zhang C, Kovacs JM (2012) The application of small unmanned aerial systems for precision agriculture: a review. Precis Agric 13:693–712

Chapter 10
Sensing for Weed Detection

S. Christensen, M. Dyrmann, M. S. Laursen, R. N. Jørgensen, and J. Rasmussen

Abstract Conventional weed control methods are based on uniform treatments of the whole field, however, weeds are not distributed uniformly within fields, which means that the uniform distribution of herbicides is inappropriate. Considerable research has been conducted on different aspects of site-specific weed management in the past three decades from fundamental studies on the spatial distribution of weeds to the applied development and testing of new technologies for weed detection and site-specific control. Despite the available technologies and knowledge, there has been little practical adoption of these technologies mainly due to the lack of automated weed detection methods. However, significant progress has been made in the past few years. This chapter describes the general principles of automated weed detection and how these principles can be used in an agricultural context. Two cases demonstrate the progress of spot-application of herbicides based on automated weed detection using RGB cameras mounted on unmanned aerial vehicles (UAV) and ground vehicles.

Keywords Site-specific weed management · Weed sensing · Weed detection · Weed mapping · Machine learning · Deep learning

10.1 Introduction

Weeds have a significant impact on crop production. Oerke (2006) estimates that the potential loss of crop yield without weed control on a global scale would be 43%. In fully industrialized countries, the most widely applied weed control method is

S. Christensen (✉) · J. Rasmussen
Department of Plant and Environmental Sciences, University of Copenhagen, Copenhagen, Denmark
e-mail: svc@plen.ku.dk; jer@plen.ku.dk

M. Dyrmann · M. S. Laursen · R. N. Jørgensen
Department of Engineering, Aarhus University, Aarhus, Denmark

R. Kerry, A. Escolà (eds.), *Sensing Approaches for Precision Agriculture*, Progress in Precision Agriculture, https://doi.org/10.1007/978-3-030-78431-7_10

herbicides, with mechanical weed control used to a lesser extent, whereas in developing countries manual weed control is still used the most widely. Increasing environmental problems and herbicide resistance caused by decades of intensive herbicide use have given rise to an urgent need to develop site-specific weed management (SSWM) strategies to reduce herbicide use.

Conventional spraying of herbicides is based on uniform broadcast or band application methods to reduce weed competition and contamination by weed seed in harvested crops. However, weeds do not tend to distribute in fields uniformly (Wiles et al. 1992; Rew and Cousens 2001; Gerhards 2010), which means that the uniform distribution of herbicides is inappropriate. Therefore, there has been an increasing number of studies on SSWM and its use as a strategy for herbicide applications (Christensen et al. 2003, 2009; Guttjahr et al. 2012; Andújar et al. 2013). Site-specific weed management represents a four-step cyclical process that includes (1) detection of the weeds, (2) decision-making, (3) precision weed control and (4) evaluation.

The ultimate technical vision for SSWM could be the detection and treatment of individual weeds, but this is technically and economically feasible in only a very limited range of conditions. To be economically feasible, weed detection has to be fast, reliable and matched to the resolution of the associated weed control system. For example, it does not make sense to use micro-sprayers to control individual plants if weeds are detected from satellite images with a spatial resolution that is far greater than the solution required to identify individual plants. Weed detection in agricultural crops is a trade-off between what is technically possible and what is feasible in an agricultural context. The value gained from weed detection has to exceed its associated costs.

In a review of the potential use of ground-based sensor technology for weed detection, Peteinatos et al. (2014) concluded that despite promising experimental results, major constraints to implementation remain because most of the significant results have been achieved under controlled conditions or following a specific protocol, which would be difficult to replicate under normal field conditions. Low adoption of SSWM technologies has indicated that relatively crude weed mapping procedures may be more feasible in current farming than more sophisticated procedures because of trade-offs between the accuracy of weed detection and cost-effectiveness. The challenge for SSWM is to develop and implement weed detection systems that are economically feasible in a practical farming context.

This chapter focuses on the integration of weed detection into an SSWM within a short timeframe. The objective is to provide a brief insight into the general principles of weed detection, demonstrate how these principles can be used in an agricultural context, and finally provide two practical case studies of SSWM based on weed detection using RGB cameras mounted on unmanned aerial vehicles (UAV) and ground vehicles.

10.2 Types of Weed Detection Sensors

A variety of sensors are available, some of which are suitable for weed detection. Sensors may be classified into optical and non-optical sensors. Non-optical sensors have hardly been used for weed detection, but Andújar et al. (2002) used an ultrasonic distance sensor to detect weeds because weeds affected canopy height. They found that weed presence or absence was predicted correctly in about 90% of the samples used in a winter wheat field. However, the method was greatly affected by the growth stage. Ehlert and Dammer (2006) used another non-optical sensor, Crop-Meter, for the assessment of crop biomass and for variable application rates of fungicides in real time. The Crop-Meter sensor is based on the pendulum principle, where the inclination angle is measured when a pendulum body passes through the crop.

In relation to weed detection, optical sensors are by far the most common sensors and they can be categorized according to the following criteria:

1. imaging *versus* non-imaging sensors
2. active *versus* passive sensor systems
3. visible light *versus* invisible light sensors
4. broad-band *versus* narrow-band sensors.

Imaging sensors create digital images based on the reflected light from objects (e.g. weeds), whereas non-imaging sensors only produce a single response value, based on measurements from a single or multiple narrow bands of the electromagnetic spectrum. The response value could be a vegetation index representing the amount of photosynthetically active biomass such as the normalized difference vegetation index (NDVI) (Heege et al. 2008). Several non-imaging sensors mounted on ground vehicles are commercially available for on-the-go patch spraying (Fernández-Quintanilla et al. 2018). They have been commercialized for weed detection between rows, in row crops (Sui et al. 2008; Andújar et al. 2011) and in fallow (Blackshaw et al. 1998).

A sensing system that is dependent on an external source of energy to irradiate the plants is called a passive system. An active sensing system possesses both a sensor and an emitting energy source, used to irradiate the objects to be detected.

The human eye is not sensitive to the near-infrared (NIR) electromagnetic spectrum, but for weed detection this spectrum is useful. Multispectral cameras are sensitive to NIR and many multispectral cameras have between three and six bands, although multispectral sensors in satellites typically have up to 15 bands. Cameras that produce images comprising a combination of visible and NIR light produce so-called false-colour images or colour-infrared (CIR) images (Fig. 10.1). The NIR is represented by a distinct colour, and thereby made visible. Consumer-grade RGB cameras can be modified to record the NIR spectrum by exchanging the colour filter of an existing channel to NIR (Rasmussen et al. 2016).

Some multispectral cameras measure light in broad bands while others use narrow bands. Hyperspectral cameras collect information from the electromagnetic

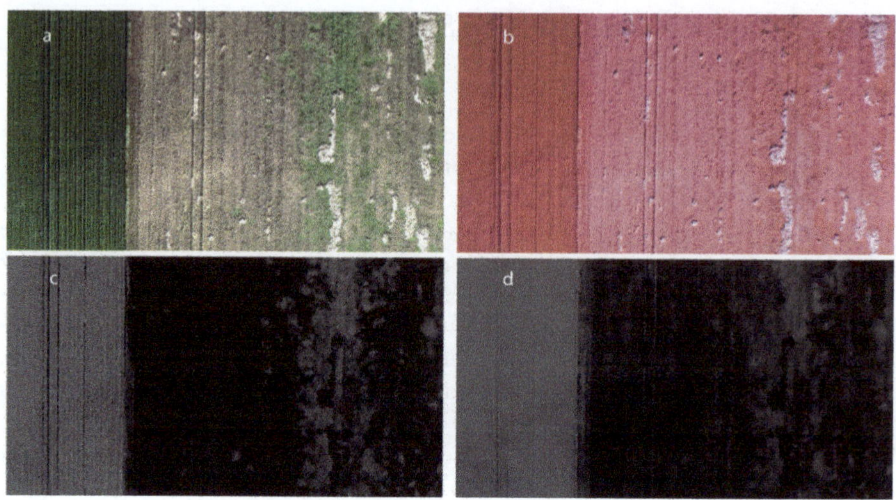

Fig. 10.1 Example image and image processing based on an RGB (**a**) and a CIR (**b**) image of the same scenery of winter wheat and a senescent grass field infested with green weeds. Green is green in the RGB image (**a**) but red in the CIR image (**b**). (**c**) is the processed ExG image and (**d**) is the processed NDVI image, both presented as a grey scale

spectrum in narrow bands (3–10 nm). Hundreds of contiguous spectral narrow bands are available in hyperspectral cameras (Adão et al. 2017).

Experimental work with spectrometers measuring laser-induced fluorescence (Hilton 2020), and stereo vision and time-of-flight (ToF) sensors (Kazmi et al. 2014) for depth images (3D) have all been investigated as weed detection sensors. The 3D methods are elaborated in Chap. 3.

10.3 Sensor Platforms

Data for weed detection can be collected using sensors mounted on satellites, piloted aircrafts, UAV and ground vehicles. These various platforms have different advantages and disadvantages in relation to weed detection.

Satellite imagery has low spatial resolution, high spectral resolution, high performance in terms of mapping capacity (ha h^{-1}) and variable frequency of image delivery because of orbits and clouds. The most popular satellite images used for agricultural research purposes are Landsat 8, which has 15 m spatial resolution, and Sentinel-2 products, which have a ground sample distance (GSD) of 10, 20 and 60 m depending on the native resolution of the different spectral bands. Sentinel-2 and Landsat 8 data are free for all types of users. There are also commercial satellites such as WorldView and GeoEye that have higher spatial resolution. Satellites have mainly been used for weed detection on rangelands because of their low spatial resolution (Thorp and Tian 2004), but experiments with weed detection in arable crops also exist. Backes and Jacobi (2006) mapped *Cirsium arvense* successfully in sugar beet at the cotyledon stage based on images from the QuickBird satellite, which had 2.62–2.90 m resolution, however, QuickBird is no longer active. New commercial satellites have even greater spatial resolution but their potential in weed detection still needs to be evaluated.

Piloted aircraft can deliver imagery with high spatial and spectral resolution. Camera size and weight are not limiting factors for piloted aircraft, which means that large cameras can be used. Flying at an altitude of around 1–2 km can result in image resolution in the range of 0.1 m to 0.5 m. Weed detection in arable crops based on piloted aircraft has been investigated. Hamouz et al. (2008) detected *C. arvense* in winter wheat with a detection accuracy of 89%, Castillejo-González et al. (2014) detected *Avena sterilis* patches in maturing wheat with an accuracy of >90% and de Castro et al. (2013) detected flowering cruciferous weed patches in winter wheat with an accuracy of >85%. More studies have been reviewed in Thorp and Tian (2004), López-Granados (2011) and Fernández-Quintanilla et al. (2018). As for satellites, the detection quality of piloted aircraft imagery should be evaluated relative to the smallest objects that can be distinguished.

Prior to 2013, few reports had been published on the use of UAV imagery for weed mapping (Rasmussen et al. 2013), but recent studies show great promise for UAV weed mapping in row crops such as maize and sunflowers (Peña et al. 2013, 2015; Torres-Sánchez et al. 2013, 2015; Pérez-Ortiz et al. 2015). Promising results have also been found for mapping perennial weeds in cereals (Olsen et al. 2017; Rasmussen et al. 2019).

Unmanned aerial vehicle (UAV) imagery offers large image resolution (about 0.01 m) and good flexibility in terms of the timing of image acquisition. The large image resolution allows the detection of low weed densities, which is important in high-input agriculture where large and dense weed patches are rare. The image acquisition flexibility allows images to be captured when the spectral differences between crops and weeds are large in the later growth stages. Small consumer UAVs (rotary wings) can cover 10–30 ha depending on flight altitude (40–100 m) in 20 min, which corresponds to the duration of a battery, approximately. Fixed-wing UAVs map around three hectares per minute, depending on the field size.

Proximal sensors mounted on in-field machines, including harvesters, tractors, ATVs and robots, can also detect weeds potentially with submillimetre resolution. However, high resolution results in sparse sampling compared to the airborne systems. Hence, any potential accuracy in cotyledon stage weed detection and classification gained may be lost in the need for interpolation between the sample points. Examples of such sparse sampling camera systems are the bispectral H-Sensor (Trengove 2016) and the RGB camera system from Dimension Agri Technologies, Norway. Several commercial spraying systems now available can detect the presence or absence of plants using non-imaging sensors, without discriminating between species. 'WeedSeeker' from Trimble and 'Amaspot' from Amazone are systems used to spot spray stubble for emerging weeds (Grunwald et al. 2020).

10.4 Weed Detection and Management Systems

In arable crops, weeds are most often controlled during the early growth stages, which challenge weed detection with the requirement of high spatial and or spectral image resolution. Detection of individual weed seedlings in crops requires a spatial resolution of at least 0.25 mm. Furthermore, procedures for advanced image analysis have to be applied if weed seedlings are growing in narrow row crops such as cereals, where the leaves of crops often overlap the weeds. This presents a far greater challenge than mapping weeds at later growth stages when the weeds are larger, and their leaf shape and colour are more distinctive.

There are two very different approaches for weed detection in relation to SSWM. One is based on sensing hours or days before the control option, the so-called map-based approach. The other approach is based on sensing immediately before the control option, the so-called on-the-go or the real-time approach.

In map-based approaches, sensor data are stored and processed off-line, with images stitched together or sampled sensor data being interpolated. Decision algorithms may integrate other types of data, such as agronomic and economic data or preferences for a strategy of weed control over several years, generating a prescription map. In on-the-go approaches, sensing, processing and the decision algorithms have to be fast and relatively simple. On-the-go systems usually involve opening and closing single nozzles or sprayer boom sections when the occurrence of weeds exceeds a predetermined threshold. The map-based approach is primarily used in dense crops such as cereals and grasslands, whereas the on-the-go approach is primarily used in row crops or in fallow. In certain situations, a map of the spatial distribution of very competitive weed species can be used to subdivide the field into zones where different herbicide dosage should be sprayed or different intensities of mechanical treatment should be applied.

10.5 Image Analysis – From Sensing to Weed Detection

The detection of weeds is based on sensor output and analysis, which are the most critical components for the adoption of SSWM. Considerable research effort has focused on image analysis. Detection of weeds in a post-emergence crop setting with narrow row distances (e.g. cereals) has proved difficult, and classification algorithms to separate weed species are still challenging.

Weed detection and classification have to be simple to be operational under field conditions. The simplest approach to weed detection is to detect green vegetation, assuming that all green plants are weeds. This approach works in real time and is straightforward in crops before emergence or when in fallow without crop plants. Common weed detection approaches are outlined briefly below.

10.6 Vegetation Indices

Vegetation indices (VIs) are the most common variable in weed detection algorithms. A vegetation index is calculated on data from two or more bands of the electromagnetic spectrum to enhance the contribution of vegetation properties. For example, the excess green index (ExG) enhances the green colour relative to the soil colour and is determined as 2G – R – B where R, G and B are the digital numbers of the red, green and blue channels in the camera, respectively. The index ExG is widely used to detect vegetation based on RGB images.

Normalized difference vegetation index (NDVI) is also a well-known VI based on red and NIR. It makes use of the inverse relation between red and NIR reflectance associated with healthy green vegetation, and is calculated as NDVI = (NIR – red)/(NIR + red). Figure 10.1 shows an RGB and a CIR picture converted into ExG and NDVI, respectively. The ExG and NDVI are presented as greyscale images in Fig. 10.1. Dark shades indicate small values and bright shades indicate large values.

Spectral reflectance of plant species is unique at specific stages and is known as the spectral signature. The spectral signature of weed species can be useful for weed identification (López-Granados et al. 2008). However, weed-species classification based on spectral signatures is not considered robust at the early growth stages because varying field conditions have a considerable impact on the spectral signal.

Weed detection with non-imaging sensors (crop sensors) is based exclusively on spectral information. In-field sensor calibrations have to be carried out to adjust the threshold separating vegetation from the background. Non-imaging sensors mounted on agricultural ground vehicles (tractors or sprayers) detect the stripes of the field in the driving direction, with the underlying assumption that the area detected represents the weed infestation along the width of the sprayer or sprayer section as a whole.

An alternative approach to weed detection by non-imaging sensors is presented by Barroso et al. (2017). They used an optical sensor, installed for on-the-go measurement of grain protein concentration in a combine harvester, to detect the presence of green plant matter in the flowing grain at harvest. The grain stream was recorded continuously during the harvest of spring wheat. Based on visual evaluations of the fields by experts, maps of the green appearance in the grain stream showed an overall agreement of 78% with reference maps.

10.7 Vegetation Cover Fraction

There are several robust algorithms to estimate the green vegetation cover fraction (VCF) (Rasmussen et al. 2007; Hu et al. 2018), which serve many different purposes in agriculture (Thorp and Tian 2004; Rasmussen et al. 2013). The VCF is important in weed detection for estimating weed extent in terms of ground cover. The VCF calculations are based on a VI and a threshold. There are algorithms that automatically estimate the threshold separating the vegetation and background using the ExG index (Rasmussen et al. 2007; Hu et al. 2018). Based on binary images, VCF can be calculated as the percentage of pixels classified as vegetation.

10.8 Object-Based Methods

The main drawback of classifying individual pixels based on spectral information is that it disregards shape and texture aspects, which are among the main clues for a human interpreter. Object-oriented image analysis involves the identification of image objects that are spatially contiguous pixels of similar texture, colour and tone. This approach allows the consideration of shape, size and context as well as spectral content. Object-oriented methods are often more effective than pixel-based methods. Considering groups of pixels as objects helps overcome the increased complexity of a high-resolution scene because of shadows, changes in vegetation density or the similar spectral signatures of dissimilar features. Blaschke (2010) and López-Granados et al. (2016) give an overview of the development of object-based methods.

10.9 Machine Learning, Convolutional Neural Networks and Deep Learning

In recent years, machine learning has emerged that creates new opportunities for analysing image data. Machine learning is a general term for automated pattern recognition in which a computer groups data based on various feature descriptors. Either these feature descriptors can be described manually or the computer can be programmed to identify useful features. Manually designed features in weed recognition could be numbers describing the leaf shape or colour. Normally, machine-learning methods will be trained on a vast number of images that are given a target label, e.g. the species. The computer's task is to map the feature descriptors to the label so that when it is presented with an unknown sample, it can map it to a known class. In the case of weed detection, the label could be the location of weeds, and in the case of weed classification, the label is the weed species.

Popular machine-learning methods include support vector machines (SVM), random forests (RF), and *k*-nearest neighbour analysis. However, as complexity and the number of classes increase, for example in the form of number of weed species and their growth stages, the latter models often fall short (Kamilaris and Prenafeta-Boldú 2018). Because of the large variation within fields, non-parametric methods with computer-designed features, known as deep learning are becoming popular. In deep learning, the computer finds patterns in the raw pixel values across all the training images. Because of the availability of cheap computer power, these deep-learning models can easily contain millions of parameters, which enable them to detect complex features while disregarding the less important content of images.

Applications of machine learning and deep learning for weed detection are becoming common. Dyrmann (2017) demonstrate a weed instance detector, which can detect the location of weeds in cereal fields despite overlapping leaves. Lottes et al. (2016) also detected weeds with machine learning in sugar beet fields in order to remove weeds mechanically. Pantazi et al. (2016) have developed a method based on hyperspectral imaging and machine learning to recognize 10 weed species in maize. Crop recognition was 100% and the correct recognition or the rate of weeds varied between 53% and 94%. However, the hyperspectral fingerprint of plants changes markedly when plants are affected by microclimate and local soil conditions, or when a species changes morphological characteristics at different growth stages as shown in Fig. 10.2. Dyrmann (2017) used convolutional neural networks to classify 18 weed species in images collected using consumer-based cameras. Accuracy of the weed classification averaged 88%.

Variation within a weed species presents a major challenge. Plants respond to environmental factors such as wind, light and nutrition, which can have an effect on the visual appearance. Some species also change significantly during their early growth stages, making them more difficult to recognize. Nevertheless, an automated system should be able to deal with the fact that many weed species look very alike during early growth stages, but may look very different across several growth stages (Fig. 10.2).

One problem in developing automated weed detection is that a vast amount of annotated training data have to be provided and that some weed species are very alike at the early growth stages and therefore difficult to annotate. Ideally, a weed classifier should be able to classify weeds at the earliest growth stages, but in a

Fig. 10.2 Examples of morphologically differences in *Veronica persica* (**a**) and *Tripleurospermum inodorum* (**b**) at the early growth stages

practical context this may be impossible if the training data provided are missing or of poor quality. Therefore, to handle all weed species the classifier could be designed to classify weeds at a higher taxonomic level, for example as dicots or monocots.

10.10 Weed Detection Accuracy

The performance of weed detection algorithms is often expressed as accuracy. Weed detection can have four possible outcomes, of which only two are accurate: true positive (TP) and true negative (TN) (Table 10.1). Based on the confusion matrix below (Table 10.1), the true detection rate (TDR), also known as the sensitivity, is TDR = TP/ (TP + FN), where FN is false negative, and the true negative rate (TNR), also known as the specificity, is TNR = TN/ (TN + FP), where FP is false positive. The accuracy (ACC) is the sum of TP and TN divided by the total detection outcomes.

When site-specific weed management is implemented, typically, the whole field is not sprayed, and there may be concerns about missed weeds (false negatives, FN). In an agricultural setting, the performance of weed detection is never 100%. In general, the greater the true detection (TP), the higher the FP. In other words, the greater is the true weed detection, the larger is the overestimate. These ratios need to be considered when designing algorithms desirable for implementation by farmers.

10.11 Case Study 1: From UAV Mission Planning to Patch Spraying of Cirsium arvense with Commercial Sprayers

10.11.1 Introduction

This case study concerns UAV mapping of *Cirsium arvense* (creeping thistle) in cereals and the potential use of *C. arvense* weed maps in site-specific weed management. The UAV mapping of *C. arvense* is relevant because *C. arvense* is a highly competitive weed (O'Sullivan et al., 1985) and it is distributed in persistent patches (Hamouz et al., 2014) (Fig. 10.3). *Cirsium arvense* is easy to detect in pre-harvest cereals because it remains green throughout the growing season and has a distinct green colour in senescent cereals (a distinct spectral signature).

Table 10.1 A confusion matrix showing the performance of a weed detection algorithm

Detection outcome	Reality	Detection
True positive (TP)	Weeds	Weeds
True negative (TN)	No weeds	No weeds
False positive (FP)	No weeds	Weeds
False negative (FN)	Weeds	No weeds

Cirsium arvense is sprayed with glyphosate products before or after crop harvest (Darwent et al. 1994) and further controlled by mechanical methods shortly after crop harvest (Graglia et al. 2006). There are different restrictions on pre-harvest spraying with glyphosate products and in some countries, it is banned. The persistence of *C. arvense* allows pre-harvest weed maps to be used for spraying with auxin analogue herbicides the following year. The objective of the case study is to describe how UAV images can be used to generate weed maps used as the basis for selective on/off patch spraying with commercial sprayers on farmers' fields.

10.11.2 Methods

10.11.2.1 Platform and Sensor

A quadcopter UAV, Phantom 4 (DJI, Shenzhen, China), was used to acquire images of the field. The Phantom 4 is equipped with an integrated RGB camera with a Sony 12.4 megapixel 1/2.3" CMOS sensor (4000 × 3000 pixels) and 20 mm focal length (35 mm equivalent). It is a user-friendly off-the-shelf UAV, which can be used by farmers and consultants without professional knowledge of UAV technology. The weed detection and mapping procedures used in this case study do not depend on a specific UAV brand, but the sensor should be an RGB camera delivering focused

Fig. 10.3 A pre-harvest spring barley field with *Cirsium arvense* patches. Some are shown with red arrows

and well-exposed images. The UAV has a rechargeable lithium polymer (LiPo) battery and a maximum flight time of 25 min per battery. It is able to cover about 0.5 ha min^{-1} at 40 m altitude and a little more than double that at 80 m altitude.

10.11.2.2 Flight Planning and Image Acquisition

An autopilot was used to control the UAV flight including the acquisition of images (Fig. 10.4). It is possible to fly the UAV using a manual remote control, but a predetermined flight plan executed by an autopilot is preferable. It helps in obtaining correct frontal and side overlaps between images. Manual control may cause issues

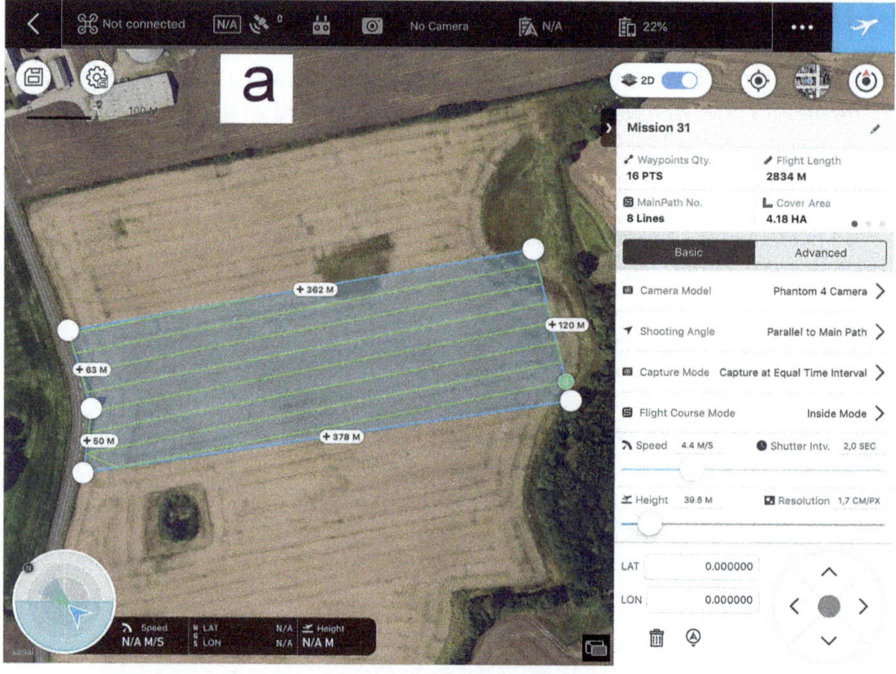

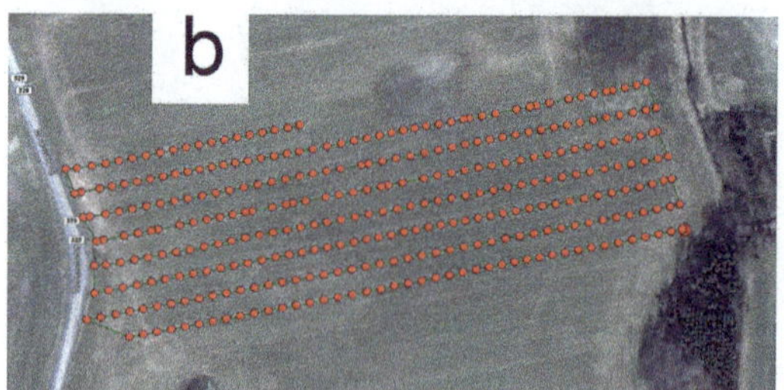

Fig. 10.4 The autopilot app (DJI Ground Station Pro) generates flight paths (**a**) and ensures that images are equidistant from each other (**b**). In (**b**), the red dots indicate the position of the captured images

for the subsequent image stitching process from irregular or limited overlap between images, which could create image gaps for the field. The autopilot app used was the DJI GS Pro (https://www.dji.com/dk/ground-station-pro). This software is free and developed to autopilot DJI UAVs. An Apple tablet (iPad) was used to run the app (Fig. 10.4). Other free downloadable apps are available for field mapping for both Android and Apple tablets (e.g. Pix4DCapture and DroneDeploy).

After wireless upload of the flight plan to the UAV, the autopilot controlled the mission. Information about the UAV's position, orientation and altitude is integrated in the autopilot system, which adjusts the course of the UAV according to the flight plan and controls the acquisition of images. Hence, the autopilot launches the UAV, controls the image acquisition and guides the UAV back to its starting position after the mission is accomplished. If the flight plan requires more than one battery (> 25 min duration), the autopilot flight mission is paused and reactivated after the battery is changed.

The flight altitude was 40 m and images were taken by the camera pointing to the nadir direction (Fig. 10.5). Image resolution was 17 mm. The frontal overlap was 80% and the lateral overlap was 70%. Although flight altitude does not generally affect the calculation of vegetation indices (VIs), it can greatly influence image segmentation, with more mixed pixels present in those images collected at higher altitudes (Rasmussen et al. 2013). For the detection of *C. arvense*, a spatial resolution of about 1.5–2 cm is suitable (Rasmussen et al. 2019).

The acquired images were stored on a memory card in the camera and downloaded after the flight. Images were stored in the JPG format with embedded GNSS-coordinates and UAV orientation (attitude) data (EXIF data). These data were used to assist with image stitching to determine the image centre position and camera orientation estimates (Smith et al. 2016). The UAV's built-in GNSS provided information on the positional, and the IMU (inertial measurement unit) and UAV compass provided the orientation data (yaw, pitch and roll).

Fig. 10.5 Image from the UAV (**a**) and a magnified view of a small part (**b**) (marked with a red rectangle in **a**) to show details

10.11.2.3 Image Processing: Stitching and Orthorectification

Prior to image analysis and weed detection, images went through a stitching process in which they were recombined into an orthoimage – also known as an orthomosaic. An orthoimage is an aerial image geometrically corrected (orthorectified) such that the image has the same lack of distortion as a map. There is a variety of software available for the photogrammetric processing of images into orthoimages, but in this case Pix4Dmapper (Pix4D, Lausanne, Switzerland) was used.

Two factors are important for achieving high global spatial accuracy of the orthoimage: a high overlap between images and accurate global coordinates giving the positions of a number of visible features in the images. Alternatively, UAVs can be equipped with RTK-GNSS, but this requires more costly UAVs than were used in this case study. Considerable overlap between images assists the photogrammetric software in identifying common points between image pairs, which is an important component in the stitching process. Such overlap also helps to minimize bidirectional reflectance effects by allowing the image processing software to extract only the central portion of each image for the image mosaic (Rasmussen et al. 2016). Strong global spatial accuracy may be achieved by using ground control points (GCPs) and their coordinates in the photogrammetric processing to position the map accurately in relation to the real world (Gómez-Candón et al. 2011; Azim et al. 2019). The GCPs are visible targets in the field (artificial or natural) with accurate positions. In this case study natural targets (stones, field constructions, road intersections and road markings) were used. The locations of 10 points were also measured in the field using a Trimble R10 GNSS receiver with TSC3 controller (Trimble, California, USA) to give the exact positions. The positioning system receives correction signals from GPSnet.dk (Geoteam, Ballerup, Denmark), a Danish network of ground-based reference stations based on Trimble's VRS technology (Trimble, California, USA). Orthoimages are in the GeoTiff format including geographical information (Fig. 10.6). The spatial accuracy of the orthoimage is within 10 cm (Azim et al. 2019).

10.11.2.4 Weed Detection

The Thistle Tool program was used to detect *C. arvense* (Rasmussen et al. 2019). The image analysis procedure in Thistle Tool is based on spectral analysis and requires *a priori* knowledge of the field vegetation because weed detection is based exclusively on colour discrimination. If other green weed species were present, these would also be detected. The Thistle Tool works in mature cereals and in late ripening growth stages when the crop colour becomes yellow.

The Thistle Tool, programmed in MATLAB (MathWorks Inc., MA, USA), processes orthoimages. It takes about 2 min to process a 1 GB orthoimage into a binary image showing weeds and cereal patches. Thistle Tool works on 1 m^2 grid squares called detection patches. A weed patch is defined as a detection patch with green plants covering at least 5% of the detection patch, whereas detection patches classified as cereals only include senesced cereals (Fig. 10.6).

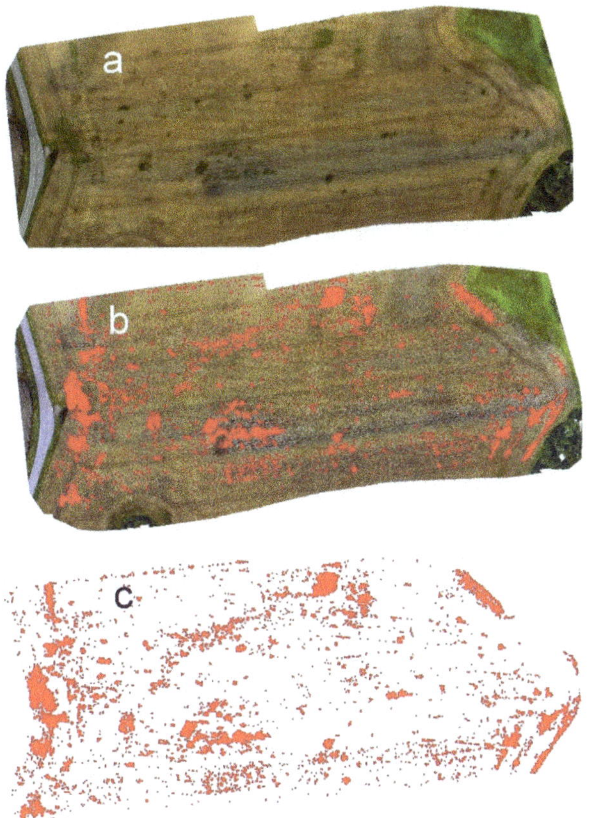

Fig. 10.6 Orthoimage (**a**) of the study area shown in Fig. 10.4, with detected *Cirsium arvense* on top of the orthoimage (**b**) and the prescription map (**c**). The total study area was 5.52 ha and the area with detected weed patches was 0.75 ha

The first step in Thistle Tool divides images into detection patches of 1 m^2 irrespective of flight altitude. In the second step, the greenness of each pixel is converted into a normalized scalar value using the excess green projection (ExG) (Rasmussen et al. 2007). The normalized ExG is calculated as follows:

$$ExG_{x,y} = \frac{2 \cdot G_{x,y} - R_{x,y} - B_{x,y}}{G + R + B} \tag{10.1}$$

where $ExG_{x,y}$ is the excessive green index for each pixel co-ordinate (x, y), and $G_{x,y}$, $R_{x,y}$ and $B_{x,y}$ are the green, red and blue intensities (0–255) for each pixel co-ordinate (x, y), respectively, and G, R and B are the average green, red and blue digital numbers (0–255) for the whole orthoimage. In the third step, a classifier called TopMax ExG is calculated. This classifier is the mean value of the p % highest $ExG_{x,y}$ values in each detection patch. If $p = 5\%$, then the largest 5% of $ExG_{x,y}$ pixel values in each detection patch are used in weed detection. In the fourth step, thresholding is performed to detect weeds. Weeds have large TopMax ExG values and senesced plants have small values. Thistle Tool allows the user to determine the segmentation threshold in two different ways: manual adjustment based on visual threshold editing, and threshold estimation based on annotated training data. In this case, thresholding was based on manual adjustment and $p = 5\%$.

In the threshold-editing mode, the threshold is adjusted interactively. If weed patches are misclassified as cereals by the default threshold, the threshold is adjusted. Thistle Tool allows the user to zoom in and out to show just a few or several patches simultaneously in the thresholding process. The threshold is adjusted until the user decides to perform the segmentation. According to Rasmussen et al. (2019), Thistle Tool classified 92–97% of the patches correctly under varying environmental conditions.

10.11.2.5 Patch Spraying

In this *C. arvense* example, the relation between the binary weed detection map and prescription map was as simple as it could be. If a detection patch was classified as a weed patch, it should be sprayed. A commercial sprayer with automatic section control was used. Such sprayers automatically turn off boom sections as they pass over previously treated areas (Larson et al. 2016). They can also target herbicide application within pre-loaded field boundaries based on RTK-GNSS positioning. This map-based function ensures that boom sections are automatically turned off when passing over areas outside the field. It also allows maps with interior patches within the field that are not to be sprayed to be loaded.

On/off patch spraying maps (prescription maps) were prepared in the GIS program QGIS Desktop 2.18.20 (https://www.qgis.org/en/site/) (Fig. 10.6). The maps consisted of field polygons indicating weed patches in a shape file format. Some sprayer consoles read shape files, others read ISOXML, and still others read both formats. Conversion between formats and the file adaptation to specific sprayer consoles require in-depth knowledge of file compatibility. The prescription map was used after crop harvest. When a boom section entered a weed polygon, the boom section automatically turned on spraying, and then turned off again when it left the polygon.

The weed patches represented 14% of the total field area (Fig. 10.6). However, this does not necessarily result in an 86% reduction in herbicide use. The reduction in herbicide depends strongly on the sprayer boom configuration and the sizes and geometry of the weed patches. If the boom section is divided into three-metre sections controlled independently, areas outside but connected to the weed patches will also be sprayed because of different resolutions of the weed detection and the sprayer. A recent study showed that if the resolution of the weed patches is 1 m^2 and the resolution of sprayer boom is 3 m^2, the sprayed area outside the weed patches would be about the same as the weed patches (Rasmussen et al. 2020). This means that a herbicide saving of 72% is reasonable if the prescription map shown in Fig. 10.6 and Fig. 10.7 is used by a sprayer with 3 m boom sections.

10.11.3 Discussion and Conclusions

This case study shows that UAV imagery can be used to detect and map *C. arvense* in pre-harvest cereals and that the maps can be used for on/off patch spraying with commercial sprayers after crop harvest. The concept presented of site-specific weed management has been verified in farm studies. All the steps from UAV mission planning to on/off patch spraying have been documented (Azim et al. 2019; Rasmussen et al. 2019, 2020). Research shows that there is still scope for improvement in weed detection due to the current inability to distinguish between weed species. However, currently, a lack of scientific knowledge or technical skills is not considered to be a major barrier to the adoption of UAV imagery in site-specific weed management. Not even economic considerations seem to hinder the adoption of UAV imagery in site-specific control of *C. arvense*. With the current technology, a rough estimate of the time spent on UAV image acquisition and image processing is about 3–5 min ha^{-1}, and small consumer-friendly UAVs with an integrated camera are affordable (1000–2000 €). However, it is important that all steps in the workflow from UAV mission planning to on/off patch application of herbicides are smooth and user-friendly.

10.12 Case Study 2: Use of Machine Learning to Identify Grass for Site-Specific Weed Management in Winter Cereals During Patch Spraying of Four Fields

10.12.1 Introduction

Previous studies have shown that if the within-field weed populations are known when selecting herbicides and dosages, herbicide application can be reduced by 20–40%. A system called 'IPMwise', which determines the optimal herbicide dosages based on weed populations (and its predecessor system Plant Protection Online), has undergone field validation trials in Denmark, Norway, Germany, Spain and Latvia. Typically, one common herbicide mixture and dose was recommended for each field and all requirements for agronomic robustness were examined (Sønderskov et al. 2014; Sønderskov et al. 2015; Montull et al. 2014; Vanaga and Zarina 2008). Thus, by performing blanket spraying with the right herbicide mixture and dose for the current weed population within the field, any sprayer, irrespective of its level of technology, may obtain 20–40% savings when using cost-effective image acquisition and automated weed identification (Somerville et al. 2019). Christensen et al. (2009) named this spatial treatment resolution as precision level 4 out of 4, where the whole field is sprayed uniformly. However, further herbicide reductions will require spatial on/off nozzle control and/or variable herbicide mixtures and dosages, moving to what Christensen et al. (2009) characterizes as resolution level 3, where weed patches or subfields with clusters of weed plants are treated individually. In this way, Timmermann et al. (2003) shows that patch spraying in cereal fields can result in 90% savings on herbicides for grass weeds and 60% savings on herbicides for broadleaf weeds.

The aim of this case study was to use machine learning to automate the identification of grass weeds within RGB images to apply the map-based approach and thereby obtain significant reductions in herbicide use.

The case study demonstrates all stages from image capture in the field to calculating herbicide rates and spraying the field. The main emphasis, however, will be on image acquisition, detection, classification principles and algorithms, followed by mapping and herbicide calculations.

10.12.2 Methods

10.12.2.1 Sensing

On 1 March 2019, 7242 images were recorded by a flash-supported machine vision camera mounted on an ATV as described in Laursen et al. (2017). The location of each 0.4 m^2 image was recorded with RTK-GNSS. A minimum sampling density of 7 m × 5 m (~285 images per hectare) within four winter rye fields with a total area of ~15 hectares was obtained while driving the ATV at an average velocity of 11.5 m s^{-1}. Unfortunately, the camera was a little out of focus due to a change in height after moving from one ATV to another, which resulted in less sharp images than usual. However, the quality was still good enough for the weed detection algorithm used to frame the different weed objects and classify them as monocot weeds or dicot weeds (see the next section). The images were uploaded to a cloud service hosted by I•GIS (www.i-gis.dk) as part of an Innovation Fund Denmark funded project called RoboWeedMaPs (www.roboweedmaps.com). This service was also used for image processing and mapping.

10.12.3 Image Processing

The images were processed by a fully convolutional neural network, which is trained to detect individual weed instances in cereal fields. Since monocots and dicots were to be treated differently, the detector was trained to distinguish monocot weeds from dicot weeds. In addition to providing the basis for herbicide selection, distinguishing monocots from dicots is beneficial from an image-processing point of view when operating in cereal fields because monocot weeds share many common features with cereals. Therefore, the detector already had to include features that could identify monocot weeds (as separate from the cereal crop). Dicot weeds are easier to distinguish from cereals, and by separating them from monocots earlier in the pipeline, it was possible to use other properties to detect the monocot weeds.

Optimally, the detector should also handle the finer classification of the weed species, but this is not practically feasible because all training images have to be fully annotated.

Training of the networks was supervized, which means that the correct labelling of the training data, i.e. the locations of the weeds in the images, had to be completed beforehand. The reason behind training the neural network to detect individual weed instances is that the system for determining the choice of optimal herbicide dosages is based on the densities of weeds in terms of plants per square meter. Therefore, individual weeds need to be detected to assess weed density.

A challenge when training the weed detector is that the training phase requires a large number of training images, ideally several thousands of them, which have to be annotated visually. To make the system robust, these images should cover the differences between fields. Such differences include weed species and growth stages, soil types, leaf occlusion, and changes to water, temperature and nutrient levels. Since the training images must be annotated to train the computer, RGB images are a natural choice in this case because they are easy to interpret visually.

Annotation of these training images requires visual marking of all plants in the images because automated detection of non-annotated plants would otherwise count as errors, which would decrease the overall accuracy of the detector.

The architecture of the weed-detecting neural network is based on the RetinaNet architecture by Lin et al. (2017). This detector's architecture consists of a classical ResNet backbone that gradually creates high-level semantic features. However, since the network gradually downscales the image to learn the concept of 'weeds', it loses the small details and thus the small weeds. Therefore, the backbone network is connected layer by layer to a feature pyramid that helps the network to detect weeds of various sizes. Finally, a bounding box regression and classification network were used to determine where in the image each monocot and dicot weed was located (Fig. 10.7).

The neural network for detecting each monocotyledon and dicotyledon weed was trained using 7975 images from cereal fields collected across several autumn and spring seasons, soil types and field treatments such as ploughing, low tillage or no tillage. In all, the training images contained 169,336 annotated weed objects.

10.12.3.1 Mapping

The mapping of the grass weeds was performed using 5 m rasterisation of an inverse distance interpolation within QGIS Desktop 3.2.0-Bonn (https://www.qgis.org/en/site/) (Fig. 10.8). Pixels outside the field boundary were removed.

Fig. 10.7 Close-up of monocot weeds (purple) and dicot weeds (blue) identified by the trained model

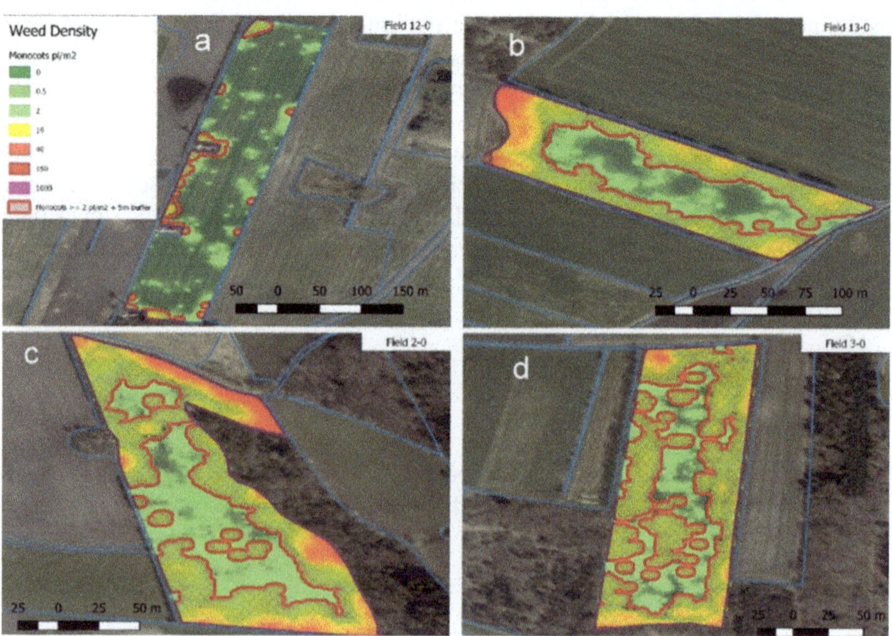

Fig. 10.8 Interpolated grass weed densities in four different winter rye fields: (**a**) 6.7 ha; (**b**) 2.3 ha; (**c**) 2.8 ha; and (**d**): 3.1 ha giving a total of 14.9 ha. The red lines indicate the border of patches that exceed two grass weeds per m^2 plus a 5 m buffer extending the area to spray (see next section and Table 10.1 for details)

10.12.3.2 Decisions

With support from the farmer and IPM Consult (dk.ipmwise.com), the grass species were determined by studying the images, and the optimal herbicide and dosage were then decided with support from the IPMwise DSS listing possible herbicide mixtures. Furthermore, the weed density threshold for turning a boom section on and off was decided after a discussion with the farmer who had a close to zero acceptance to grass weeds because he had a grass seed production year in his crop rotation.

Without knowing the current weed composition, the farmer's consultant advised that the field should be sprayed with a 0.6 l ha^{-1} of Cossack OD based on their common historical knowledge about the expected weed composition of the field. (https://middeldatabasen.dk/product.asp?productID=51055). This herbicide was also in the listed possibilities of herbicide based on manually inspecting the collected weed images containing grass weeds with exception of the dose. Hence, IPM Consult advised that the grass patches be sprayed with a 0.5 l ha^{-1} dose of Cossack OD when the grass weed density exceeded two grass plants per square metre. To minimize the likelihood of missing sparse grass patches, a 5 m buffer zone was added to the boundary lines enclosing the detected grass weed patches.

10.12.3.3 Prescription Maps and Spraying

The prescription maps were generated by Agrinavia (http://agrinavia.com) ensuring that the spray controller adapted shape files containing the grass weed polygons illustrated in Fig. 10.9 by the outer red lines. In this case, the shape file was created for an AgLeader InCommand 1200 sprayer controller.

Patch spraying of the four fields was performed using a Danfoil AirBoss 24 m sprayer with seven boom sections that can be turned on and off independently by the AgLeader InCommand 1200 sprayer controller equipped with an RTK-GNSS receiver. Each of the four outermost sections was three metres wide, while the three innermost sections were four metres wide.

In all, 5.82 ha were to be patch sprayed with 0.5 l ha^{-1} Cossack OD (see Table 10.2 in the next section). The farmer was informed of this because he had a tank mix sprayer and therefore needed to know the dosage beforehand. This dosage was a 61% planned reduction in the area to be sprayed compared to the consultant's advice. However, to be sure of having enough herbicide, the farmer mixed a tank volume covering 50% of the total field area of 14.94 ha, resulting in an actual herbicide reduction of 58% compared to the consultant's suggested blanket spraying. To use up all of the mixed herbicide in the tank, the farmer was forced to spray an additional area in field 13–0 (Table 10.2). The logfiles from the sprayer were used to create the application maps to evaluate the precision of the patch- spraying operation.

A study of the patches to be sprayed relative to the areas that were actually sprayed according to the log file in Fig. 10.9 showed that the sprayer missed

Fig. 10.9 Prescription map of Field 3–0 including logged areas sprayed. The red inner contour line filled with red squares are the grass patches with two or more grass weed per hectare to be sprayed without buffer zones. The red contour lines which include the latter areas plus a 5 m buffer zone exceed the patches to be sprayed with less than two grass weeds per hectare. The blue colour is the areas actually sprayed according the sprayer log file. See Table 10.1 for details

Table 10.2 Case study overview with a focus on potential and actual herbicide savings with and without 5 m buffer zones. The herbicide savings calculated as the difference between patch spraying with a dose of 0.5 l ha^{-1} and blanket spraying with a dose of 0.6 l ha^{-1}. The patch-spraying dose was based on integrated pest management (IPM) principles. The blanket-spraying dose was based on the advice of a consultant

				Monocots > = 2					
				No buffer		5 m buffer			
Field	Field area	Images / ha	Sample ratio area (%)	Spray area	Savings (%)	Planned spray area	Savings (%)	Logged spray area	Savings (%)
12–0	6.66	281	1.1	0.19	98	0.48	94	0.56	93
13–0	2.33	338	1.4	0.94	66	1.40	50	–	–
2–0	2.83	566	2.3	1.19	65	1.88	45	1.99	41
3–0	3.12	955	3.8	0.93	75	2.06	45	2.17	42
Overall	14.94	485	1.9	3.25	82	5.82	68	–	–

boom-wide strips in the top right, bottom left and bottom right corners of the field. These areas also had relatively high grass densities, indicating that this may be a reoccurring sprayer error that was also present previously when the field was sprayed. This is because the sprayer controller will not open the nozzles until a minimum velocity is reached to avoid herbicide application hotspots.

10.12.4 Discussion and Conclusions

The actual herbicide saving obtained was 58%, but it could have been a lot closer to 68% if the farmer had had an injection-based sprayer system that does not need to know the exact area to spray in advance.

Additional savings could be obtained by reducing the width of the buffer, as shown in Table 10.2. In addition, the dose may be adjusted in accordance with the grass counts per square metre, as undertaken by Timmermann et al. (2003), thereby achieving 90% herbicide savings in grass weeds. However, the limited abilities of the current sprayer controller and sprayer systems complicate the use of variable herbicide dosing. To compensate for this, Somerville et al. (2019) suggest a possible method to overcome this problem, but this requires the locations of the tramlines to be known in advance.

10.13 Conclusions for the Chapter

The two case studies show that it is technically possible to detect and patch spray a perennial weed, *C. arvense*, pre- or post-harvest based on UAV mapping with RGB cameras and to patch spray monocot grass weeds in cereals based on an RGB camera mounted on a ground vehicle (ATV). In the first case the whole study area was covered by images that were stitched together into an orthoimage and in the second case, the map was created after interpolation based on images located 5–7 m apart. In the first case, the weed detection was exclusively based on colour segmentation and in the second case, an automated neural network was used. Both cases indicate that SSWM may be practical and feasible in an agronomic context, but adoption barriers have not yet been studied.

References

Adão T, Hruška J, Pádua J et al (2017) Hyperspectral imaging: a review on UAV-based sensors, data processing and applications for agriculture and forestry. Remote Sens 9(11):1110

Andújar D, Weis M, Gerhards R (2002) An ultrasonic system for weed detection in cereal crops. Sensors 12:17343–17357

Andújar D, Escolà A, Dorado J et al (2011) Weed discrimination using ultrasonic sensors. Weed Res 51:543–547

Andújar D, Ribeiro A, Fernández-Quintanilla C et al (2013) Herbicide savings and economic benefits of several strategies to control Sorghum halepense in maize crops. Crop Prot 50:17–23

Azim S, Rasmussen J, Nielsen J et al (2019) Manual geo-rectification to improve the spatial accuracy of ortho-mosaics based on images from consumer-grade unmanned aerial vehicles (UAVs). Precis Agric 20:1199–1210

Backes M, Jacobi J (2006) Classification of weed patches in QuickBird images: verification by ground truth data. In: EARSel (European Association of Remote Sensing Laboratories)

eProceedings, EARSel, Warsaw, Poland. Available at: http://www.eproceedings.org/static/vol05_2/05_2_backes1.html
Barroso J, McCallum J, Long D (2017) Optical sensing of weed infestations at harvest. Sensors 17(10):2381
Blackshaw RE, Molnar LJ, Lindwall CW (1998) Merits of a weed-sensing sprayer to control weeds in conservation fallow and cropping systems. Weed Sci 46:120–126
Blaschke T (2010) Object based image analysis for remote sensing. ISPRS J Photogramm Remote Sens 65:2–16
Castillejo-González IL, Pena-Barragán JM, Jurado-Expósito M et al (2014) Evaluation of pixel- and object-based approaches for mapping wild oat (*Avena sterilis*) weed patches in wheat fields using Quick Bird imagery for site-specific management. Eur J Agron 5:57–66
Christensen S, Heisel T, Walter A et al (2003) A decision algorithm for patch spraying. Weed Res 43:276–284
Christensen S, Søgaard HT, Kudsk P et al (2009) Site specific weed control technologies. Weed Res 49:233–241
Darwent AL, Kirkland KJ, Baig MN et al (1994) Preharvest applications of glyphosate for Canada thistle (*Cirsium arvense*) control. Weed Technol 8:477–482
De Castro AI, López-Granados F, Jurado-Expósito M (2013) Broad-scale cruciferous weed patch classification in winter wheat using QuickBird imagery for in-season site-specific control. Precis Agric 14:392–413
Dyrmann M (2017) Automatic detection and classification of weed seedlings under natural light conditions, Thesis PhD, University of Southern Denmark
Ehlert D, Dammer K-H (2006) Widescale testing of the crop-meter for site-specific farming. Precis Agric 7:101–115
Fernández-Quintanilla C, Peña JM, Andújar D et al (2018) Is the current state of art of weed monitoring suitable for site-specific weed management in arable crops? Weed Res 58:259–272
Gerhards R (2010) Spatial and temporal dynamics of weed populations. In: Oerke EC, Gerhards R, Menz G, Sikora R (eds) Precision crop protection – the challenge and use of heterogeneity. Springer, Dordrecht
Gómez-Candón D, López-Granados F, Caballero-Novella JJ et al (2011) Geo-referencing remote images for precision agriculture using artificial terrestrial targets. Precis Agric 12:876–891
Graglia E, Melander B, Jensen RK (2006) Mechanical and cultural strategies to control *Cirsium arvense* in organic arable cropping systems. Weed Res 46:304–312
Grunwald LC, Belyaev VI, Meinel T (2020) Improving efficiency of crop protection measures. A technical contribution for better weed control, less pesticide use and decreasing soil tillage intensity in dry farming regions exposed to wind erosion. In: Frühauf M, Guggenberger G, Meinel T, Theesfeld I, Lentz S (eds) KULUNDA: climate smart agriculture, Innovations in landscape research. Springer, Cham
Guttjahr C, Sokefeld M, Gerhards R (2012) Evaluation of two patch spraying systems in winter wheat and maize. Weed Res 52:510–519
Hamouz P, Novakova K, Soukup J et al (2008) Detection of *Cirsium arvense* L. in winter wheat using a multispectral imaging system. J Plant Dis Prot 21:167–170
Hamouz P, Hamouzová K, Holec J, Tyšer L (2014) Effect of site-specific weed management in winter crops on yield and weed populations. Plant Soil Environ 60:518–524
Heege HJ, Reusch S, Thiessen E (2008) Prospects and results for optical systems for site-specific on-the-go control of nitrogen-top-dressing in Germany. Precis Agric 9:115–131
Hilton PJ (2020) Laser-induced fluorescence for discrimination of crops and weeds. Proc SPIE 4124:223–231
Hu JB, Dai MX, Peng ST (2018) An automated (novel) algorithm for estimating green vegetation cover fraction from digital image: UIP-MGMEP. Environ Monit Assess 190:687
Kamilaris A, Prenafeta-Boldú FX (2018) Deep learning in agriculture: a survey. Comput Electron Agric 147:70–90

Kazmi W, Foix S, Alenyà G et al (2014) Indoor and outdoor depth imaging of leaves with time-of-flight and stereo vision sensors: analysis and comparison ISPRS. J Photogramm Remote Sens 88:128–146
Larson JA, Velandia MM, Buschermohle MJ et al (2016) Effect of field geometry on profitability of automatic section control for chemical application equipment. Precis Agric 17(1):18–35
Laursen MS, Jørgensen RN, Dyrmann M et al (2017) RoboWeedSupport-sub millimeter weed image acquisition in cereal crops with speeds up till 50 km/h. In 19th international conference on precision agriculture. Kyoto, Japan, pp. 27–28
Lin TY, Goyal P, Girshick R et al (2017) Focal loss for dense object detection. In: IEEE International Conference on Computer Vision (ICCV), pp 2980–2988
López-Granados F (2011) Weed detection for site-specific weed management: mapping and real time approaches. Weed Res 51:1–11
López-Granados F, Pena-Barragan JM, Jurado-Exposito M et al (2008) Multispectral classification of grass weeds and wheat (Triticum durum) using linear and nonparametric functional discriminant analysis and neural networks. Weed Res 48:28–37
López-Granados F, Torres-Sánchez J, De Castro AI et al (2016) Object-based early monitoring of a grass weed in a grass crop using high resolution UAV imagery. Agron Sustain Dev 36(4):36–67
Lottes P, Horferlin M, Sander S et al (2016) Effective vision-based classification for separating sugar beets and weeds for precision farming. J Field Rob 34(6):1160–1117
Montull JM, Sønderskov M, Rydahl P et al (2014) Four years validation of decision support optimising herbicide dose in cereals under Spanish conditions. Crop Prot 64:110–114
O'Sullivan PA, Weiss GM, Kossatz VC (1985) Indices of competition for estimating rapeseed yield loss due to Canada thistle. Can J Plant Sci 65:145–149
Oerke EC (2006) Crop losses to pests. J Agric Sci 144:31–43
Olsen SI, Nielsen J, Rasmussen J (2017) Thistle detection. In: Sharma P, Bianchi FM (eds) Scandinavian conference on image analysis 2017, Tromsø, Norway, Part II, Lecture notes in computer science. Springer, Basel
Pantazi X, Moshou D, Bravo C (2016) Active learning system for weed speciesrecognition based on hyperspectral sensing. Biosyst Eng 146:193–202
Peña JM, Torres-Sánchez J, de Castro AI et al (2013) Weed mapping in early-season maize fields using object-based analysis of unmanned aerial vehicle (UAV) images. PLoS One 8(10):e77151
Peña JM, Torres-Sánchez J, Serrano-Pérez A et al (2015) Quantifying efficacy and limits of unmanned aerial vehicle (UAV) technology for weed seedling detection as affected by sensor resolution. Sensors 15:5609–5626
Pérez-Ortiz M, Peña JM, Gutiérrez PA et al (2015) A semi-supervised system for weed mapping in sunflower crops using unmanned aerial vehicles and a crop row detection method. Appl Soft Comput 37:533–544
Peteinatos G, Weis M, Ujar D (2014) Potential use of ground-based sensor technologies for weed detection. Pest Manag Sci 70:190–199
Rasmussen J, Nørremark M, Bibby BM (2007) Assessment of leaf cover and crop soil cover in weed harrowing research using digital images. Weed Res 47:299–310
Rasmussen J, Nielsen J, Garcia-Ruiz F et al (2013) Potential uses of small unmanned aircraft systems (UAS) in weed research. Weed Res 53:242–248
Rasmussen J, Ntakos G, Nielsen J, Svensgaard J et al (2016) Are vegetation indices derived from consumer-grade cameras mounted on UAVs sufficiently reliable for assessing experimental plots? Eur J Agron 74:75–92
Rasmussen J, Nielsen J, Streibig JC et al (2019) Pre-harvest weed mapping of *Cirsium* arvense in wheat and barley with off-the-shelf UAVs. Precis Agric 20:983–999
Rasmussen J, Azim S, Nielsen J et al (2020) A new method to estimate the spatial correlation between planned and actual patch spraying of herbicides. Precis Agric 21:713–728
Rew LJ, Cousens RD (2001) Spatial distribution of weeds in arable crops: are current sampling and analytical methods appropriate? Weed Res 41:1–18

Smith MW, Carrivick JL, Quincey DJ (2016) Structure from motion photogrammetry in physical geography. Prog Phys Geogr 40:247–275
Somerville GJ, Joergensen R, Boyer OM et al (2019) Marrying futuristic weed mapping with current herbicide sprayer capacities. In: Stafford J (ed) The 12th European conference on precision agriculture, pp 231–237
Sønderskov M, Kudsk P, Mathiassen S et al (2014) Decision support system for optimized herbicide dose in spring barley. Weed Technol 28(1):19–27
Sønderskov M, Fritzsche R, de Mol F et al (2015) DSSHerbicide: weed control in winter wheat with a decision support system in three South Baltic regions – field experimental results. Crop Prot 76:15–23
Sui R, Thomasson JA, Hanks J, Wooten J (2008) Ground-based sensing system for weed mapping in cotton. Comput Electron Agric 60:31–38
Thorp KR, Tian LF (2004) A review on remote sensing of weeds in agriculture. Precis Agric 5:477–508
Timmermann C, Gerhards R, Kühbauch W (2003) The economic impact of site-specific weed control. Precis Agric 4(3):249–260
Torres-Sánchez J, López-Granados F, De Castro AI et al (2013) Configuration and specifications of an unmanned aerial vehicle (UAV) for early site specific weed management. PLoS One 8(3):e58210
Torres-Sánchez J, López-Granados F, Peña JM (2015) An automatic object-based method for optimal thresholding in UAV images: application for vegetation detection in herbaceous crops. Comput Electron Agric 114:43–52
Trengove S (2016) Weed mapping. Dev Demos 12(2):20–22
Vanaga I, Zarina L (2008) Evaluating herbicides in a range of doses for integrated plant protection in Latvia. Zemdirbyste-Agric 95:22
Wiles LJ, Oliver GW, York AC et al (1992) Spatial distribution of broadleaf weeds in North Carolina soybean (*Glycine max*) fields. Weed Sci 40:554–557

Chapter 11
Applications of Sensing to Precision Irrigation

Yafit Cohen, George Vellidis, Carlos Campillo, Vasileios Liakos, Nitsan Graff, Yehoshua Saranga, John L. Snider, Jaume Casadesús, Sandra Millán, and Maria del Henar Prieto

Abstract Precision irrigation aims at improving productivity and sustainability by addressing spatial as well as temporal variability of soil and crop water status. This chapter presents three case studies from south-eastern USA, Israel and Spain which relate to different attributes of precision irrigation: (1) type of precision: whether it

---

Yafit Cohen and Nitsan Graff: Introduction and Case Study 11.2
George Vellidis and Vasileios Liakos: Case Study 11.1
Yehoshua Saranga and John L. Snider: Case Study 11.2
Carlos Campillo, Jaume Casadesús, Sandra Millán and Maria del Henar Prieto: Case Study 11.3

Y. Cohen (✉)
Agricultural Engineering Institute, Agricultural Research Organization, Volcani Institute, Rishon LeZion, Israel
e-mail: yafitush@volcani.agri.gov.il

G. Vellidis · V. Liakos · J. L. Snider
Crop and Soil Sciences Department, University of Georgia, Athens, GA, USA
e-mail: yiorgos@uga.edu

C. Campillo · S. Millán · M. d. H. Prieto
Centro de Investigaciones Científicas y Tecnológicas de Extremadura (CICYTEX), Finca La Orden, Junta de Extremadura. Guadajira, Badajoz, Spain
e-mail: carlos.campillo@juntaex.es

N. Graff
Agricultural Engineering Institute, Agricultural Research Organization, Volcani Institute, Rishon LeZion, Israel

The Robert H. Smith Faculty of Agriculture, Food and Environment, The Hebrew University of Jerusalem, Rehovot, Israel

Y. Saranga
The Robert H. Smith Faculty of Agriculture, Food and Environment, The Hebrew University of Jerusalem, Rehovot, Israel

J. Casadesús
Institut de Recerca i Tecnologia Agroalimentàries (IRTA), Centre UdL IRTA, Lleida, Catalonia, Spain

R. Kerry, A. Escolà (eds.), *Sensing Approaches for Precision Agriculture*, Progress in Precision Agriculture, https://doi.org/10.1007/978-3-030-78431-7_11

targets the spatial variability by using variable-rate irrigation (VRI) system or the temporal variability by using automatic triggering of the irrigation, (2) type of irrigation system, and (3) type of data the system uses for irrigation decisions. Each of the case studies addresses a unique combination of attributes and together they draw a more complement picture of precision irrigation. All case studies have shown that well-timed data provides decision support for VRI management, i.e. soil moisture sensor-data (south-eastern USA and Spain) and thermal aerial imagery (Israel). The three case studies showed that precision irrigation treatments performed better than uniform irrigation. Yield increased by 4.3–12% and water-use efficiency (WUE) was improved by 14–40%. The reliance on point-sensor data in the case studies from south-eastern USA and Spain dictated the use of pre-determined irrigation management zones (IMZ), yet, enabled adaptive in season irrigation management. In season remotely sensed images can be further used for adaptive IMZ, i.e. to modify their boundaries, yet it currently suits VRI in drip irrigation. From these case studies, it can be seen that full VRI implementation, which adapts for spatial and temporal changes, faces "site-specific" challenges, i.e. every irrigation system is unique, and thus requires tailored solutions.

Keywords Variable-rate-irrigation · Automatic-irrigation · Soil-sensors · Thermal-imagery · Decision support system

11.1 Introduction

Irrigation accounts for more than 70% of total water withdrawals on a global basis (The World Bank; https://www.worldbank.org/en/topic/water-in-agriculture). The inevitable competition between agriculture and other users of limited water resources will require that farmers become more efficient at producing crops with a finite water supply. In addition, because irrigated agriculture provides about 40% of the global food supply on 20% of the total cultivated land, the pressure to produce even more food on irrigated land will also intensify as global population increases. Productivity per unit of water consumed would increase by implementation of crop location strategies (optimal soil and climate attributes), conversion to crops with higher economic value, and adoption of alternate drought-tolerant crops as well as by the implementation of precision irrigation (Evans and Sadler 2008). These high-technology tools will not only allow adjustments of the water schedule depending on the needs of the crops at each moment and the characteristics of the soil, but will also increase the farmer's expertise and be more attractive to new generations of young farmers. The term precision agriculture (PA) in general and precision irrigation in particular is perceived differently by stakeholders from research, extension, industry and by the farmers themselves. From the perspective of the research community, PA aims at improving productivity and sustainability by addressing spatial and temporal variation within fields. Automated irrigation scheduling is recognized as precision irrigation, and even though it does not address spatial variation, it addresses temporal variation. This chapter presents three case studies from south-eastern USA, Israel and Spain which relate to different attributes of precision irrigation: (1) type of

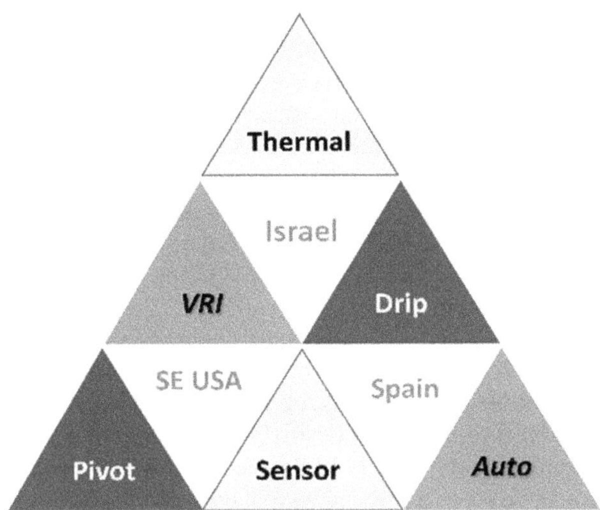

Fig. 11.1 Shared and complementary attributes of the case studies. White represents country; Outlined light grey represents type of input data; Dark grey represents the type of irrigation system and light grey (no outline) the type of precision irrigation

precision: whether it targets the spatial variation using a variable-rate irrigation (VRI) system or temporal variation by automatic triggering of the irrigation, (2) type of irrigation system and (3) type of data the system uses for irrigation decisions. Each of the case studies addresses unique combinations of attributes and together they draw a more complete picture of precision irrigation. Yet they are not totally separated from each other because they share some attributes which connect them (Fig. 11.1). The first case study from south-eastern USA describes VRI with centre pivots based on soil sensors, the second case study from Israel deals with variable-rate drip irrigation (VRDI) based mainly on thermal remote sensing, and the third case study from Spain presents an automatic drip irrigation system triggered by soil sensors. Figure 11.1 summarizes the shared and complementary attributes.

11.2 Case Study 11.1. Variable-Rate Irrigation with Centre Pivot in South-Eastern USA

11.2.1 Introduction

Precision irrigation has its roots in VRI technology developed for centre pivot irrigation systems by the University of Georgia (UGA) Precision Agriculture team in 2001 (Perry et al. 2002; Perry and Podcknee 2003). The UGA Precision Agriculture team recognized that variable-rate application of irrigation water was a key enabling technology for the adoption of PA in south-eastern USA. This was because fields in this region have very variable soil type and texture, moisture holding capacity and slope. If site-specific water needs are disregarded while attempting to vary other inputs such as fertilizers, then this would not result in the desired gains in efficiency theoretically possible with PA. An additional reason was that in the south-eastern USA, irrigation of agronomic crops such as cotton (*Gossypium* spp.), maize (*Zea mays*), peanut (*Arachis hypogaea*) and soya bean (*Glycine max*), is now done almost exclusively with centre pivot irrigation systems.

Conventional centre pivots apply the same rate of water along the entire length of the pivot and cannot account for within-field variability or non-farmed areas. The UGA VRI technology was commercialized by FarmScan, an Australian electronics company. The large pivot manufacturers began offering their own VRI systems once the original patent expired in 2012.

The VRI allows centre pivots to vary water application rates along the length of the pivot using electronic controls to cycle sprinklers and control pivot speed. Sprinklers are controlled individually or together typically in groups of 2–10 depending on the level of resolution desired by the farmer. Each group or bank of sprinklers represents a grid with a 1–10-degree arc in which the application rate of irrigation water can be set as a percentage of the normal application rate; for example from 0 to 200% of normal (Fig. 11.2c). The number of degrees in the arc is determined by the level of resolution desired. A 50% application rate is half the normal rate and is achieved by cycling the sprinklers on and off every 30 s. A 150% application rate is achieved by leaving the sprinklers on continuously while decreasing the travel speed of the pivot. If other grids along the length of the pivot require smaller rates of application, the VRI controller adjusts the sprinkler cycling pattern within those grids accordingly.

The VRI can be installed retroactively on most existing pivots. Installation costs vary widely by brand and are also a function of the length of the pivot and the level of resolution desired by the farmer to address the variability of the field. Application rates are determined from an application or prescription map. A short video describing VRI is available at (https://www.youtube.com/watch?v=DgexX_IToI0).

11.2.2 Prescription Maps

The prescription map for a field is developed jointly by the farmer and VRI dealer on desktop software (Fig. 11.2a, b, c) and then downloaded to the VRI controller on the pivot. The field is divided into irrigation management zones (IMZs) and application rates assigned to each of the IMZs using whatever information is available. In almost all VRI applications, the prescription maps are static. In other words, they are typically developed once and used thereafter. Static prescription maps do not respond to environmental variables such as weather patterns and other factors that affect soil moisture conditions and rates of crop growth. Although VRI offers a great leap forward in improving water-use efficiency (WUE), the system can be greatly enhanced with real-time information on crop water needs to control the application rates. One approach for creating dynamic prescription maps is to use soil moisture sensors to estimate the amount of irrigation water needed to return each IMZ to an

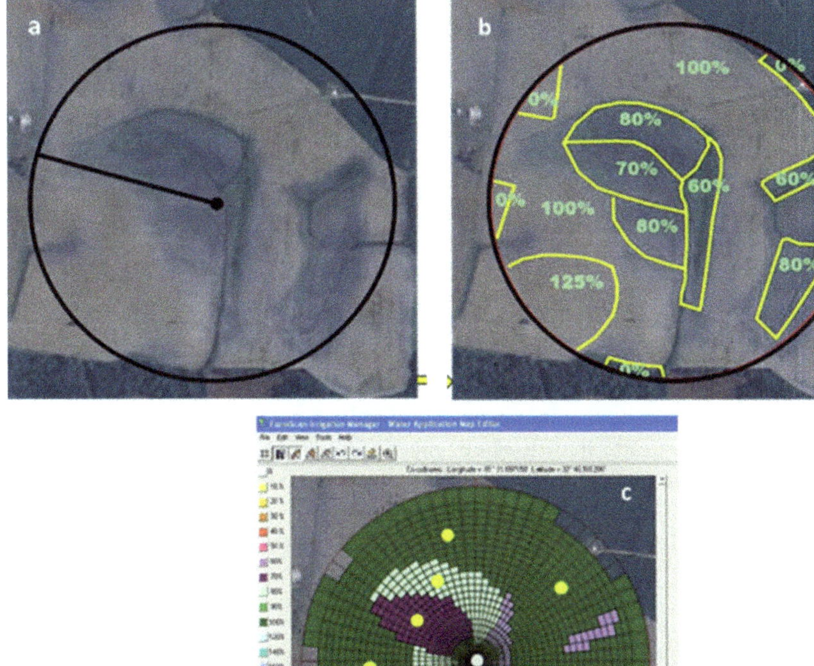

Fig. 11.2 The process of creating a VRI prescription map for a 49-ha field in Georgia. A bare-soil image is used as a base map to delineate IMZs (**a**). The percentages in (**b**) indicate the percent of the normal application rate assigned to each IMZ. Grid cells in (**c**) represent discreet areas that can receive unique application rates. Cells grouped into the same colour represent the IMZs delineated in (**b**). The yellow circles represent potential locations of soil moisture sensor nodes

ideal soil moisture condition (Fig. 11.2c). This case study describes the on-farm application of a *dynamic* VRI control system developed by the UGA Precision Agriculture Team. The control system couples real-time soil moisture sensing networks, an irrigation scheduling decision support system (DSS), and VRI.

11.2.3 A Dynamic VRI System

The operational paradigm of the UGA dynamic VRI control system is that the field is divided into IMZs and a soil moisture sensing network with a high density of sensor nodes is installed to monitor soil conditions within the IMZs. Between one and three nodes are installed in each IMZ depending on the zone's size. Soil moisture data are uploaded hourly to a web-based user interface.

At the web-based user interface, the soil moisture data are used by the DSS running in the background to develop irrigation scheduling recommendations for each IMZ. The recommendations are then approved by the user (farmer) and downloaded wirelessly to the VRI controller on the pivot as a precision irrigation prescription. When the pivot is engaged by the farmer, the pivot applies the recommended rates. Figure 11.3 shows the dashboard of the UGA dynamic VRI system.

11.2.3.1 Real-Time Soil Moisture Sensing Network

A key requirement of a soil moisture sensor based dynamic VRI system is a dense network of sensor nodes that is inexpensive, reliable, wireless, energy efficient, easy to install and remove, and that does not interfere with farming operations. These types of networks are only now becoming commercially available. The UGA smart sensor array (UGA SSA) soil moisture sensor network was specifically developed for the dynamic VRI system described here.

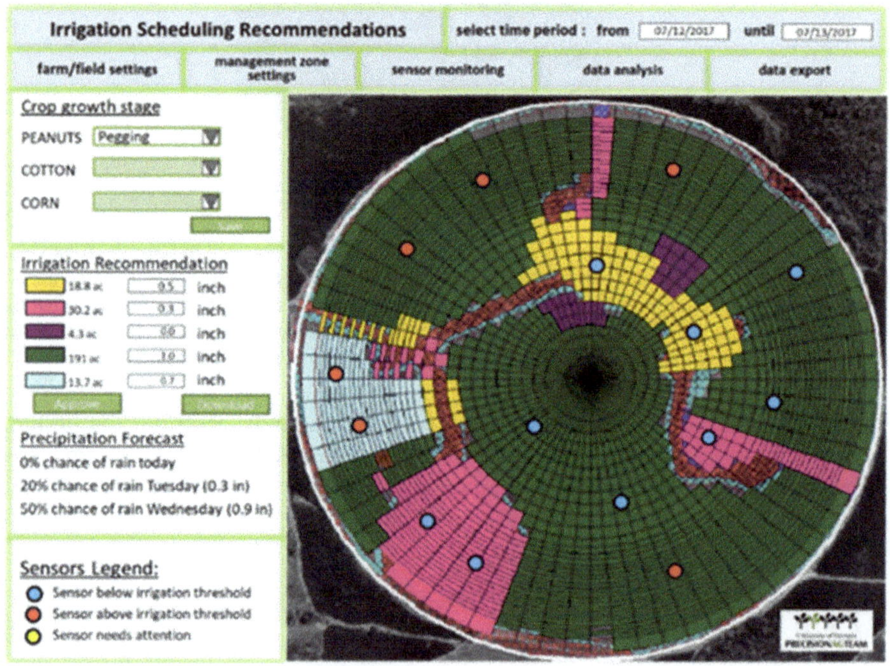

Fig. 11.3 Dashboard of the UGA dynamic VRI system as seen by the user. The system requires the user to approve and then download the daily prescription map to the VRI controller located on the pivot. Real-Time Soil Moisture Sensing Network

The UGA SSA consists of smart sensor nodes and a base station. The term sensor node refers to the combination of electronics and sensor probes installed within a field at a single location. The electronics include a circuit board for data acquisition and processing, and a radio frequency (RF) transmitter. The UGA SSA uses Watermark® soil moisture sensors that measure soil moisture in terms of soil water tension (SWT)—the absolute value of soil matric potential in units of kilopascals (kPa). Each soil moisture probe integrates up to three Watermark® sensors. In addition, each node supports two thermocouples for measuring soil and or canopy temperature. For deep-rooted field crops like cotton or maize, the sensors on the probe are arranged so that when installed they are at 20, 40 and 60 cm below the soil surface although any combination of depths is possible. For shallow-rooted crops like peanut, 10, 20 and 40-cm depths are used. Users access the data through a dedicated web-based user interface.

11.2.3.2 Web-Based User Interface and Decision Support System

The purpose of the web-based interface is to allow users to visualize their soil moisture data and to make irrigation recommendations for each IMZ. A PHP (Personal Home Page) and JavaScript programming languages were used to create different visualizations of the soil moisture data (Fig. 11.4). The different visualizations provide users with the opportunity to view instantaneous SWT, time series of SWT and IMZ delineation of their fields. The user interface was smartphone compliant.

The web-based user interface also included a DSS which offers irrigation recommendations for each IMZ (Liakos et al. 2015). The DSS uses a modified Van Genuchten model to convert SWT data to volumetric water content (Liang et al. 2016). The strength of the method is that it uses soil properties readily available from the United States Department of Agriculture – Natural Resources Conservation Service (USDA-NRCS) Web Soil Survey (https://websoilsurvey.sc.egov.usda.gov/) to translate measured SWT into irrigation recommendations that are specific to the soil moisture status of each IMZ. The DSS uses mean SWT data collected at 07:00 h from all nodes within an IMZ to calculate the volume of irrigation water needed to bring the soil profile back to the desired soil moisture condition, which could be field capacity or a percentage of field capacity (for example 75% of field capacity) (Fig. 11.5). Each node's SWT value is a weighted average of the SWT values of the three Watermark® sensors of the node as shown in Eq. (11.1) where α, β and γ are weighting factors based on the phenological stage of the crop. Early in the growing season when the root system is not fully developed, more weight is given to the shallow sensors. As the root system develops, the weighting factors change accordingly. For peanut, at maturity, α, β and γ were 0.5, 0.3 and 0.2 respectively. Weighting factors may differ by crop.

$$\text{Weighted SWT} = \alpha \times \text{SWT}_{10\text{cm}} + \beta \times \text{SWT}_{20\text{cm}} + \gamma \times \text{SWT}_{40\text{cm}} \quad (11.1)$$

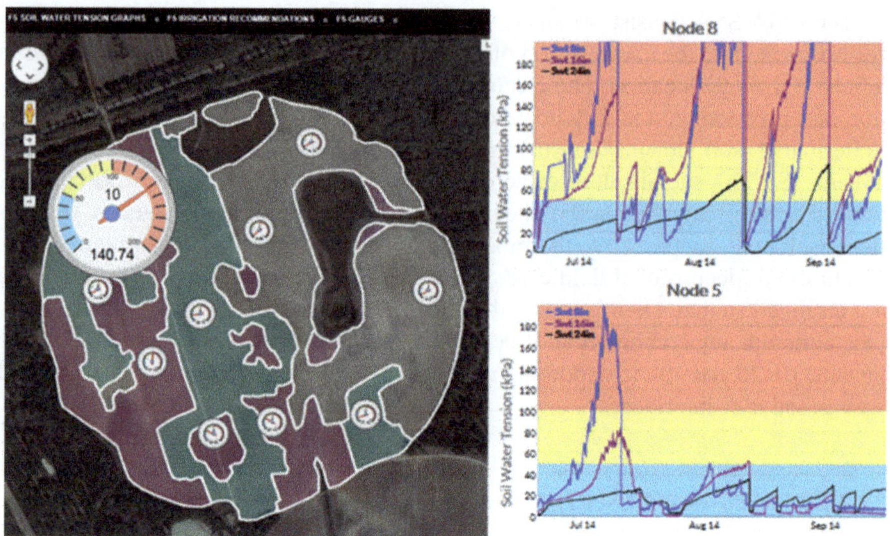

Fig. 11.4 Two different visualizations of UGA SSA data. On the left is current SWT displayed through colour coded gauges. Touching the gauges with the cursor or finger enlarges them. On the right are SWT curves for three depths for the entire growing season. Note the difference in response between nodes in the same field

Irrigation recommendations may use the same SWT threshold across all of the crop's phenological stages or thresholds could be adjusted by phenological stage. For example, UGA cotton physiologists recommend a 70-kPa threshold before flowering and a 40-kPa threshold after flowering (Meeks et al. 2017).

11.2.3.3 On-Farm Testing of the UGA Dynamic VRI Control System

The UGA dynamic VRI control system was field-tested on commercial farms with funding from the United States National Peanut Board's Southern Peanut Research Initiative (SPRI). Because of the funding source, the work was conducted with peanut (*Arachis hypogaea* L.). A farmer who already operated several VRI-enabled centre pivots was recruited to participate in the project. Before this project, the farmer had used his VRI systems only to turn sprinklers off over non-farmed areas in the field. He had not varied application rates over farmed areas. The evaluation compared the performance of the dynamic VRI control system to the farmer's standard method for scheduling irrigation on large fields. The case study below describes the on-farm evaluation conducted in 2017.

The 94-ha field was divided into IMZs using soil electrical conductivity, digital elevation models (DEMs), hydrography and the Management Zone Analyst geostatistical software (MZA) (Fridgen et al. 2004). Eight alternating 120–row wide (109 m) conventional irrigation and precision irrigation strips were then

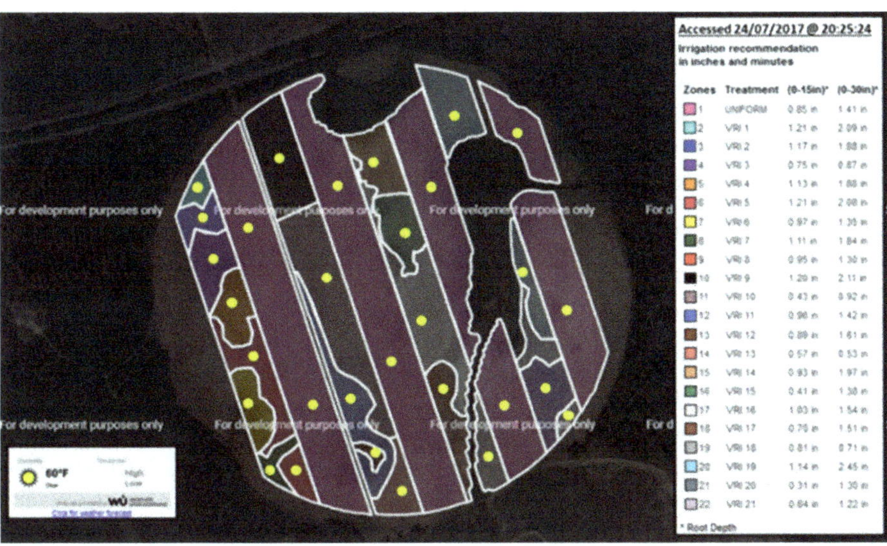

Fig. 11.5 The experimental design of the field used in the 2017 case study showing four pairs of parallel VRI and conventional strips (brown). The VRI strips maintain the delineated IMZ. The yellow circles indicate the location of 30 UGA SSA sensor nodes. The table to the right indicates the depth of water that is recommended for each IMZ by the DSS. The recommendations are provided for shallow and deep-rooted crops

superimposed over the IMZs. The precision irrigation strips retained the IMZ boundaries while the conventional strips did not (Fig. 11.5). After planting, UGA SSA sensor nodes were installed in each of the IMZs in the precision irrigation strips. Sensor nodes were located close to the centre of each IMZ to avoid boundary effects created by changing the rates of water application as the VRI system transitioned from one IMZ to another. Three sensor nodes were also installed in each of the conventional strips. A total of 30 nodes, 22 in IMZs and 8 in the conventional strips, were installed. Each probe contained Watermark® sensors at 10, 20 and 40-cm depths below the soil surface.

The UGA SSA sensor probes installed in the conventional irrigation strips were used only to monitor soil moisture conditions. Each conventional strip was irrigated uniformly by the farmer using Irrigator Pro (Davidson Jr. et al. 2000) to make irrigation decisions. Irrigator Pro is a public domain peanut crop growth model and irrigation scheduling tool developed by USDA that uses soil temperature, ambient temperature and precipitation to provide yes/no irrigation decisions for peanuts. The user decides how much water to apply. The farmer installed his own soil thermometers and a rain gauge in the field and manually collected data two to three times a week. He then entered the data into the Irrigator Pro model running on his personal computer to make irrigation decisions. Each uniform strip was irrigated uniformly.

All irrigation decisions and the amount of irrigation water applied for the precision irrigation strip IMZs were made using the DSS. At each irrigation event, the DSS used mean weighted SWT from each IMZ to calculate the irrigation water needed to return the soil profile of the zone to 75% of field capacity. The irrigation recommendations for each IMZ were then transferred manually to the prescription map that was downloaded wirelessly to the pivot VRI controller. In this field, approximately 72 h were required for the pivot to irrigate the field. Because of this, a new prescription map was downloaded every morning during an irrigation event to address soil moisture changes that occurred over the past 24 h.

11.2.3.4 Yield and Water-Use Efficiency

Yield was aggregated by strip. Water-use efficiency (WUE) was calculated by dividing yield by the irrigation applied to each strip. The 2017 peanut growing season was relatively dry with 272 mm of precipitation, which is about half of the normal precipitation received in this area. Since each IMZ within the VRI strips received different amounts of irrigation, the irrigation assigned to each VRI strip was weighted by the area of each IMZ. Table 11.1 presents a summary of the results. Every VRI strip had greater WUE than the conventional uniformly irrigated strips. Three of the VRI strips had larger yields than the uniform strips. Overall, the VRI strips resulted in a 39.7% increase in WUE and 4.3% more yield.

11.2.4 Conclusions

Pivot manufacturers have observed the outcomes of this study and dynamic VRI experiments conducted by other research groups and have begun to develop their own solutions to this issue. Lindsay Corporation which manufactures the Zimmatic brand centre pivot irrigation systems is offering its version of dynamic VRI under the trade name of FieldNET® Advisor (https://www.lindsay.com/usca/en/irrigation/brands/fieldnet/resources/). This system uses FAO-56 type evapotranspiration models to estimate daily crop water use in each IMZ. This estimate and soil properties from the NRCS Web Soil Survey are used to calculate a daily soil water balance. When the soil water balance reaches a predetermined threshold, the IMZ is irrigated. Data from the model automatically populate a prescription map that is downloaded wirelessly to the pivot's panel. This is likely to be a more adoptable dynamic VRI solution until sensor costs are greatly reduced and installation and removal requires less effort than current sensor solutions. However, models might not

Table 11.1 Comparison of results between the VRI and Uniform treatments for the entire field

Treatment	Area (ha)	Yield (kg ha^{-1})	Avg Irrig (mm)	WUE (kg $ha^{-1}mm^{-1}$)	% Diff Yield	% Diff WUE
VRI	22.5	5983	91	66	4.3	39.7
Uniform	24.2	5735	122	47		
Totals	47					

account accurately for soil differences within IMZs that may be accounted for with a high-density sensor network.

11.3 Case Study 11.2. Variable-Rate irrigation with Drip irrigation in Israel

11.3.1 Introduction

Approximately half of the agricultural land in Israel is irrigated and the majority of that (around 60%) is irrigated with drip systems. The widespread implementation of drip irrigation has made a major contribution to the increase in irrigation water-use efficiency (WUE) in Israel, in terms of production per irrigation unit. Irrigation WUE in Israel has increased steadily since the 1960s and doubled in the period 2000–2010. Yet, it hardly changed after 2011. Evans and Sadler (2008) argue that emerging computerized GNSS-based precision irrigation technologies for self-propelled sprinklers and micro-irrigation systems will enable growers to apply water and agrochemicals more precisely and site-specifically to match soil and plant status and needs using wireless sensor networks. While VRI technology for centre pivot irrigation systems goes back to the 2000s and is already implemented by farmers (case study 1 from south-eastern USA), it is still under development for micro irrigation systems like sprinklers and drip systems.

There are two primary challenges for variable-rate drip irrigation (VRDI) development: a lack of mobility and a lack of variable-rate emitters. A few recent studies have applied VRDI in vineyards (e.g. McClymont et al. 2012; Nadav and Schweitzer 2017). To overcome the lack of mobility, the vineyards were subdivided into management zones each with valves and piping necessary for autonomous irrigation. The abovementioned studies differed in zonation methodology. McClymont et al. (2012) assumed relative stability of the IMZ condition over time, i.e. a zone with less vigour in 1 year was assumed likely to exhibit less vigour in other years. Accordingly, the vineyard was designed to irrigate a few pre-determined management zones that differed in size and shape based on multi-year NDVI maps and a canopy temperature map. In this way, the number of valves and hence the cost of the system is reduced. In contrast to this, several studies have shown that IMZs are dynamic and can change from year to year and vary even during a single season (O'Shaughnessy et al. 2015; Scudiero et al. 2018). Sanchez et al. (2017) and Bahat et al. (2019) used a 'blind' zonation based on regular polygons of equal sizes allowing greater flexibility and adaptation to in-season spatial changes of IMZ. Despite differences in zonation approach, and how irrigation decisions were made for each zone, all have shown a decrease in spatial variation in yield or in water status. Currently, there are no off-the-shelf VRDI systems, but prototypes have already been installed and tested in a few orchards.

This section described how precision irrigation studies in Israel have evolved over the last 20 years following three central themes, i.e. (1) thermal imaging collection and analysis to map the variation in water status, (2) creation of irrigation prescription maps and (3) implementing variable-rate drip irrigation to address variability in a commercial field.

11.3.2 Thermal Imaging for Mapping Variation in Water Status

Drip irrigation is very efficient compared to other irrigation systems, therefore, most of the precision drip irrigation studies in Israel have focused on crops that it would benefit most: cotton and vineyards. Both cotton and wine grape (*Vitis vinifera*) vineyards are grown under controlled regimes of water stress, therefore, the best irrigation practices involve monitoring of plant water status. This section focuses on cotton. In Israel, irrigation scheduling is predetermined and water stress corrections are made by modifying irrigation amounts.

Quantities of irrigation water are estimated according to the equation:

$$\text{Irrigation quantity} = \alpha \times K_c \times \text{ET}_0 \tag{11.2}$$

where ET_0 is the reference evapotranspiration, (K_c) is the crop coefficient (FAO56; Allen et al. 1998) and α is the water status (or water stress) coefficient. Real-time daily values of ET_0 for many locations in Israel are available from the Ministry of Agriculture and Rural Development's (MARD) Agro-meteorology website (meteo.co.il). The K_c tables are also available on the web, For example, this is the link for cotton: https://www.moag.gov.il/shaham/professionalinformation/documents/daily_exhalation_user_guide_2014.pdf. In addition, an app was developed for most crops by MARD's Extension Service (https://play.google.com/store/apps/details?id=il.gov.moag.shaham.irrigationapp). The K_c of cotton while in its vegetative growth stage is further adjusted according to the rate of plant growth and in the boll-filling period to a measure of plant water status, leaf water potential (LWP). For these, growers measure, among other properties, height and LWP of a few plants in the field once or twice a week. Based on target curves of these properties that were developed by the cotton board they can estimate crop water status, α, and further correct irrigation quantities. Direct plant measures are reasonably reliable and reflect individual plant water status, but they do not represent the spatial variability of water status in the plot and are laborious and therefore expensive.

Thermal infrared (TIR) images have been used in most of the precision irrigation studies as alternatives for estimating and mapping water status in the productive periods. Initial studies focused on correlating the crop water stress index (CWSI) (Idso et al. 1981) extracted from thermal images and leaf or stem water potential (Cohen et al. 2005; Möller et al. 2007). The CWSI is defined as follows:

$$\mathrm{CWSI} = \frac{T_{\mathrm{canopy}} - T_{\mathrm{wet}}}{T_{\mathrm{dry}} - T_{\mathrm{wet}}} \tag{11.3}$$

where T_{canopy} is leaf or canopy temperature, and T_{wet} and T_{dry} are approximations of minimum and maximum canopy temperatures, respectively. Correlations between CWSI calculated using ground-based thermal images and LWP/SWP in both cotton and vines were strong (r = 0.85–0.92). Further studies examined whether robust relationships exist between the two measures throughout a cotton growing season, for different varieties, across years and in different geographical areas (different climates and soils) (Cohen et al. 2015). To test this, a dataset from three cotton growing seasons and from different geographical areas was created. A linear CWSI–LWP relationship was found to be appropriate, and had with a high coefficient of determination ($r = 0.84$).

To upscale the use of thermal imaging to commercial application, aerial thermal images over commercial cotton fields were used to map the variation in LWP and to produce prescription maps according to water status levels (Cohen et al. 2017b). Transformation of raw aerial thermal images into CWSI to be used further to map LWP requires a simple, rapid and yet accurate procedure to extract T_{wet}. An extensive comparison between various wet baselines that were used to calculate CWSI in cotton have shown the superiority of bio-indicator wet baselines including the average temperature of the coolest 5–10% of the canopy pixels in the overall field (Alchanatis et al. 2010; Rud et al. 2014; Gonzalez-Dugo et al. 2013). Furthermore, the statistical bio-indicator for wet-baseline (T_{wet}) has minimal requirements (measurements of air temperature) compared to other existing alternatives, thus paving the way towards commercialization.

Based on bio-indicators, prescription-like maps which were based on the LWP recommended curve for several commercial fields were produced on three dates during the season (Fig. 11.6 presents maps for one of the fields). The variation in water status was not constant (Cohen et al. 2017b). The maps show the importance of in-season mapping of the variation for rational irrigation management. This means that to improve VRI, in-season IMZ's should be used.

11.3.3 Thermal-Based Water Status Maps for Irrigation Management

11.3.3.1 Irrigation Management Experiment in Cotton

Cotton is an important crop in the crop rotation widely practiced in Israel. Cotton yield and quality are very dependent on an adequate supply of water and in the boll-filling growth stage they are grown under controlled water stress to allow balance between vegetative growth and cotton production. While it has been proved efficient,

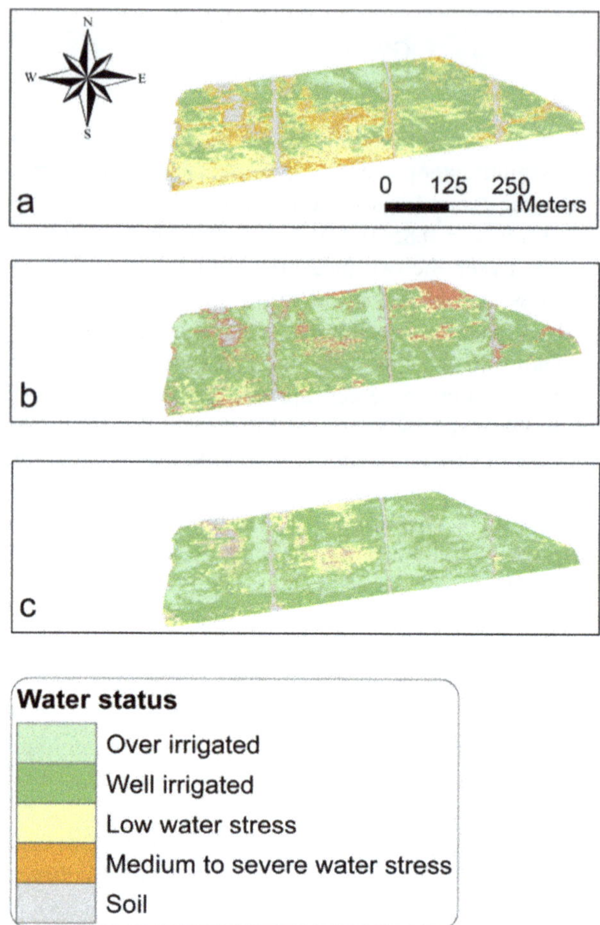

Fig. 11.6 Multi-temporal LWP variability maps of a 15 ha commercial cotton fields in Yavne, Israel. The LWP maps are based on aerial thermal images acquired on 10/7/11 (**a**), 27/7/11 (**b**), 18/8/11 (**c**)

adoption of LWP as a complementary tool for irrigation management has been limited because of the time and manpower needed to obtain manual measurements. In addition, these measurements do not necessarily represent the entire field or its variability. The experiment described in this section compared thermal-imaging based irrigation management to LWP-based management (Rosenberg et al. 2014).

Field Plots and Irrigation Treatments:

Field measurements were conducted in the summer of 2013 at Kibbutz Givat Brener, Israel. Experimental plots were planted with cotton, *Gossypium hirsutum* × *Gossypium barbadense* hybrid ("Acalpi") at the beginning of April and were drip-irrigated with one drip line every other row. Each plot was 18 m × 19 m. Plots were irrigated with four different treatments and with six replicates (a total of 24 sub-plots). Irrigation amounts were determined based on ET_0 and crop coefficients, and stress factor (α) corrections were made according to height measurements in the vegetative period and LWP in the boll-filling period. The treatments differed by the estimation stress factor α and thus the irrigation amounts. In one treatment, denoted Reg-LWP, α was estimated according to the commercial practice, i.e. based on a few direct LWP measurements conducted twice a week and irrigation decisions were made based on a recommended LWP curve. In the three other treatments, denoted Reg-TIR, Def-TIR (deficit irrigation) and Oir-TIR (over-irrigation) α was estimated based on LWP calculated by thermal images acquired once a week. Different LWP curves were used for the different treatments. In Reg-TIR, like the Reg-LWP, irrigation decisions were based on a recommended LWP curve. In Def-TIR and Oir-TIR, irrigation decisions were based on water saving and over-irrigation LWP curves, respectively.

Image Acquisition:
Oblique thermal images were acquired above the experimental plots using uncooled infrared thermal cameras (SC655 and SC2000, FLIR systems, Oregon, USA) that were attached to a vertical 20 m mast (mounted to a tractor). The two cameras are sensitive in the thermal range of 7.5–13 μm and have measurement sensitivity and accuracy of 0.1 °C and ± 2 °C, respectively. Image acquisition was carried out on 4 days in the boll-filling period: 04/08/13, 11/08/13, 18/08/13 and 25/08/13. The thermal images were acquired a few hours before irrigation, around solar noon (11:00–15:00 h daylight saving Israel Standard Time, GMT +3).

Processing of Thermal Images and LWP Calculation:
To estimate T_{canopy} separation between vegetation and background pixels in the thermal images was first done empirically following Meron et al. (2010). Second, the mean canopy temperature was extracted for each sub-plot. The CWSI was calculated according to Eq. (11.3). The T_{dry} was used in its empirical form, i.e. T_{air} + 5 °C which gave a reasonable estimate of the maximum leaf temperature for cotton (Alchanatis et al. 2010). To estimate T_{wet} temperature, a reference strip adjacent to the experimental plot was double irrigated using a drip-line every row. The temperature of this strip was extracted each time from the thermal images similar to the extraction of the T_{canopy}. Last, LWP was determined from a linear CWSI–LWP relationship (Eq. 11.4) based on a multi-year database (Cohen et al. 2015).

$$\mathrm{LWP} = -1.77\mathrm{CWSI} - 1.28 - K \tag{11.4}$$

where K is a transformation constant between the LWP [MPa] measurement methodology and other methodologies (Cohen et al. 2017b).

The LWP Measurements:
Parallel to image acquisition, 4–6 leaves from each replicate were sampled and their LWP was measured with a pressure chamber (model ARIMAD 1, Mevo Hama Instruments, Israel), as described by Meron et al. (1987).

11.3.4 Results

The CWSI-based and measured LWP were well correlated with $r = 0.88$ and RMSE of 0.14 MPa. In addition, the slope of the regression model was not significantly different from 1.

Until day 30 from onset (21/07/2013) the entire field received homogeneous irrigation of 316.4 mm. From 31 days after onset, treatments were irrigated differently as LWP measurements or calculations were used for K_c corrections. From 22/7/2013 until 28/8/2013 Def-TIR and Oir-TIR received 70.6 and 156.1 mm, respectively while Reg-LWP and Reg-TIR received very similar irrigation amounts (96.5 and 90.0 mm, respectively). Figure 11.7 shows the yield and irrigation water-use efficiency (WUE) of the four treatments. The WUE here was not calculated for the seasonal irrigation amounts but for the irrigation amounts given during the irrigation experiment (during the reproductive stage; from 31 days after onset onwards). The yields of the different treatments were very similar (*one-way ANOVA*, $F = 0.26$, $p = 0.85$). In terms of irrigation amounts, both regular treatments (Reg-LWP and Reg-TIR) were larger than the deficit irrigation treatment and smaller than the over-irrigation treatment. Significant differences were found in WUE (*one-way ANOVA*, $F = 169.9$, $p < 0.01$; $n = 24$). With a standard error of the difference (SED) of 1.3 and least significance difference (LSD) of 2.7 it can be said that WUE of Reg-TIR (88.3 kg mm^{-1} ha^{-1}) was significantly larger than that of Reg-LWP (79.1 kg mm^{-1} ha^{-1}) (Fig. 11.7). The results suggest that remotely sensed thermal data can replace direct pressure chamber measurements, while maintaining at least the same yield with similar amounts of water application. Furthermore, it has the potential to improve WUE. It is assumed that despite the imperfect estimation of LWP measurements by the thermal images, the images can represent most if not all the plants leading to sound irrigation decisions.

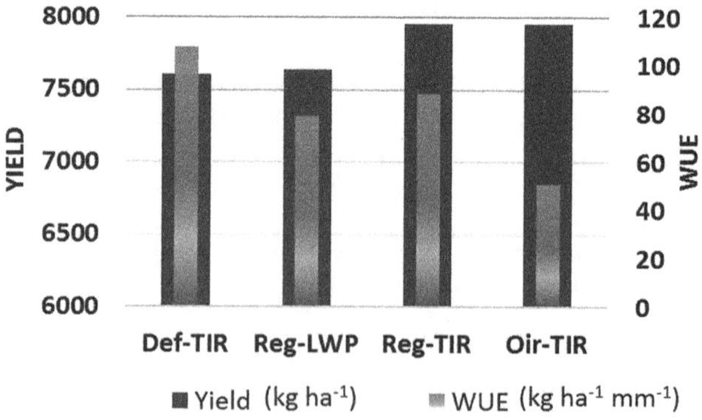

Fig. 11.7 Yield and irrigation water-use efficiency (WUE) for every irrigation treatment. The WUE was calculated from the irrigation given at the reproductive stage only

11.3.4.1 Addressing the Natural Variation in Water Status in a Commercial Cotton Field

The experiment in Givat Brener, described above, demonstrated the ability of thermal based LWP maps to aid irrigation decisions by correcting K_c. As mentioned above, variable-rate drip irrigation is challenging and its efficiency should be justified by either increasing yield and or WUE and or reducing variability. For that, an experiment was conducted in a highly variable zone in a commercial cotton field in Bnei-Darom farm near to the coastline plain of southern Israel, a well-known area for cotton cultivation. This experiment aimed at demonstrating the implementation of VRI in a drip irrigation system and its efficiency compared to unified commercial irrigation. The northern part of the field is characterized by considerable natural variability in soil type and water holding capacity. Figure 11.8 shows there is a patch of sandy soil in the northern part of the field (white arrow). The experimental area was divided into three treatments, one commercial and two VRIs, denoted COM, VRI, TIR (Fig. 11.8a) with 8 blocks (Fig. 11.8b) assuming each block to have similar soil texture characteristic. Every treatment was divided into eight cells (24 m wide by 30–40 m long), giving a total of 24 cells. From each, a soil sample was taken for texture. In addition, the height of three plants was measured twice a week in every cell from May to July, and LWP of four leaves was measured once a week from July to the end of August (boll-filling period). All treatments were irrigated similarly by the farmer until the boll-filling period (beginning of July). In general, the farmer followed the cotton board recommendations, but he added 10% to the recommended irrigation amounts relying on his experience that this area is sandy and requires more water. From the boll-filling period, the commercial treatment (COM) was irrigated by the farmer as for the Reg-LWP described in the previous section. The VRI and TIR were irrigated as for Reg-TIR, i.e. LWP was calculated from CWSI extracted from weekly aerial thermal images. The two treatments were differentiated by the size of their management cell (Fig. 11.8a). In TIR, the eight cells were grouped into two irrigation units (outlined in green in Fig. 11.8a), i.e. the northern and southern four cells were irrigated individually based on their averaged calculated LWP. In general, the northern cells are sandier (62% on average) than the southern cells (41% on average). In VRI, each of the eight cells (outlined in black in Fig. 11.8a) were irrigated individually based on their averaged calculated LWP. Figure 11.8a presents the map of accumulated irrigation amounts for each irrigation unit. Focusing on both VRI treatments (VRI and TIR), it can be seen that the northern areas were irrigated more than the southern areas, which are less sandy and experienced less water stress (as indicated by their calculated LWP in different dates, data not shown).

The unified irrigated strip (COM) was irrigated with 476 mm and 386 mm in the northern and southern irrigation units, respectively. The northern and the southern irrigation units of TIR treatment were irrigated with 440 mm (−8% vs COM) and 347 mm (−10% vs COM), respectively. Irrigation in the VRI strip ranged between 329 mm (−15% vs COM) and 590 mm (+24% vs COM). Table 11.2 presents a summary of the yield (before ginning) and WUE results. Yield for the whole

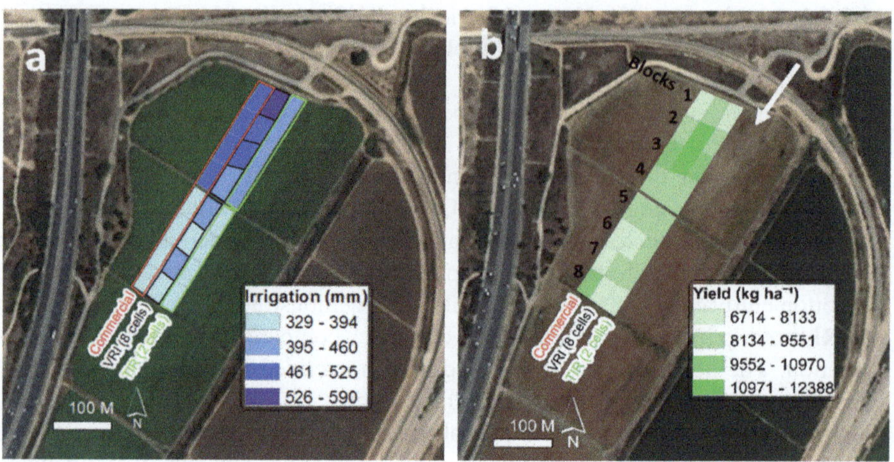

Fig. 11.8 Treatment borders and accumulated seasonal irrigation (**a**) and yield (**b**) for every cell. The white arrow marks a patch of sandy soil

experiment area ranged between 6714 and 12 388 kg ha^{-1} and WUE ranged between 15.7 kg ha^{-1} mm^{-1} and 26.6 kg ha^{-1} mm^{-1}. No significant differences were found between blocks in both yield and WUE and no significant differences were found between treatments (*Two-way ANOVA, no repetition, $p > 0.05$*). Four of the VRI blocks had larger yields than uniform irrigation (COM) while three blocks had smaller yields. Additionally, in four of the VRI blocks greater WUE was observed compared to the conventional uniformly irrigation (COM). Overall, the VRI treatments resulted in 6% more yield and 4% increase in WUE. The TIR treatment showed lower performance in yield (−3.2%) while better performance in WUE (6.1%). It might be related to the precision scale. The TIR treatment represents a lower resolution of VRI having only two irrigation units compared with eight irrigation units of the VRI. A similar experiment was conducted in the 2019 season in which VRI was practiced throughout the whole irrigation period. In the 2019 experiment seven out of eight blocks had more yield and WUE than the uniform irrigation treatment (COM). The VRI treatments resulted in 15.5% more yield (not significant) with 17.4% larger WUE (significant; data not shown).

11.3.5 Conclusions

Multi-year studies on thermal-remote sensing for water status mapping and for irrigation management have been summarized briefly. The results suggest that remotely sensed thermal data can replace manual measurements of water status, with the same or higher yields and similar or less amounts of water application. Thus, thermal imaging has the potential to improve WUE. It is assumed that despite the imperfection in the estimation of LWP by thermal imaging, the latter has the advantage of

Table 11.2 Comparison of results between the VRI, TIR and Uniform (COM) treatments by blocks

	Block	Com	TIR	VRI	% Diff TIR	% Diff VRI
Yield (kg ha^{-1})	1	7997	7609	9282	−5	16
	2	9263	9392	12388	1	34
	3	10398	10467	12108	1	16
	4	9103	8765	9199	−4	1
	5	9531	8674	8718	−9	−9
	6	7794	8674	7276	11	−7
	7	7013	9104	9444	30	35
	8	10593	6714	7519	−37	−29
	Avg.	**8961**	**8675**	**9492**	**-1.4**	**7.2**
WUE (kg $ha^{-1}mm^{-1}$)	1	18.68	17.30	15.73	3	−6
	2	21.64	21.35	26.65	10	37
	3	24.29	23.79	25.19	9	15
	4	22.27	19.92	20.58	4	8
	5	22.27	21.85	22.09	1	−10
	6	18.21	21.85	19.28	24	−4
	7	16.38	22.93	21.99	44	21
	8	24.75	16.91	22.87	−30	−17
	Avg.	**20.94**	**20.74**	**20.55**	**1.2**	**4.0**

measuring many more of the plants, leading to better irrigation decisions. Another step towards the implementation of this technique was a field irrigation experiment in which natural variation of water status in a commercial cotton field was addressed by variable-rate drip irrigation. Larger, though not significant, yields were obtained. Since this is the first attempt in using VRDI in field crops, more experiments should be conducted to prove its added value. Thermal imaging, however, can be used not only for VRI but also for better estimation of the field water status and to improve irrigation decisions. Because of its relatively high cost, to assimilate thermal imaging for commercial use more cost effective methods should be developed (Cohen et al. 2017a).

11.4 Case Study 11.3. Automatic Irrigation of Orchards Using Soil Moisture Sensors (IRRIX Model)

11.4.1 Introduction

At the scientific level, the processes that determine the water demand of crops are well known, and various methodologies have been developed to improve irrigation efficiency. The water balance method, in which the water inputs of the soil-plant system must be balanced to the expected outputs (Allen et al. 1998), is the most widely used system to determine the irrigation needs of a crop. One of the drawbacks of the water balance method is that any inaccurate parameterization will produce a systematic error that will accumulate throughout the crop cycle due to deviations that occur between estimated values and actual consumption. Irrigation control based on feedback from moisture sensors is a viable alternative to the water balance method, having as its main advantage the ability to adjust irrigation to the needs of a specific plot (Casadesús et al. 2012). However, there are also drawbacks to irrigation control based only on sensors such as the risk of sensor breakdown, variability in the measurements (Nolz and Kammerer 2017) and difficulty in interpreting sensor results, especially for drip irrigation where soil water distribution is heterogeneous.

In general terms, as both the water balance and sensor-based irrigation control methods have their pros and cons, combining both approaches seem to be the best way to improve the efficiency of irrigation in agricultural systems. This can be done by determining the irrigation dose from a water balance model and then using sensors to adjust that model to the real situation of each plot. Information and communication technologies (ICTs) can be used to attain this objective. In the last decade, most of the studies that have been published on automatic irrigation controllers have focused on regulating soil water content (SWC) or soil water tension (SWT) with feedback-based on/off strategies (Luthra et al. 1997; Miranda et al. 2005; Cáceres et al. 2007; Tahar et al. 2011). These devices are relatively inexpensive and easy to use, but ground water measurements imply certain limitations: they require a large number of sensors and do not consider plant status and response. Xiang (2011) and Zhu and Li (2011) published a study on irrigation controllers that used a combination of SWC and weather data to control drip irrigation. Romero et al. (2012) concluded that the approach of combining weather data with soil moisture signals could increase irrigation efficiency in almond trees. O'Shaughnessy and Evett (2010) and Peters and Evett (2008) proposed irrigation controllers aimed at regulating canopy temperature instead of SWC sensors.

Software tools and web applications are of fundamental importance when it comes to determining when and how much irrigation should be applied in response to crop development, crop type and environmental conditions (Casadesús et al. 2012). Various computer software packages have been developed to monitor soil properties and irrigation scheduling over a wide range of irrigation systems (Hess 1996; Abreu and Pereira 2002). Kim and Evans (2009) developed decision support software to collect information from wireless sensor networks (WSN) and control a site-specific linear-move irrigation system on a malting barley (*Hordeum vulgare* L.) field. A similar approach was described in the first case study of this chapter. There are also researchers who have developed software tools that allow operation in combination with the water balance approach and crop evapotranspiration (ET_c) estimation using soil or plant humidity sensors to enable subsequent readjustment of the ET_c estimation (Bacci et al. 2008; Casadesús et al. 2012; Osroosh et al. 2016). In addition, there are researchers who used a DSS that executed a pre-established irrigation schedule in which regulated deficit irrigation (RDI) was applied without human intervention in a Japanese plum crop *(Prunus salicina)* (Millán et al. 2019) and a hedgerow olive orchard (Millán et al. 2020) combining the water balance method with soil moisture sensors.

11.4.2 Semi Commercial Testing of an Automatic Irrigation System

To demonstrate the use of an automatic irrigation system (IRRIX), an experiment was conducted during 2018 (FERTINNOWA Project) on a commercial farm called "Finca El Chaparrito". The farm belongs to the company Haciendas Bio S.A., and is in the municipality of Pueblonuevo del Guadiana, Badajoz (latitude 38°5 6′ 13.59″ N, longitude 6° 45′ 21.98″ W, WGS84), Spain. Three different varieties of early-maturing peach *(Prunus persica)* were present in the measurement area (4.5 ha): Kay-sweet (1.5 ha), Almaneb (1.5 ha) and UFO-4 (1.5 ha). All trees were planted in the spring of 2009 at a spacing of 5 m × 3 m, in an east-west row orientation. The trees were irrigated daily using a drip system with a single lateral line per tree row located 0.5 m from the base of the tree and on-line emitters with discharge rates of 2.2 l h^{-1}, spaced at 0.5 m. The irrigation sector was separated into two parts for each variety, thus allowing the establishment of two irrigation management systems for each variety.

11.4.2.1 Selection of the Testing Zone

An analysis of soil and plant spatial variation was carried out to select the best area to install the IRRIX system and enable comparison between it and the irrigation system carried out by the farmer (FARMER). For this purpose, the most homogeneous zone possible was selected. To determine soil heterogeneity, the apparent electrical conductivity (EC_a) of the soil was measured with a Dualem-1S non-contact sensor (Dualem, Inc., Milton, Ontario, Canada), equipped with a global positioning system (GPS) antenna. The entire field was surveyed for EC_a, obtaining both shallow (0–50 cm) electrical conductivity (EC_s) and deep (0–150 cm) electrical conductivity (EC_d) values. The EC_s data were used as most root activity occurs in the first 50 cm. The final data set consisted of 3386 measurements for EC_s. Figure 11.9 shows the EC_s values, with the green line indicating the division of irrigation sectors. This line separates each variety into different irrigation sectors for IRRIX installation and allows comparison with the application carried out by the farmer. The data obtained with the soil sensor indicate a less variation in soil characteristics between the corresponding irrigation sectors in the zone of the Kay-sweet (red) than between those of the Almaneb (purple) and UFO-4 (brown) varieties.

Sentinel-2 (ESA) satellite images were used to study crop variability. The normalized difference vegetation index (NDVI) values were used to characterize spatial variation within the different varieties and sectors (Fig. 11.10). The maximum and minimum NDVI values for the different irrigation sectors of the varieties show different values in the sectors for each variety (Table 11.3). In the pre-harvest period, the greatest variability occurred in the Kay-sweet variety.

The soil and crop values indicated that the best location to develop the trial was in the Kay-sweet area because the soil was less variable and crop development was similar, especially in the post-harvest period. The area occupied by the Kay-sweet variety was delimited by a central pipeline controlled by two solenoid valves that allowed two types of irrigation in two zones. In one of the zones (FARMER), traditional irrigation was carried out by the farmer following his experience in irrigation scheduling, replacing ET_c in pre-harvest and a deficit irrigation strategy at 75% of ET_c post-harvest. Irrigation doses were applied using a general irrigation programmer (Agronic 2500, Progrés, Spain) installed on the farm. In the other zone, an IRRIX system was installed to automate irrigation decisions. The IRRIX treatment was applied through the IRRIX application, and the seasonal plan that was introduced in IRRIX was the same as in the FARMER treatment.

11.4.2.2 Automatic Irrigation System

The IRRIX system was calibrated in a national project (RTA2013-00045-C04) funded by the Spanish Agrarian and Food Research Institution (INIA) with different crops. The IRRIX system comprised two components: (a) sensors installed in the field and (b) a cloud-hosted web platform (IRRIX) which uses a control algorithm

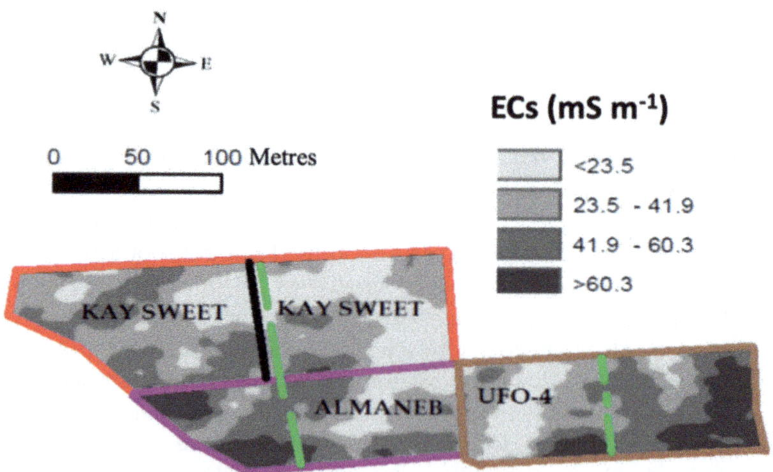

Fig. 11.9 Shallow soil apparent electrical conductivity (EC_s) (mS m^{-1}) values in trial areas. The dashed green lines indicate the division of irrigation sectors and the continuous red, purple and brown lines indicate the areas corresponding to the different varieties Kay-sweet, Almaneb and UFO-4 varieties, respectively

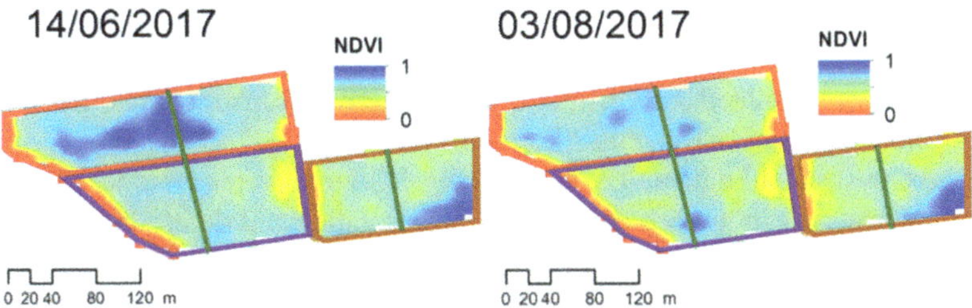

Fig. 11.10 Normalized difference vegetation index (NDVI) values in pre- (14/06/2017) and post-harvest (03/08/2017) periods using Sentinel-2 satellite images of the study zone. The green lines indicate the division of irrigation sectors and the continuous red, purple and brown lines indicate the areas corresponding to the different varieties Kay-sweet, Almaneb and UFO-4 respectively

Table 11.3 Maximum and minimum NDVI values for different varieties in the pre- and post-harvest period

	Kay-Sweet			Almaneb			UFO-4		
NDVI	Min	Max	SD	Min	Max	SD	Min	Max	SD
Pre-harvest	0.50	0.70	0.04	0.49	0.58	0.02	0.47	0.61	0.02
Post-harvest	0.45	0.63	0.02	0.50	0.62	0.03	0.47	0.64	0.03

Min minimum, *Max* maximum, *SD* standard deviation

combines a water-balance-based estimate of crop water needs with readjustment based on sensor readings:

(a) Various sensors and other devices were installed in the field: (1) Soil moisture sensors at three selected control points (Cp). At each Cp, five 10HS soil moisture sensors (Decagon Devices Inc., Pullman, WA, USA) were installed at different depths and locations in relation to dripper position (two sensors at a depth of 30 cm and located under the dripper; one sensor at a depth of 60 cm and located under the dripper; one sensor at a depth of 30 cm and located between drippers; one sensor at a depth of 60 cm and located between drippers; Fig. 11.11). The 10HS sensors used the general calibration proposed by the manufacturer for mineral soils, (2) an air temperature sensor (CS2015, Campbell Scientific Inc., Logan, UT, US), (3) a solenoid valve, (4) a water meter and (5) a relay. All sensors were connected to a CR1000 data logger (Campbell Scientific Inc., Logan, UT, US). All data were collected every 5 minutes and downloaded to the IRRIX server four times a day.

Other soil moisture measurement points were selected in the farm: (1) Observation points (Op) were also used to monitor soil moisture, but not to control IRRIX. Op1 was installed in the FARMER zone using the Hidrosoph® system with soil moisture sensors every 10 cm to a maximum depth of 80 cm; and (2) Op2 was installed in the IRRIX zone using the same configuration as Cp above.

(b) A cloud-hosted web platform (IRRIX) carried out the following daily tasks for water scheduling:

1. Data collection from sensors installed in the field. IRRIX downloads sensor data at periodic intervals during the day. The reference evapotranspiration (ET_0) is estimated daily from the air temperature sensor using the Hargreaves equation (Hargreaves and Allen 2003).
2. Analysis of all data and calculation of irrigation water volumes. The IRRIX analyses all incoming data to detect anomalies and detect if any important event has occurred in the system (irrigation, rain). IRRIX interprets the soil moisture sensors by focusing on the trend, between consecutive days, of the driest measurement of each day (SWCd). Then, to reduce variability between sensors, IRRIX normalizes these values specifically for each sensor through Eq. (11.4).

$$\mathrm{NSWCd} = \left(\mathrm{SWCd} - \mathrm{SWCWP}\right) / \left(\mathrm{SWCFC} - \mathrm{SWCW}\right) \tag{11.4}$$

where soil water content wilting point (SWCWP) and soil water content field capacity (SWCFC) correspond to the values that this sensor would record under wilting point and field capacity conditions, respectively. In practice, the SWCFC is taken at the beginning of the campaign from actual measurements of the system under field conditions, while the value assigned to the SWCWP is the expected SWC value under wilting point conditions for this type of soil. To obtain a single value to summarize the state of an irrigation sector, IRRIX performs a weighted average of the values obtained with the 15 sensors installed in a sector where the weight of each sensor is the product of its reliability and representativeness. The reliability of a sensor is related to the quality of the data that a sensor records, in other words, whether the sensor is operating correctly or not due to some unexpected anomaly. IRRIX automatically estimates sensor reliability on a daily basis with a scale of values ranging from 0 to 1. All sensors have an initial value of 1, and each time an anomaly is detected the reliability value is decreased (multiplied by a value ranging

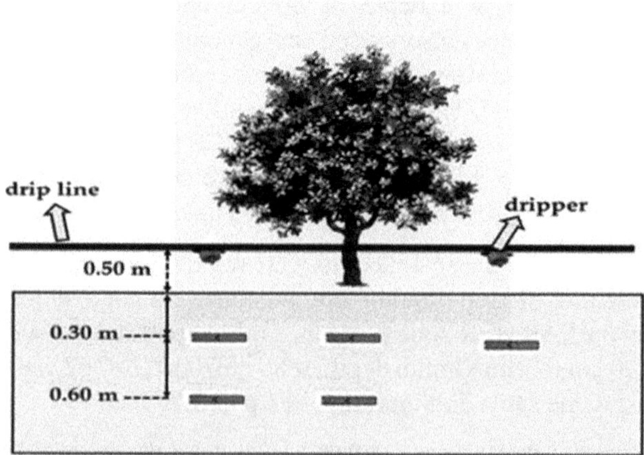

Fig. 11.11 Location of soil moisture sensors

between 0 and 0.5) depending on the seriousness of the anomaly. If the sensor stops working, IRRIX automatically removes it from the system (value of 0) and does not take the data from the sensor into account in its estimates. The representativeness of each sensor refers to whether the information provided by the sensor is relevant or not for the decision-making process in the irrigation control system. The level of representativeness is not set automatically by IRRIX but by the user according to his or her criteria. In practice, representativeness takes the value 1. The automatic system and its control algorithms are described in Millán et al. (2019). Then, IRRIX analyses the set of data to determine and adjust the irrigation dose according to the information provided by the soil sensors.

3. Irrigation scheduling. IRRIX sends the updated irrigation doses to the data logger. Then, the IRRIX system orders the solenoid valve to be opened and closed in order to apply the corresponding irrigation dose as indicated by the water meter.
4. Interaction with users. IRRIX is an autonomous system whose main objective is to free the user from work. The main function of the user is to check that the system has worked correctly, and the irrigation campaign has been undertaken as expected. Any anomaly in the system must also be resolved manually.
5. In addition, before starting the irrigation season, the user must input a seasonal plan to IRRIX that includes a rough forecast of how the water will be distributed throughout the irrigation season. For this purpose, a set of curves has to be defined, and the automated control system positioned between those curves (limits of the system) to ensure it has maximum and minimum cumulative irrigation values. Between these points, IRRIX can modify the irrigation schedule on the basis of the data provided by the soil sensors.

The Cp and Op were located in relation to the soil texture characteristics of the farm. The EC_s map was used to select 46 points to measure the surface (0–30 cm) and deep (30–60 cm) soil properties. Soil sample texture was analyzed using the method introduced by Gee and Bauder (1986). Computation of variograms for ordinary kriging is unreliable with so few data points (Webster and Oliver 1992), therefore, regression kriging with the dense EC_s data was used. Goovaerts and Kerry (2010) showed that the relationship between soil and dense ancillary data can account for a large proportion of the spatial variation and thus, a smaller sample size (around 50 points) can be used when a multi-variate geostatistical approach like regression kriging is used. Regression kriging involves regression between, for example, the sand and EC_s data, computation of a variogram of the residuals and then kriging of the residuals (Millán et al. 2019) (Fig. 11.12). Regression kriging was conducted using the Geostatistical Analyst extension of the ArcGIS software (version 10.3, ESRI, Inc., Redlands, Cal.). The sand and clay content values measured on the map allowed us to identify similar areas between the two zones (Fig. 11.12 a, b). The area next to the dividing line of the sector (black line) was identified as the most representative area for both zones. The three Cps were installed in this area in the IRRIX zone (right-hand side of map), with the distance between the different control points for automatic irrigation limited by the maximum possible cable distance to maintain the electrical signal with sufficient quality. The Op1 was installed in the

FARMER zone (left-hand side of map). Another Op (Op2) was installed in the IRRIX zone to allow monitoring of another point location at the highest elevation for potential pipe leakage problems (Fig. 11.12c), and to carry out checks with respect to the Cp, but becoming part of decision-making process.

Figure 11.13 shows the average soil moisture values at the Cp and Op sites integrated in the root zone (first 60 cm of soil depth). Information from the sensors indicated water application above irrigation needs during the pre-harvest period at all points. This is traditional practice to avoid a loss of fruit size during the pre-harvest period. In this period, the IRRIX system had minor correction limits, determined in the seasonal plan by the farmer, with irrigation doses above crop needs. During the post-harvest period, the farmer uses RDI strategies and so the system limits are larger and allow RDI-based adjustment. With Cp-sourced soil sensor data (Fig. 11.13 a, b, c), the IRRIX management system enabled a significant reduction of irrigation water during the post-harvest period, thereby reducing the large water content in the soil (below the stress line). However, in the case of FARMER management, the decrease was less, resulting, as can be seen in Fig. 11.13e (Op1), in the crop maintaining its water status within the buffer zone of the system (solid and dashed lines). This indicates that the crop was not under the desired stress (dashed line) and that the farmer's goals were not being achieved during this phase. This was subsequently confirmed with the measured stem water potential values (data not shown). The Op2 (Fig. 11.13d), installed at another point of the IRRIX zone at a

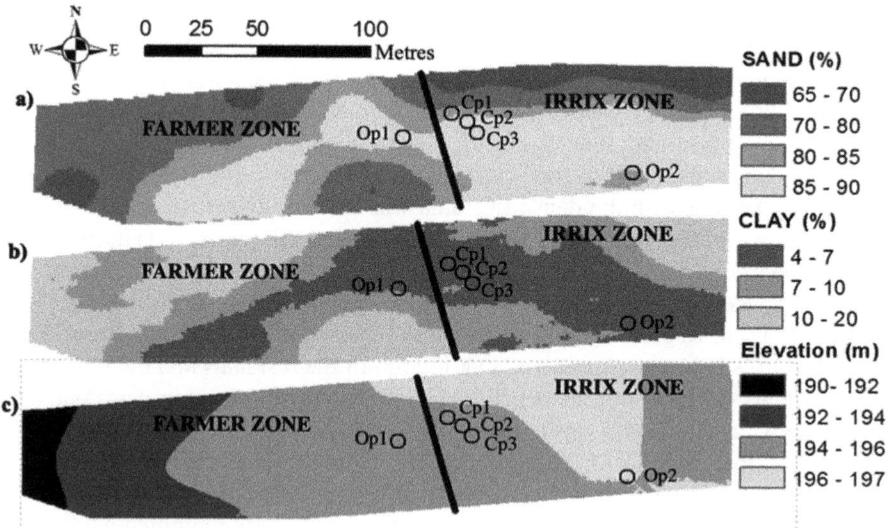

Fig. 11.12 Prediction maps of soil properties (**a**) sand (%), (**b**) clay (%) according to the method described by Millan et al. (2020) and elevation (**c**) elevation (metres) in trial part of the farm. The Cp (control point) and Op (observation point) location shown in the map. The black line is the zone where the sector was divided into farmer and IRRIX zones

higher elevation, showed a lower soil water content (due to different water pressure) closer to the stress situation (dashed line) intended by the farmer.

The IRRIX system not only allowed automation of the irrigation of a commercial plot, but also adjustment of the irrigation dose, reducing the application of water during the post-harvest period and resulting in water savings of 20–25% for all of the irrigation periods (Fig. 11.14a), especially in the post-harvest period. With respect to the impact of the IRRIX strategy on the following year's yield. The yield was evaluated for 12 trees in each of the zones (FARMER and IRRIX). No losses were observed (Fig 11.14b), with a similar commercial yield between treatments (no significant differences).

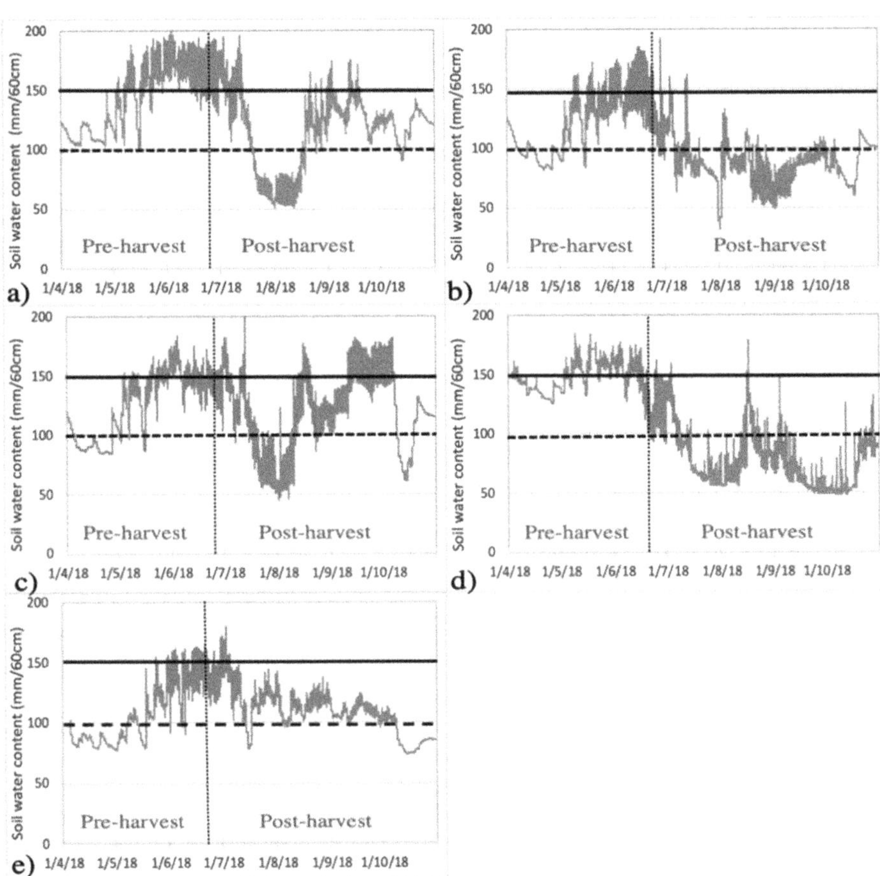

Fig. 11.13 Evolution of soil moisture, average value of sensors in root zone (0–60 cm) at the different control and observation points: Cp1 (**a**), Cp2 (**b**), Cp3 (**c**), Op2 (**d**) and Op1 (**e**). Solid line indicates field capacity and dashed line indicates the start of crop stress. The vertical line marks the harvest date

11.4.3 Conclusions

For efficient management of fruit crop irrigation and the possible implementation of water-saving irrigation strategies, soil moisture control systems provide important benefits with knowledge of the amount of water available to the plant in each crop phase; they are cheap, easy-to-install and can measure soil moisture content continuously. However, these systems also have some drawbacks, with the trickiest issues involving sensor positioning, possible errors and inconsistencies in sensor readings, the treatment given to the sensor readings, interpretation of the data obtained and the decision making process. In this sense, advances in computer science combined with the development of ICTs and improvement of communication systems have revolutionized the possibilities for the integration of crop monitoring sensors in DSS. Automatic irrigation systems such as IRRIX allow the integration of DSS with sensors in the field, enabling early estimation of irrigation needs. Having a mechanism of readjustment based on feedback control from soil sensors allows selection of the most reliable data. Such interpretation of the results is very useful for decision making.

The correct installation of the soil moisture sensors in a representative area of the farm is an important factor to obtain useful data for the water management objectives set by the farmer. The use of PA techniques to characterize spatial variation in the soil and plants and selection of the correct zone to install the sensors has also been a major advance. This enables representative data to be obtained of the hydraulic state of the soil that can be applied to the whole plot and helps to avoid irrigation programs that influence each zone in a different way.

The IRRIX system could be very useful to farmers in the application of RDI strategies, saving water and reducing vegetative growth (pruning) while at the same time maintaining yield and fruit quality.

11.5 Conclusions for the Chapter

Three precision irrigation case studies have been described above that address a range of challenges associated with incorporating precision irrigation. The case study from south-eastern USA, presented the use a commercial VRI for pivot irrigation system, the case study from Spain demonstrated an automatic irrigation system for a drip irrigated orchard and the one from Israel described the first steps towards VRDI implementation. Until recently, precision irrigation was based on predetermined IMZ, essentially based on historical and soil data and predetermined irrigation scheduling based primarily on weather data. Adaptive irrigation control strategies can use both historical data and (near) real-time quantitative measurements of crop status, weather and soil, either singly or in combination, to adjust the irrigation application locally, as required, to account for temporal and spatial variation in the field (McCarthy et al. 2010). All three case studies have shown that

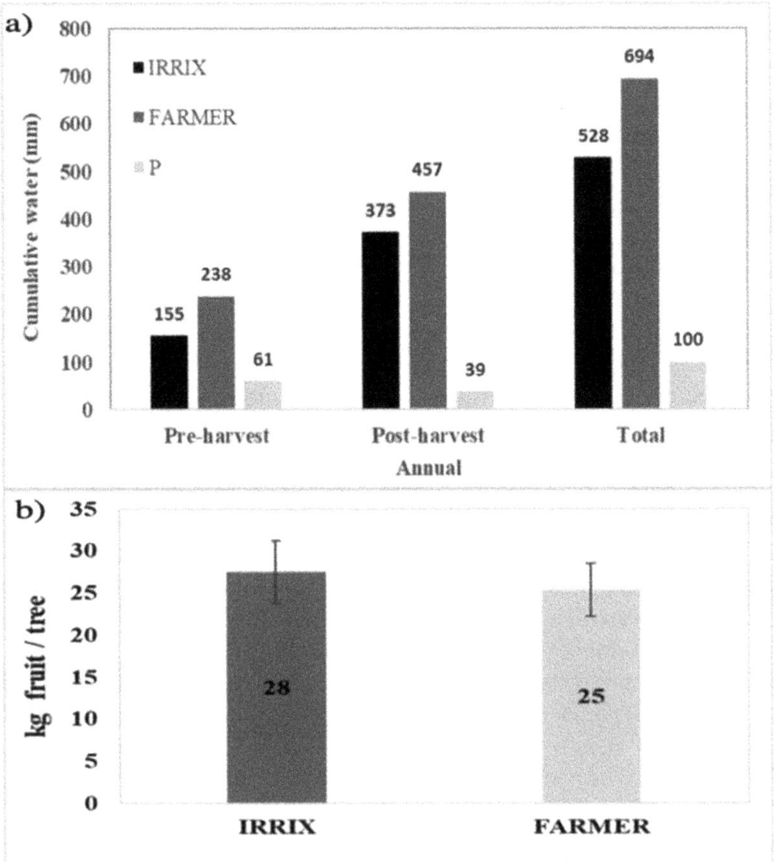

Fig. 11.14 Automatic (IRRIX) and traditional (FARMER) irrigation and rainfall (p) in the different water scheduling programs in pre- and post-harvest period (**a**) and average yield of 12 trees from each zone in the following growing season (**b**). The error bars stand for the standard error

timely data provide decision support for VRI management, i.e. soil moisture sensor data (south-eastern USA and Spain) and thermal aerial imagery (Israel). The three cases studies showed that precision irrigation treatments performed better than uniform irrigation. Yield increase was 4.3, 6–15, and 12% in south-eastern USA, Israel and Spain, respectively. In addition, in all cases but one (in Israel, 2018) WUE was improved by 14, 25 and 40% in Israel, Spain and southern USA, respectively.

The reliance on point-sensor data in the case studies from south-eastern USA and Spain dictated the use of pre-determined IMZ, yet, enabled adaptive in-season irrigation management. In-season remotely sensed images can be used further for adaptive IMZ, i.e. to modify their boundaries (Fontanet et al. 2020). The use of satellite imagery and to some extent aerial imagery, however, currently have a relatively long revisit time which limits their use with VRI pivot irrigation systems. Irrigation of a field with VRI pivot irrigation systems is lengthy, thus a snapshot image does not adequately represent the variation in plant water status that is relevant to irrigation decisions. To overcome that, a sensor network mounted on the lateral of an irrigation system (O'Shaughnessy and Evett 2010) was suggested. The case study from Israel, by comparison showed that remotely sensed imagery might depict the variation in water status well in drip irrigated fields, enabling the delineation of dynamic management zones. In this case, however, the implementation of precision irrigation is limited by the current lack of VRI for drip irrigation. From these case studies, it can be seen that full VRI implementation, which adapts for spatial and temporal changes, faces 'site-specific' challenges, i.e. every irrigation system is unique, and requires unique solutions.

Precision irrigation is studied widely, but is still in its infancy in terms of assimilation and commercialization. Similar to other PA disciplines, PI management is mainly based on: (1) data acquisition mostly from remote and proximal sensing, (2) data processing, (3) modelling and development of spatial decision support systems to manage within-field variability and create prescription maps, and (4) variable-rate application (VRA) devices. Previous scientific efforts regarding PI have concentrated on data acquisition and processing. Until recently, industries have focused mainly on the development of VRI systems. Both scientific and industrial communities currently invest increasing effort to develop advanced decision-making methods and tools (e.g. Navarro-Hellín et al. 2016; McCarthy et al. 2014). More effort should be invested in developing irrigation decision support systems that integrate atmospheric data with timely data from soil sensors and in-season spatial plant data from remote sensing. Moreover, to shift from purely responsive irrigation control that relies on past and 'near real time' soil and plant sensing data, forecasts of soil and plant water status that are based on crop models should also be incorporated into DSS.

References

Abreu VM, Pereira SL (2002) Sprinkler irrigation systems design using Isadim. St. Joseph, MI, USA

Alchanatis V, Cohen Y, Cohen S et al (2010) Evaluation of different approaches for estimating and mapping crop water status in cotton with thermal imaging. Precis Agric 11(1):27–41

Allen RG, Pereira L, Raes D et al (1998) Crop evapotranspiration: guidelines for computing crop water requirements: FAO irrigation and drainage paper 56. Rome: Food and Agriculture Organisation

Bacci L, Battista P, Rapi B (2008) An integrated method for irrigation scheduling of potted plants. Sci Hortic 116(1):89–97

Bahat I, Netzer Y, Ben-Gal A et al (2019) Comparison of water potential and yield parameters under uniform and variable rate drip irrigation in a cabernet sauvignon vineyard. In: Stafford JV (ed) 12th European conference on precision agriculture, Montpelier, France. Wageningen Academic Publishers, pp 125–131

Cáceres R, Casadesús J, Marfà O (2007) Adaptation of an automatic irrigation-control tray system for outdoor nurseries. Biosyst Eng 96(3):419–425

Casadesús J, Mata M, Marsal J et al (2012) A general algorithm for automated scheduling of drip irrigation in tree crops. Comput Electron Agric 83:11–20

Cohen Y, Alchanatis V, Meron M et al (2005) Estimation of leaf water potential by thermal imagery and spatial analysis. J Exp Bot 56(417):1843–1852
Cohen Y, Alchanatis V, Sela E et al (2015) Crop water status estimation using thermography: multi-year model development using ground-based thermal images. Precis Agric 16(3):311–329
Cohen Y, Agam N, Klapp I et al (2017a) Future approaches to facilitate large-scale adoption of thermal based images as key input in the production of dynamic irrigation management zones. Adv Anim Biosci 8(2):546–550
Cohen Y, Alchanatis V, Saranga Y et al (2017b) Mapping water status based on aerial thermal imagery: comparison of methodologies for upscaling from a single leaf to commercial fields. Precis Agric 18(5):801–822
Davidson JI Jr, Lamb MC, Sternitzke DA (2000) Farm suite-irrigator pro: peanut irrigation software and USER'S guide. The Peanut Foundation, Alexandria
Evans RG, Sadler EJ (2008) Methods and technologies to improve efficiency of water use. Water Resour Res 44(7)
Fontanet M, Scudiero E, Skaggs TH et al (2020) Dynamic management zones for irrigation scheduling. Agric Water Manag 238:106207
Fridgen JJ, Kitchen NR, Sudduth KA et al (2004) Management zone analyst (Mza). Agron J 96(1):100–108
Gee GW, Bauder JW (1986) Particle-size analysis. In: Klute A (ed) Methods of soil analysis part 1. Soil Science Society of America, Madison, pp 383–411
Gonzalez-Dugo V, Zarco-Tejada P, Nicolas E et al (2013) Using high resolution Uav thermal imagery to assess the variability in the water status of five fruit tree species within a commercial orchard. Precis Agric 14(6):660–678
Goovaerts P, Kerry R (2010) Using ancillary data to improve prediction of soil and crop attributes in precision agriculture. In: Oliver M (ed) Geostatistical applications for precision agriculture. Springer, Dordrecht
Hargreaves GH, Allen RG (2003) History and evaluation of Hargreaves Evapotranspiration Equation. J Irrigation Drainage Eng 129(1):53–63
Hess T (1996) A microcomputer scheduling program for supplementary irrigation. Comput Electron Agric 15(3):233–243
Idso SB, Jackson RD, Pinter PJ Jr et al (1981) Normalizing the stress-degree-day parameter for environmental variability. Agric Meteorol 24(C):45–55
Kim Y, Evans RG (2009) Software design for wireless sensor-based site-specific irrigation. Comput Electron Agric 66(2):159–165
Liakos V, Vellidis G, Tucker M et al (2015) A decision support tool for managing precision irrigation with center pivots. In: Stafford JV (ed) 10th European conference on precision agriculture, Rishon-LeZion, Israel. Wageningen Academic Publishers, pp 677–683
Liang X, Liakos V, Wendroth O et al (2016) Scheduling irrigation using an approach based on the Van Genuchten model. Agric Water Manag 176:170–179
Luthra SK, Kaledonkar MJ, Singh OP et al (1997) Design and development of an auto irrigation system. Agric Water Manag 33(2):169–181
McCarthy AC, Hancock NH, Raine SR (2010) Variwise: a general-purpose adaptive control simulation framework for spatially and temporally varied irrigation at sub-field scale. Comput Electron Agric 70(1):117–128
McCarthy AC, Hancock NH, Raine SR (2014) Simulation of irrigation control strategies for cotton using model predictive control within the variwise simulation framework. Comput Electron Agric 101:135–147
McClymont L, Goodwin I, Mazza M et al (2012) Effect of site-specific irrigation management on grapevine yield and fruit quality attributes. Irrig Sci 30(6):461–470
Meeks CD, Snider JL, Porter WM et al (2017) Assessing the utility of primed acclimation for improving water savings in cotton using a sensor-based irrigation scheduling system. Crop Sci 57:2117–2129

Meron M, Grimes DW, Phene CJ et al (1987) Pressure chamber procedures for leaf water potential measurements of cotton. Irrig Sci 8(3):215–222

Meron M, Tsipris J, Orlov V et al (2010) Crop water stress mapping for site-specific irrigation by thermal imagery and artificial reference surfaces. Precis Agric 11(2):148–162

Millán S, Moral FJ, Prieto MH et al (2019) Mapping soil properties and delineating management zones based on electrical conductivity in a hedgerow olive grove. Trans ASABE 62(3):749–760

Millán S, Campillo C, Casadesús J et al (2020) Automatic irrigation scheduling on a hedgerow olive orchard using an algorithm of water balance readjusted with soil moisture sensors. Sensors (Basel) 20(9)

Miranda FR, Yoder RE, Wilkerson JB et al (2005) An autonomous controller for site-specific management of fixed irrigation systems. Comput Electron Agric 48(3):183–197

Möller M, Alchanatis V, Cohen Y et al (2007) Use of thermal and visible imagery for estimating crop water status of irrigated grapevine. J Exp Bot 58(4):827–838

Nadav I, Schweitzer A (2017) Vrdi – variable rate drip irrigation in vineyards. Adv Anim Biosci 8(2):569–573

Navarro-Hellín H, Martínez-del-Rincon J, Domingo-Miguel R et al (2016) A decision support system for managing irrigation in agriculture. Comput Electron Agric 124:121–131

Nolz R, Kammerer G (2017) Evaluating a sensor setup with respect to near-surface soil water monitoring and determination of in-situ water retention functions. J Hydrol 549:301–312

O'Shaughnessy SA, Evett SR (2010) Canopy temperature based system effectively schedules and controls center pivot irrigation of cotton. Agric Water Manag 97(9):1310–1316

O'Shaughnessy SA, Evett SR, Colaizzi PD (2015) Dynamic prescription maps for site-specific variable rate irrigation of cotton. Agric Water Manag 159:123–138

Osroosh Y, Peters RT, Campbell CS et al (2016) Comparison of irrigation automation algorithms for drip-irrigated apple trees. Comput Electron Agric 128:87–99

Perry C, Podcknee S (2003) Development of a variable-rate pivot irrigation control system. In: Georgia water resources conference, The University of Georgia, 23–24 April 2003

Perry C, Pocknee S., Hansen O, Kvien C, Vellidis G, Hart E. (2002) Development and testing of a variable rate pivot irrigation control system. (ASABE Paper No. 022290). St. Joseph: ASAE

Peters RT, Evett SR (2008) Automation of a center pivot using the temperature-time-threshold method of irrigation scheduling. J Irrig Drain Eng 134(3):286–291

Romero R, Muriel JL, Garcia I et al (2012) Improving heat-pulse methods to extend the measurement range including reverse flows. In: VIII International symposium on sap flow, Leuven, Belgium, Vol. 951. International Society for Horticultural Science (ISHS), Acta Horticulturae, pp 31–38

Rosenberg O, Alchanatis V, Cohen Y et al (2014). Are thermal images adequate for irrigation management? In: 12th International Conference on Precision Agriculture, Sacramento, California, USA

Rud R, Cohen Y, Alchanatis V et al (2014) Crop water stress index derived from multi-year ground and aerial thermal images as an indicator of potato water status. Precis Agric 15(3):273–289

Sanchez LA, Sams B, Alsina MM et al (2017) Improving vineyard water use efficiency and yield with variable rate irrigation in California. Ad Anim Biosci 8(2):574–577

Scudiero E, Teatini P, Manoli G et al (2018) Workflow to establish time-specific zones in precision agriculture by spatiotemporal integration of plant and soil sensing data. Agronomy 8(11):253

Tahar B, Abdellah A, Abdulkhaliq A et al (2011) Evaluation of the effectiveness of an automated irrigation system using wheat crops. Agric Biol J N Am 2(1):80–88

Webster R, Oliver MA (1992) Sample adequately to estimate variograms of soil properties. J Soil Sci 43(1):177–192

Xiang X (2011) Design of fuzzy drip irrigation control system based on Zigbee wireless sensor network. In: Computer and computing technologies in agriculture IV. Berlin Heidelberg: Springer, pp 495–501

Zhu HL, Li X (2011) Study of automatic control system for irrigation. Adv Mater Res 219–220:1463–1467

Chapter 12
Applications of Optical Sensing of Crop Health and Vigour

James A. Taylor, Evangelos Anastasiou, Spyros Fountas, Bruno Tisseyre, Jose P. Molin, Rodrigo G. Trevisan, Hongyan Chen, and Marcus Travers

Abstract This chapter presents case studies that focus on canopy sensing using proximal and unmanned aerial vehicle (UAV)-mounted optical sensors, rather than satellite-based optical sensing applications. The potential use of optical canopy sensing for crop quality and quantity is explored across four varied case studies. The case studies have been chosen to represent a diversity of crops, countries and stages of sensor development and translation (from emerging research to near commercial applications). In

---

James A. Taylor: Introduction, Case Study 12.2 and Case Study 12.4
Evangelos Anastasiou and Spyros Fountas: Case Study 12.1
Bruno Tisseyre: Introduction and Case Study 12.2
Rodrigo G. Trevisan: Case Study 12.2
Jose P. Molin: Case Study 12.3
Hongyan Chen and Marcus Travers: Case Study 12.4

J. A. Taylor (✉)
ITAP, Univ Montpellier, INRAE, Institut Agro, Montpellier, France

School of Natural and Environmental Sciences, University of Newcastle, Newcastle-upon-Tyne, UK
e-mail: james.taylor@irstea.fr

E. Anastasiou · S. Fountas
Agricultural University of Athens, Athens, Greece
e-mail: sfountas@aua.gr

B. Tisseyre
ITAP, Univ Montpellier, INRAE, Institut Agro, Montpellier, France
e-mail: Bruno.Tisseyre@supagro.fr

J. P. Molin · R. G. Trevisan
USP, Piracicaba, SP, Brazil
e-mail: jpmolin@usp.br

H. Chen
School of Natural and Environmental Sciences, University of Newcastle, Newcastle-upon-Tyne, UK

M. Travers
Soil Essentials, Hilton of Fern, Brechin, Scotland, UK

R. Kerry, A. Escolà (eds.), *Sensing Approaches for Precision Agriculture*, Progress in Precision Agriculture, https://doi.org/10.1007/978-3-030-78431-7_12

each case study, optical sensing is shown to be relevant to assessing productivity, either directly or through an indicator of crop health. It represents a powerful tool for crop management; however, across all the case studies, the optical sensing solution could only be used directly to address local issues. A clear message is that the suitability and adaptability of this technology to a variety of end-uses in cropping systems depends on local calibration and interpretation. The need for these is a limitation to technology adoption despite the widespread potential applications of optical sensors.

Keywords Viticulture · Cotton · Potatoes · Crop circle · UAV · Vegetative indices · LiDAR · Ultrasonic sensors · Multispectral · Proximal sensors

12.1 Introduction

This chapter presents four contrasting applications of crop sensors to various cropping systems including table grapes in Greece, wine-grapes in France, potato production in the United Kingdom and cotton production in Brazil. In each case, one or more canopy or crop sensors are used to assess crop vigour and to relate it to crop production, in terms of quantity and or quality. Collectively, they illustrate how information on plant vigour can be connected directly to production attributes or can be used as an ancillary variable to assess another primary crop production attribute. They also illustrate a diversity of platforms (terrestrial, unmanned aerial vehicle [UAV] and aerial) and a diversity of sensors (light detection and ranging [LiDAR], ultrasonic, optical infrared and optical visible) to obtain data on vigour. The case studies highlight the potential breadth of applications of crop vigour data, either directly or indirectly, in crop management, but this does not suggest that this is the full extent of potential uses. The case studies move from emerging research questions (such as the use of cumulative canopy reflectance responses over the season) to clear commercial applications (such as the incorporation of sensing into potato agronomy services).

A key message across the case studies is that sensor data help to identify spatial patterns in crop health and vigour, but also require local, site-specific interpretation and the ability to relate them directly to a production attribute and to a management operation. Ultimately this is the key objective when starting to collect spatio-temporal information on crop vigour.

The first case study describes a simple application of a proximal (tractor-mounted) visible–near-infrared (Vis–NIR) optical sensing system to table grapes. It explores the diversity of spectral data that can be derived from even simple multispectral systems and investigates which of the vegetative indices are the most useful for predicting table grape yield and quality. The second case study continues in the area of viticulture with a Vis–NIR optical sensor. However, it focuses on wine grapes and on data obtained from an aerial platform. It does not directly relate the information on vigour to production attributes (as in case study 1), but rather uses the vigour data to identify and rank fields according to the amount of observed spatial variation in canopy vigour. This information is then used to sample intelligently for information related to vine water stress and to identify fields, and zones in fields, where crop vigour relates to vine water stress, and therefore to final grape quality.

Case study 3 concerns cotton production systems in Brazil. It compares optical sensors focused on crop vigour with sensors of crop height for mapping crop yield and productivity, especially at early developmental stages. Commercial optical sensing systems were robust and effective for spatial crop management. However, the alternative sensors were shown to work well in the given conditions. The LiDAR system, while experimental, was a particularly effective method of estimating crop biomass from crop height. The opportunities for sensor data-fusion and issues with sensor resolution versus input application resolution are discussed.

The final case study investigates potato production in the UK. This case study relies only on a simple RGB camera. It explores how cheap camera-based systems mounted on UAV platforms can give relevant and timely information on crop emergence and canopy development, even if the vigour of the canopy is not measured directly. Information on crop emergence and on canopy development, expressed as percentage ground cover, is used to update a local crop model. Scenarios run with the model in its native state and with the addition of different levels of information derived from the UAV images are used to demonstrate the value of basic colour images when used intelligently.

12.2 Case Study 12.1. Health and Vigour for Table Grapes in Greece

12.2.1 Introduction

The world market for table grapes is large with production estimated at 27.3 million tonnes in 2018 (OIV 2019). Farming practices and field conditions have a strong effect on the quantity and quality characteristics of table grapes. This is important because specific characteristics of table grapes will influence their commercial value. These quality characteristics are total soluble solids and berry diameter, as well as the berry sugar/acid ratio, which is strongly related with storability and consequently with shelf life (Sen et al. 2016; Tehrani 2016). For this reason, high resolution vineyard information is desirable that relates to either or both the quantity or quality of berries in a field and can be used for management.

Proximal canopy sensing in vineyards is used to assess crop growth and yield variables in a non-destructive way and may include the use of radar, optical or multispectral sensors (Henry et al. 2019). The latter type of sensor (multispectral sensors) records information in different spectral bands that are being combined through mathematical equations for calculating the spectral vegetation indices (VIs) (Xue and Su, 2017). More specifically for vineyards, there are numerous VIs that have been used for assessing important crop variables such as vigour, yield and quality characteristics (e.g. total soluble solids and titratable acidity), as well as for pest and disease infestation and water stress (Hall et al. 2002; Tisseyre et al. 2007). The normalized difference vegetation index (NDVI) (Rouse et al. 1974) is the most common VI in agriculture (Badr et al. 2015). Although many studies have already found a significant relationship between a VI recorded at a single crop stage with vine vigour, yield and quality attributes, the studies do produce different degrees of correlation at different crop stages (Anastasiou et al. 2018). This has prompted the use of cumulative VIs that aggregate the values of the single crop stage VIs for providing better crop yield and quality assessments. The advantage of cumulative VIs is that the variable nature of the growing season is taken into account, resulting in a better correlation with crop yield and quality (Sun et al. 2017; Mirasi et al. 2019). However, there are no studies to date on the use of cumulative VIs in vineyards. Therefore, the main aim of this case study was to assess the relationship between five different cumulative vegetation indices (VIs) and yield and quality attributes in a vineyard that had a large number of viticulture operations.

12.2.2 Materials and Methods

The study site was a commercial table grape vineyard (*Vitis vinifera* cv. Thompson seedless) near Athens, Greece with data collected over three consecutive years (2015–2017). Within the vineyard, there are two distinct soil texture types, a sandy clay loam and a clay loam. Each year the vineyard received approximately 2400 mm ha^{-1} of water (irrigation + precipitation), 16 spray applications of foliar fertilizers, pesticides and crop growth regulators and 4 pruning operations to control the vine growth.

Canopy properties at six different developmental stages (fruit set [FS], pea-size berries [PS], majority of berries touching [BT], beginning of veraison [BV], middle of veraison [MV] and harvest [H]) were measured using a Crop Circle sensor (ACS-470, Holland Scientific Inc., Lincoln, NE, USA). The sensor was mounted at 1.5 m height above the soil surface and pointed horizontally at the vines from a distance of 1.2 m to scan the side 'curtain' (canopy) area of the vine. This particular canopy sensor has the ability to change filters and the wavelengths sensed according to user requirements. For the purposes of this experiment, six different filters with the same characteristics in terms of focal length were used, specifically, 532 nm (green), 550 nm (green), 670 nm (red), 700 (red edge), 730 nm (red edge) and 760 nm (NIR). The measurements were used for developing five VIs, the NDVI, two versions of the green normalized difference vegetation index (GNDVI) calculated with different wavelengths in the green region of the spectrum (550 nm and 532 nm) and two versions of the red edge normalized difference vegetation index (NDRE), using two different wavelengths (730 nm and 700 nm) in the red edge area of the spectrum. Details of the formulae for the five VIs are given in Table 12.1. These VIs have previously shown good correlations with crop properties such as nitrogen uptake, biomass and leaf area index (LAI) in cereal crops (Rodriguez et al. 2006; Wang et al. 2007). Many researchers have proposed cumulative VIs for mapping and monitoring important crop variables such as net primary productivity, LAI, biomass and yield (Ricotta et al. 1999, Kross et al. 2015, Zhou et al. 2017).

Cumulative VIs were calculated by aggregating average VI values collected pre-veraison (FS, PS, BT), post-veraison (BV, MV, H) and at all stages (FS, PS, BT BV,

Table 12.1 Formulae and references for the spectral vegetation indices (VIs) used in the current work

Spectral Vegetation Index	Equation	References
NDVI	NDVI = $(\rho_{760} - \rho_{670})/(\rho_{760} + \rho_{670})$	Rouse et al. (1974)
GNDV1	GNDVI1 = $(\rho_{760} - \rho_{550})/(\rho_{760} + \rho_{550})$	Lymburner et al. (2000)
GNDVI2	GNDVI2 = $(\rho_{760} - \rho_{532})/(\rho_{760} + \rho_{532})$	Lymburner et al. (2000)
NDRE1	NDRE1 = $(\rho_{760} - \rho_{730})/(\rho_{760} + \rho_{730})$	Hunt et al. (2013)
NDRE2	NDRE2 = $(\rho_{760} - \rho_{700})/(\rho_{760} + \rho_{700})$	Hunt et al. (2013)

MV, H) for each plot. This was done to assess the effect of the variation in vine vigour before and after veraison on correlations with final yield and quality characteristics.

For this study a regular grid of 36 cells (~10 m × 25 m) was established across the vineyard to facilitate field sampling. This methodology follows the approach used by Tagarakis et al. (2013 and 2018). Samples of 50 berries were taken from each vineyard cell before harvest. Berry diameter was computed using image analysis in Image J (National Institutes of Health, USA). The berries were then crushed to produce juice using a juicing machine and the juice total soluble solids (TSS) (°Brix) measured by refractometry in an SR400 digital refractometer, the titratable acidity (TA) measured in a Fruit Acidity Meter GMK-708 (G-won Hitech Co., Seoul, South Korea) and the pH measured with an AD8000 multi-parameter meter (Adwa Hungary Kft., Szeged, Hungary). Harvest was performed manually on the 2–3 September 2015, the 21–22 August 2016 and the 16–17 August 2017. Table grape yield was estimated at harvest by measuring the number of bins per cell and multiplying it by the average bin weight.

Maps of yield and berry quality for the 3 years of study were produced using ArcGIS 10.2 software (ESRI Inc., Redlands, CA, USA). To assess the relationship between the cumulative VIs and table grape yield and quality, descriptive statistics, Pearson's correlation and regression analysis were performed. Step-wise linear regression was only performed for the cumulative VIs (cVI) that had the strongest correlation for each variable. The statistical analysis was performed with the statistical software Statgraphics 16 (StatPoint Technologies Inc., Warrenton, VA, USA).

12.2.3 Results and Discussion

There were variable weather conditions during the study. In this region the weather conditions from March to June are usually characterized by a considerable amount of precipitation and moderate temperatures, while July and August have higher temperatures and dry weather (Fig. 12.1). In 2016, however, precipitation was very limited and temperatures were higher during most crop stages. In contrast, in 2017 there was more precipitation that continued late into the growing season (July).

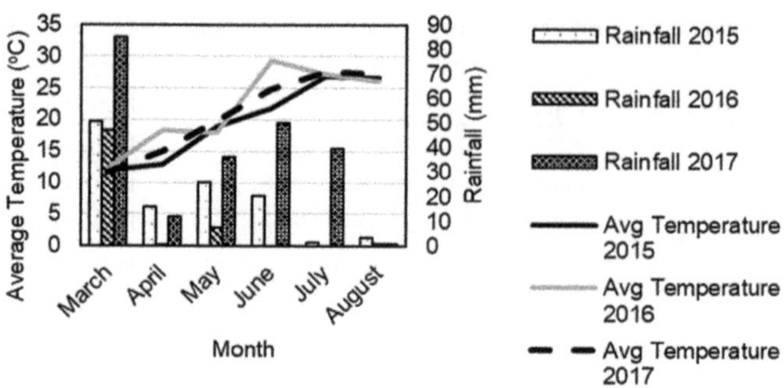

Fig. 12.1 Monthly average temperature and precipitation from March to August of 2015, 2016 and 2017 table grape seasons

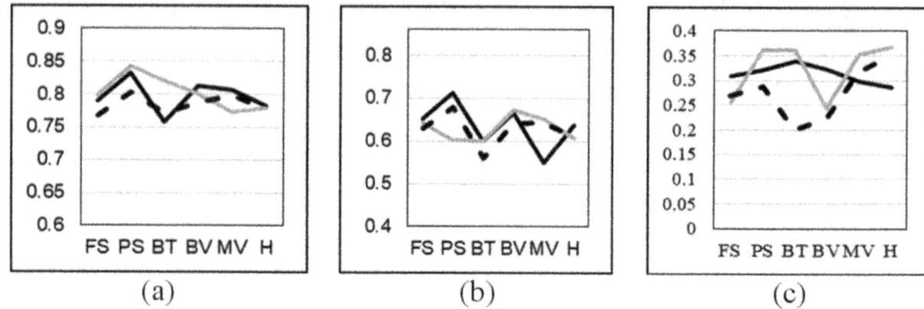

Fig. 12.2 Temporal profiles of NDVI (**a**), GNDVI1 (**b**) and NDRE1 (**c**) averaged over the field at each sampling time (fruit set [FS], pea-size berries [PS], majority of berries touching [BT], beginning of veraison [BV], middle of veraison [MV] and harvest [H]) in 3 years (2015 – solid black; 2016 – solid grey; 2017 – dashed black line). These are shown as examples of the difference in response over time between the vegetation indices (VIs)

The development of example VIs by crop stage and by year are shown in Fig. 12.2. The VIs showed more variation during pre-veraison crop stages than for post-veraison stages. This variation can be explained by the fact that weather conditions can greatly affect vine development and therefore yield (Cheng et al. 2014; Oczkowski, 2016; Ollat et al. 2016).

There was spatial variation of yield attributes between the different years (Fig. 12.3). Specifically, the south and east part of the field had the smallest values in terms of yield, berry diameter and titratable acidity for the 3 years of the study indicating that in this part of the field different management practices can be applied. In addition, the spatial pattern for other quality attributes such as TSS and pH was not stable. This variation can be explained by the different weather conditions that occurred among the different years of the study.

For pH the strongest correlation was with pre-veraison cNDVI ($r = -0.55$), for TSS with post-veraison cGNDVI2 ($r = 0.42$), TA with whole-season cGNDVI1 ($r = -0.25$), berry diameter with whole-season cGNDVI2 PrV ($r = 0.736$) and yield with pre-veraison cGNDVI2 PrV ($r = 0.58$). The strongest correlations were with cGNDVI and not cNDVI or cNDRE. These findings accord with other studies that

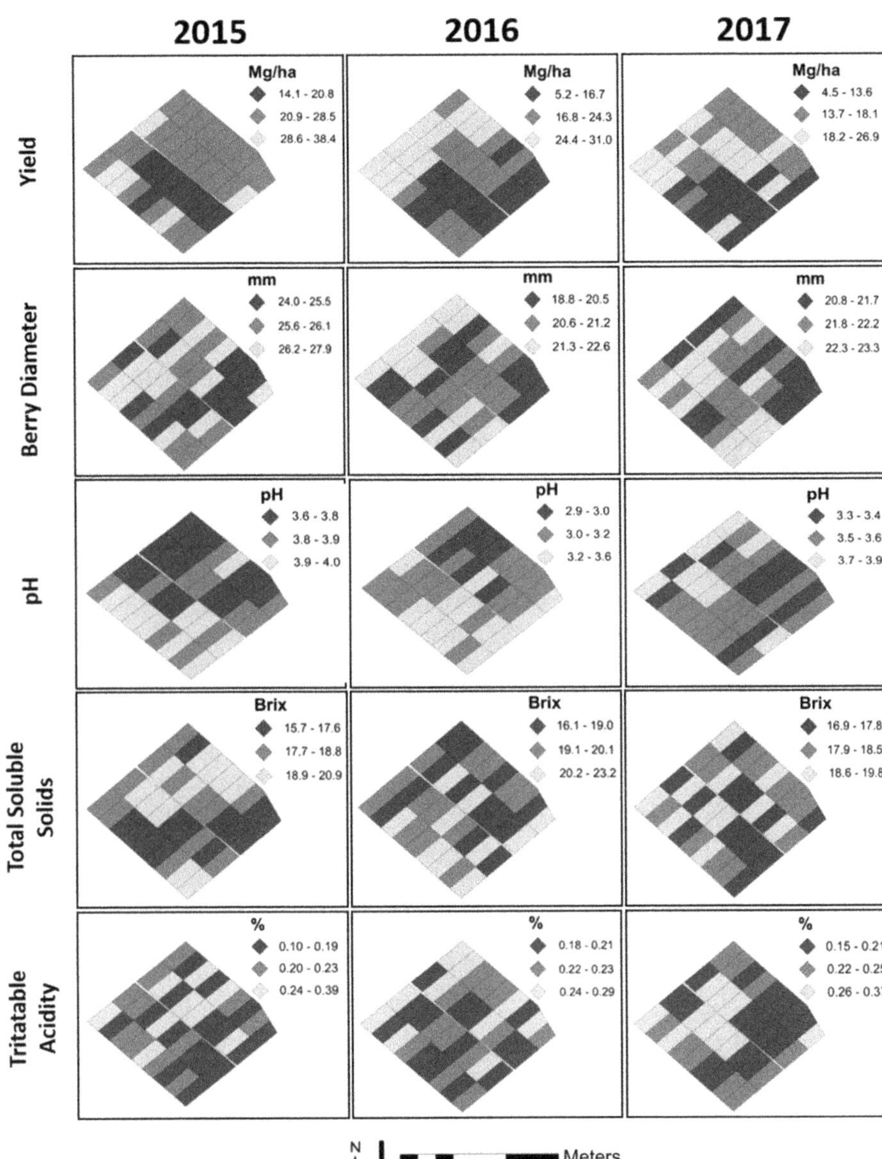

Fig. 12.3 Thematic maps of Yield (top row), Berry Diameter (second row), pH (third row), Total Soluble Solids (fourth row) and Titratable Acidity (bottom row) for 2015 (first column), 2016 (second column) and 2017 (third column) to illustrate similarities and differences in variable response between years and between variables within years

indicate that the performance of other VIs might be better when compared with classical spectral VIs, such as NDVI (Hall and Wilson 2013; Anastasiou et al. 2018). Moreover, the strongest correlations with yield characteristics were for different cumulative VIs (Table 12.2). The latter agrees with the result of Anastasiou et al. (2018) who found that different crop stages have different accuracy for estimating yield characteristics of table grapes.

The linear regression models of table grape yield and quality attributes with proximally sensed cVIs showed that the model fit explained between 20 and 60% of the variance in grape quality and quantity attributes (Table 12.3). The berry diameter regression model had the best fit (adjusted $R^2 = 0.60$) and TSS the poorest (adjusted $R^2 = 0.20$). The yield estimation model explained 45% of the variance (Table 12.3). These results contradict those of Anastasiou et al. (2018) who found larger coefficients of determination when modelling table grape quality attributes, such as BD and TSS, with VIs from only one crop growth stage. This can be explained by the fact that specific crop stages, and the vigour conditions at these stages, will have different effects on final production (Sipiora and Granda, 1998; Intrigliolo and Castel, 2010). Quality in particular may be more aligned with conditions at specific stages, rather than an integration over the entire season. In contrast, absolute yield might be better explained through a cumulative approach.

The various cumulative VIs used in this study resulted in different performances with the cumulative VIs based on GNDVI having the strongest correlations with the

Table 12.2 Pearson's correlation matrix between cumulative VIs and table grape yield and quality attributes

VI	pH	TSS (°Brix)	TA (%)	Berry diameter (mm)	Yield (kg ha^{-1})
Whole-season					
cNDVI	**−0.29**	0.16	**−0.23**	0.01	**0.37**
cGNDVI1	0.06	0.06	***−0.25***	0.13	0.08
cGNDVI2	**0.30**	0.06	**−0.23**	**0.63**	**0.52**
cNDRE1	**−0.28**	0.11	**−0.19**	0.07	**0.34**
cNDRE2	−0.18	**0.31**	−0.18	0.11	**0.33**
Pre-Veraison					
cNDVI	**−0.55**	0.19	−0.12	**−0.25**	**0.26**
cGNDVI1	**0.31**	**−0.20**	−0.27	0.43	**0.27**
cGNDVI2	**0.41**	−0.08	**−0.23**	***0.74***	***0.59***
cNDRE1	−0.18	0.02	−0.15	0.17	**0.37**
cNDRE2	**−0.26**	**0.21**	**−0.21**	0.11	**0.44**
Post-Veraison					
cNDVI	***0.34***	−0.01	**−0.21**	**0.39**	**0.24**
cGNDVI1	**−0.32**	**0.36**	−0.02	**−0.38**	**−0.24**
cGNDVI2	**−0.23**	***0.42***	−0.08	−0.12	−0.03
cNDRE1	**−0.34**	**0.27**	−0.16	**−0.27**	−0.03
cNDRE2	0.04	**0.27**	−0.03	0.04	−0.03

Bold values indicate significance at p = 0.05 level. Italics indicate VIs used subsequently (Table 12.3)

Table 12.3 Linear regression models of table grape yield and berry quality with cVIs. The cVIs were selected based on correlation analysis

Regression Model	adjusted R^2	RMSE
Yield = − 5764.59 + 11679.57 × cGNDVI2 PrV	0.45	3.64 mg ha^{-1}
$BD = \sqrt{(253.84 + 51.86 \times cGNDVI2PrV^2)}$	0.60	1.2 mm
TSS = 11.39 + 3.65 × cGNDVI2PoV	0.20	0.7 °Brix
TA = 0.56 – 0.09 × cGNDVI1	0.20	0.02%
pH = 1/(0.951 – 0.551 × ln cNDVI PrV)	0.42	0.01

PrV = pre-veraison; PoV = post-veraison

most variables (yield, BD, TSS and TA). This accords with Anastasiou et al. (2018) who found that both proximally- and satellite-derived single crop stage GNDVI produced better results than NDVI. According to their study, this can be explained by the saturation of NDVI values, which resulted in smaller estimates of crop yield characteristics. Furthermore, the cumulative VIs had poorer accuracies than single stage VIs from the same study (Anastasiou et al. 2018). This is because the various agricultural operations (e.g. irrigation, canopy trimming) that take place during the crop growth periods decrease the within-field spatial variation and smooth the values of the cumulative VIs. Finally, this study accords with that of Sun et al. (2017) and Mirasi et al. (2019) who found that cumulative VIs can assess crop properties better, such as yield, when compared to single crop stage measurements.

12.2.4 Conclusions

In this study, an assessment of proximal sensed cumulative VIs for the estimation of crop yield and quality characteristics was conducted during three crop growing seasons (2015, 2016 and 2017) on a commercial table grape vineyard in Greece cultivated with Thompson Seedless variety grapes.

The cumulative VIs based on GNDVI presented higher correlations with yield, BD, TSS and TA when compared with other cumulative VIs. Moreover, the results indicated that cumulative VIs have the potential to become a tool for assessing table grape characteristics and thus being used in management zone delineation for selective harvesting. However, cumulative VIs must be carefully used as they might not be suitable for estimating specific attributes under intensive management conditions when frequent agricultural operations occur in the field. In this situation, single stage VIs may exhibit stronger relations and be more useful for management.

12.3 Case Study 12.2. Airborne Multispectral images as a Tool to Characterize the Spatial Variation of vine Water Status: Application to a Non-irrigated Mediterranean Vineyard

12.3.1 Introduction

The evolution of vine water status (VWS) throughout the vineyard growth cycle has a direct effect on grape composition and harvest quality through its influence on fruit growth, yield, and fruit metabolism (Dry and Loveys 1998; Ojeda et al. 2002). Therefore, it is important to monitor VWS, to predict either expected harvest quality or as a source of critical information for vineyard management, under both irrigated and non-irrigated conditions (Naor et al. 1997; Choné et al. 2001).

Several reference methods have been proposed to measure VWS; leaf water potential (LWP), stem water potential (SWP) and predawn leaf water potential (PLWP) (Sibille et al. 2007). These methods are widely used, but they are laborious manual techniques, requiring specific instrumentation and a certain level of skill when making measurements (Ojeda et al. 2002; Sibille et al. 2007). These constraints make systematic spatio-temporal (S-T) VWS measurements with these methods time-consuming and difficult to perform. For this reason, these reference measurements have been used mainly to monitor the temporal change in VWS at only a few locations within vineyards. The VWS observations therefore tend to be reported at a low spatial resolution and often under an assumption that VWS is homogeneous over the measurement area (i.e. the block or set of blocks are assumed to behave similarly).

However, VWS can be highly variable within blocks, within vineyard estates (Taylor et al. 2010) and across viticulture regions (Baralon et al. 2012). So, characterizing the spatial variation of VWS is key to improving the positioning of reference measurements to enable managerial decisions based on the observed spatial patterns (Acevedo-Opazo et al. 2008a). Availability of soil water, because of differences in soil depth and soil physical properties is a key influence on VWS. Under non-irrigated Mediterranean conditions, water restriction decreases the vegetative growth of the vine (Celette and Gary, 2013), so vine vigour should constitute an indicator of the spatial variation in VWS.

This case study illustrates the use of a vegetation index, obtained from multispectral aerial images, to assess the spatial variation of VWS under non-irrigated Mediterranean conditions. The aim was to verify that remotely sensed vegetative indices are a useful auxiliary data layer for understanding and managing viticulture practices according to the spatial variation of VWS.

12.3.2 Materials and Methods

12.3.2.1 Description of the Vineyard

The study vineyard is at Pech Rouge (Gruissan, Aude, France; 43° 08′ 30″ N, 3° 07′ 30″ E, WGS84). It has 28 blocks with a total area of ~32 ha. The vineyard is representative of vineyard diversity in southern France in terms of training systems (vertical shoot positioning and gobelet), the age of vines (from 2 to >50 years old) and varieties (*Vitis vinifera* cv Syrah, Grenache, Chardonnay, Petit Verdot, Muscat, Mourvedre and Carignan).

Management practices (pruning, fertilization, trimming, mechanical weeding, etc.) were very similar for all blocks. The vineyard is non-irrigated and has a Mediterranean climate with a hot dry summer. Precipitation occurs mainly in autumn and spring. A large evaporative demand usually leads to considerable vine water restrictions in summer. Examples of average water constraint over the vineyard, estimated by PLWP, were – 0.75 MPa in August 2003 (a very dry year) and – 0.60 MPa in August 2006 (a wet year) (Taylor et al. 2010).

12.3.2.2 Multispectral Image

Aerial images (1 m^2 pixels) were collected by l'Avion Jaune (Montpellier, Hérault, France) at veraison (August, 2006). The spectra included were: (i) blue (445–520 nm), (ii) green (510–600 nm), (iii) red (632–695 nm) and (iv) near-infrared (757–853 nm). The NDVI was derived for each image after the 1 m^2 pixels were aggregated into 3 m^2 pixels using the methodology outlined in Acevedo-Opazo et al. (2008b). This approximates the 'mixed pixel' row spacing approach of Lamb et al. (2004). Data mapping was performed using QGIS 2.18. Three classes of NDVI were created; high, medium and low relative to the tertiles in each individual block.

12.3.2.3 Spatial Variation and vineyard Block Selection

Not all blocks had the same magnitude of spatial variation of NDVI. Vineyard managers will focus on blocks that show considerable variation and with spatial patterns large enough to justify the use of site or zone-specific practices. To determine blocks that would be most appropriate for site-specific management, the technical opportunity index (TOi) (Tisseyre and McBratney, 2008) was used to rank them. It was computed for each block with the GeoFIS freeware (Leroux et al. 2018). The TOi ranks were used to meet two objectives: 1) to identify blocks that could justify zone-specific management and 2) to target blocks to collect additional data from.

12.3.2.4 Sampling Scheme

Sampling sites were selected based on the NDVI information. Only blocks with a large potential for zone-specific management (large TOi values) were considered. For the selected blocks, two zones were selected corresponding to a relatively high and a relatively low NDVI response, with a constraint that the selected zones presented a significant area in the block (> 100 m^2). This last criterion was considered mainly for practical reasons, to ensure that the number of vines in each zone was relevant for further analysis. A measurement site of 40 m^2 was randomly located within the selected high and low NDVI zones (2 sites per block). Zones of medium NDVI values were considered transition zones and not sampled.

12.3.2.5 Vine Measurements

Two types of measurement were performed to verify the origin of the difference in NDVI values: vine vegetative expression, using the external canopy area (ECA) per vine (m^2 pl^{-1}), and PLWP (MPa) measured at veraison, when the water stress was assumed to be at its strongest for the season. The ECA was estimated manually over five vines along the row using measurements of canopy height (m) and width (m). The PLWP was measured over the same five vines between 3:00 and 5:00 a.m. using a pressure chamber (Scholander et al. 1965).

To identify the possible uses of the NDVI in relation to VWS, a more intensive temporal survey of PLWP coupled with harvest measurements was performed on the two NDVI zones in Block 5. This block was chosen as it had a large TOi value and was considered a block with potentially high fruit quality by the manager. Zone-specific management may therefore be important to improve yield and quality. The PLWP was measured on seven dates during the season. At harvest, variables were measured to characterize production (yield $vine^{-1}$) and berry quality. Quality measurement from both the high and low NDVI zones were based on 10-bunch samples (from different vines). Soluble solid concentrations (using a thermo-compensated refractometer) (°Brix), total acidity (g L^{-1} of sulphuric acid) and pH were measured at berry maturity. Total extractable anthocyanins and total polyphenols index were assessed at harvest using the methodology proposed by Iland (2000).

12.3.3 Results

The 3-class NDVI map shows that all blocks have spatial patterns corresponding to low, medium or high NDVI values (Fig. 12.4). However, some blocks have larger, more contiguous NDVI zones. It is possible to identify geometric patterns (rectangles, squares) that correspond to experiments within blocks related to different rootstocks and or different training systems. Other blocks (with no experiments) had zones with more complex, irregular shapes. These zones are likely to be related to

Fig. 12.4 Vineyard blocks at Pech Rouge with the NDVI layer classified into three equal classes (tertiles) within-blocks

environmental characteristics (soil, topography, and so on). The vineyard manager wanted to focus on these blocks to consider different practices for better control of yield and berry quality at harvest. Recall that in this non-irrigated vineyard, the assumption is that the variation in the NDVI values makes it possible to estimate the variation of vegetative expression induced by differences in VWS.

The TOi opportunity index was calculated to rank the blocks according to the within-block variation in NDVI and to identify three classes of blocks (Fig. 12.5). There were 13 blocks that had strong and structured spatial variation in NDVI (in white on Fig. 12.5). The vineyard manager's expertise reduced this to 11 blocks as two blocks (Blocks 12 and 13; Fig. 12.5) showed significant variation caused by experiments related to the training system or the rootstock. The rest of this case study focused on these 11 blocks, whose variation was assumed to originate from environmental factors.

With aerial NDVI collected over 3 years, the 11 blocks showed a relationship between the EC_a ground measurements and the NDVI up-scaled to 3 m^2 pixel ($R^2 = 0.64$, $n = 33$, data not shown). Similar findings have been found in California, USA (Johnson et al., 2003). It confirmed that NDVI was a reliable indicator of the variation in vegetative expression within blocks in these specific production conditions.

The variation highlighted by the NDVI was mainly explained by water access and the resulting PLWP (Fig.12.6). Of the 11 selected blocks, the observed PLWP at veraison was systematically smaller in the zones of low NDVI. The differences were particularly significant ($p < 0.05$) for Blocks 3, 4, 5, 6, 7, 8 and 9 on calcareous soils. The differences are less, but still significant ($p < 0.05$), for Blocks 10 and 11 on colluvial soils. For Blocks 1 and 2 on sandier soils, the differences in PLWP were small ($p > 0.05$) between NDVI zones. However, for these last two blocks, the

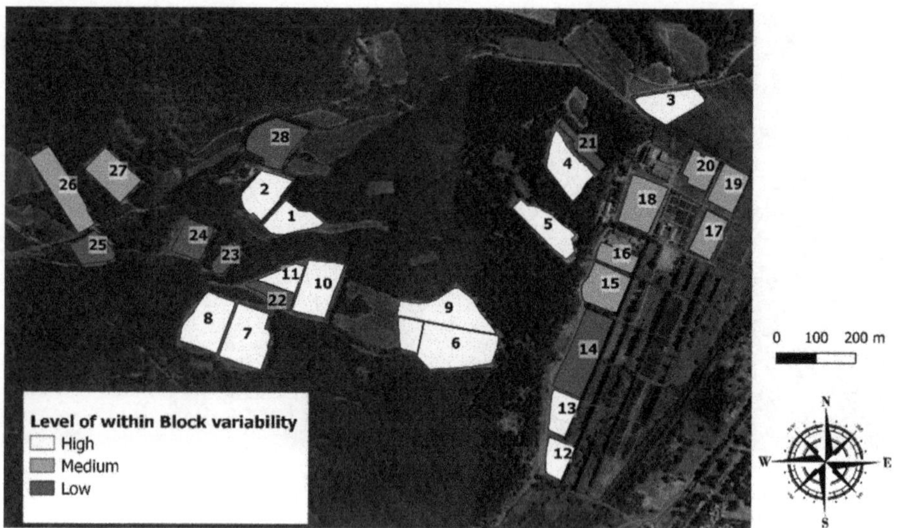

Fig. 12.5 Classification of the blocks according to their variability in NDVI response as assessed using the TOi metric. Block IDs are shown that relate to Block IDs referred to in the text and in Fig. 12.6

general trend was followed, with smaller PLWP (more water stress) associated with smaller NDVI values.

Although NDVI proved to be interesting auxiliary information for mapping the spatial variation of VWS, Fig. 12.6 emphasizes that any analysis must be supported by expert knowledge of the soil type and block conditions. For example, it would have been irrelevant to propose a linear model of water stress as a function of the NDVI at the whole vineyard level given the different responses observed on different soil types.

Figure 12.7 shows the VWS results from Block 5 where the PLWP was monitored simultaneously in both NDVI zones (low and high) on 7 dates throughout the season. During the summer, evaporation demand was large, but water consumption by the vines was not compensated for by either irrigation or rainfall. However, water restriction remained moderate in the high NDVI zone where the soil water capacity must be greater. In contrast, in the low NDVI zone, considerable water restriction was observed from early in the season, which can be explained by a relatively smaller soil water capacity. The NDVI zones therefore made it possible to highlight likely differences in soil water capacity. However, the depth of vine roots in these non-irrigated plots (> 5 m) and the difficult soil conditions in which to dig makes it difficult to support this hypothesis with more objective observations. These technical difficulties argue in favour of the use of a simple observation, like NDVI, to integrate and explain varying vine growth conditions.

In Fig. 12.7 the differences in VWS between the high and low NDVI zones were apparent from the beginning of the season and increased as the season progressed.

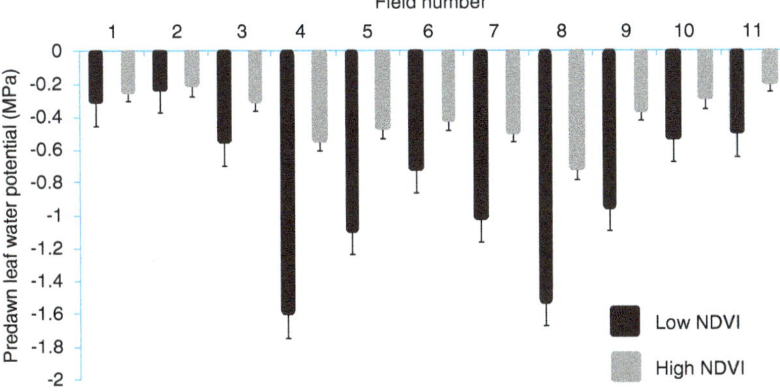

Fig. 12.6 Predawn Leaf water potential observed at within block sites corresponding to high and low NDVI values over 11 blocks (veraison). Field numbers refer to blocks indicated in Fig. 12.5

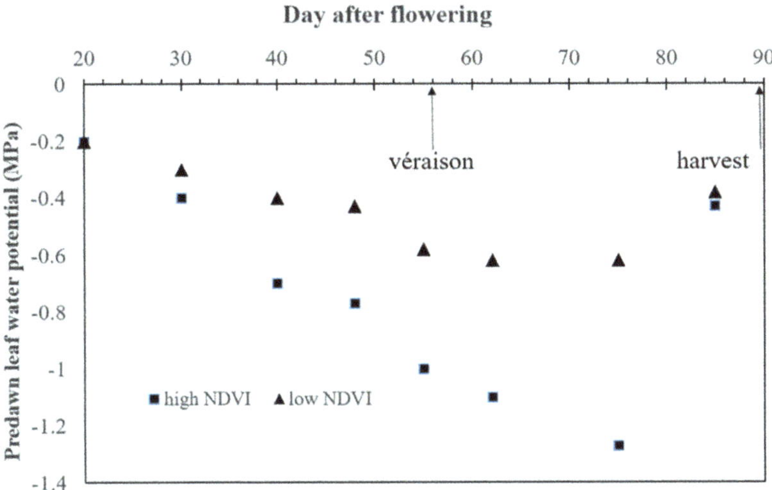

Fig. 12.7 Evolution of vine water stress (measured as predawn leaf water potential (MPa)) over a season for two sites stratified between a low (▲) and a high (■) NDVI zone in Block 5. Arrows and labels indicate key crop stages

Since it has an impact throughout the season, from flowering to grape maturation, this difference affected the growth of the vines (which explains differences in NDVI values) at the beginning of the season and the quality of the harvest during ripening (Table 12.4). The low NDVI zone had less sugar, less phenolics and more acid at harvest than the high NDVI zone. All three conditions are known to reduce the potential quality of any wine produced from the grapes. These values indicated that the low NDVI zone had a problem achieving crop maturity, resulting in poorer berry quality.

Table 12.4 Mean berry quality attributes at harvest for two NDVI zones (High and Low) in Block 5

Zone	Sample size (n)	Sugar (g l^{-1})	pH	Acidity (g l^{-1})	Total polyphenol index (TPI)
High NDVI	29	22.55	3.81	3.54	36.9
Low NDVI	20	20.89	3.78	3.99	33.66

The knowledge of within-block variation explained by water stress and highlighted by NDVI zoning justified the implementation of several applications in the vineyard: i) targeted sampling to estimate yield, maturity and harvest date better, ii) some experiments to assess site-specific mechanical weeding in the inter-row to decrease competition for water and iii) irrigation combined with appropriate nitrogen fertilization on blocks with very variable NDVI. Differential harvest has not been implemented because of logistical constraints.

12.3.4 Conclusions

In a non-irrigated Mediterranean context, this case study showed the value of NDVI in mapping vegetative expression and explaining the variation in VWS because access to water is the main factor affecting vine growth. The aerial NDVI was an 'easy-to-acquire' data set that made it possible to highlight zones of different water stress easily and with high spatial resolution. The case study also highlighted the limits of the approach. The NDVI/water stress relation shown here is only relevant in this pedoclimatic system and the results cannot be extrapolated directly to all vineyards. This case study is a reminder that NDVI remains an integrative data source for summarizing how environmental factors and operations affect the canopy. Its interpretation for decisions must be systematically based on knowledge of the ecophysiology of the vine and with local soil expertise. For significant water constraints, the study shows that the variety effect is negligible compared to the effect of environmental factors. However, the interpretation of NDVI values to quantify differences in water stress remains difficult without prior expertise on soil conditions and especially without taking into account the characteristics of the field (training system, planting density, grass cover, etc.) likely to affect NDVI values.

This research has contributed to the launch of commercial services specifically dedicated to remote sensing in viticulture: in particular the Oenoview service (Terranis, ICV, France). Although constantly increasing, the use of remote sensing in viticulture remains marginal in France, it represents about 1% of the area planted with vines (Lachia et al. 2019). The most important use (in terms of area) concerns cooperatives faced with a great diversity in vineyard blocks spread over a large area. For these structures, the benefit from maps of vegetative expression at veraison is an interesting source of information for the qualitative selection of blocks for improving yield and quality estimates before harvest. In this respect, in the vineyards of southern France, knowledge of the within–field spatial variation controlled by water restriction is relevant to rationalize and optimize the positioning of observations in the blocks before harvest.

12.4 Case Study 12.3. Proximal Sensing for Cotton Management

12.4.1 Introduction

Cotton (*Gossypium* sp.) is one of the most important fibre crops, with approximately 35 million ha grown worldwide, making it one of the 20 most important global commodities in terms of its value in 2016 (http://www.fao.org/faostat). Among the top producers, production in China, India and Pakistan is mainly by small farmers in labour intensive crop systems, whereas production is highly mechanized in the USA, Brazil and Australia. In these latter countries, the cropping systems are capital intensive and require many interventions throughout the season.

Cotton is a perennial plant that is cultivated as an annual crop. It has a growing season of approximately 180 days. If the crop goes through severe biotic or abiotic stresses in this period, the plant will drop most of its reproductive structures as a self-preservation mechanism (Stewart et al. 2009). If, however, conditions are very favourable, the (perennial) plant will prioritize vegetative over reproductive development, a scenario known as rank growth. Although domestication, selection and breeding have produced cultivars with characteristics more aligned with growers' needs, crop management still plays an important role in fibre quality and yield. The main factors controllable by growers that affect productivity are irrigation, fertilizer (mainly nitrogen) and plant growth regulators (PGR). The PGRs are used to control growth, promote higher yields, better fibre quality and permit mechanical harvest.

The optimal rate of PGR application depends on crop height, biomass, and growth rate and these are influenced by variable environmental conditions. Given known variability, there is a great interest in using remote and proximal sensors to monitor crop development and to guide variable-rate application of inputs. Active and passive optical reflectance sensors mounted on various platforms can detect infield crop variation (Sui et al., 2012; Trevisan et al., 2018). Sensors have been used to predict biomass, plant height, height-to-node ratio, nitrogen nutrition status, crop maturity and lint yield (Portz et al., 2014; Arnall et al., 2016). The performance of sensors can vary throughout the crop season because of crop canopy changes and management practices. Some limitations of optical sensors have been observed when used later in the season, related to the known effect of signal saturation with dense canopies and changes in the spectral signature of the plants (Mutanga & Skidmore, 2004). Unlike optical readings, crop height and volume observations have shown good relations with crop biomass through the entire season, without saturation effects. The aim of this case study is to evaluate the potential uses of several proximal sensors as tools to monitor cotton development and yield.

12.4.2 Material and Methods

Observations were taken from 10 fields over 4 years in two states in Brazil (Goiás in 2013 and 2014 and Mato Grosso in 2015 and 2016). The soil in the fields ranges from clay Oxisols to sandy loam Quartzipsamments (Leão, 2016), with different levels of spatial variation in clay content.

Information on soil spatial variation was provided by an apparent soil electrical conductivity (EC_a) survey of the 0–0.3 m layer using a Veris 3100 system (Veris Technologies, Salina, Ka, USA) on 12 m swaths and georeferenced using real-time kinematic (RTK) Global Navigation Satellite System (GNSS) receivers. The EC_a data were acquired in January (in the rainy season), between the soya bean harvest and cotton planting. Data processing consisted of interpolating the EC_a and elevation data by ordinary kriging. A digital elevation model was used to extract the topographic derivatives slope, aspect, curvature and topographical wetness index. Ten points representing the full range of EC_a variation in each field were chosen for soil sampling. The EC_a values at these points were compared with laboratory soil particle size analysis to generate clay and sand maps for each field.

In-season monitoring of crop vigour was done by two types of proximal optical sensor systems. The first optical sensor system, OPS1 (N-Sensor™ ALS, Yara International ASA, Dülmen, Germany), was installed above the vehicle cabin and readings were taken from an oblique position. The N-sensor measures canopy reflectance in the red edge (730 nm) and near-infrared (NIR) (760 nm), and a scaled logarithmic difference of reflectance at the two wavelengths was used as the vegetation index in all comparisons (Jasper et al., 2009). The second optical sensor system, OPS2 (Crop Circle ACS-430, Holland Scientific, Lincoln, NE, USA), was mounted to take readings directly above the crop canopy. The OPS2 was integrated with the GEOSCOUT GLS-420 (Holland Scientific, Lincoln, NE, USA) for data acquisition. The CropCircle sensor measures canopy reflectance at three wavelengths, but only the red edge (730 nm) and NIR (780 nm) were used to calculate the normalized difference red edge index (NDRE) (Horler et al. 1983), which was used in all comparisons.

In-season proximal monitoring of plant height was done using a georeferenced ultrasonic system, US1 (HC-SR04, generic sensor), mounted in the same position as OPS1. The system was developed for this research, using low-cost hardware commonly used in automation projects and data acquisition based on an Arduino® Mega 2560 (Arduino, Ivrea, Italy). The time-of-flight principle was used to measure the distance between the top of the canopy and the sensor (at a fixed height above the ground). Finally, a terrestrial 2D light detection and ranging scanner system (LiDAR–LMS200, Sick, Waldkirch, Germany) was mounted at the front of the vehicle 3.0 m above ground level. Customized data acquisition software was developed to collect georeferenced LiDAR data. The LiDAR sensor was programmed to collect data with an angular resolution of 1°, an angular range of 180° and a distance range of 80 m. The distance resolution was set to 0.01 m. The acquisition rate was 75 Hz, achieved by configuring a 500 kbps baud rate communication through a

RS422 serial interface. Returns further than 24 m were excluded, and distances and angles were used to calculate the UTM coordinates of each return point. The data were subsampled to obtain a constant spatial density because the raw data were denser closer to the machine path. The point cloud was used to extract the ground and canopy levels, defined as the fifth and 95th percentile of the values within 1 m. Plant height was then obtained by subtraction.

All sensors were installed on a high-clearance vehicle operating with a swath width of 24 m, at a maximum travel speed of 6.4 m s^{-1} (23 km h^{-1}). All sensor data acquisition was made simultaneously in a single machine pass in each field. Cotton was harvested using a JD 7760 cotton picker (John Deere, Moline, IL, USA), with the Harvest Doc™ yield monitoring system. All data were georeferenced using RTK-GNSS. All sensor datasets were filtered to remove local extreme values (Spekken et al. 2013). The point data were interpolated to a common grid with 1 m spatial resolution using inverse distance weighted. This method was preferred as the data were already dense and it required less computational effort. All maps were prepared using QGIS software (QGIS Development Team 2018). All statistical procedures were performed using the R programming language (R Core Team 2018).

12.4.3 Results and Discussion

12.4.3.1 General Observations

Data acquisition was successful, and there were no major problems with the sensors. A few problems were reported with the custom-built systems, mainly wiring and some software problems. The commercially available optical systems operated consistently over the four years of data acquisition. The robustness of the sensors is one of the main concerns of growers when considering the adoption of new technologies, therefore this was a very positive result.

The performance of the sensors was better when the variability in crop development had longer spatial ranges and less short-scale variation. Usually the large regions with larger or smaller values have similar patterns, but the pass to pass differences in one map did not match the same short-scale variation in the others (Fig. 12.7). The practical implications of this result is that choice of the best sensor and prescription protocol will depend on the resolution at which any variable-rate prescription can be applied for each particular input. There was also a temporal trend in the proportion of short-scale variation, which was greater in the early stages of the crop because of uneven crop emergence. This made the footprint of the sensor and the ability to average short-scale variation more important at this stage. The OPS1 had an advantage in these early stages (Trevisan et al. 2018) due to its oblique angle of operation enhancing crop reflectance and minimizing soil reflectance compared to sensing the crop from a top-view angle (OPS2). The temporal stability of the spatial variability is also related with its scale. In general, the long-range variability was more stable over long periods of time and different crops in the same field, while short-scale variation was largely affected by operational quality and the interaction with pests and diseases.

12.4.3.2 Example Field

We chose one field from the 2015 crop season to describe the sensor results. The field was 70 ha with a sandy loam soil and clay content ranging from 50 to 150 g kg^{-1}. This was not the most representative field for cotton production in the region, as usually soils with higher clay contents (> 350 g kg^{-1}) are preferred. Nevertheless, considerable crop variability at both short and long ranges made it a good choice to compare sensor performances. The sandy soils are more fragile and small differences in organic matter and nutrient availability can have a large influence on crop development. The EC_a map gives an idea of soil variability over the field (Fig. 12.7a). The density of points collected was approximately 200 points ha^{-1}. The EC_a was very small, reflecting the low cation exchange capacity of this soil. The eastern half of the field had predominantly smaller EC_a and sandier soils, while the western half had larger EC_a and clay content.

The final crop height measured by US1 (Fig. 12.8b) showed an overall similarity with the EC_a map. The density of points was approximately 1800 points ha^{-1}. However, they were all concentrated around the machine path, representing an effective sample of only four crop rows in every 30 rows. The correlation between

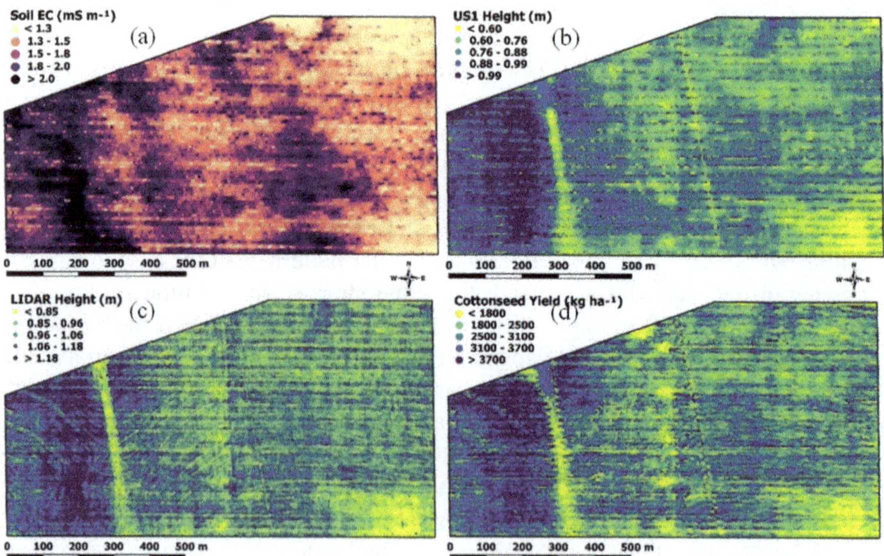

Fig. 12.8 Spatial distribution of the soil electrical conductivity from the Veris 3100 sensor (**a**), final crop height measured by an ultrasonic sensor system (US1) (**b**), final crop height measured by a LiDAR sensor system (c) and cottonseed yield (d)

crop height measured by sensors and by traditional hand-based methods was strong when evaluating small plots (Sui et al. 2012; Trevisan et al. 2018). There was a clear trend of taller plants in the western half of the field and smaller plants in the centre and eastern parts. The stripe with small plants on the western side was associated with the topography of the field. This region has a depression and was the preferred path for run-off water. Heavy rain events (> 100 mm h^{-1}) caused soil erosion and crop lodging in this area. The narrow strip of small plants in the centre was created artificially one month before harvest because of the need to install a pipeline across the field, which required plants to be removed.

The errors observed in the US1 were higher at the early crop stages because of the machine path error, making positioning of the sensors directly over the canopy more difficult (Trevisan et al. 2018). There were also systematic errors observed related to changes in the machine dynamics. The air suspension and tyre compression in the soil gradually changed as the input tank was emptied, altering the relative height of the fixed mounted sensor to the canopy. Irregularities in the terrain topography and machine inclination could also contribute to the error. These challenges would be greater if the sensors were installed on a boom with additional difficulties in calculating the distance to the ground from a moving (swaying) boom.

The same general patterns of crop height observed for the US1 sensors can be seen in the LiDAR map (Fig. 12.7c). The density of points is larger, approximately 600,000 points ha^{-1} in the raw data and 5000 points ha^{-1} in the final height map. The greater resolution allows better identification of short-scale variation and visualization of some patterns that were not clear in the US1 map. The presence of shorter plants following the contour levels of the terrain were probably related to earthworks used to control soil erosion. The stripes along the direction of seeding are related to unequal emergence associated with planter effects at sowing.

The LiDAR system outperformed the ultrasonic system in mapping crop height. There were always sufficient points at survey times to determine the ground level, therefore the absolute position of the sensor was not needed to calculate crop height. It may become a limitation if the canopy achieves full cover and the laser is unable to determine the ground level. The cost of the LiDAR system (hardware and processing requirements) is an order of magnitude higher than the US1, therefore the economic advantage of this system will depend on the use of the data. Currently, the management of spatial variation is usually limited by the resolution of the variable-rate applicators rather than the sensing capabilities, which makes it more difficult to take advantage of the higher resolution data generated by the LiDAR.

Similar results and resolution could be obtained with a UAV and photogrammetric tools. The main challenges at the moment for the use of UAV imagery are in the general logistics of collecting UAV data, the time needed to process images and to make the data available for prescription applications, the accuracy of the GNSS receiver used and the need for ground control points (Feng et al. 2020).

The yields were generally small (Fig. 12.8d), compared to the national average of 3800 kg ha^{-1} (CONAB 2015), mainly because of the sandier soils and poor crop establishment. There was a reasonable positive correlation (> 0.5) between yield and crop height measured by both sensors (Table 12.5). This correlation will not always hold true and will depend on local growth characteristics.

The correlation of EC_a with yield is a good indicator of the proportion of variability that is likely to be stable over the years, since EC_a patterns tend to be stable (Guo 2018). The EC_a map could be used to adjust seeding rates according to variable soil conditions. The general recommendation is to apply larger seed rates in sandier soils and reduced rates in heavier (clay) soils. This promotes some compensation as the plants in sandier soils tend to have a reduced development, and therefore a smaller number of reproductive structures per plant.

The stronger correlation of yield with the LiDAR height compared with the US1 height is mostly explained by the sources of errors previously discussed in the US1 data collection. Although the sensor resolution may be important to obtain better estimates of infield variability, the best economic results will be observed when the sensing resolution and the application resolution are matched (Amaral et al. 2018). The LiDAR data showed significant short-scale and row-to-row variation. Variable-rate application at this resolution is still not practical because of the lack of resolution in the appropriate equipment, being either too expensive or too complicated to be adopted by farmers.

Other research has demonstrated that when fields can be divided into management zones and long-range variation dominates, low-cost systems of variable-rate application have produced good results. In this scenario, PGRs can be applied effectively, resulting in more uniform cotton plant height and yields within fields (Baio et al. 2018). To show the economic viability of investing in a more expensive system is difficult. The economic return of investment of each technology will depend on the degree of variation in each field and the ability to take advantage of it with site-specific applications. It is usually easier to show the benefits with product savings than with significant differences in crop yields. It is even more challenging to show consistent differences in fibre quality (Trevisan et al. 2018).

The main challenge that remains for the use of proximal sensors in cotton management is related to the lack of agronomic knowledge and algorithms to convert sensor readings into decisions and prescriptions. The sensors provide information that is well correlated to the spatial variation of crop development or nutritional status. But to make the right decisions information on the crop response to the input that will be applied is required. For example, cotton biomass in the early stages may be more affected by population counts than by nitrogen deficiency. These confounding factors make it difficult to make the right decision based on a single evaluation from a crop canopy sensor. Combining multiple sensors and readings from different

Table 12.5 Correlation matrix for soil and crop attributes in a cotton field

	Yield	EC_a*	US1 Height
EC_a	0.38		
US1 height	0.51	0.52	
LiDAR height	0.61	0.46	0.57

EC_a: Apparent soil electrical conductivity; US1: Ultrasonic sensor system crop stages can certainly improve the prescriptions (Sharma et al. 2016). The use of calibration strips and on-farm precision experiments can further improve the results, accounting for the temporal and regional variability of crop response to any input (Arnall et al. 2016; Kindred et al. 2017).

Further evidence of this challenge is the wider adoption of solutions as services instead of products (sensors). Even though having the sensors installed on the agriculture machinery has a lower operational cost than using a dedicated platform to collect the data, the services approach has been more successful because it allows more flexibility and less disruption of the field operations. It is also easier to combine more sources of information and agronomic knowledge when the prescription is done as a separate step rather than in real-time (Trevisan et al. 2018).

Furthermore, the variable-rate application of PGR alone is not enough to manage spatial variation when there are large soil differences (Trevisan et al. 2018). The use of variable-rate seeding, nitrogen and PGR applications needs to be planned in an integrated manner to maximize the economic results in every part of the field. However, combining multiple sources of information to guide decisions about multiple inputs under many uncertainties is not a trivial task. This requires better software and more complex models. Sensor fusion and autonomous decision-making techniques are likely to benefit from machine learning methods that have been applied to many domains (Chlingaryan et al. 2018).

12.4.4 Conclusions

Crop sensors showed good performance for monitoring within-field variability in cotton height and vigour (biomass) in fields. There are sensors available with different resolutions and working principles, which might perform better in different cases. Integrating temporal information and multiple sensors is important for improving results. Local farmers have access to sensors available in the market and some already use them for nitrogen management. Challenges remain with agronomic knowledge and algorithms to convert the sensor readings into decisions and prescriptions. The high value and large-scale production of cotton associated with innovation-oriented producers in Brazil make this crop ideal for developing and experimenting with new technologies.

The availability of variable-rate applicators that match the scale of variation is also important. Managing all the inputs in an integrated framework allows for better crop management. Spraying solutions remain a challenge, not because of a lack of technology, but because of the cost associated with this technology for the variable-rate application of multiple products required in certain situations. Further research is needed to evaluate the return on investment of different technologies.

12.5 Case Study 12.4. Integration of UAV Imagery into Potato Crop Modelling Services

12.5.1 Introduction

The increased availability of UAVs has significantly increased the ability of growers to acquire imagery of their crops. Previously, growers had to rely on satellite imagery (with spatial resolution, over-pass time and cloud cover issues) or aerial surveys, which are often expensive, to obtain imagery. The availability of UAVs has allowed growers greater flexibility with high-resolution image acquisition. They can be operated by growers to collect information when needed. A UAV system still has limitations. However, there are some limitations that initially revolved around converting multiple images of a field automatically into a 'mosaicked' image of the field. Algorithms have advanced recently and growers can now acquire timely, mosaicked, geo-rectified images of whole fields without needing any specific image processing skills. This means that UAV imagery is now a viable information source for modern commercial agriculture (Tsouros et al. 2019, Cucho-Padin et al. 2019). The next major limitation is the cost of the system, particularly the camera. A UAV itself is just a platform, like a tractor, airplane or satellite, onto which a sensor can be mounted. The more advanced the sensing system, the more expensive it is. It is possible to mount multispectral (MS), hyperspectral (HS) and thermal sensors onto UAVs, however, for most growers, especially small-holder farmers, the cost of these sensors is prohibitive (Cucho-Padin et al. 2019). The 'sensor' that is accessible, is a simple colour (RGB) camera. Ideally, a MS sensor with a NIR band would be preferable for mapping vigour in a similar way to MS sensors mounted on satellites or tractors (as illustrated in previous case studies in this chapter). It is likely that such MS camera systems will become available at a competitive price point for the majority of growers; however, in the interim the use of colour imagery and of very high-resolution images is challenging for a UAV-mounted RGB camera to deliver.

Another advance in precision agriculture is the use of crop models for short-term spatial prediction (Chen et al. 2017). This shifts traditional (semi-) deterministic models from long-term strategic applications (e.g. climate change predictions) to short-term tactical application for variable-rate management. This transformation requires a change of input information, and in some cases a change in parameters, within the crop model to allow newly accessible digital information, like UAV imagery, to be used. This case study illustrates how basic colour UAV imagery, has been incorporated into an existing potato model to

(a) improve functioning of the model, and
(b) allow the model to be spatialized within a field

The research and extension was performed and developed within commercial systems and led primarily by industry.

12.5.2 *The Crop Model*

The growth model in this study simulates daily crop growth of a single potato plant and its interaction with the local environment. It predicts both potential crop yield and the yield under restricted water conditions with the assumption of adequate, non-limiting nutrients and correct crop protection practices. It originates from experimental work done at the James Hutton Institute (MacKerron and Waister, 1985 and MacKerron, 1985). The model is mechanistic, modelling the sequential stages of crop sprouting, leaf expansion, dry matter accumulation and partitioning, and crop senescence of a potato plant as well as their effects on crop yield. Plant development is primarily influenced by weather factors (temperature and solar radiation interception) and soil moisture conditions (MacKerron et al. 2004). Plant density information is then used to scale the plant model to a field response. The flow of the crop modelling process, and the input and intermediate parameters and variables, is shown in Fig. 12.9.

As the model is deterministic, errors at early stages of the model related to crop development will have an effect, and in some cases a large effect on final yield prediction. For example, dry matter accumulation may be up to 1 t ha^{-1} day^{-1} mid-late season. Errors in modelling the emergence date can therefore affect the predicted yield at crop burn-down by effectively shortening or lengthening the growing season within the model. For strategic, long-term uses, these errors are not critical. The

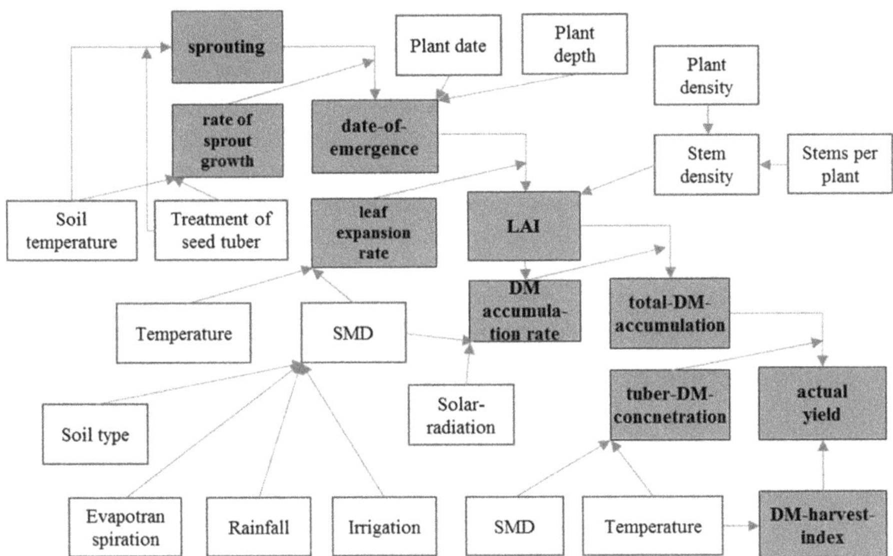

Fig. 12.9 Schematic diagram of a potato crop model. The white blocks show initial conditions and other model inputs for model simulation. The grey blocks represent the progression of crop growth and development from sprouting, emergence, vegetative development to actual yield. LAI = Leaf Area Index, SMD = Soil Moisture Deficit, DM = Dry Matter

model is only being used for scenario-testing and the emergence date error would be a constant. However, for short-term tactical uses, yield predictions need to be as accurate and precise as possible, and a 4–7 day difference in emergence date (model vs observed) will introduce a large error in yield prediction.

12.5.3 Adapting the Model Using UAV Imagery

The model is essentially a dry matter accumulation model, therefore, having the correct canopy size (leaf area index or LAI) at any given period is critical to good model predictions. The ability to operate UAVs and process imagery at high temporal resolutions provides an opportunity to gather relevant ground cover (%) information to validate and correct the model. The model has been adjusted in three fundamental ways with observed ground cover to align the model outputs more closely with local conditions;

1. Early-season imagery allows identification and counting of individual plants to provide spatial plant density,
2. Multi-temporal early-mid season imagery can be used to either identify emergence data or to model emergence date, and
3. Multi-temporal early-mid-season imagery can be used to identify the date of canopy closure or the observed level of canopy closure (if not full)

These three factors adjust spatial solar radiation interception, dry matter accumulation and ultimately the spatial estimation of potato yield within a field. This depends on two factors. First, availability of the relevant information derived from the UAV imagery (see below), and second, the method by which this information is incorporated into the model. New, observed information can be incorporated into the model in several different ways. The two most common approaches are by a 'forcing' strategy or a 're-initialization or re-parameterization' strategy (Moulin et al. 1998). There are advantages and disadvantages to both of these approaches. To develop this commercial service, an approach based on both re-initialisation and re-parameterisation of the potato crop model was developed to enable the integration of spatial crop canopy cover information from UAV-mounted RGB cameras.

1. **Spatial plant density information**

Within the model, plant density affects the rate of canopy development and LAI. Higher densities have faster rates of canopy development. The grower at planting sets plant density, and stems per plant is a constant in the model. The assumption is therefore that the planter produces a uniform plant and stem density in the field.

Early season imagery provides an opportunity to count plants automatically as they emerge and to observe local plant densities. At very early stages of growth, new, green plants are contrasted against the soil background. Figure 12.10(a–c) shows part of a UAV image taken around emergence. This colour image is converted to greyscale and a threshold used to generate a binary image (plant or non-plant)

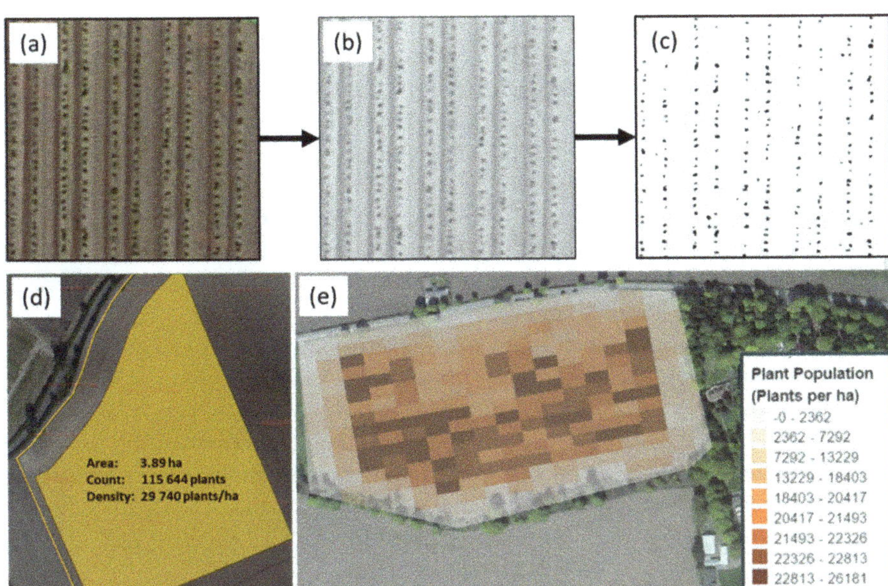

Fig. 12.10 An example of a very early season high resolution image of part of a potato field (~100 m^2) that shows individual potato plants emerging (**a**). The image is converted into greyscale (**b**), and then a binary image using a threshold algorithm (**c**). The discrete elements in the binary image relate to the number of plants and can be summed for a whole field (**d**) or sub-field/'pixel' scales (**e**) to generate observed plant density maps

and a count of the number of discrete elements in the threshold image. These processes are standard and use common image processing techniques. The result is that for any given area the number of observed, emerged plants can be calculated and mapped (Fig. 12.10 d–e). In the first instance, this provides the grower with clear information about emergence and establishment within the field, and possible areas for improvement. It also provides spatial information on plant density that can be entered directly into the model to improve LAI simulation.

Plant counting does have some limitations; mainly that it depends on each plant being an individual, discrete object, i.e. the plants cannot overlap. If plants do overlap, then multiple plants are considered as only one individual. Growth rates early in the season are often very fast, so there is only a short window of opportunity between emergence and the start of overlapping, a window in which imagery needs to be collected for the counting to be effective.

2. **Improved estimation of emergence date**

The original model calculated emergence date based on seed quality, soil temperature and planting depth. However, within a field, different areas can have different rates of crop sprouting (because of variation in soil conditions, planting depth or seed quality). Regular early season flights can provide information on emergence and the actual, or likely, date where 50% of plants have emerged. This direct approach depends on regular (1–2 day) flights around emergence. The plant counting algorithm used for plant density can be run at each time step, the maximum density found and an estimate made of the date when 50% emergence occurred. However, such high-frequency flying might be problematic, especially where regulations on UAV operation are strict, when flights cannot be fully automated, and when labour is scarce. An alternative is to take weekly early season flights to observe changes in ground cover over time and to back-predict the emergence date.

Percentage ground cover from UAV imagery can be obtained in a way similar to that of plant counts, i.e. the ratio of green to non-green pixels is the percentage ground cover in the image (Fig. 12.11). If ground cover (%) is recorded at several (usually 3+) time periods, then simple mathematical models of plant growth can be used to predict emergence. For most annual crops, canopy development follows a sigmoidal response and fitting a relevant equation allows prediction of the date when ground cover began to develop. This modelling approach has the advantage that fewer UAV flights are needed, and the timing of these flights is less critical, i.e. flights immediately pre and post-emergence are not required. Whether directly observed or modelled from early season imagery, a more accurate and spatially varying emergence date can be generated for a field.

3. **Observed LAI and canopy closure date**

The same information, multi-temporal ground cover (%) images during canopy development (Fig. 12.11) and the same sigmoidal canopy development model can also be used to provide a forward prediction of the likely date of canopy closure as

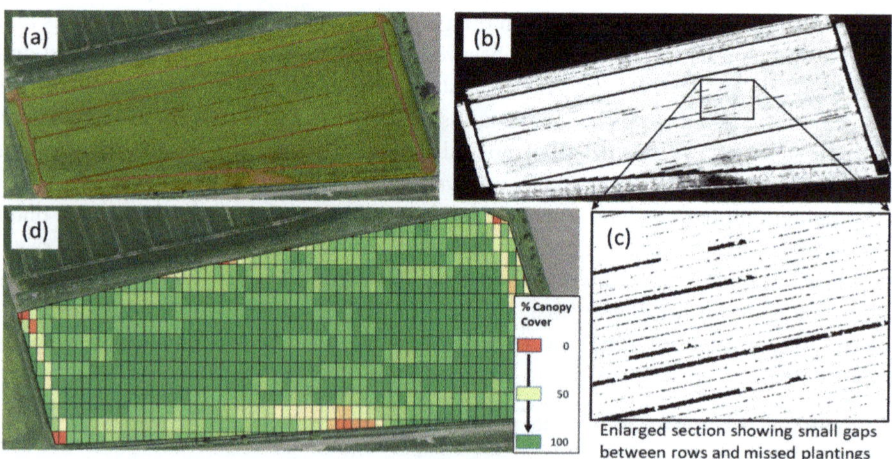

Fig. 12.11 A mid-season (after expected canopy closure) UAV colour image of a potato field (**a**) again converted into a binary image (**b**) to determine % ground cover in the field. An enlarged section, illustrating missed plants and incomplete canopy fill (lower production potential) is shown (black areas) (**c**). Operational ‘pixels’ can be superimposed and the local canopy cover (%) calculated to form a map for spatial modelling and or management (**d**)

well. The estimated date of canopy closure (or full canopy development) can be verified by another UAV image around or after the expected canopy closure date.

This is particularly useful in fields (or parts of fields) where local conditions restrict canopy development and canopy closure may not be achieved. Information on full canopy development (date where closure is definite or the maximum ground cover % is realized) permits the canopy development model to be updated and to reflect actual, local conditions. This provides more accurate information for potential solar radiation interception and on dry matter accumulation and yield.

12.5.4 *Examples of Improvement in Model Performance*

To illustrate the value of incorporating the UAV imagery into the potato crop model, the model was run in several different modes in a commercial seed potato field in Scotland in 2016.

The four approaches were:

(a) the original model at the whole-field scale (using data sourced only from the grower, i.e. no UAV or in-season observations and emergence date simulated by the crop model.)
(b) the model with the emergence date adjusted (back-prediction from early-season canopy observations)
(c) the model with LAI adjusted to fit the mid-season canopy development (but retaining the original simulated emergence date from (a))
(d) the model with both emergence date and mid-season LAI adjusted according to in-season field observations

To calibrate and to validate the four approaches, 100 sites were selected in the field where manual canopy (ground cover %) and yield observations were made (see Taylor et al. 2018 for details on the sampling and data collected) and complemented with mid-season UAV flights. In this example, the emergence date and canopy information were averaged and for all cases (a–d) the model was run at the whole-field scale. Weather and soil information were constant for all four approaches. Table 12.6 shows the simulated yields for the different approaches while Fig. 12.12

Table 12.6 Results from simulating potato yield using the original crop model (a) and with adjustments derived from in-season UAV imagery (b-d) to correct emergence date and LAI

Approach	Change(s) made	Observed yield (t ha^{-1})	Predicted yield (t ha^{-1})	Error (%)
a	Original model	37.13	52.23	40.67
b	Emergence date adjusted		45.26	21.90
c	LAI adjusted to mid-season observation		41.99	13.09
d	Emergence date and LAI adjusted		40.42	9.86

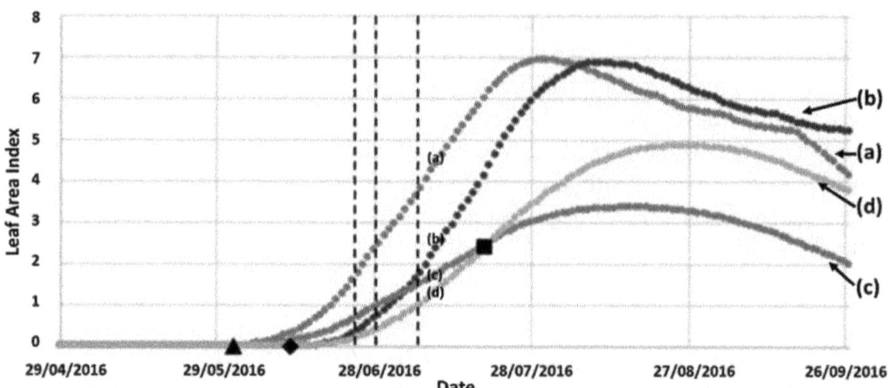

Fig. 12.12 Plots of the evolution of LAI as an intermediate value within the crop model over time for the four approaches. Labels (a–d) relate to the approaches in Table 12.6. The simulated emergence date (crop model-based) is indicated by ▲, the emergence date based on early season ground cover observation (dates indicated by vertical dashed lines) is indicated by ◆ and the LAI from a UAV-derived groundcover image (July 19th) is indicated by ■

shows the change in LAI within the model as the emergence date and the mid-season canopy data were included.

For all four approaches, the crop model over-estimated the final yield. This might be due to the soil conditions, especially soil moisture, being incorrectly specified in the model, even though the best available data were used. The soil data were kept constant, therefore any error should be constant. The native, deterministic model approach (a) over-estimated yield by 40%. Adjusting the emergence date (b) or the mid-season LAI (c) information reduced this error to 22 and 13%, respectively. Adjusting both the emergence date and the mid-season LAI within the model reduced the error in yield prediction to <10% (3.29 t ha^{-1}). Simulating canopy development and size correctly appears to be more important than observing emergence correctly. Plotting the daily LAI model values (Fig. 12.12) clearly illustrated the effect of incorporating in-season observations on the simulation.

The simulated emergence date (based on soil conditions and planting depth) was much earlier than expected, which led to earlier canopy development, greater dry matter accumulation and effectively a longer production season. The early season canopy observations produced an estimated emergence date 11 days later than the simulated model emergence date, resulting in a 7 t ha^{-1} reduction in yield prediction. However, canopy development was still simulated to be very strong in the field (LAI > 5), whereas a UAV image in mid-July revealed that full canopy closure had not occurred. When this information was included in the model, the effective LAI, dry matter accumulation and yield potential decreased, and the model output moved closer to the observed yield values.

12.5.5 Conclusions

Optical sensing for crop vigour has traditionally used the response within red and near-infrared wavelengths to interpret crop vigour and biomass. This case study has shown that it is not always necessary to have these wavelengths to generate good agronomic information. The increasing availability of platforms (UAV or terrestrial) that provide high-resolution imagery, from even relatively cheap camera systems, is changing the way that optical systems are being deployed and the information used. The likelihood is that best agronomic practices will be obtained with imagery from both visible and near-infrared systems within models and decision support systems that are formulated and designed with these data sets in mind.

12.6 Conclusions for the Chapter

These examples illustrate that reflectance in the red and near-infrared wavelengths from the canopy of all crops provides both spatial and temporal information on the canopy condition. As agriculturists, we can link this reflectance to productivity and to crop health. However, the canopy response is generic and it is not possible to extract directly which specific biotic and or abiotic effects control the canopy response. Consequently, a wide variety of different agronomic services have been proposed using optical imagery in crop systems. However, under an assumption of good crop management (good pest and disease control) the majority have been aimed at identifying fertilizer rates. A case study on nitrogen fertilizer was not included here because this application is already widely accepted (and described elsewhere in this book). Instead, the case studies presented above show how the use of multispectral optical sensing in agriculture is much broader than variable-rate fertilizer application and is expanding, for example, to account for temporal changes over the season or to assist in high-resolution crop modelling. Multispectral optical sensing has been, and will continue to be, an important information source for all aspects of site-specific crop management.

However, as illustrated in these examples, it is not necessarily the type of multispectral sensor that is important, but the agronomic service that it is incorporated into. The sensors work; but must deliver agronomic solutions to be adopted. The same sensor, a CropCircle™, was used for both the table grape (Greece) and cotton (Brazil) case studies, but delivered different information for different services. Both cases are valid applications of the same sensor. It is also critical that any agronomic services are developed to account for any advantages (or disadvantages) associated with a particular sensing system. Even though similar sensors can be tractor-mounted or UAV-mounted, it does not follow that data from a UAV-mounted system can be incorporated directly into a decision support system built on tractor-mounted data acquisition. In a similar vein, services and decisions must be tailored to suit the local agronomy, even if the same sensing system is used. Fertilizer decisions are always linked to current crop vigour (making canopy sensing useful), however for wheat in northern Europe these decisions will be different from those for wheat in southern Europe even if the same satellite or tractor-mounted system is used because the production system needs are different.

The multispectral systems featured here are current mainstays of commercial optical sensing systems. They are affordable, robust and easy to use, but generate a generic response. Increasing the number of bands and the sensitivity (width) of the bands generates more specific information (and more expensive sensors). These sensors are termed hyperspectral (typically >20 bands and often >100). For optical crop sensing systems, hyperspectral sensors will eventually overtake multispectral systems because of the additional information that can be collected and used to separate biotic and abiotic stresses in the crop. However, these sensors are prohibitively costly at the moment for wide commercial use. Adoption is also limited by the availability of commercial services that can make use of the additional information in the hyperspectral data. Without unlocking this potential, end-users will probably continue with the cost-effective and proven multispectral systems.

Finally, optical sensing, even hyperspectral sensing, is unlikely to be the 'holy grail' and provide the perfect solution for all agronomic decisions. These optical sensors need to be incorporated into decision systems with other types of sensors (as illustrated in the cotton example) or with other tools (e.g. the crop models in the potato case study) to optimize production systems.

References

Acevedo-Opazo C, Tisseyre B, Ojeda H et al (2008a) Is it possible to assess the spatial variability of vine water status? OENO One 42(4):203–219

Acevedo-Opazo C, Tisseyre B, Guillaume S et al (2008b) The potential of high spatial resolution information to define within-vineyard zones related to vine water status. Precis Agric 9(5):285–302

Amaral LR, Trevisan RG, Molin JP (2018) Canopy sensor placement for variable-rate nitrogen application in sugarcane fields. Precis Agric 19:147–160

Anastasiou E, Balafoutis A, Darra N et al (2018) Satellite and proximal sensing to estimate the yield and quality of table grapes. Agriculture 8(7):94

Arnall DB, Abit MJM, Taylor RK et al (2016) Development of an NDVI-based nitrogen rate calculator for cotton. Crop Sci 56(6):3263–3271

Badr G, Hoogenboom G, Davenport J et al (2015) Estimating growing season length using vegetation indices based on remote sensing: a case study for vineyards in Washington state. Trans ASABE 58(3):551–564

Baio FHR, Neves DC, Souza HB et al (2018) Variable rate spraying application on cotton using an electronic flow controller. Precis Agric 19(5):912–928

Baralon K, Payan JC, Salançon E et al (2012) SPIDER: spatial extrapolation of the vine water status at the whole denomination scale from a reference site. J Int Sci Vigne Vin 46(3):167–175

Celette F, Gary C (2013) Dynamics of water and nitrogen stress along the grapevine cycle as affected by cover cropping. Eur J Agron 45:142–152

Chen H, Leinonen I, Marshall B et al (2017) Conceptual Spatial Crop Models for Potato Production Advances in Animal Biosciences 8(2):678–683

Cheng G, He YN, Yue TX et al (2014) Effects of climatic conditions and soil properties on cabernet sauvignon berry growth and anthocyanin profiles. Molecules 19(9):13683–13703
Chlingaryan A, Sukkarieh S, Whelan B (2018) Machine learning approaches for crop yield prediction and nitrogen status estimation in precision agriculture: a review. Comput Electron Agr 151:61–69
Choné X, Van Leeuwen C, Chery PH et al (2001) Terroir influence on water status and nitrogen status of non-irrigated Cabernet Sauvignon (Vitis vinifera) Vegetative development, must and wine composition (example of a Medoc top estate vineyard, Saint Julien area, Bordeaux, 1997) S Afr J of Enol Vitic 22(1):8–15
CONAB (2015) Acompanhamento da safra brasileira (Brazil crop production), 2(12):1–134
Cucho-Padin, G, Loayza, H, Palacios, S. et al (2019) Development of low-cost remote sensing tools and methods for supporting smallholder agriculture. Appl Geomatics
Dry PR, Loveys BR (1998) Factors influencing grapevine vigour and the potential for control with partial rootzone drying. Aust J Grape Wine R 4(3):140–148
Feng A, Zhou J, Vories ED, Sudduth KA et al (2020) Yield estimation in cotton using UAV-based multi-sensor imagery. Biosyst Eng 193:101–114
Guo W (2018) Spatial and temporal trends of irrigated cotton yield in the southern High Plains. Agronomy 8(12):298
Hall A, Wilson MA (2013) Object-based analysis of grapevine canopy relationships with winegrape composition and yield in two contrasting vineyards using multitemporal high spatial resolution optical remote sensing. Int J Remote Sens 34:1772–1797
Hall A, Lamb DW, Holzapfel B et al (2002) Optical remote sensing applications in viticulture-a review. Aust J Grape Wine R 8(1):36–47
Henry D, Aubert H, Véronèse T (2019) Proximal radar sensors for precision viticulture. IEEE T Geosci Remote 57(7):4624–4635
Horler DNH, Dockray M, Barber J (1983) The red edge of plant leaf reflectance. Int J Remote Sens 4(2):273–288
Hunt ER Jr, Doraiswamy PC, McMurtrey et al (2013) A visible band index for remote sensing leaf chlorophyll content at the canopy scale. Int J Appl Earth Obs 21:103–112
Iland P (2000) Techniques for chemical analysis and quality monitoring during winemaking. Patrick Iland Wine Promotions, Campbelltown, S.A, Australia
Intrigliolo DS, Castel JR (2010) Response of grapevine cv.'Tempranillo'to timing and amount of irrigation: water relations, vine growth, yield and berry and wine composition. Irrigation Sci 28(2):113
Jasper J et al (2009) Active sensing of the N status of wheat using optimized wavelength combination: impact of seed rate, variety and growth stage. In: van Henten EJ, Goense D, Lokhorst C (eds) Precision Agriculture '09. 7th European Conference on Precision Agriculture, Wageningen, July 2009. Wageningen Academic Publishers, Wageningen, p 23
Johnson LF, Roczen DE, Youkhana SK et al (2003) Mapping vineyard leaf area with multispectral satellite imagery. Comput Electron Agr 38(1):33–44
Kindred DR, Milne AE, Marchant B et al (2017) Spatial variation in Nitrogen requirements of cereals, and their interpretation. 303–307. Adv Anim Biosci 8(2):303–307
Kross A, McNairn H, Lapen D et al (2015) Assessment of RapidEye vegetation indices for estimation of leaf area index and biomass in corn and soybean crops. Int J Appl Earth Obs 34:235–248
Lachia N, Pichon L, Tisseyre B (2019) A collective framework to assess the adoption ofprecision agriculture in France: description and preliminary results after two years.In: Stafford JV (ed) Precision agriculture '19. 12th European conference on Precision agriculture, Montpellier, 8-11 July 2013. Wageningen Academic Publishers, Wageningen, p 851
Lamb DW, Weedon MM, Bramley RGV (2004) Using remote sensing to predict grape phenolics and colour at harvest in a cabernet sauvignon vineyard: timing observations against vine phenology and optimising image resolution. Aust J Grape Wine R 10(1):46–54
Leão TP (2016) Particle size distribution of Oxisols in Brazil. Geoderma Reg 7(2):216–222

Leroux C, Jones H, Pichon L et al (2018) GeoFIS: an open source, decision-support tool for precision agriculture data. Agriculture 8(6):1–21
Lymburner L, Beggs PJ, Jacobson CR (2000) Estimation of canopy-average surface-specific leaf area using Landsat TM data. Photogramm Eng Rem S 66(2):183–192
MacKerron DKL (1985) A simple model of potato growth and yield. II. Validation and external sensitivity. Agric For Meteorol 34:285–300
MacKerron DKL, Waister PD (1985) A simple model of potato growth and yield. I. Model development and sensitivity analysis. Agric For Meteorol 34:241–252
MacKerron DKL, Marshall B, McNicol JW (2004) MAPP and the underlying functions that it contains. In: MacKerron DKL, Haverkort AJ (eds) Decision support systems in potato production: bringing models to practice. Wageningen Academic Publishers, Wageninen
Madni A, Madni C, Lucero S et al (2019) Leveraging digital twin Technology in Model-Based Systems Engineering. Systems 7(1):7
Mirasi A, Mahmoudi A, Navid H et al (2019) Evaluation of sum-NDVI values to estimate wheat grain yields using multi-temporal Landsat OLI data. Geocarto Int:1–16
Moulin S, Bondeau A, Delecolle R (1998) Combining agricultural crop models and satellite observations: from field to regional scales. Int J Remote Sens 19(6):1021–1036
Mutanga O, Skidmore AK (2004) Narrow band vegetation indices overcome the saturation problem in biomass estimation. Int J Remote Sens 25:3999–4014
Naor A, Gal Y, Bravdo B (1997) Crop load affects assimilation rate, stomatal conductance, stem water potential and water relations of field-grown Sauvignon blanc grapevines. J Exp Bot 48(9):1675–1680
Oczkowski E (2016) The effect of weather on wine quality and prices: an Australian spatial analysis. J Wine Econ 11(1):48–65
OIV 2019 Statistical report on world Vitiviniculture. International Organisation of Vine and Wine Intergovernmental Organisation. Available via http://www.oiv.int/public/medias/6782/oiv-2019-statistical-report-on-world-vitiviniculture.pdf. Accessed 14 Feb 2020
Ojeda H, Andary C, Kraeva E et al (2002) Influence of pre-and postveraison water deficit on synthesis and concentration of skin phenolic compounds during berry growth of Vitis vinifera cv. Shiraz Am J Enol Viticult 53(4):261–267
Ollat N, Touzard JM, van Leeuwen C (2016) Climate change impacts and adaptations: new challenges for the wine industry. J Wine Econ 11(1):139–149
Portz G, Vilanova Jr NDS, Trevisan RG et al (2014) Cotton field relations of plant height to biomass accumulation and n-uptake on conventional and narrow row systems. In: proceedings of the 12th international conference on precision agriculture. Sacramento, USA, 20-23 July 2017
QGIS Development Team. (2018) QGIS geographic information system, Version 32 http://qgisorg Accessed 14 Feb 2020
R Core Team. (2018) R: a language and environment for statistical computing. Vienna, Austria https://wwwr-projectorg/ Accessed 14 Feb 2020
Ricotta C, Avena G, De Palma A (1999) Mapping and monitoring net primary productivity with AVHRR NDVI time-series: statistical equivalence of cumulative vegetation indices. ISPRS J Photogramm 54(5–6):325–331
Rodriguez D, Fitzgerald GJ, Belford R et al (2006) Detection of nitrogen deficiency in wheat from spectral reflectance indices and basic crop eco-physiological concepts. Aust J Agric Res 57:781–789
Rouse JW, Haas RH, Schell JA et al (1974) Monitoring vegetation systems in the Great Plains with ERTS. NASA special publication 351:309
Sen F, Oksar RE, Kesgin M (2016) Effects of shading and covering on "Sul-tana seedless" grape quality and storability. J Agric Sci Technol 18:245–254
Sharma LK, Bu H, Franzen DW et al (2016) Use of corn height measured with an acoustic sensor improves yield estimation with ground based active optical sensors. Comput Electron Agr 124:254–262

Sibille I, Ojeda H, Prieto J et al (2007) Relation between the values of three pressure chamber modalities (midday leaf, midday stem and predawn water potential) of 4 grapevine cultivars in drought situation of the southern of France. Applications for the irrigation control. In: Proceedings of the XVth conference of Groupe d'Etude des Systèmes de Conduite de la Vigne (GESCO). Porec, Croatia, 20-23 June 2007
Sipiora MJ, Granda MJG (1998) Effects of pre-veraison irrigation cut-off and skin contact time on the composition, color, and phenolic content of young cabernet sauvignon wines in Spain. Am J Enol Vitic 49(2):152–162
Spekken M et al (2013) A simple method for filtering spatial data. In: Stafford JV (ed) Precision Agriculture '13. 9th European Conference on Precision Agriculture, Lleida, July 2013. Wageningen Academic Publishers, Wageningen, p 259
Stewart JM, Oosterhuis D, Heitholt JJ et al (eds) (2009) Physiology of cotton. Springer Science & Business Media, Dordrecht
Sui R, Fisher DK, Reddy KN (2012) Cotton yield assessment using plant height mapping system. J Agr Sci 5(1):23–31
Sun L, Gao F, Anderson MC et al (2017) Daily mapping of 30 m LAI and NDVI for grape yield prediction in California vineyards. Remote Sens-Basel 9(4):317
Tagarakis AC, Liakos V, Fountas S et al (2013) Management zones delineation using fuzzy clustering techniques in grapevines. Precis Agric 14:18–39
Tagarakis AC, Koundouras S, Fountas S et al (2018) Evaluation of the use of LIDAR laser scanner to map pruning wood in vineyards and its potential for management zones delineation. Precis Agric 19:334–347
Taylor JA, Acevedo-Opazo C, Ojeda H et al (2010) Identification and significance of sources of spatial variation in grapevine water status. Aust J Grape Wine R 16(1):218–226
Taylor JA, Chen H, Smallwood M et al (2018) Investigations into the opportunity for spatial management of the quality and quantity of production in UK potato systems. Field Crops Res 229:95–102
Tehrani MM, Kamgar- Haghighi AA, Razzaghi, F et al (2016) Physiological and yield responses of rainfed grapevine under different supplemental irrigation regimes in Fars province, Iran. Sci Hortic-Amsterdam 202:133–141
Tisseyre B, McBratney AB (2008) A technical opportunity index based on mathematical morphology for site-specific management: an application to viticulture. Precis Agric 9(1–2):101–113
Tisseyre B, Ojeda H, Taylor J (2007) New technologies and methodologies for site-specific viticulture. J Int Sci Vigne Vin 41(2):63–76
Trevisan R, Vilanova N Jr, Eitelwein M et al (2018) Management of Plant Growth Regulators in cotton using active crop canopy sensors. Agriculture 8(7):101
Tsouros DC, Bibi S, Sarigiannidis PG (2019) A review on UAV-based applications for precision agriculture. Information 10
Van Leeuwen C, Seguin G (1994) Incidences de l'alimentation en eau de la vigne, appréciée par l'état hydrique du feuillage, sur le développement de l'appareil végétatif et la maturation du raisin (Vitis vinifera variété Cabernet franc, Saint-Emilion, 1990). J Int Sci Vigne Vin 28(2):81–110
Wang FM, Huang JF, Tang YL et al (2007) New vegetation index and its application in estimating leaf area index of Rice. Rice Sci 14(3):195–203
Xue J, Su B (2017) Significant remote sensing vegetation indices: a review of developments and applications. J Sensors
Zhou X, Zheng HB, Xu XQ et al (2017) Predicting grain yield in rice using multi-temporal vegetation indices from UAV-based multispectral and digital imagery. ISPRS J Photogramm 130:246–255

Chapter 13
Applications of Sensing for Disease Detection

Ana Isabel de Castro Megías, Claudia Pérez-Roncal, J. Alex Thomasson, Reza Ehsani, Ainara López-Maestresalas, Chenghai Yang, Carmen Jarén, Tianyi Wang, Curtis Cribben, Diana Marin, Thomas Isakeit, Jorge Urrestarazu, Carlos Lopez-Molina, Xiwei Wang, Robert L. Nichols, Gonzaga Santesteban, Silvia Arazuri, and José Manuel Peña

Abstract The potential loss of world crop production from the effect of pests, including weeds, animal pests, pathogens and viruses has been quantified as around 40%. In addition to the economic threat, plant diseases could have disastrous consequences for the environment. Accurate and timely disease detection requires the use of rapid and reliable techniques capable of identifying infected plants and providing

---

Ana Isabel de Castro Megías: Introduction and Case Study 13.2
J. Alex Thomasson, Tianyi Wang, Curtis Cribben, Thomas Isakeit, Xiwei Wang and Robert L. Nichols: Case Study 13.1
Reza Ehsani and José Manuel Peña: Case Study 13.2
Claudia Pérez-Roncal, Ainara López-Maestresalas, Chenghai Yang, Carmen Jarén, Diana Marin, Jorge Urrestarazu, Carlos Lopez-Molina, Gonzaga Santesteban and Silvia Arazuri: Case Study 13.3

A. I. de Castro Megías (✉)
National Institute for Agricultural and Food Research and Technology (INIA-CSIC), Madrid, Spain
e-mail: ana.decastro@inia.es

C. Pérez-Roncal · A. López-Maestresalas · C. Jarén · D. Marin · J. Urrestarazu
C. Lopez-Molina · G. Santesteban · S. Arazuri
School of Agricultural Engineering and Biosciences, Public University of Navarre (UPNA), Pamplona, Spain

J. A. Thomasson
Mississippi State University, Mississippi State, MS, USA
e-mail: thomasson@tamu.edu

R. Ehsani
University of California, Merced, CA, USA

C. Yang
USDA Agricultural Research Service, College Station, TX, USA

R. Kerry, A. Escolà (eds.), *Sensing Approaches for Precision Agriculture*, Progress in Precision Agriculture, https://doi.org/10.1007/978-3-030-78431-7_13

the tools required to implement precision agriculture strategies. The combination of suitable remote sensing (RS) data and advanced analysis algorithms makes it possible to develop prescription maps for precision disease control. This chapter shows some case studies on the use of remote sensing technology in some of the world's major crops; namely cotton, avocado and grapevines. In these case studies, RS has been applied to detect disease caused by fungi using different acquisition platforms at different scales, such as leaf-level hyperspectral data and canopy-level remote imagery taken from satellites, manned airplanes or helicopter, and UAVs. The results proved that remote sensing is useful, efficient and effective for identifying cotton root rot zones in cotton fields, laurel wilt-infested avocado trees and esca-affected vines, which would allow farmers to optimize inputs and field operations, resulting in reduced yield losses and increased profits.

Keywords Crop disease · Image analysis · Spectral analysis · Multispectral imaging · Hyperspectral imaging · Prescription map

13.1 Introduction

The potential yield of agricultural crops can be affected by biotic and abiotic stress factors that can reduce the quality and quantity of production. It has been estimated that around 40% of world crop production is lost due to the impact of pests, including weeds, animal pests, pathogens and viruses (Oerke and Dehne 2004). Moreover, in terms of the efficacy of actual crop protection practices, the control of diseases caused by fungi and bacteria is considerably less than protection obtained for other pests (Oerke and Dehne 2004). In addition, plant diseases are not only an economic threat, but could also have disastrous consequences for the environment, as new diseases and the re-emergence of controlled ones are developing at an alarming rate in crops around the world with transfers between hosts, global climate change and the use of some intensive management practices to increase productivity (Howden et al. 2007). Precision disease control is therefore a challenging goal in agriculture that could assist growers in decision-making to improve crop yields and reduce economic costs and environmental risks.

T. Wang
Texas A&M AgriLife Research, Dallas, TX, USA

C. Cribben
Bridgestone Americas, Inc., Burlington, NC, USA

T. Isakeit
Texas A&M University, College Station, TX, USA

X. Wang
Nanjing Forestry University, Xuanwu, China

R. L. Nichols
Cotton Incorporated, Cary, NC, USA

J. M. Peña
Institute of Agricultural Sciences (ICA-CSIC), Madrid, Spain

Traditionally, diagnostic methods consist of visual inspection of suspicious trees, collecting symptomatic plants and laboratory analyses, including microscopic evaluation, and molecular, serological and microbiological diagnostic techniques. These methods are costly and time-consuming, especially when disease symptoms are similar to those caused by abiotic stress such as nutrient and water deficiency, making visual diagnosis of the disease very complicated. In addition, visual estimation is subject to the individual's experience and, therefore, to human bias (Mahlein 2016). Accurate and timely disease detection requires the use of rapid and reliable techniques to collect and process data, based on spatial and temporal information on the crop in the field. In that sense, techniques capable of detecting infected plants before they show symptoms noticeable to the human eye would allow better crop management and mitigate the spread of disease (De Castro et al. 2015a).

The need for robust and timely indicators of disease is increasing the focus of the agricultural sector to find technological advances for monitoring plant health. The main objective of such advances is to provide growers with tools to implement precision agriculture (PA) methods that reduce the economic and environmental costs related to agricultural activity and adapt to the social demands for improved food security and the sustainability of agricultural production systems. In this context, remote sensing (RS) systems play a key role through the application of powerful technologies, such as new terrestrial (tractor, carriers, robots) or remote platforms (satellite, manned aircraft and unmanned aerial vehicles -UAVs), and multispectral, hyperspectral and thermal sensors, which have immense potential for monitoring the health status of crops (Zhang et al. 2019). Sensors can provide dense information for the whole field with less expense and can also provide information at wavelengths not visible to the human eye making earlier disease detection possible. In addition, the development of increasingly powerful algorithms for image and data analysis (e.g. multivariate analysis, machine learning and deep learning) enables the discovery of hidden patterns and unknown correlations between the factors involved in the disease development. However, the data analysis necessary can be time-consuming and requires automation once sound approaches have been developed. The combination of suitable remote sensing data and advanced analysis algorithms makes it possible to develop prescription maps for site-specific pest management programs in sustainable crop production.

In this chapter, case studies on the use of remote sensing techniques in arable crops, horticulture and viticulture, accounting for some of the world's major crops; namely cotton, avocado and grapevines, are described. In all the cases, RS has been applied to detect disease caused by fungi using data from several types of RS platform. Images from satellite, manned plane and UAVs were used to map cotton root rot (CRR) for site-specific application of fungicide using tractor pulled variable-rate (VR) control systems. In the case of avocado, the study describes the spatial and spectral properties for the diagnosis of laurel wilt (Lw) using spectral information and remote images from helicopter flights at low altitude. In addition, another widespread avocado disease, i.e. Phytophthora root rot, and abiotic factors causing similar symptoms were evaluated. In the grapevine case study, a hyperspectral (HS) imaging system was employed on *esca* diseased leaves to distinguish between visually asymptomatic and symptomatic leaves at the laboratory scale using multivariate data analysis and several pre-processing imaging techniques.

13.2 Case Study 13.1. Detecting Cotton Root Rot disease for Precision Fungicide Application

13.2.1 Introduction

Cotton Root Rot (CRR), caused by the soilborne fungus *Phymatotrichopsis omnivora*, has been a major disease in cotton crops in the southwestern USA (mainly Texas and Arizona) since first described by Pammel (1888). From 2002 to 2011, roughly 6% of the Texas cotton crop was lost to CRR annually (NCC 2013). Roots of infected plants rot, and then the plants wilt and die quickly with the leaves still attached. Symptoms in a cotton crop begin during vegetative growth but typically are more noticeable during flowering and fruit development. The CRR tends not to affect entire fields; instead it begins at various points in a field and typically spreads from these foci throughout the growing season (Smith et al. 1962) in irregular, mainly circular patterns (Lyda 1978). Diseased areas (Fig. 13.1a) range in size from less than a square metre to several hectares and expand as the season progresses, especially in rainy years. Moreover, CRR tends to occur from year to year in virtually the same areas within fields (Yang et al. 2016). Therefore, remote sensing (RS) imagery recorded late in one growing season and used to detect CRR zones can be useful to predict their occurrence in future seasons.

Over many decades, several treatments have been evaluated for disease control with little or no success. But in 2008 Topguard fungicide (FMC Corp., Philadelphia, PA, USA), containing the active ingredient flutriafol, was found in a research trial to be effective at controlling CRR (Isakeit et al. 2009) and has been available to growers since 2012. Growers who applied the fungicide on their fields achieved lower CRR incidence, larger yields, and better fibre quality on affected fields (Drake et al. 2013; Yang et al. 2014). The most effective method of fungicide application is during planting, which is many weeks before the appearance of symptoms in plants growing in an infested field.

13.2.2 Studies, Methods, and Results

Now that a successful treatment for CRR has been found, fields with a history of CRR are commonly treated uniformly, even though the fungicide is expensive (up to $125 USD per ha in 2019) and the grower is aware that only a portion of the field is infected. Uniform treatment ensures that all existing and potential new infected areas are treated because it is not known whether infected areas will expand from year to year. Furthermore, growers historically have not had ready access to site-specific application equipment and prescription maps that could potentially be developed from RS imagery. The following descriptions illustrate the use of three types of RS data in field studies to map areas of diseased cotton for site-specifically applying fungicide, as well as describing the advantages and disadvantages of each method, and potential future advances in the technology.

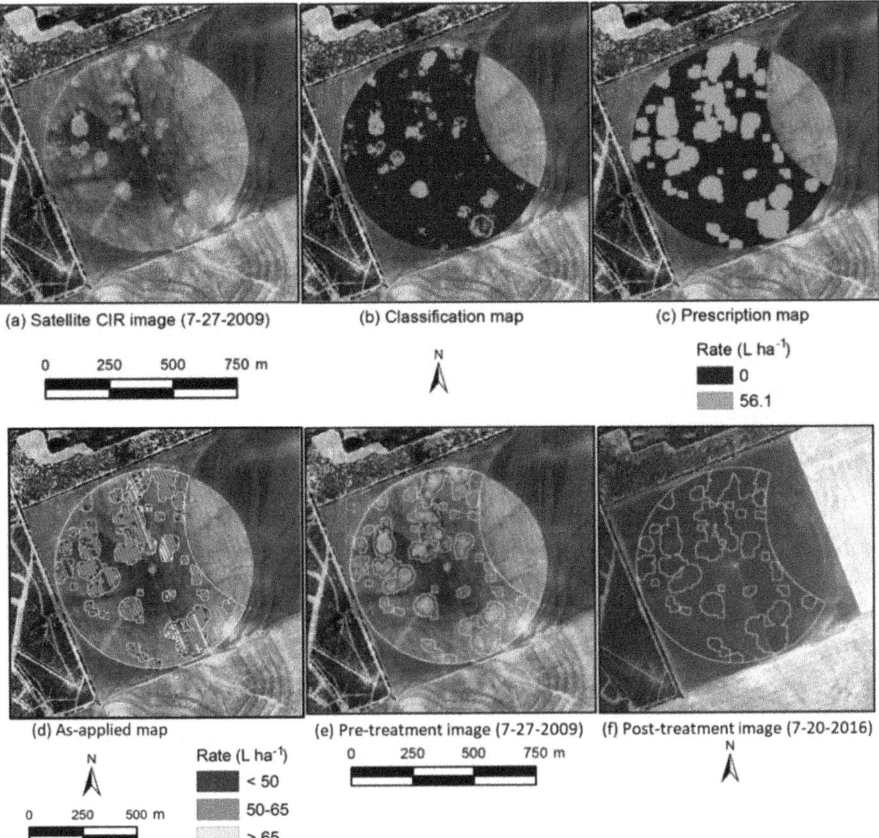

Fig. 13.1 (**a**) A 2009 GeoEye-1 satellite colour-infrared (CIR) image of a 41 ha cotton field in southern Texas with areas of CRR, (**b**) a two-zone classification map of the field, in which light-grey spots are classified as infected and the dark-grey area is classified as healthy, (**c**) a prescription map of the field in which only the light-grey spots were treated with fungicide, (**d**) the as-applied map for actual fungicide applied during planting, (**e**) the original satellite image of the field with fungicide-application zones delineated, and (**f**) the post-treatment image, showing that virtually no disease is evident in the field in 2016 after the precision fungicide application. (Adapted from Yang et al. 2018. Used with permission)

13.2.2.1 Satellite Remote Sensing

Until recently, available satellite data (e.g. Landsat) did not have adequate spatial resolution to map CRR precisely in a field. Newer satellite systems, however, are reaching a level of resolution (1.0 to 2.0 m per pixel) that can be useful in mapping CRR. Yang et al. (2018) used a GeoEye (DigitalGlobe, Inc., Longmont, CO, USA) satellite scene acquired on July 27, 2009, to detect CRR in a southern Texas field with a history of CRR (Fig. 13.1a). Image-acquisition timing was late in the growing season when the full extent of CRR was expressed, but prior to pre-harvest defoliation so that green leaves remained on the live plants. The scene included four image bands (red, green, blue and near-infrared; or RGB plus NIR) at fairly high resolution (2 m).

A prescription map based on the satellite image was created in eight steps. (1) A field boundary was defined for the field to create an area of interest (AOI). (2) A normalized difference vegetation index (NDVI) image was created for the AOI. Pixels with smaller NDVI values were generally associated with CRR-infected zones, while those with larger values were associated with non-infected zones. (3) The NDVI images were classified with Imagine software (Erdas Inc., Norcross, GA, USA) into CRR-infected and non-infected zones by ISODATA (iterative self-organizing data analysis) unsupervized classification (Campbell 2002). The ISODATA method was initiated with two classes having arbitrary class means based on the NDVI image statistics. Each pixel was assigned to the class closest to the NDVI mean. Once all image pixels were assigned to a class, the NDVI means of the two classes were recalculated and used for subsequent iteration. The process was repeated until the number of iterations reached a prescribed limit or the percentage of changed pixels was within a small prescribed tolerance of 0% between iterations. (4) Once the iterative process was complete, the classification maps contained many polygons representing CRR infection. Some of the smaller polygons represented actual CRR infection, whereas some were artefacts of the classification procedure. The CRR polygons with areas less than or equal to 4 m^2 were filtered out where they were sparse or merged together where they were dense with ArcInfo GIS software (ESRI Inc., Redlands, CA, USA). About 11% of the field was classified as having CRR (Fig. 13.1b) at this point in the classification process, and it occurred at various locations around the field. (5) Ground observations were made to verify that field areas classified as CRR were related to CRR. In general, ground observations supported the classification of field areas into CRR zones based on analysis of the satellite imagery. However, some field areas classified as CRR were related to anomalies such as planter skips and human-caused artefacts that needed to be removed manually from the prescription map. (6) To account for possible spatial variation of CRR from year to year, a buffer of 10 m was added around the CRR areas on the classification map to become part of the treatment area in the prescription map (Fig. 13.1c). The buffer zones significantly increased the treatment area and tended to connect some of the CRR areas, making the prescription map more practical for site-specific application. After this step about 37% of the field was prescribed as treatment area. (7) The polygons in the prescription map were assigned as Spray for CRR areas and No-Spray for non-CRR areas. (8) The prescription map was converted to an ESRI shapefile for use by the variable-rate (VR) application equipment.

The prescription map described in the previous paragraph and based on 2009 satellite data was loaded onto a tractor's VR control system to apply Topguard Terra at planting in 2016. The map was converted to the appropriate format for the variable-rate (VR) control system installed on a John Deere 8230 tractor (Deere & Company, Moline, IL, USA) owned by the cotton grower. Planting of half the field occurred on March 18, 2016 and the other half on March 23. The application rate of Topguard Terra was 0.585 L ha^{-1} (full prescribed rate) mixed with 56.1 L ha^{-1} of water. The liquid was applied during planting with the modified in-furrow method. Liquid flow was distributed to the shanks of a 12-row planter with row spacing at 0.965 m. Rainfall occurred a day after the first half of the field was planted resulting in a poor stand, so replanting of that half of the field was required on April 6. No fungicide was applied at replanting.

Two tasks were carried out to evaluate the results of the satellite-based prescription map and VR fungicide application. (A) An as-applied map for the fungicide was recorded during planting (Fig. 13.1d), and it consisted of rectangular regions with fixed width equivalent to the effective swath of the planter (11.58 m). The map data included target and actual rates for comparison between actual application and the prescription. The application system missed some small areas and did not turn on and off exactly when entering and exiting treatment zones, respectively. The actual treatment area was 6% smaller than the prescribed area, and the actual application rate was 0.4% higher than the prescribed rate, but with such small deviations from the prescription map, the VR application was considered adequate for evaluation of overall efficacy. (B) A two-camera aerial imaging system on a manned aircraft, capturing RGB plus NIR images, was used in 2016 to collect post-application images to evaluate the efficacy of the site-specific fungicide application. An image acquired late in the season at 1220 m above ground level (AGL) with a pixel size of 0.30 m was used to detect CRR areas and assess the efficacy of the site-specific application. Compared to the map of fungicide-application zones based on the 2009 image (Fig. 13.1e), the 2016 image (Fig. 13.1f) made it clear that the fungicide effectively controlled the disease in the treated areas, although CRR appeared in a few treated areas towards the end of the growing season. This late CRR manifestation had little effect on yield, because most cotton bolls were fully developed by that time.

13.2.2.2 Manned Aircraft Remote Sensing

High-resolution satellite imagery has shown potential for CRR detection and creating fungicide prescription maps, but manned-aircraft images have been used repeatedly for this purpose on numerous fields (Yang et al. 2014). Aerial imagery has advantages over satellite imagery including higher spatial resolution, more flexibility in timing of data acquisition and the ability to collect data on cloudy days. Yang et al. (2018) used a four-camera system (Yang 2012) to collect RGB plus NIR images of two fields with a history of CRR in different cotton-growing regions of Texas. Field 1 was in southern Texas, and Field 2 was in western Texas. Images were collected from a single-engine aircraft shortly before harvest to aid application in 2010, when CRR was fully expressed for the season. Image acquisition occurred at 2740 m AGL, giving a pixel resolution of 0.90 m, which was resampled to 1 m.

Prescription maps based on the aerial images of Fields 1 and 2 were created in eight steps, similar to those described in the section on satellite remote sensing. Minor differences are noted here. After step (4), about 33% of Field 1 and 37% of Field 2 were classified as CRR area. In step (6) a 5 m buffer (instead of 10 m) was added around the CRR areas on the classification maps to become part of the treatment area in the prescription maps. After this step about 57% of Field 1 and 63% of Field 2 were prescribed as treatment areas.

These prescription maps based on 2010 aerial image data were loaded onto tractors' variable-rate (VR) control systems to apply fungicide at planting in 2015, the next growing season when access to the field was possible and when cotton was being grown in the rotation. For Field 1 a John Deere VR control system was installed on a John Deere 8230 tractor owned by the grower. Topguard Terra fungicide was applied at planting on May 1, 2015, about six weeks later than the usual planting date because of excessive rainfall. The application rate was 0.292 L ha^{-1} of product (half the prescribed rate) mixed with 56.1 L ha^{-1} of water. Liquid flow was distributed to the shanks of a 12-row planter with row spacing of 0.965 m and applied with the modified in-furrow method. For Field 2 a Trimble Field-IQ spray control system was installed on a John Deere 8210 tractor owned by the grower. The older Topguard fungicide formulation (approved under EPA Section 18) was applied at planting on June 3, 2015. The application rate was 2.34 L ha^{-1} of product (full prescribed rate) mixed with 46.8 L ha^{-1} of water. Liquid flow was distributed to the shanks of an 8-row planter with row spacing of 1.016 m and applied with the T-band method.

Similar to the assessment with satellite imagery, two tasks were carried out to evaluate the results of the aerial-image based prescription map and VR fungicide application. Minor differences are noted here. In task (A), the effective swaths of the planters were 11.580 m in Field 1 and 8.128 m for field 2. The actual treatment area was 1.5% smaller than prescribed for Field 1 and 1.4% larger for Field 2. The actual application rate was 4.1% higher than prescribed for Field 1 and 1.5% lower for Field 2. In task (B), evaluation of post-application images, a two-camera aerial imaging system on a manned aircraft was used in 2015 to collect post-application images. The images were acquired late in the growing season at 1070 m AGL with a pixel size of 0.35 m and used to detect CRR areas. These 2015 post-application images were compared to the 2010 prescription maps to determine efficacy. In Field 1, the fungicide applied at half rate was able to control the disease for most of the growing season, but late-season CRR infection may have had negative effects on cotton yield and quality. In Field 2, the fungicide effectively controlled the disease in the treated areas, although CRR occurred at a few treated areas toward the end of the season but had little effect on the crop.

13.2.2.3 UAV Remote Sensing

Unmanned aerial vehicles (UAVs) have been used extensively for RS in agricultural research over the last few years. While they have disadvantages including large volumes of data and challenges in pre-processing of images, their advantages include higher spatial-resolution imagery, more flexibility in timing of data acquisition and lower cost of data acquisition. With respect to fungicide application for CRR, technological advances suggest that future VR systems will possibly be capable of precision spraying to the level of an individual seed at planting. For example, state-of-the-art planting systems with real-time kinematic (RTK) Global Navigation Satellite System (GNSS) receivers are currently capable of precisely planting individual seeds at a known location, accurate to within 2 cm, and auxiliary state-of-the-art spraying systems are currently capable of applying starter fertilizer adjacent to each seed planted. The high resolution of UAV imagery, therefore, offers the possibility of prescription maps with extremely high precision, potentially capable of fungicide application on a seed-by-seed basis during planting, and limiting fungicide application to a small area adjacent to each seed planted in a small CRR zone. Research with UAVs for CRR detection was undertaken first to replicate the generation of prescription maps possible with imagery from manned aircraft and satellites, and second to pursue development of prescription maps at the single-plant level. One study (Thomasson et al. 2018) was conducted on a 32.9 ha field in central Texas with a history of CRR. On August 22, 2015, image data in the green, red and NIR bands were acquired with a Lancaster (PrecisionHawk Corp., Raleigh, NC, USA) fixed-wing UAV flown at 120 m AGL, giving 3.7 cm pixels. Another study (Wang and Thomasson 2019) was conducted on a 5.7 ha field with a history of CRR, also in central Texas. In this study RGB plus NIR and red edge band image data were acquired on August 20, 2017, with a UAV Mapper fixed-wing UAV (Tuffwing LLC, Boerne, TX, USA) flown at 120 m AGL, giving 7.6 cm pixels. Images in both studies were acquired with a minimum of 70% image overlap (forward and sideward) to enable generation of a high-quality mosaic.

Regional Classification

The methods used previously to classify satellite and manned-aircraft images for CRR have been regional methods, classifying fields into zones of multiple square metres to match the precision of current VR equipment. Regional classification can be based on traditional image-analysis techniques and is relatively computationally efficient. To show the capability of UAVs for practical RS tasks, regional classification of UAV imagery was used to develop CRR prescription maps to demonstrate efficacy, in essence mimicking what has been done previously with manned-aircraft and satellite RS. An image mosaic of the 2015 images of the 32.9 ha field was created with Pix4D software (Lausanne, Switzerland) and resampled in ENVI software (Harris Geospatial, Boulder, CO, USA) to a resolution of 1.0 m per pixel. Support vector machine (SVM) classification was applied to the mosaic to classify it into CRR and non-CRR areas. Based on the classified image data, prescription maps were developed in ArcGIS (ESRI, Redlands, CA, USA). The proportion of CRR area was 5.52% at this stage in the classification process. Ground observations were made to verify CRR areas of various sizes along the western edge and in the southeastern corner of the field, and that the classified CRR areas were actually CRR. In general, most were classified correctly, but a few small areas were misclassified because of, for example, planter skips. To accommodate the potential expansion and temporal variation of the disease, a 5 m buffer zone was created around the infected areas as part of the treatment areas in the prescription map.

The 2015 UAV images were used for VR fungicide application in 2017. The classified image was converted into a shapefile-based prescription map in ArcMap (ESRI, Redlands, CA, USA). The polygons in the prescription map were assigned Spray for CRR areas and No-Spray for non-CRR areas. A Trimble Field IQ system was integrated with a Trimble FM 1000 monitor and a Trimble RTK-GNSS receiver installed on a CASE IH Puma 210 Tractor owned by the grower. At planting on April 25, 2017, Topguard Terra was applied with the T-band method at a rate of 0.585 L ha^{-1} (full prescribed rate) with 57.1 L ha-1 of water from a 12-row planter with row spacing of 0.762 m.

An as-applied map (A) was not evaluated in this study, but based on the work with satellite and manned-aircraft imagery the fungicide application was expected to conform closely to the prescription map. Similar to the assessment with satellite and manned-aircraft imagery, one additional task (B) was carried out to evaluate the results of the UAV-based prescription map and VR fungicide application. The UAV images of the 32.9 ha field were collected to determine the efficacy of the UAV-based prescription map based on regional classification. On August 20, 2017, RGB plus NIR and red edge image data were acquired with a Micasense RedEdge camera on a UAV Mapper flown at 120 m AGL, giving 7.64 cm pixel resolution. As with the 2015 image data, the 2017 images were collected with minimum 70% image overlap (forward and sideward) so that a mosaic could be created with Pix4D and processed and classified into CRR and non-CRR areas. These 2017 post-application images were compared to the 2015 prescription maps to determine efficacy. The proportion of CRR area in the field was reduced from 5.52% in 2015 to 0.55% in 2017, giving a strong indication of the efficacy of the UAV-based prescription map in mitigating CRR, similar to the results with manned-aircraft and satellite imagery.

Plant-by-Plant Classification

In the future it may be desirable to apply fungicide precisely at a greater level of detail than 5 m × 5 m, potentially even at the single-plant level. To take full advantage of the capability of UAV remote sensing, one must utilize the high resolution inherent in the images. Two methods have been developed to approach plant-by-plant (PBP) classification. The first method, a custom row-searching algorithm, identified individual crop rows and then scanned each row, applying a plant-size mask to the image data, to enable PBP classification. A global (i.e. to be used consistently across an entire mosaic) algorithm was developed to exploit the fact that straight rows in a field share the same angle between row direction and latitude lines, so that angles need to be measured at only one representative location in a given field. The algorithm automatically pre-processed the mosaic and then measured the row angle based on two-dimensional gradients. Then a morphological operation, tailored to the angle of the rows was used with a customized structural element set to the specific row spacing in the field. Row centre lines were constructed with this process, and then collinear rows were connected across breaks resulting from the presence of long planting skips or streaks of dead plants along the rows. The complete row centre lines were then overlaid onto an image mosaic. Plant-size zones were classified into three categories (live plant, dead plant, and no plant) by applying a plant-size mask from one end to the other on each row. The row-searching algorithm was applied to the mosaic of the 2017 image data from the 32.9 ha field (Fig. 13.2a) and performed well, providing generally accurate identification of live and dead plants. The algorithm was efficient when it was restricted to searching for linear crop rows, requiring only a few minutes of processing on a PC for the entire field. Computation time with curved rows (e.g. in fields irrigated by centre pivot) would probably be significantly longer, and programming improvements would be needed to achieve acceptable processing times.

The second method for PBP CRR classification, a superpixel algorithm, used simple linear iterative cluster (SLIC) superpixel segmentation, a state-of-the-art object-based image classification technique. The superpixel algorithm was applied to the 2015 image mosaic of the 5.7 ha field and 'seeded' (i.e. the iteration process was initiated) with a large number that closely resembled the number of plants expected to be in the field based on planting density. The image of the field was segmented into roughly that number of small pieces (superpixels), each based on the colour, shape and texture of the original image data, and assigned spectral and spatial statistics based on the constituent pixels. The *k*-means clustering was applied automatically to the superpixel image to generate a classification map. The overall algorithm was efficient, taking about two minutes for the 5.7 ha field. The superpixel algorithm provided accurate classification (93.5%) of individual plant zones (Fig. 13.2b), with small errors of omission and commission, and it was faster than the more detailed row-searching algorithm.

13.2.3 Conclusions

In summary, remote sensing has proved to be useful for developing prescription maps to enable precision application of fungicide to protect cotton plants against CRR disease. High-resolution satellite and manned-aircraft images have been shown to be useful for delineating zones of disease incidence in fields. Images from UAVs can also be used for this purpose, but the extremely high resolution of UAV images also allows for the possibility of plant-by-plant fungicide application. Two studies have shown how individual cotton plants can be located and classified into diseased and healthy categories. Therefore, fungicide applied during planting may, in the future, be placed very precisely next to the seed, further reducing cost and environmental risk associated with over-application.

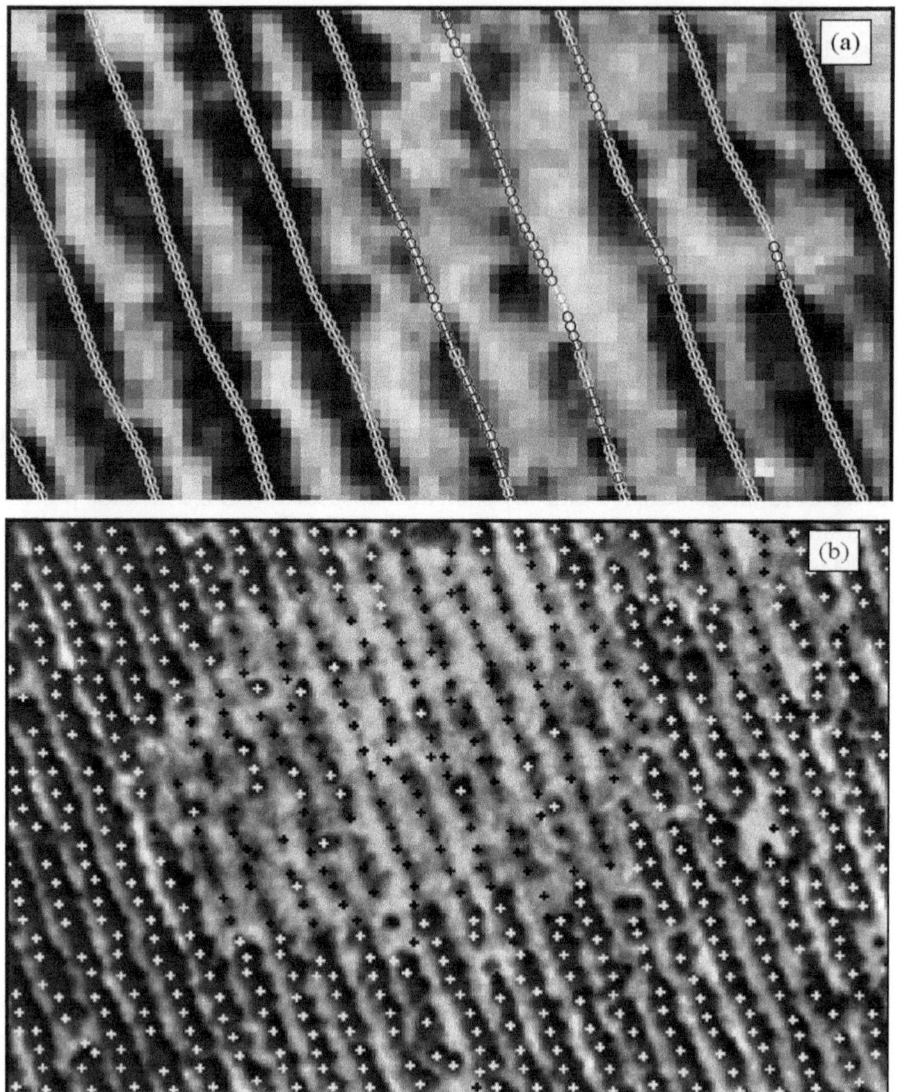

Fig. 13.2 (**a**) Plant-by-plant labelling of live plants (white circles) and dead plants (black circles) and soil (light grey) along cotton rows based on a custom row-searching algorithm. Centre lines of rows recognized by the algorithm are marked with white lines. (**b**) Plant-by-plant labelling of live plants (white crosses) and dead plants (black crosses) based on superpixel classification

13.3 Case Study 13.2. Detection of Laurel Wilt Disease in Avocado: A Case Study for Avocado Production in Florida

13.3.1 Introduction

Avocado (*Persea americana*) is important to the agricultural economy of Florida. The avocado industry is second in importance after citrus; it represents approximately 60% of the tropical fruit crop in Florida (2800 ha) (Evans et al. 2015). It provides an economic benefit of approximately $100 million per year (Evans et al. 2015). In addition, this economic importance reflects that the United States is one of the main producers and importers of avocados in the world (Statista 2018).

However, the avocado industry in Florida is under severe threat because of the invasion of an exogenous pathogen, the Asian fungus *Raffaelea lauricola*, and its original vector, the redbay ambrosia beetle *Xyleborus glabratus*, which causes the lethal vascular disease laurel wilt (Lw) (Ploetz et al. 2011). This complex disease has spread rapidly along the southeastern seaboard of the United States because of the natural dissemination of *X. glabratus*, the great susceptibility of the native *Persea* spp. and their attractiveness to *X. glabratus,* the substantial amounts of inoculum that most females of *X. glabratus* carry and the human transport of infested wood (Hanula et al. 2008; Ploetz et al. 2017a). Moreover, the disease appears to spread through interconnected root systems, which allows the movement of the pathogen without the aid of vectors (Ploetz et al. 2017b). Consequently, Lw has caused the loss of 300 million redbay trees throughout the coastal forests from North Carolina to Florida and over 25 000 avocado trees since its migration into Florida (Ploetz et al. 2017b; Mendel et al. 2018). Furthermore, it is difficult to prevent the spread of the disease as there is no effective fungicide-based control strategy. Sanitation, which consists of identifying affected trees and destroying them before new generations of vectors emerge and colonize new host trees, is the only available control measure (Ploetz et al. 2011; De Castro et al. 2015b).

Laurel wilt impairs xylem function as soon as three days after inoculation, impeding the flow of water and nutrients into affected trees, which soon causes external symptoms of wilting and foliar necrosis in affected portions of the tree, and full defoliation within 2–3 months of symptomatic onset (Ploetz et al. 2011). Before the appearance of external symptoms consisting of leaves changing from an oily green to brown colour, internal symptoms of increased tree temperature that arise from water and nutrient blockage occur. This results in a reduced amount of chlorophyll in the leaves and damaged cell structure (De Castro et al. 2015b). Those variations in leaf plant pigment and temperature make it possible to detect diseased trees, even in the early stages of disease development, with remote sensing tools including those that are spectroscopic and imaging-based.

Laurel wilt symptoms are very similar to those caused by other vascular diseases or factors such as frost damage, Phytophthora root rot, Verticillium wilt, nutrient deficiencies, salinity and fruit stress (overbearing), and consequently their visual discrimination is very difficult (De Castro et al. 2015b). Despite this, diagnostic practice consists of visual inspection of suspect trees, collection of wood and laboratory analyses, which is time-consuming, labour intensive, expensive and requires symptomatic trees. Once plants display external symptoms, it is then complicated to manage the disease because by that time significant colonization of the host by the fungus has already occurred; at which point the best choice is sanitation, i.e. elimination of affected trees (Ploetz et al. 2011). Therefore, a methodology to detect Lw before the external symptoms appears, i.e. at an early stage, and its discrimination from other biotic and abiotic stress factors is highly desirable. This case study describes the required spatial and spectral properties for the rapid and accurate diagnosis of Lw at an early stage using remote sensing tools, long recognized as suitable for the fast monitoring of large areas and reducing the costs of extensive field campaigns. In addition, another widespread avocado disease (Phytophthora root rot caused by *P. cinnamomi*) as well as abiotic factors (salinity and nitrogen and iron nutrient deficiencies), which cause similar symptoms, were evaluated.

13.3.2 Materials and Methods

An effective mapping system begins with an evaluation of the spectral signature at the leaf level of diseases and factors affecting avocado. Once a suitable sensor is selected based on spectral requirements, the study should be scaled up to the canopy level to evaluate other aspects related to image analysis, such as flight altitude, spatial resolution, pre-processing and image algorithm.

13.3.2.1 Spectral Requirements: Spectral Data Analysis

First, the feasibility of discriminating healthy plants from damaged plants due to biotic and abiotic stressors at an early stage with spectral information was determined. Next, the optimal wavebands and hence the sensor for affected and healthy plant discrimination was selected.

Multivariate analysis tools are considered as one of the most suitable and advanced techniques for the detection of spectral difference (De Castro et al. 2012). Among these tools, neural networks have received great attention from the remote sensing community because of their flexibility and adaptability to the results, tolerance of noisy data and errors, fast computation processing speed, and ability to explore correlations or models that could not be detected by traditional statistical procedures (Han et al. 2012).

Spectral Data Collection

Spectral data were taken from avocado leaves under controlled laboratory conditions using a handheld spectroradiometer (SVC HR-1024, Spectra Vista Corp., Poughkeepsie, NY, USA) placed at a height of 50 cm above the leaf. Five reflectance spectra per leaf were taken at the range of 400 to 950 nm with a 10 nm spectral resolution based on published recommendations and the noise of the remainder of the spectral range (Fig. 13.3).

Four asymptomatic and slightly affected leaves in the early stage of stress development, i.e. just beginning to lose turgidity, were selected from each abiotic factor- and disease-affected plant. These leaves were taken from potted 'Simmonds' variety avocado trees grown in a greenhouse at the University of Florida's Tropical Research and Education Center (TREC) in Homestead, FL, USA. The experiment consisted of 10 plants for each class, and all showed symptoms like those caused by Lw. In addition, healthy (H) leaves were obtained from potted plants grown in full sun. The disease induction and symptom development were performed as follows:

- Laurel wilt (Lw): Conidial suspensions of *R. lauricola* with a concentration of 30,000 colony forming units (CFUs) mL^{-1} were introduced in four small holes 5 cm above the soil level around each trunk circumference, resulting in a total of 3000 CFUs per plant. Early symptoms of Lw began to develop by 14 days after inoculation.
- Phytophthora root rot (Prr): 6 g of wheat seed colonized with *P. cinnamomi* were used for the inoculation. Early symptoms, i.e. yellowing of some leaves, appeared after 14 days.

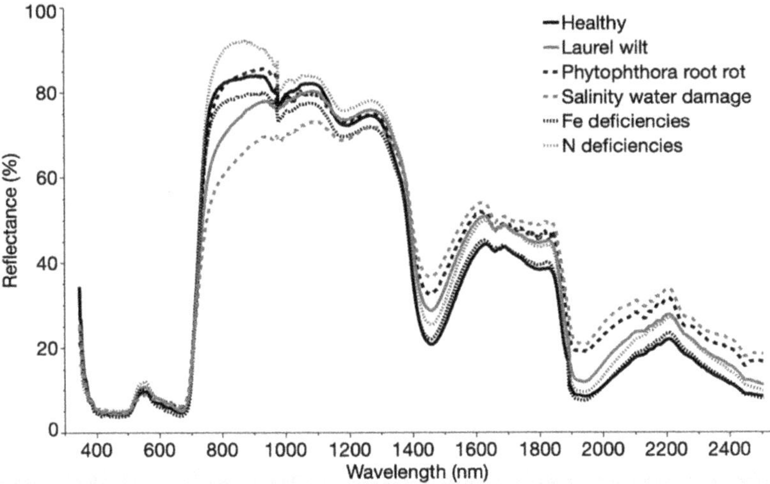

Fig. 13.3 Mean reflectance spectra of leaves representing healthy, laurel wilt, Phytophthora root rot, salinity, water damage, and Fe and N deficiencies of avocado trees. All leaves are typical of the early stage of symptom development. (The figure has been adapted from De Castro et al. 2015a and Abdurhina et al. 2018)

- Salinity (Sln): One litre of a salt solution with a similar concentration to that of sea water from the east coast of Florida, i.e. 36 g L^{-1}, was applied to each tree. Early browning symptoms were found in leaves after seven days.
- Nitrogen (N) deficiency: avocado plants growing in a nutrient-free matrix composed of sand and perlite received a modified Hoagland solution with all essential nutrients except N once a week. Early symptoms occurred 60 days after the beginning of the procedure.
- Iron (Fe) deficiency: these avocado samples grew under the same conditions as N deficient samples, although the applied Hoagland solution contained all the minerals with the exception of Fe. The first symptoms occurred in the same time as the previous case.

The spectral dataset was calibrated using a barium sulphate standard reflectance panel (Spectralon®, Labsphere Inc., North Sutton, NH, USA) in the presence of two portable 500 W halogen work lamps used as extra light source.

Spectral Data Analysis

The 10 nm averaged spectral measurements were analysed statistically with the multilayer perceptron (MLP) neural network to identify the best waveband for discrimination of H, Lw and other stressors such as Prr, salinity, and N and Fe deficiencies.

As a multilayer feed-forward neural network, MLP creates an analytically adjusted model based on supervized training with a back-propagation algorithm that minimizes the prediction error (Han et al. 2012). The weight, bias and typology parameters of the network are adjusted by learning the relation between inputs (spectral information in this case) and outputs data (health status class in this case). The MLP comprises an input layer, in this case a 10 nm averaged spectral data set, a hidden layer of neurons to compute the data and create the model, and an output layer consisting of the classes to which the samples are classified (H, Lw, and other disease or abiotic factors). The validation of the MLP algorithm was performed by a hold-out cross-validation procedure, where $3n/4$ of the full data set was used to train the model and $n/4$ was used as a test set to provide the generalization accuracy; n was the number of units in the full dataset in every analysis.

The statistical analyses were performed using SPSS software (IBM Statistical Package for Social Science, SPSS Inc., Microsoft Corp., Redmond, WA).

13.3.2.2 Image Specifications: Image Data Analysis

Multispectral image acquisition and spatial requirements. A user-configurable bandpass filter camera was selected for the experiment. The Tetracam mini-MCA-6 (Tetracam, Inc., Chatsworth, CA, USA) multispectral camera is a lightweight sensor of six individual digital channels with independent optics, each holding a

1.3-megapixel CMOS sensor (1280 × 1024 pixels) with a focal length of 9.6 mm and FOV of 43.7° × 35.6°. Each unit independently stored the data in compact flash cards embedded in the camera. The images were taken in the presence of avocado experts from the Florida Avocado Committee in a commercial avocado production field in Miami-Dade County containing healthy and Lw-infested trees. Experts identified diseased trees that were shortly after confirmed as such by a diagnostic DNA test. No other damaging biotic agents or disturbances were found in or around this field.

The remote images were acquired by a helicopter flight at a height of 250 m, considered to be the optimum height according to the size of typical avocado trees. The pixel size obtained using this sensor at this flight height was 15 cm, large enough to identify a standard avocado tree with a canopy diameter of 7–9 m. Moreover, the average avocado orchard size ranges from 0.4 ha to 2 ha (Evans et al. 2015), which was covered by images taken with the MCA-camera from a height of 170 m. A lower flight altitude would involve more flight time and cost, and may require an extra mosaicking process to cover the entire field.

Ground truth data. Healthy and Lw avocado trees at early stage were located in the images and manual digitalization was conducted to extract the digital information of the affected portion of the trees and healthy plants. The ground truth data consisted of 21 Lw-infested and 12 healthy trees.

13.3.3 Results

13.3.3.1 Spectral Analysis-Leaf Level

The six channels of the Mini-MCA were selected according to the results obtained in the spectral analysis (Table 13.1), where values ranging from 96% to 100% accuracy were obtained in all the classifications. Table 13.1 shows the wavelengths that contributed to the greater specific weights in the neural network algorithm, which all used one hidden layer with a similar number of neurons.

The most frequently selected 10 nm wavelengths were 740 nm and 750 nm, which were also among the first variables entered into the MLP model in all the cases, indicating that they are crucial in discriminating between infested and healthy avocado plants. An extra filter was selected in the red edge region (760 nm) because wavelengths around that value were chosen in several MLP algorithms and also the importance of that part of the spectrum to detect vascular diseases in plants (De Castro et al. 2015b). The Lw plugs the xylem, blocking the flow of water and increasing the tree temperature. Consequently, leaf chlorophyll concentration and photosynthesis decrease while carotenoid production increases, affecting the reflectance values in the green, red edge and near-infrared regions (Chappelle et al. 1992). For these reasons, it was appropriate to add a filter in the NIR region. Taking into account the most frequently selected wavelengths in that part of the region, economic reasons and commercial availability, a band with a centre wavelength at

Table 13.1 Accuracy assessment on 10 nm bandwidth data classification for healthy (H), laurel wilt-infested plants (Lw) and other stressors such as Phytophthora root rot (Prr), salinity (Sln) and N and Fe deficiencies, using MLP neural networks

Analysed classes	Selected wavelengths[a] (nm)	Accuracy (%)
H *vs* Lw *vs* Prr	740, 750, 830, 760	100
H *vs* Lw *vs* N vs Fe	840, 930, 750, 720, 830, 740	100
H *vs* Lw *vs* Sln	720, 750, 740, 526, 950, 770	96

[a]Wavelengths selected to account for the greater specific weights in the neural network algorithm

850 nm and 40 nm full-width was added to the camera. In addition, because of those changes in the vegetal pigment concentrations, the absolute difference between the red edge and NIR region with the green one decreases in diseased plant spectral data, making the ratios between bands useful to separate plants affected by damaging agents from healthy ones. Therefore, an additional filter was selected in the green region (580–10 nm full-width). Finally, another 10 nm full-width (650 nm) was added in the red region because the large variety of narrow-band vegetation indices (VIs) obtained from remote sensing data to assess plant health rely on the combination of NIR and red reflectance.

13.3.3.2 Image Processing-Canopy Level

After the suitable filters were selected and attached to the camera, images were taken to assess the feasibility to detect infested avocado plants at an early stage of symptom development.

Multispectral Band Alignment and Image Radiometric Calibration

Both processing steps are required before image analysis. The alignment process reduces geometric differences between the bands and groups the six images saved in each channel. This was carried out by Tetracam PixelWrench 2 (PW2) software (Tetracam Inc.) that provides a band-to-band registration file. During this process, the vignetting parameters were also adjusted.

Radiometric correction was conducted using two calibration targets (black of 3% and white of 82%) and an empirical line calibration method with ENVI software (ENVI®, Research Systems Inc., Boulder, CO, USA).

Image Data Analysis

The mean digital information extracted from pixels of Lw–infested and healthy trees was used to calculate and evaluate a large pool of VIs calculated from the six bands of the customized MCA camera. The VIs have been widely used in physiological stress detection (Lu et al. 2017, 2018) as they magnify the differences in spectral signatures, thus making the identification of infested plants easier (Mahlein et al. 2012).

The *M*-statistic was applied to quantify the histogram separation of vegetation indices and to establish their potential for spectral discrimination. The *M*-statistic evaluates the mean (μ) difference of the class 1 and class 2 histograms normalized by the sum of their standard deviations (σ) (Kaufman and Remer 1994) (Eq. 13.1). The larger is the M value, the better is the spectral separation. Values less than 1 indicate poor separation.

$$M = \frac{\mu_{\text{class1}} - \mu_{\text{class2}}}{\sigma_{\text{a}} + \sigma_{\text{b}}} \tag{13.1}$$

The resulting *M* values varied according to the vegetation indices, suggesting that the separation capacity depends largely on the spectral region analysed. Only *M* values >1.5 were considered as indicators of strong discriminatory power here. The best results were obtained with red edge/G, GRVI, VIgreen and GNDVI, where any of the bands related to the red edge region (740, 750 and 760 nm) of the Tetracam camera were used (Table 13.2). These VIs work by combining digital values in the green, red edge and near-infrared region of the spectrum and are related to changes in vegetal pigment concentration and cellular damage, both of which occur in Lw-infested plants due to xylem blockage. These results confirm the importance of proper band selection early in the procedure because their use made it possible to identify Lw-infested trees at an early stage of disease development with minimal symptoms, i.e. leaves are still green and have barely begun to lose turgidity.

Therefore, the analysis of images obtained from the camera with the attached filters using the selected VIs can overcome the challenge of early detection of Lw, which represents a great advance in preventing the spread of this lethal avocado disease.

Table 13.2 The M values obtained in the analysis of digital data of laurel wilt-infested trees at the early stage of symptom development and those of healthy avocado trees using remote sensed data

Vegetation Index	Equation	Adapted from	Bands used	*M* value
R/G	Redge_x/G	–	Redge_{740}	1.8
			Redge 50	1.8
			Redge_{760}	2.1
Green ratio vegetation index	$\text{GRVI} = \text{NIR}_x/G$	–	NIR_{850}	1.9
Green vegetation Index	$\text{VIgreen} = \frac{G - R_x}{G + R_x}$	Gitelson et al. 2002	Redge_{740}	1.8
			Redge_{750}	1.8
			Redge_{760}	2.1
Green normalized difference vegetation index	$\text{GNDVI} = \frac{\text{NIR}_x - G}{\text{NIR}_x + G}$	Gitelson et al. 1996	NIR_{850}	1.8

Redge_x in this form represents the filters in the red edge region of the MCA-camera used to calculate the VI. i.e., 740, 750 or 760 nm

13.3.4 Conclusions

The spatial and spectral specifications for the quick and accurate diagnosis of Lw at an early stage, as well as the possibility to separate it from other abiotic and biotic factors that cause similar symptoms, were evaluated in this case study. Therefore, once suitable sensor and flight planning requirements have been defined, an automatic algorithm based on aerial system imaging, such as UAV, may be developed for early and rapid Lw detection in further research. The early detection of Lw will prevent the spread of the disease and facilitate the implementation of disease control precision strategies, such as targeted sanitation, in the context of PA.

13.4 Case Study 13.3. The Use of Hyperspectral Imaging for Esca Detection in a Vineyard

13.4.1 Introduction

Hyperspectral (HS) imaging systems are one of the most currently used image-based phenotyping methods in modern agriculture due to their inherent advantages. They include the possibility of acquiring data in a non-destructive and non-invasive way, being amenable to automation and allowing in-field sample analyses (ElMasry and Sun 2010). For these reasons, HS systems represent a promising tool for plant disease diagnosis, together with the fact that not only can an infection be identified successfully, but also its location within the plant can be detected (Mutka and Bart 2015; Rançon et al. 2019). The HS techniques generally work in the near-infrared (NIR) region of the electromagnetic spectrum because the spectral signature of vegetation is characterized by high reflectance in this region (Rodríguez-Pérez et al. 2007). It is particularly relevant for disease detection, as symptoms can sometimes be detected before the naked eye is able to do so (Di Gennaro et al. 2016). Thus, HS systems may have the potential to enable diagnosis of plant diseases that have no visible symptoms at the early stages of their development, as in the case for esca, a grapevine fungal trunk disease.

Currently, grapevine trunk diseases are one of the main concerns of viticulture worldwide because they are responsible for substantial economic loss to the wine industry (Levasseur-Garcia et al. 2016). They result in a decrease in crop productivity and, in many cases, the early decay of plants (Laveau et al. 2009). Among these diseases the most prominent in the Mediterranean countries is esca (Fischer 2002). It was considered to be a problem in older vineyards only, and it was relatively easily controlled with fungicides (Graniti et al. 2000). However, the use of sodium arsenite – the main fungicide tool against it – was banned at the beginning of the twenty-first century in many countries which, together with other changes in growing techniques, led to a considerable increase in esca incidence worldwide (Bertsch et al. 2009).

Esca is a complex disease, mainly caused by the ascomycete fungi *Phaeomoniella chlamydospore* and *Phaeoacremonium aleophilum* and the basidiomycete fungus *Formitiporia mediterranea* (Di Marco et al. 2011). It usually affects adult plants aged above 10 years, either causing foliar discoloration or sudden wilting of the entire vine (apoplexy) which kills the plant within a short period (Mugnai et al. 1999). Affected leaves generally show a 'tiger-stripe' pattern (Surico et al. 2008), while a characteristic spotting, described as 'black measles' in the USA, is observed on berries (Mugnai et al. 1999). Foliar symptoms may or may not be observed in consecutive years, but affected plants generally end up dying from apoplexy (Hofstetter et al. 2012). Currently, in the absence of chemical methods of control of proven efficiency against esca, any treatment should be preventive and various cultural and crop management measures are recommended, including good pruning practices and the use of a high-quality plant material (García-Jiménez et al. 2010). Once the vine is affected, alterations to the cells arise at leaf level before symptoms become visible (Valtaud et al. 2009). Therefore, a technique capable of detecting infected vines before the symptoms become visible would allow better crop management and decision-making.

The present case study shows the potential of a near-infrared hyperspectral system (NIR-HSI) to distinguish between visually asymptomatic grapevine leaves, picked from esca-affected vines, and symptomatic leaves, collected from the same vines, at a laboratory scale. This methodology opens up an area of research aiming to apply it at the field scale through the development of sensors that could help growers to detect disease presence early, before the symptoms become visually noticeable.

13.4.2 Materials and Methods

13.4.2.1 Plant Material

In this study, grapevine leaves of cv. Tempranillo (*Vitis vinifera* L.) picked from an experimental vineyard belonging to the Viticulture and Enology Station of Navarra (EVENA) and located in Olite (Navarra, Spain) were used. Two leaf categories were selected visually, identified and handpicked from the field at a growth stage close to harvest (September 20, 2018). A total of 60 samples were collected: 30 asymptomatic leaves from esca-affected vines, named Esca 1 (E1), and 30 symptomatic leaves from the same esca-affected vines of class E1 and designated Esca 2 (E2). Samples were kept in cold storage at 3 °C until analysis. The measurements were made approximately 24 h later. Before hyperspectral image acquisition, a reference RGB image was obtained for each leaf.

13.4.2.2 Hyperspectral Imaging

Hyperspectral Image Acquisition

Hyperspectral images were recorded using an NIR-HSI system consisting of an NIR InGaAs camera with 320 × 256 pixel resolution (Xeva 1.7–320, Xenics, Leuven, Belgium) coupled to a spectrograph (ImSpector N17E, Specim, Spectral Imaging Ltd., Oulu, Finland), both sensitive in the range 900–1700 nm. This line-scanning imager was mounted 400 mm above a linear translation stage (LEFS25, SMC Corporation, Tokyo, Japan) that allowed samples to be moved under the field of view of the camera. Four 46 W halogen lamps and a black cover enclosing the entire set-up were used for stable lighting conditions of the scene. A computer equipped with Xeneth 2.5 and ACT Controller software was used to control the camera and the translation stage and to record the leaf images.

One hyperspectral image of the adaxial leaf side was acquired per sample with a spatial resolution of 0.75 mm per pixel (320 pixels per line) and a spectral resolution of about 3 nm (256 spectral bands). Detector saturation was avoided by optimizing the integration time at 2 ms. In addition, white reference with standard reflectance of 99% (Teflon white calibration tile, Specim, Spectral Imaging Ltd., Oulu, Finland) and dark reference (camera lens covered by an opaque black cap) images were taken for reflectance calibration.

Image Processing

The first step in image processing consisted of forming the three-dimensional data cube (hypercube) by stacking the raw leaf images. Then, reflectance calibration was performed to convert the raw intensity values in hyperspectral images into relative reflectance (*R*) values by using Eq. 13.2 (Geladi et al. 2004):

$$R = \frac{I_{\text{Raw}} - D}{\text{W} - \text{D}}, \tag{13.2}$$

where I_{Raw} is the raw irradiance intensity acquired on the sample, *D* is the intensity acquired for the dark reference and *W* is the intensity acquired on the white reference.

At the next step, images were segmented to separate the region of interest, in this case the whole leaf, from the saturated areas and background. In this study segmentation was accomplished following the algorithm presented in Lopez-Molina et al. (2017). Moreover, data between 900–1000 nm were removed as spectral noise was observed within that region.

Finally, the relevant spectral data were extracted by unfolding the 3-dimensional hypercube into a 2-dimensional data matrix of the leaf pixel reflectance values at the selected wavelengths (224 bands). In this case, the dataset was divided randomly into calibration and validation groups, comprising 60 and 40% leaves of each class, respectively. For each leaf that composed the calibration group (18 images per class), 10 pixels were manually selected using the graphical user-friendly interface HYPER-Tools (Mobaraki and Amigo 2018) and taking the RGB images as a reference. For class E1, pixels were selected from one external and one internal leaf ring (5 pixels per ring), while for class E2 only the pixels corresponding to leaf zones with visible esca symptoms were selected. The resulting X matrix consisted of 360 rows and 224 columns (180 rows per class), and was used as the calibration set to form classification models. In the remaining 12 images per class, the unfolding process was performed automatically, and one matrix including the leaf pixels contained in the segmented mask was obtained for each leaf sample for validation purposes.

Image processing was performed in MATLAB R2016b (The MathWorks, Natick, MA, USA).

13.4.2.3 Multivariate Data Analysis

Data processing and qualitative analysis were performed using the PLS_Toolbox (Eigenvector Research Inc., Wenatchee, WA) within MATLAB® computational environment.

Spectral Pre-processing

Prior to model building, spectral data were pre-processed to correct light scattering and system noise effects. The following pre-processing techniques were tested individually and combined: standard normal variate (SNV), multiplicative scatter correction (MSC), detrending, smoothing, and first and second derivatives (1st Der and 2nd Der, respectively). Smoothing was performed using the Savitzky–Golay algorithm, on a total window of 15 points and a zero-order polynomial, while derivatives were calculated using the Savitzky–Golay method by second order polynomial and a 15-point window. The effect of no pre-processing (None) was also analysed.

Leaf Pixel Classification

A partial least squares discriminant analysis (PLS-DA) method was used to create a two-class classifier to differentiate pixels belonging to class E1 (asymptomatic leaves from esca-affected vines) from those belonging to class E2 (symptomatic leaves). The PLS-DA is a supervized classification technique in which a PLS regression is carried out to predict class membership (Barker and Rayens 2003). For that reason, a **Y** matrix consisting of 0 s and 1 s needs to be formulated to indicate class membership (1) or non-membership (0). In this case, the spectral information (**X** matrix) was linked with the category the samples belonged to (E1 or E2) (**Y** matrix).

As stated above, 60% of samples (36 leaves) of each class were randomly selected for calibration and cross-validation (CV; Venetian blinds cross-validation method with 10 data splits), while the remaining 40% (24 leaves) were used as a validation group.

The performance of PLS-DA models was evaluated in terms of the percentage of correctly classified (%CC) pixels, and the sensitivity and specificity in CV, together with the percentage of correctly predicted pixels per class obtained on each sample in the validation.

13.4.3 Results and Discussion

Figure 13.4 shows the mean spectra of the selected pixels of each of the two classes, E1 (asymptomatic) and E2 (symptomatic), in the calibration group. Considerable differences in the magnitude of reflectance were observed between the two classes along the selected spectral range (1000–1700 nm). A deep dip in the spectrum is evident at around 1450 nm because of the first overtone of the OH-stretching band (Osborne et al. 1993). As can be seen in Fig. 13.4 the reflectance of class E1 at 1450 nm is lower and thus, absorbance was higher, than that of class E2. Since the strong water absorption bands near this wavelength change according to the water content status of foods (Büning-Pfaue 2003), it is hypothesized that this difference occurs because of the greater water content in the asymptomatic leaves than symptomatic ones where esca has already caused desiccation of some leaf areas. This

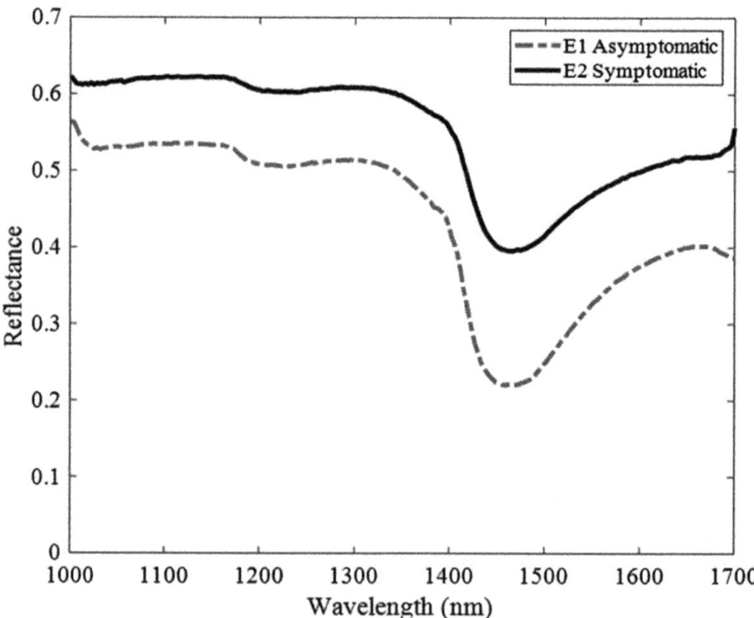

Fig. 13.4 Mean spectra of classes E1 and E2 in the calibration group

statement accords with the findings of Büning-Pfaue (2003), who observed that the absorption band at around 1400 nm of sliced pear flesh decreased in intensity at the same time as dehydration increased.

Table 13.3 presents the % CC pixels in the calibration and the CV groups obtained with the different pre-processing methods applied. The number of samples (*n*) (after elimination of outliers) and the number of latent variables (LVs) used to develop the PLS-DA models are also included. Good classification results were obtained with all of the pre-processing techniques, achieving more than 85% CC pixels. However, the best results were achieved when applying smoothing, with more than 94% of pixels correctly classified in the CV group.

Table 13.4 shows the confusion matrix and the sensitivity and specificity values obtained for the CV group after the smoothing pre-process. Class E1 has a higher sensitivity value than class E2, indicating that pixels belonging to E1 were classified better into their corresponding group (97.2% CC *versus* 92.2%).

This is an interesting result, since quite the opposite was expected, i.e. that symptomatic pixels would have been classified better than asymptomatic ones. However, this highlights the capability of HS systems to identify vines potentially affected by esca, but without visual symptoms.

Regarding the results obtained for the validation group (24 leaves) (data not shown), in most cases, a larger proportion of pixels was classified into the class they belonged to. In total, 84% of the pixels from the 12 leaves of class E1 were correctly

Table 13.3 Number of LVs and % CC samples obtained in the PLS-DA models with the different pre-processing

Pre-processing	*n*	LVs	% CC_{Cal}	% CC_{CV}
None	360	3	91.9	91.1
SNV	354	3	91.0	90.4
MSC (mean)	358	2	90.8	90.5
Detrending	360	2	90.8	89.7
Smoothing	360	5	**95.3**	**94.7**
1st Der	360	4	90.6	90.6
2nd Der	358	3	86.6	85.2
Smoothing+2nd Der	360	3	89.7	88.6
Smoothing+MSC	360	4	90.6	90.0
Smoothing+SNV	358	4	90.5	89.9
Smoothing+1st Der	360	2	89.7	90.0
1st Der + MSC	359	6	93.6	93.0
1st Der + SNV	360	4	90.8	90.6

Values in bold correspond to the highest % CC pixels in the PLS-DA models

Table 13.4 Confusion matrix and sensitivity and specificity values of CV group after smoothing

		Actual class (%)		Sensitivity	Specificity
Predicted class (%)	E1	97.2	7.8	0.972	0.928
	E2	2.8	92.2	0.928	0.972
	Not assigned	0	0		

Fig. 13.5 Classification of pixels in the validation (grey: E1 class; black: E2 class) leaf samples (**a**,**b**) E1; (**c**,**d**) E2 obtained by HS system (b,d) and their corresponding RGB images (a,c)

classified into their corresponding class (asymptomatic), whereas 76% of the pixels from the 12 leaves of class E2 were correctly labelled as symptomatic.

Figure13.5 displays the classification of pixels from two leaves of the validation group belonging to classes E1 (a,b) and E2 (c,d), respectively. Images in Fig. 13.5a and c correspond to the RGB images taken as reference and images in Fig. 13.5b and d are those obtained by the HS system. In sample E1 (Fig. 13.5b) 77.5% of pixels were correctly assigned as class E1 (grey pixels), whereas in sample E2 (Fig. 13.5d) 76.8% of pixels were classified as class E2 (black pixels). Fig. 13.5d also shows that most of the black pixels were at the edges of the leaf, matching the most esca-affected areas as shown in the equivalent RGB image (Fig. 13.5c).

13.4.4 Conclusions

The feasibility of NIR hyperspectral imaging, combined with multivariate analysis, to differentiate between asymptomatic and symptomatic leaves from esca-affected vines was evaluated in this case study. Good classification rates (above 85% CC in CV) were obtained when applying different pre-processing techniques in PLS-DA models. More accurate discrimination of asymptomatic (E1) and symptomatic (E2) pixels was achieved after the smoothing pre-process (94.7% CC). Furthermore, a pixel-based prediction accuracy above 75% was obtained in the validation group. Class E1 was classified better than class E2 suggesting that HS systems could be used for esca diagnosis at early stages of infection.

13.5 Conclusions for the Chapter

Based on the importance of having early and accurate indicators of disease infestation in crops for timely and proper disease control management, case studies in cotton, avocado and grape vines using remote sensing technology have been illustrated. Different acquisition platforms were evaluated, such as leaf-level hyperspectral data and canopy-level remote imagery taken from manned airplanes or helicopter and UAVs, as well as from satellites. The results proved that remote sensing is very useful, efficient and effective for identifying CRR zones in cotton field, laurel wilt-infested avocado trees and esca-affected vines. The use of powerful analytical algorithms on remotely-sensed data enables the challenge of detecting infested plants at an early stage to be overcome, i.e. with minimal symptoms, discriminating them from asymptomatic plants and from plants affected by other biotic and abiotic factors that cause similar symptoms, and for developing prescription maps. Therefore, the combination of suitable remote-sensing data and advanced algorithms are presented as robust tools for rapid and accurate disease detection, offering major savings compared to traditional diagnostics such as visual inspection, which is costly, time-consuming, and subject to human bias. The choice between the remote-sensing platforms and analysis techniques depends on the agronomic goal, the cost and availability of data and their ease of analysis, the computing power required and the overall ease of use.

The early identification of infested plants could assist growers in the decision-making process and in developing proper and timely site-specific disease management strategies to control the spread of these important diseases. In addition, the use of disease and prescription maps would allow farmers to optimize inputs and field operations, resulting in reduced yield losses and increased profits. Consequently, the environmental impact would be lessened with fewer and targeted inputs. Further research should be aimed at developing automatic algorithms applied at the plant level to control the evolution of these diseases in a robust, fast and accurate way.

Acknowledgments The research presented here was partly financed by the USDA Specialty Block Grant No. 019730 (Florida Department of Agriculture and Consumer Services, USA), AGL2017-83325-C4-1R and AGL2017-83325-C4-4R Projects (Spanish Ministry of Science, Innovation and Universities and AEI/EU-FEDER funds), Public University of Navarre postgraduate scholarships (FPI-UPNA-2017, Res.654/2017), Project DECIVID (Res.104E/2017, Department of Economic Development of the Navarre Government-Spain), and the Spanish MINECO project TIN2016-77356-P (AEI, Feder/UE). The authors thank Don Pyba and Sherrie Buchanon for their helpful assistance, as well as the Viticulture and Enology Station of Navarra-Spain (EVENA) for providing the samples and for their valuable support.

References

Abdulridha J, Ampatzidis Y, Ehsani R, de Castro A (2018) Evaluating the performance of spectral features and multivariate analysis tools to detect laurel wilt disease and nutritional deficiency in avocado. Comput Electron Agric 155:203–211

Barker M, Rayens W (2003) Partial least squares for discrimination. J Chemometr 17(3):166–173

Bertsch C, Larignon P, Farine S et al (2009) The spread of grapevine trunk disease. Science 324(5928):721
Büning-Pfaue H (2003) Analysis of water in food by near infrared spectroscopy. Food Chem 82(1):107–115
Campbell JB (2002) Introduction to remote sensing, 3rd edn. Guilford Press, New York
Chappelle EW, Kim MS, McMurtrey JE (1992) Ratio analysis of reflectance spectra (RARS): an algorithm for the remote estimation of the concentrations of chlorophyll A, chlorophyll B, and carotenoids in soybean leaves. Remote Sens Environ 39(3):239–247
De Castro AI, Jurado-Expósito M, Gómez-Casero MT et al (2012) Applying neural networks to hyperspectral and multispectral field data for discrimination of cruciferous weeds in winter crops. Sci World J 630390
De Castro AI, Ehsani R, Ploetz R et al (2015a) Optimum spectral and geometric parameters for early detection of laurel wilt disease in avocado. Remote Sens Environ 171:33–44
De Castro AI, Ehsani R, Ploetz RC et al (2015b) Detection of laurel wilt disease in avocado using low altitude aerial imaging. PLoS One 10(4):e0124642
Di Gennaro SF, Battiston E, Di Marco S et al (2016) Unmanned Aerial Vehicle (UAV)-based remote sensing to monitor grapevine leaf stripe disease within a vineyard affected by esca complex. Phytopathol Mediterr 55(2):262–275
Di Marco S, Osti F, Calzarano F et al (2011) Effects of grapevine applications of fosetyl-aluminium formulations for downy mildew control on "esca" and associated fungi. Phytopathol Mediterr 50(4):S285–S299
Drake DR, Minzenmayer RR, Multer WL et al (2013) Evaluation of farmer applications of Topguard (flutriafol) for cotton root rot control in the first Section 18 exemption year. In: Proccedings of the Beltwide Cotton Conf. National Cotton Council of America, Cordova
ElMasry G, Sun DW (2010) Principles of hyperspectral imaging technology. In: Sun DW (ed) Hyperspectral imaging for food quality analysis and control. Academic Press, San Diego, pp 3–43
Evans EA, Bernal Lozano I (2015) Sample avocado production costs and profitability analysis for Florida. Electronic data information source (EDIS) FE837. Gainesville, FL: Food and Resource Economics Department, University of Florida. https://edis.ifas.ufl.edu/dosearch.html. Accessed 23 March 2018
Fischer M (2002) A new wood-decaying basidiomycete species associated with esca of grapevine: *Fomitiporia mediterranea* (Hymenochaetales). Mycol Prog 1(3):315–324
García-Jiménez J, Raposo R, Armengol J (2010) Enfermedades fúngicas de la madera de la vid. In: Jiménez-Díaz RM, Montesinos Seguí E (eds) Enfermedades de las plantas causadas por hongos y oomicetos: naturaleza y control integrado. SEF-Phytoma España, pp 161–189
Geladi P, Burger J, Lestander T (2004) Hyperspectral imaging: calibration problems and solutions. Chemom Intell Lab Syst 72(2):209–217
Gitelson AA, Merzlyak MN (1996) Signature analysis of leaf reflectance spectra: algorithm development for remote sensing of chlorophyll. J Plant Physiol 148(3–4):494–500
Gitelson AA, Kaufman YJ, Stark R et al (2002) Novel algorithms for remote estimation of vegetation fraction. Remote Sens EnvironRemote Sens Environ 80(1):76–87
Graniti A, Surico G, Mugnai L (2000) Esca of grapevine: a disease complex or a complex of diseases? Phytopathol Mediterr 39(1):16–20
Han J, Kamber M, Pei J (2012) Data mining. Concepts and techniques, 3rd edn. Morgan Kaufmann Publishers, Waltham
Hanula JL, Mayfield AE III, Fraedrich SW et al (2008) Biology and host associations of redbay ambrosia beetle (*Coleoptera: Curculionidae: Scolytinae*), exotic vector of laurel wilt killing redbay trees in the southeastern United States. J Econ EntomolJ Econ Entomol 101(4):1276–1286
Hofstetter V, Buyck B, Croll D et al (2012) What if esca disease of grapevine were not a fungal disease? Fungal Divers 54:51–67

Howden SM, Soussana JF, Tubiello FN et al (2007) Adapting agriculture to climate change. Procceding of the Natl Acad Sci USA 104(50):19691–19696
Isakeit T, Minzenmayer R, Sansone C (2009) Flutriafol control of cotton root rot caused by Phymatotrichopsis omnivora. In Procceding of the Beltwide Cotton Conf. 130–133. Cordova, Tenn.: National Cotton Council of America
Kaufman YJ, Remer LA (1994) Detection of forests using mid-IR reflectance: an application for aerosol studies. IEEE Trans Geosci Remote Sens 32(3):672–683
Laveau C, Letouze A, Louvet G et al (2009) Differential aggressiveness of fungi implicated in esca and associated diseases of grapevine in France. Phytopathol Mediterr 48(1):32–46
Levasseur-Garcia C, Malaurie H, Mailhac N (2016) An infrared diagnostic system to detect causal agents of grapevine trunk diseases. J Microbiol Methods 131:1–6
Lopez-Molina C, Ayala-Martinez D, Lopez-Maestresalas A et al (2017) Baddeley's Delta metric for local contrast computation in hyperspectral imagery. Prog Artif Intell 6:121–132
Lu JZ, Ehsani R, Shi YY et al (2017) Field detection of anthracnose crown rot in strawberry using spectroscopy technology. Comput Electron Agric 135:289–299
Lu JZ, Ehsani R, Shi YY et al (2018) Detection of multi-tomato leaf diseases (late blight, target and bacterial spots) in different stages by using a spectral-based sensor. Sci Rep 8:2793
Lyda SD (1978) Ecology of *Phymatotrichum omnivorum*. Annu Rev Phytopathol 16:193–209
Mahlein AK (2016) Plant disease detection by imaging sensors – parallels and specific demands for precision agriculture and plant phenotyping. Plant Dis 100(2):241–251
Mahlein AK, Steiner U, Hillnhütter C et al (2012) Hyperspectral imaging for small-scale analysis of symptoms caused by different sugar beet diseases. Plant Methods 8:3
Mendel J, Burns C, Kallifatidis B et al (2018) Agri-dogs: using canines for earlier detection of laurel wilt disease affecting avocado trees in South Florida. HortTechnology 28(2):109–116
Mobaraki N, Amigo JM (2018) HYPER-Tools. A graphical user-friendly interface for hyperspectral image analysis. Chemom Intel Lab Syst 172:174–187
Mugnai L, Graniti A, Surico G (1999) Esca (black measles) and brown wood-streaking: two old and elusive diseases of grapevines. Plant Dis 83(5):404–418
Mutka AM, Bart RS (2015) Image-based phenotyping of plant disease symptoms. Front Plant Sci 5:734
NCC (2013) Disease Database (2011). National Cotton Council of America, Cordova. Available at: http://www.cotton.org/tech/pest/index.cfm. Accessed 20 February 2013
Oerke EC, Dehne HW (2004) Safeguarding production - losses in major crops and the role of crop protection. Crop Prot 23:275–285
Osborne BG, Fearn T, Hindle PH (1993) Practical NIR spectroscopy with applications in food and beverage analysis. Longman Scientific and Technical, Harlow
Pammel LH (1888) Root rot of cotton, or "cotton blight". Texas Agric Exp Station Ann Report 1:50–65
Ploetz RC, Harrington T, Hulcr J et al (2011) Recovery plan for laurel wilt of avocado (caused by *Raffaelea lauricola*). National Plant Disease Recovery System. Homeland Security Presidential Directive Number 9 (HSPD-9). http://www.ars.usda.gov/research/docs.htm?docid=14271 accessed 20 April 2013
Ploetz RC, Konkol JL, Narvaez T et al (2017a) Presence and prevalence of *Raffaelea lauricola*, cause of laurel wilt, in different species of ambrosia beetle in Florida USA. J Econ Entomol 110(2):347–354
Ploetz RC, Kendra PE, Choudhury RA et al (2017b) Laurel wilt in natural and agricultural ecosystems: understanding the drivers and scales of complex pathosystems. Forest 8(2):48
Rançon F, Bombrun L, Keresztes B et al (2019) Comparison of SIFT encoded and deep learning features for the classification and detection of esca disease in Bordeaux vineyards. Remote Sens (Basel) 11(1):1–26
Rodríguez-Pérez JR, Riaño D, Carlisle E et al (2007) Evaluation of hyperspectral reflectance indexes to detect grapevine water status in vineyards. Am J Enol Vitic 58(3):302–317

Smith HE, Elliot FC, Bird LS (1962) Root rot losses of cotton can be reduced. Pub. No. MP361. Texas A&M Agricultural Extension Service, College Station

Statista (2018) Import value of avocados worldwide in 2017, by leading country (in million U.S. dollars). Source: UN Comtrade; 2017. https://www.statista.com/statistics/938571/major-importers-avocado-import-value/ Accessed 06 November 2018

Surico G, Mugnai L, Marchi G (2008) The esca disease complex. In: Ciancio A, Mukerji KG (eds) Integrated management of diseases caused by fungi, phytoplasma and bacteria. Integrated management of plant pests and diseases, vol 3. Springer, Dordrecht, pp 119–136

Thomasson JA, Wang T, Wang X et al (2018) Disease detection and mitigation in a cotton crop with UAV remote sensing. In Proccedings of the autonomous air and ground sensing Systems for Agricultural Optimization and Phenotyping. Bellingham, Wash.: SPIE

Valtaud C, Larignon P, Roblin G et al (2009) Developmental and ultrastructural features of *Phaeomoniella chlamydospora* and *Phaeoacremonium aleophilu*m in relation to xylem degradation in esca disease of the grapevine. J Plant Pathol 91(1):37–51

Wang T, Thomasson JA (2019) Plant-by-plant level classifications of cotton root rot by UAV remote sensing. In Proccedings of the Autonomous Air and Ground Sensing Systems for Agricultural Optimization and Phenotyping. Bellingham, Wash.: SPIE

Yang C (2012) A high-resolution airborne four-camera imaging system for agricultural remote sensing. Comput Electron Agric 88(1):13–24

Yang C, Odvody GN, Fernandez CJ et al (2014) Monitoring cotton root rot progression within a growing season using airborne multispectral imagery. J Cotton Sci 18(1):85–93

Yang C, Odvody GN, Thomasson JA et al (2016) Change detection of cotton root rot infection over 10-year intervals using airborne multispectral imagery. Comput Electron Agric 123(1):154–162

Yang C, Odvody GN, Thomasson JA et al (2018) Site-specific management of cotton root rot using airborne and high-resolution satellite imagery and variable-rate technology. Trans ASABE 61(3):849–858

Zhang J, Huang Y, Pu R et al (2019) Monitoring plant diseases and pests through remote sensing technology: a review. Comput Electron Agric 165:104943

Chapter 14
Conclusions: Future Directions in Sensing for Precision Agriculture

Ruth Kerry and Alexandre Escolà

Abstract In this final chapter, we provide an overall conclusion to this book on the use of sensors in precision agriculture (PA) based on the conclusions of the individual chapters concerning key themes and future research needs. The authors highlighted aspects related to the need to improve sensor resolutions (spatial, temporal and spectral), increase accuracy and simplify the process of calibration, when required. The ability to obtain gigabytes or even terabytes of data is complicated by the need to store, process and analyse them. Although computing power is increasing continuously, automated data processes are also required to ease the adoption of new sensing systems. In addition, some barriers to the widespread adoption of sensing approaches in PA are identified. Most important are economics and training. Gathering, processing and analysing data from sensing systems should lead farmers to make more informed management decisions and that is only possible if the information derived helps them to increase profits in a sustainable way.

Keywords Spatial resolution · Temporal resolution · Spectral resolution · Accuracy · Adoption · Barriers · Computing issues · Economic viability

R. Kerry (✉)
Department of Geography, Brigham Young University, Provo, UT, USA
e-mail: ruth_kerry@byu.edu

A. Escolà
Department of Agricultural and Forest Engineering, Research Group on AgroICT & Precision Agriculture – GRAP, Universitat de Lleida/Agrotecnio-CERCA Center, Lleida, Catalunya, Spain

R. Kerry, A. Escolà (eds.), *Sensing Approaches for Precision Agriculture*, Progress in Precision Agriculture, https://doi.org/10.1007/978-3-030-78431-7_14

14.1 Introduction

This book has covered a wide range of sensors that are currently routinely used by precision agriculture (PA) practitioners as well as sensors that are still in the research stages and not yet commercially available. Some common themes that arise in the book from the first review chapter (Chap. 2) on satellite remote sensing are the key issues of spatial, temporal and spectral resolution. Broadly speaking, spatial resolution is relevant to all sensing approaches if it is seen as analogous to the density of spatially sensed data. Within the broader context of general sensing, issues of temporal resolution can encompass the timeliness of data acquisition and processing to determine if real-time on-the-go sensing and application approaches are feasible compared with static map-based approaches. Finally, in the broader context of general sensing, spectral resolution can encompass not only optical sensing in different parts of the electromagnetic spectrum but sensing of different characteristics and with different technologies. This book shows that the future of PA is likely to be based on advances that involve sensing with increased spatial, temporal and spectral resolution based on the broad definitions used above. Other key issues that will affect the future of sensing in PA relate to the accuracy and calibration of sensors and computing issues. As sensors improve their spatial, temporal and spectral resolutions there will be increasingly large volumes of data that need to be stored, processed and analyzed swiftly. Increased computing power, automation of data analysis and the development of decision support systems as well as fully automated decision making are all key issues for the future of sensing in PA. Finally, the crux for PA practitioners involves economic aspects that encourage or hinder uptake of new sensing approaches by those practicing commercial PA.

14.2 Spatial Resolution

The focus in PA is primarily within fields, therefore, the spatial resolution or sampling density on the ground should be high for most activities. In Chap. 2, the satellite remote sensing products that show the best promise for PA are identified as those with a ground sample distances of between 0.5 m and 30 m, or even less. Several products with such characteristics are currently available or scheduled to be commissioned. Some are or will be free to use, therefore, there is the potential for much future work using satellite remote sensing. When higher spatial resolutions than those available from satellites are required, users will need to get closer to the crop with manned or unmanned airborne sensing systems. However, when it comes to very high spatial resolution, proximal sensing should probably be used. In Chap. 3, light detection and ranging (LiDAR)-based systems are shown to provide very dense, high resolution point clouds with several hundreds or even thousands of points per square metre on the ground. The red, green and blue depth (RGB-D) cameras are also shown to be a good low-cost solution for 3D sensing of crops. Both

LiDAR and other mobile or portable sensing approaches usually lack adequate resolution for detecting fine objects such as small fruitlets, thin branches or leaves in field conditions. It is clear from Chaps. 8, 10 and 13, especially, that for activities such as identifying fruit, weeds and individual infected plants, sensors with a higher spatial resolution (0.5–0.01 m) are key to current and future activities in PA. In Chap. 7, the influence of the time delay associated with transport of grain through the combine threshing drum on the accuracy of yield and protein maps is mentioned. This time delay affects the spatial resolution and location accuracy of yield sensors, one of the most established sensors used in PA. Dynamic grain flow models show promise for over-coming this, but are still in the development stages and need further investigation.

Increasing the spatial resolution of application machinery will also be key for PA in the future as there is little benefit in obtaining very detailed data if the machinery still has a large spatial footprint for applying water, fertilizers and pesticides. An example of this are current field sprayer booms. At the very beginning they were only able to control the whole boom with working widths of 12–36 m and even more. Then the manufacturers divided the booms into smaller sections of 3–5 m. Currently, it is possible to control the individual nozzles, that is, every 0.5 m. Case study 2 in Chap. 10 highlights this issue noting that an approximate 10% increase in savings could have been made had the spatial resolution of the sprayer used in the study been higher.

14.3 Temporal Resolution

Traditionally, PA has focused on plant nutrient needs which are assessed infrequently and often determined just once in a growing season. However, increasing emphasis on crop moisture needs, temporal monitoring and forecasting of biomass and yield and assessing weed and disease infestations within seasons has made satellite remote sensing products with greater temporal resolution and revisit frequency more desirable. For temporal resolution, the most interesting solution might be the use of stationary sensors installed in the field. They may have low spatial resolution because of their unit cost, but they can provide high temporal resolution data that can be transmitted using point-to-point communication solutions or by being part of a wireless sensor network. Stationary sensors installed in a small number of management zones are frequently used in precision irrigation applications because irrigation timing is as important as determining spatial irrigation zones. The case studies in Chap. 11 illustrated the importance of timely data and the move towards adaptive rather than static irrigation management zones. Currently, there are several satellites with revisit frequencies of just one day, but they are not usually free to access. This increasing need for data with a high temporal resolution has fuelled the use of sensors attached to masts in fields (Chap. 11), attached to ground vehicles (all terrain vehicles (ATVs) in Chap. 10) or attached to irrigation central pivots (Chap. 11), and this has also been the reason for the almost exponential increase in the use of sensors deployed on unmanned aerial vehicles (UAVs), as described in Chap. 9. It is clear from the current state of research in PA that the use of UAVs will remain popular and probably increase further as more cost-effective systems are developed and the need for timely data continues to grow.

14.4 Spectral Resolution and Range of Sensor Approaches

Chapter 2 indicates that identifying specific soil conditions and crop stresses are more likely with narrow bands from hyperspectral imagery, yet currently the only satellite source of hyperspectral imagery is Proba 1 CHRIS. However, there are a few hyperspectral platforms that will soon be launched. In addition to hyperspectral imagery, sun-induced fluorescence sensors (SIF) are a powerful, yet under-utilized tool for identifying crops stress. Furthermore, synthetic aperture radar (SAR) sensors have great potential for providing insight into soil surface moisture problems and can still be used at night or when there is cloud cover. However, the current SIF and SAR platforms suffer from too coarse a spatial resolution to be useful in PA unless they are down-scaled using finer scale imagery from multispectral satellites. This is clearly possible but is a time-consuming and computationally intensive activity to do successfully.

Multispectral and hyperspectral sensors, however, do also occur in UAV and in proximal sensing solutions. Multispectral devices provide broader spectral bands with more generic data, whereas hyperspectral devices produce tens or even hundreds of narrow bands with more detailed and specific data on the target. Nevertheless, the complexity and cost of the latter makes them difficult to transfer for use by farmers. Chapter 3 notes that currently, commercially available LiDAR-based equipment provides only the intensity of returns. However, some researchers have already designed LiDAR systems using light for both ranging and analysing spectral response of the measured objects in different spectral bands. Including this capability in LiDAR and other types of 3D sensing systems will complement the data obtained from crops. Chapter 4 indicated that proximal soil sensing approaches tend to produce more accurate predictions of the more permanent properties than of nutrients. Case study 4 showed that determining nutrient concentrations was possible for soils developed under markedly different conditions, and the introduction to Chap. 4 suggested that predicting nutrient levels is more feasible with sensor fusion and information from several sensors at once. Multi-sensor platforms for sensing soil are likely to play a bigger role in the future of PA. Chapter 6 suggests that nose sensors hold the most promise for detection of the severity of plant injury by insect feeding. Although current PA research contains many papers on hyperspectral imaging for disease control, Chaps. 6 and 13 mention that it is still in the development stages. The reasons are the cost of equipment, the need for sensing under controlled laboratory conditions for the best results and the labour-intensive application of increasingly complex numerical algorithms for data analysis. The spatial and temporal resolution of remotely sensed data from UAVs (Chap. 9) have been greatly increased, but if the cost of cameras with wider ranges of wavelengths is reduced it would enable more hyperspectral imaging from UAVs. Also required are longer battery life to avoid flight interruption and lighter cameras or the increased ability of UAVs to hold larger weights and therefore more sensors. Hyperspectral data is also used in precision horticulture (Chap. 8), in weed detection and classification (Chap. 10) and to detect the health and vigour status of crops (Chap. 12).

14.5 Accuracy and Calibration of Sensors

Another important message arising from this book is that many sensors, such as proximal soil sensors (Chap. 4), and remote sensing data measure bulk properties of soils and plants rather than individual properties. This means that the sensors need to be calibrated to gain information about individual properties. Although it is economically desirable to calibrate with as few samples as possible, accurate information is also needed, otherwise the management decisions made using the sensor will be poor and profits and efficiency will be less. A recurring issue in Chap. 4 was the number of soil samples needed to calibrate proximal soil sensors because they need to be calibrated locally. Furthermore, accuracy is far less when samples are not prepared and observed uniformly, therefore, spectroscopic techniques are often applied in the laboratory which is more expensive than *in-situ* observation in the field. Future PA needs more sensors that can be used *in situ* and on-the-go without sample preparation. Chapter 3 notes that changes in lighting affect the ability to apply photogrammetric sensors in the field compared with the laboratory under standardized conditions. Chapter 8 also notes that optical sensors based on reflectance, fluorescence and emission are effective for detecting problems in crops such as plants under stress. However, most optical sensors need calibration for a given site or crop variety with appropriate reference data. The future of such sensing in PA needs the development of local calibrations that can be incorporated into the software used to deploy the sensors.

14.6 Computing Issues

Large data sets from sensors need to be analyzed, and the issue of data storage and computational intensity came up in most chapters of this book. There are currently trade-offs between having high spatial, temporal and spectral resolution data and the inability to analyze them with available techniques and computing power. Another issue is the lack of training of PA practitioners with the increasingly complex computational methods needed to analyse large sets of sensed data.

A key to adoption of sensing approaches seems to be the ease of use of new sensing methods. This brings to light the importance of automated software interfaces for processing data from sensors, and output that can be easily used. Decision support systems that are largely automated are also required. Chapter 7 suggests the need for the development of software to explore multi-year yield data and Chap. 8 notes that the development of agronomic computer models to enable sensor data to support management processes is needed and is the next logical step. Chapter 4 notes the potential of sensor fusion and artificial intelligence approaches to obtain accurate estimates of soil properties with minimal soil calibration data in the future, but this will come only after much research.

In most chapters, real-time, on-the-go applications rather than map-based applications were brought up as the future. Again, making this a widely adopted reality in commercial PA will involve a great deal of computing research. The volume of data to be processed is a major problem, especially when it has to be analyzed manually. In such cases, improving the spatial, temporal and spectral resolutions of sensed data could incur costs for data-processing that are similar in magnitude to traditional sampling and laboratory analysis expenses. This would defeat the purpose of sensing approaches, therefore, there is a need for automated complex analysis of big data sets. Automated UAV data processing services are currently available and this technology has been quite widely adopted, but such services will need to continue to develop as cameras with greater spectral resolution are used (Chap. 9). Chapter 3 notes the merits of deep- and machine-learning for processing large point clouds. Also, neural networks could be key for classifying different parts of plants' anatomy (Chap. 3), distinguishing crops from soil (Chaps. 2 and 3), fruit from leaves (Chap. 8), weeds from crops (Chap. 10) and infected plants from uninfected plants (Chap. 9). Finally, Chap. 13 emphasizes the need for advanced algorithms to analyse automatically large amounts of hyperspectral data. We mention in the introduction (Chap. 1) of this book the need for a book focused on numerical methods in this Springer Precision Agriculture book series, and we reiterate that need here. Indeed, a book on modelling is currently being compiled for the book series.

Chapter 13 emphasizes that currently map-based approaches to disease control are used but there needs to be more on-the-go sensing and application software for plant protection products. Indeed, some of the greatest needs for precision irrigation are related to computer integration of weather, satellite and soil sensor data to create efficient near real-time data analysis and advanced automated decision support software to control precision irrigation strategies (Chap. 11). Chapter 5 notes that wireless sensor networks are not yet mainstream because the set up for each application is highly customized. Furthermore, efficient techniques of fault monitoring need to be developed as well as more secure ways to transmit data. Chapter 3 discussed how crop geometry and structure sensor information needs to become accurate and timely enough to allow real-time adjustment of pesticide spraying rates, cutting height, and so on. However, LiDAR-based sensing systems may generate several gigabytes of data in few minutes or even seconds and also require high capacity workstations to process them. Chapter 7 suggests the need for the development of software to explore multi-year yield data and Chap. 8 notes that the development of agronomic computer models to enable sensor data to drive management processes is needed and is the next logical step.

14.7 Issues Influencing the Adoption of Sensing Approaches in PA: Economic and Training Issues

Throughout the book, barriers to widespread adoption of sensing approaches in PA were raised as well as reasons why the use of some sensors has stayed within the realm of research rather than commercial availability. These barriers to adoption of sensing approaches primarily revolve around two issues: economics and training. Many chapters indicate that although research can provide good results from various sensing approaches, if the cost of the sensor and man-power needed to process sensor data outweigh or equal the savings made from the sensor, it will not be adopted by commercial farmers. Far more studies that assess the economic viability of sensing approaches are needed to illustrate the benefits of adopting such strategies. Several chapters showed the relative cost of using different sensors and more research like this is needed. Chapter 7 notes that the development of yield mapping software that displays profit on the screen given production costs and expected prices is needed and is a potentially attainable goal in the near future. Also, analysis of multi-year data could help commercial practitioners to understand the most profitable parts of their fields.

Chapter 2 indicates the need for imagery with high enough spatial, temporal and spectral resolutions for PA applications, but this is often off-set by the cost of imagery with the highest resolutions. Therefore, freely available Sentinel-2 imagery is likely to continue to be one of the most utilized platforms in PA because of its good spatial resolution for field crops (10–20 m), high spectral resolution and relatively frequent revisit period of five days. Chapter 6 noted that one of the most widely adopted sensing approaches was used to identify N stress, and that it was used in Oklahoma, USA, in 10% of the cereal crop area. This adoption success was due to consistent extension efforts and field demonstrations. Chapter 6 showed that farmers are more comfortable with time-saving robotics such as self-steering tractors, than with software that automatically determines N applications. Farmer response here is logical because if a time-saving robot mal-functioned it would be unlikely to have an adverse economic consequence, but if N applications were incorrect it could potentially have a severe impact on annual income. This preference towards the use of robots over automated decision support tools is a key point given the emphasis in much of the book about the need for automated decision support software. There may also be an issue here between map-based operations and real-time PA. The extra time spent on the former would allow farmers to supervise automated decisions while that would not be possible for the latter. Nevertheless, there is always the possibility to supervise as-applied maps, when available if needed. Clearly, extension services and companies that sell sensors commercially need to provide more demonstrations to convince farmers that such technologies work and will save them money or increase their profits. In addition, environmental impact should also be part of the equation to increase resource-use efficiency because of direct positive effects on the environment near farms. There is also the need for research on the economic return on investment for PA machinery and sensors.

Chapter 7 emphasizes the training needed for collection, analysis and interpretation of sensed data as a barrier to adoption. Chapter 6 also mentioned that the use of available technology should not require undue effort by the user to re-program present equipment, require purchase of expensive peripherals, or need additional tools and connectors that might be a barrier to implementation. In addition, and possibly most importantly, equipment suppliers or consulting services that understand and service the system and can guide individual farmers in their initial use of a sensor should provide support and be readily available to answer questions and troubleshoot any issues that arise with the technologies. Sensor and controller connecting protocols need to become standard to be able to mix equipment from different manufacturers without problems.

The key role of extension services is emphasized in Chap. 8. Local calibration data and reference materials should be developed by extension services or technical service providers for the calibration of optical sensors, otherwise farmers might use such sensors sub-optimally and negatively affects their profits. Any adverse experiences with sensors by farmers will make them less willing to adopt new technology in the future. Calibration and reference materials could also be provided by sensor developers who sell their services and would allow the generation of large local datasets that could be used further in future local calibrations. Each field site has local and unique characteristics that challenge the adoption of precision irrigation technologies. For example, Chap. 11 notes that each field presents a unique and complex irrigation problem that needs to be solved. This is one reason for limited precision irrigation uptake along with the cost of accurate soil moisture sensors. However, the case studies from the University of Georgia's centrally controlled irrigation program (Chaps. 5 and 11) where experts determine the best precision irrigation approach for farmers from a central control area show promise, suggesting similar approaches should be used in the future.

Although not mentioned widely throughout the book, Chap. 8 highlights the economic advantage of smart phone apps. Given their relatively low cost, their ubiquitous availability and the familiarity of the general public with smart phones, and downloading and using new apps, their use is likely to be a growth area in PA. Economic studies, for example that compared the accuracy of apps like the field scout green index app (currently costing about ~US $100) with traditional handheld normalized difference vegetation index (NDVI) metres, which currently cost in the few hundred to few thousand US $ range, are needed.

In a similar vein, UAVs have several desirable qualities. The case study in Chap. 9 highlighted the issue that Sentinel-2 data was sufficient to advise zonal management of crop vigour within fields, but that a UAV based approach could in some cases be more expensive because of the more complex image processing. However, this may be the opposite when it comes to precision horticulture where row crops need higher spatial resolution data to classify canopy pixels from ground cover to obtain more accurate results. Chapter 10 provided a brief economic assessment of UAV use for weed detection and suggested that this is a cost-effective approach. Clearly, economic studies of UAV use for different purposes need to be undertaken.

14.8 The Way Forward

Figure 14.1 shows a word cloud constructed from the text of the conclusions to each of the chapters in the book. It, along with our earlier discussion in this chapter, shows some of the key areas of research and focus for the future of sensing in PA. Our discussion and this diagram point to a bright future for PA with sensing capabilities with increased spatial, temporal and spectral resolutions. Computing speeds continue to improve, and the cost of sensors seems to be decreasing, therefore, it is vital that a bottle-neck of data that cannot be used does not build up. Farmers require guidance and information to avoid investing in sensors that they do not know how to use or know what to do with the resultant data. There needs to be an increasing focus on computing for automated complex analysis of big data and decision-support. Proof of concepts training on using such automated software and extension education programs that also help by providing local calibration data for sensors will also be key to improving adoption rates of new sensing approaches. According to the International Society for Precision Agriculture (ISPA) definition of PA (https://www.ispag.org/about/definition), it is not only a matter of gathering data compulsively, it is a matter of gathering, processing and analysing data to turn them into useful information to make management decisions with a purpose: improve resource use efficiency, productivity, quality, profitability and sustainability of agricultural production.

Fig. 14.1 Word cloud constructed from the text of the conclusions to each of the chapters in this book. The size of words is proportional to their count

Index

R. Kerry, A. Escolà (eds.), *Sensing Approaches for Precision Agriculture*, Progress in Precision Agriculture, https://doi.org/10.1007/978-3-030-78431-7

T

U